Handbook of Microalgal Culture: Applied Phycology and Biotechnology

Handbook of Microalgal Culture: Applied Phycology and Biotechnology

Editor:

Preethi Kartan

Delve Publishing

www.delvepublishing.com

Handbook of Microalgal Culture: Applied Phycology and Biotechnology

Editor: Preethi Kartan

708 3rd Avenue, 6th Floor
New York, NY 10017
United States of America
www.delvepublishing.com

ISBN: 978-1-68095-850-8

Library of Congress Control Number: 2016951603

About the Editor

Preethi Kartan

Preethi is a postgraduate in Biotechnology from University of Leeds, UK. She is currently working as a Scientific Associate at one of the CRO's in India and her interest lie in life sciences related writing.

Contents

List of Contributors

Rathinam Raja
IBB/CGB-Faculty of Sciences and Technology, Food Science Laboratory, University of Algarve, Campus de Gambelas, Faro 8005-139, Portugal

Hemaiswarya Shanmugam
IBB/CGB-Faculty of Sciences and Technology, Food Science Laboratory, University of Algarve, Campus de Gambelas, Faro 8005-139, Portugal

Venkatesan Ganesan
Acme ProGen Biotech (India) Private Limited, Balaji Nagar, Salem-636 004, India

Isabel S. Carvalho
IBB/CGB-Faculty of Sciences and Technology, Food Science Laboratory, University of Algarve, Campus de Gambelas, Faro 8005-139, Portugal

Matteo Marsullo
Department of Industrial Engineering, University of Padova, Padova, Italy

Alberto Mian
Industrial Process and Energy System Engineering Group (IPESE), École Polytechnique Fédérale de Lausanne, Lausanne, Switzerland

Adriano Viana Ensinas
Industrial Process and Energy System Engineering Group (IPESE), École Polytechnique Fédérale de Lausanne, Lausanne, Switzerland
Universidade Federal do ABC, Santo Andre, Brazil

Giovanni Manente
Department of Industrial Engineering, University of Padova, Padova, Italy

Andrea Lazzaretto
Department of Industrial Engineering, University of Padova, Padova, Italy

François Marechal
Industrial Process and Energy System Engineering Group (IPESE), École Polytechnique Fédérale de Lausanne, Lausanne, Switzerland

M. Huesemann
Pacific Northwest National Laboratory, Marine Sciences Laboratory, 1529 West Sequim Bay Road, Sequim, WA 98382, United States

B. Crowe
Pacific Northwest National Laboratory, Marine Sciences Laboratory, 1529 West Sequim Bay Road, Sequim, WA 98382, United States

P. Waller
The University of Arizona, Department of Agricultural and Biosystems Engineering, Tucson, AZ 8521, United States

A. Chavis
Pacific Northwest National Laboratory, Marine Sciences Laboratory, 1529 West Sequim Bay Road, Sequim, WA 98382, United States

S. Hobbs
Pacific Northwest National Laboratory, Marine Sciences Laboratory, 1529 West Sequim Bay Road, Sequim, WA 98382, United States

S. Edmundson
Pacific Northwest National Laboratory, Marine Sciences Laboratory, 1529 West Sequim Bay Road, Sequim, WA 98382, United States

M. Wigmosta
Pacific Northwest National Laboratory, Hydrology Group, Richland, WA 99352, United States

Peter R. Mooij
Department of Biotechnology Delft University of Technology

Lisanne D. de **Jongh**
Department of Biotechnology Delft University of Technology

Mark C. M. van Loosdrecht
Department of Biotechnology Delft University of Technology

Robbert Kleerebezem
Department of Biotechnology Delft University of Technology

Michiel H. A. **Michels**
Bioprocess Engineering, AlgaePARCWageningen University
HZ University of Applied Sciences

Atze Jan van der **Goot**
Food Process EngineeringWageningen University

Marian H. Vermuë
Bioprocess Engineering, AlgaePARCWageningen University

René H. Wijffels
Bioprocess Engineering, AlgaePARCWageningen University
Faculty of Biosciences and AquacultureUniversity of Nordland

Yoshihiko Sano
Department of Mechanical Engineering, Shizuoka University, 3-5-1 Johoku, Naka-ku, Hamamatsu, 432-8561, Japan

Akihiko Horibe
Graduate School of Natural Science and Technology, Okayama University, 3-1-1 Tsushima-naka, Kita-ku, Okayama, 700-8530 Japan

Naoto Haruki
Graduate School of Natural Science and Technology, Okayama University, 3-1-1 Tsushima-naka, Kita-ku, Okayama, 700-8530 Japan

Yugo Okino
Graduate School of Natural Science and Technology, Okayama University, 3-1-1 Tsushima-naka, Kita-ku, Okayama, 700-8530 Japan

Tim de Mooij
Bioprocess Engineering, AlgaePARC, Wageningen University, PO box 16, 6700 AA Wageningen, The Netherlands

Guus de Vries
Bioprocess Engineering, AlgaePARC, Wageningen University, PO box 16, 6700 AA Wageningen, The Netherlands

Christos Latsos
Bioprocess Engineering, AlgaePARC, Wageningen University, PO box 16, 6700 AA Wageningen, The Netherlands

René H. Wijffels
Bioprocess Engineering, AlgaePARC, Wageningen University, PO box 16, 6700 AA Wageningen, The Netherlands
Biosciences and Aquaculture, Nordland University, Bodø 8049, Norway

Marcel Janssen
Bioprocess Engineering, AlgaePARC, Wageningen University, PO box 16, 6700 AA Wageningen, The Netherlands

Jana BB
Centre for Environmental Protection (Kalyani Shine India), B-10/289, Kalyani-741235, West Bengal, India

Kundu D
International Centre for Ecological Engineering, University of Kalyani, Kalyani-741235, West Bengal, India

Datta D
International Centre for Ecological Engineering, University of Kalyani, Kalyani-741235, West Bengal, India

Lahiri S
International Centre for Ecological Engineering, University of Kalyani, Kalyani-741235, West Bengal, India

Bag S
International Centre for Ecological Engineering, University of Kalyani, Kalyani-741235, West Bengal, India

Sarkar D
International Centre for Ecological Engineering, University of Kalyani, Kalyani-741235, West Bengal, India

Bhakta JN
International Centre for Ecological Engineering, University of Kalyani, Kalyani-741235, West Bengal, India

Renil Anthony
Department of Biological Systems Engineering, University of Wisconsin-Madison, Madison, Wisconsin 53706, United States

Troy Runge
Department of Biological Systems Engineering, University of Wisconsin-Madison, Madison, Wisconsin 53706, United States

Luis G Ramirez-Merida
Applied Biotechnology Center, Department of Biology, University of Carabobo, Venezuela
Food Science and Technology Department, Federal University of Santa Maria, Brazil

Leila Queiroz Zepka
Food Science and Technology Department, Federal University of Santa Maria, Brazil

Cristiano Ragagnin de Menezes
Food Science and Technology Department, Federal University of Santa Maria, Brazil

Eduardo Jacob-Lopes
Food Science and Technology Department, Federal University of Santa Maria, Brazil

Josef Taucher
Institute of Bioprocess Engineering, Friedrich-Alexander-Universität Erlangen-Nürnberg, Erlangen, Germany

Sascha Baer
Institute of Bioprocess Engineering, Friedrich-Alexander-Universität Erlangen-Nürnberg, Erlangen, Germany

Philipp Schwerna
Institute of Bioprocess Engineering, Friedrich-Alexander-Universität Erlangen-Nürnberg, Erlangen, Germany

Dominik Hofmann
Institute of Bioprocess Engineering, Friedrich-Alexander-Universität Erlangen-Nürnberg, Erlangen, Germany

Magdalena Hümmer
Institute of Bioprocess Engineering, Friedrich-Alexander-Universität Erlangen-Nürnberg, Erlangen, Germany

Rainer Buchholz
Institute of Bioprocess Engineering, Friedrich-Alexander-Universität Erlangen-Nürnberg, Erlangen, Germany

Anna Becker
Institute of Bioprocess Engineering, Friedrich-Alexander-Universität Erlangen-Nürnberg, Erlangen, Germany

Karthik R
Department of Marine Biotechnology and Centre for Marine Bioprospecting, AMET University (Declared as deemed to be University U/S 3 of UGC Act 1956),Kanathur, Chennai–603112, Tamil Nadu, India

Ramalingam K
Department of Marine Biotechnology and Centre for Marine Bioprospecting, AMET University (Declared as deemed to be University U/S 3 of UGC Act 1956),Kanathur, Chennai–603112, Tamil Nadu, India

Yuvaraj D
Department of Biotechnology, Vel Tech High Tech Engineering College, Avadi, Chennai–600 062, Tamil Nadu, India

Vanitha MC
Department of Marine Biotechnology and Centre for Marine Bioprospecting, AMET University (Declared as deemed to be University U/S 3 of UGC Act 1956),Kanathur, Chennai–603112, Tamil Nadu, India

Muthezhilan R
Department of Marine Biotechnology and Centre for Marine Bioprospecting, AMET University (Declared as deemed to be University U/S 3 of UGC Act 1956),Kanathur, Chennai–603112, Tamil Nadu, India

Suantika G
Microbial Biotechnology Research Group, School of Life Sciences and Technology, Institute Technology Bandung (ITB), Jalan Ganesha No. 10 Bandung 40132, Indonesia

Lumbantoruan G
Microbial Biotechnology Research Group, School of Life Sciences and Technology, Institute Technology Bandung (ITB), Jalan Ganesha No. 10 Bandung 40132, Indonesia

Muhammad H
Microbial Biotechnology Research Group, School of Life Sciences and Technology, Institute Technology Bandung (ITB), Jalan Ganesha No. 10 Bandung 40132, Indonesia

Azizah FFN
Microbial Biotechnology Research Group, School of Life Sciences and Technology, Institute Technology Bandung (ITB), Jalan Ganesha No. 10 Bandung 40132, Indonesia

Aditiawati P
Microbial Biotechnology Research Group, School of Life Sciences and Technology, Institute Technology Bandung (ITB), Jalan Ganesha No. 10 Bandung 40132, Indonesia

Sibi G
Department of Biotechnology, Indian Academy Degree College, Centre for Research and Post Graduate Studies, Bangalore, India

Mustapha Benmoussa
Whistler Center for Carbohydrate Research, Department of Food Science, Purdue University, 745 Agriculture Mall Drive, West Lafayette, IN 47907-2009, USA

Hong JW
Advanced Bio-resource Research Center, Kyungpook National University, Daegu 41566, South Korea
Department of Biology, Kyungpook National University, Daegu 41566, South Korea

Kim OH
Advanced Bio-resource Research Center, Kyungpook National University, Daegu 41566, South Korea

Kim H
Advanced Bio-resource Research Center, Kyungpook National University, Daegu 41566, South Korea

Jo SW
Department of Energy Science, Kyungpook National University, Daegu 41566, South Korea

Cho HW
Department of Biology, Kyungpook National University, Daegu 41566, South Korea

Yoon HS
Advanced Bio-resource Research Center, Kyungpook National University, Daegu 41566, South Korea
Department of Biology, Kyungpook National University, Daegu 41566, South Korea
Department of Energy Science, Kyungpook National University, Daegu 41566, South Korea

Yoshiaki Maeda
Division of Biotechnology and Life Science, Institute of Engineering, Tokyo University of Agriculture and Technology, 2-24-16, Naka-cho, Koganei, Tokyo 184-8588, Japan

Takuma Tateishi
Division of Biotechnology and Life Science, Institute of Engineering, Tokyo University of Agriculture and Technology, 2-24-16, Naka-cho, Koganei, Tokyo 184-8588, Japan

Yuta Niwa
Division of Biotechnology and Life Science, Institute of Engineering, Tokyo University of Agriculture and Technology, 2-24-16, Naka-cho, Koganei, Tokyo 184-8588, Japan

Masaki Muto
Division of Biotechnology and Life Science, Institute of Engineering, Tokyo University of Agriculture and Technology, 2-24-16, Naka-cho, Koganei, Tokyo 184-8588, Japan
JST, CREST, Sanbancho 5, Chiyoda-ku, Tokyo 102-0075, Japan

Tomoko Yoshino
Division of Biotechnology and Life Science, Institute of Engineering, Tokyo University of Agriculture and Technology, 2-24-16, Naka-cho, Koganei, Tokyo 184-8588, Japan

David Kisailus
Department of Chemical and Environmental Engineering, University of California, Riverside, Room 343 Materials Science and Engineering Building, Riverside, CA 92521, USA

Tsuyoshi Tanaka
Division of Biotechnology and Life Science, Institute of Engineering, Tokyo University of Agriculture and Technology, 2-24-16, Naka-cho, Koganei, Tokyo 184-8588, Japan
JST, CREST, Sanbancho 5, Chiyoda-ku, Tokyo 102-0075, Japan

Mariam Al hattab
Department of Process Engineering and Applied Sciences, Faculty of Engineering, Dalhousie University, Halifax, Nova Scotia, Canada

Abdel Ghaly
Department of Process Engineering and Applied Sciences, Faculty of Engineering, Dalhousie University, Halifax, Nova Scotia, Canada

Amal Hammouda
Department of Process Engineering and Applied Sciences, Faculty of Engineering, Dalhousie University, Halifax, Nova Scotia, Canada
Food Technology Research Institute, Agricultural Research Center, Giza, Egypt

Melanie Oey
Institute for Molecular Bioscience, The University of Queensland, St Lucia, Qld, Australia

Anne Linda Sawyer
Ruhr-University Bochum, Bochum, Germany

Ian Lawrence Ross
Institute for Molecular Bioscience, The University of Queensland, St Lucia, Qld, Australia

Ben Hankamer
Institute for Molecular Bioscience, The University of Queensland, St Lucia, Qld, Australia

Preface

Study of algae is known as Phycology. Algae are of two types - macro algae and micro algae. Microalgae are unicellular organisms that consist of both prokaryotic and eukaryotic forms. Microalgae represent a sustainable source of natural products. Over 15,000 novel compounds have been identified in microalgae. Microalgae include cyanobacteria, (blue-green algae) green, brown and red algae. Most of micro algae grow in aquatic environments, but there are a few species that can grow in the most extreme environments. Microalgae are mostly autotrophic (grow through photosynthesis) but few are heterotrophic (algae that grows in dark using sugar or starch), or even mixotrophic (combine both growth modes). Micro algae are most promising sources of biomass and oils for fuel, food, feed, and other co-products. Some of the products produced by micro algae include Carotenoids, fatty acids, antioxidants, polymers, sterols, enzymes and biofuels. Due to their fast growth ability, consumption of CO2 and ability to store significant amounts of energy rich compounds micro algae are gaining importance in recent times. Today cultured microalgae is used as direct feed for humans and land-based farm animals, and as feed for cultured aquatic species such as molluscs and the early larval stages of fish and crustaceans. It is a potential candidate for biofuel production. Microalgae can grow 20 or 30 times faster than traditional food crops, and has no need to compete for arable land. Since microalgal production is central to so many commercial applications, there is a need for production techniques which increase productivity and are economically profitable.

The production of algae for high-value markets (aquaculture, food supplements, nutraceuticals, pharmaceuticals) will be developed through the search for, isolation and cultivation of new algal strains endowed with the activity of interest. Algal biomass might become an important source of biofuels, especially if its production will be carried out in low-cost photobioreactors and associated with wastewater treatment and greenhouse gas abatement. Handbook of Microalgal Culture: Applied Phycology and Biotechnology highlights some of the parameters established, models, techniques and methods developed for the improvement of micro algae cultivation and micro algae application. The state of the art of microalgae biotechnology, particularly focusing on new culture techniques and actual and potential uses of microalgae in human and animal nutrition, in cosmetics and pharmaceutics, and for environmental applications, is described. There should not be any doubt that microalgal culture has an exciting future. In comparison with terrestrial agriculture we are just beginning. There has been a rapid expansion in capability over the last 50 years and every reason to think it will continue to expand. There are many compounds of commercial potential in microalgae and finding economically viable methods of producing them will challenge future generations.

Editor

Preethi Kartan

INTRODUCTION

Microalgae are microscopic algae, typically found in freshwater and marine systems living in both the water column and sediment. They are unicellular species which exist individually, or in chains or groups. Depending on the species, their sizes can range from a few micrometers (µm) to a few hundreds of micrometers. Unlike higher plants, microalgae do not have roots, stems, or leaves. They are specially adapted to an environment dominated by viscous forces. Microalgae, capable of performing photosynthesis, are important for life on earth; they produce approximately half of the atmospheric oxygen and use simultaneously the greenhouse gas carbon dioxide to grow photoautotrophically. Microalgae are the base of the food web and provide energy for all the trophic levels about it. Microalgae biomass is often measured with chlorophyll a concentrations and can provide a useful index of potential production. The standing stock of microphytes is closely related to that of its predators. Without grazing pressures the standing stock of microphytes dramatically decreases.

As species of aquatic organisms such as seaweed are macroalgae, aquatic organisms such as spirulina, chlorella, and blue-green algae are known as microalgae. Microalgae are microscopic, single-cell organisms that exist in fresh water and marine environments and exist at the bottom of the food chain. Microalgae are a source of a wide variety of nutrients such as omega-3 fatty acids, and are a particularly large source of protein. Microalgae are consumed by prey fish such as sardines and herring, which in turn are consumed by bigger fish. The omega-3 fatty acids from a fish like salmon are not in fact from the salmon, but from the presence of microalgae in the food chain.

Biotechnology is technology based on biology, especially when used in agriculture, food science, and medicine.

Any technological application that uses biological systems, living organisms, or derivatives thereof, to make or modify products or processes for specific use. Algal biotechnology is a technology developed using algae.this can be further divided into microalgae technology and macroalgae technology.

The objective of the Micro algal Biotechnology Group is to further the understanding of the ecology of microalgae. This will assist with the development of commercial-scale micro algal culturing techniques for the production of bioactive compounds, aquaculture feed, fine chemicals, and renewable fuels. Additionally, environmental applications such as CO_2 bioremediation, control of excessive algal growth and development of management strategies for water supply managers are investigated.

Transgenesis in algae is a complex and fast-growing technology. Selectable marker genes, promoters, reporter genes, transformation techniques, and other genetic tools and methods are already available for various species and currently ~25 species are accessible to genetic

transformation. Fortunately, large-scale sequencing projects are also planned, in progress, or completed for several of these species; the most advanced genome projects are those for the red alga Cyanidioschyzon merolae, the diatom Thalassiosira pseudonana, and the three green algae Chlamydomonas reinhardtii, Volvox carteri and Ostreococcus tauri.

The main methods by which algae are cultivated and harvested are described. The first section deals with the environmental factors affecting mass culture and the principles governing the design and operation of mass cultivation systems. The second section gives the main current and potential uses of algae: in wastewater treatment, a source of food and feed, an energy source, and in the production of common and fine chemicals, such as polysaccharides, lipids, glycerol, pigments, and enzymes.

Principles Microalgal Culturing

A culture can be defined as an artificial environment in which the algae grow. In theory, culture conditions should resemble the alga's natural environment as far as possible; in reality many significant differences exist, most of which are deliberately imposed.

A culture has three distinct components:

- a culture medium contained in a suitable vessel;
- the algal cells growing in the medium; and
- air, to allow exchange of carbon dioxide between medium and atmosphere.

Extensive measures must be taken to keep pure unialgal cultures chemically and biologically clean. Chemical contamination may have unquantifiable, often deleterious, and therefore undesirable effects on algal growth. Biological contamination of pure algal cultures by other eukaryotes and prokaryotic organisms in most cases invalidates experimental work, and may lead to the extinction of the desired algal species in culture throughout-competition or grazing. In practice, it is very difficult to obtain bacteria-free (axenic) cultures, and although measures should be taken to minimize bacterial numbers, a degree of bacterial contamination is often acceptable.

For an entirely autotrophic alga, all that is needed for growth is light + CO_2 + H_2O + nutrients + trace elements. By means of photosynthesis the alga will be able to synthesize all the biochemical compounds necessary for growth. Only a minority of alga seem, however, to be entirely autotrophic; many are unable to synthesize certain biochemical compounds (certain vitamins, for example) and will require these to be present in the medium. This condition is known as auxotrophy.

Based on their growth characteristics, two kinds of cultures can be defined.

- In limited volume (batch) cultures, resources are finite. When the resources present in the culture medium are abundant, growth occurs according to a sigmoid curve, but once the resources have been utilised by the cells, the cultures die unless supplied with new medium. In practise this is done by subculturing, i.e. transferring a small volume of existing culture to a large volume of fresh culture medium at regular intervals.
- In continuous cultures, resources are potentially infinite: cultures are maintained at a chosen point on the growth curve by the regulated addition of fresh culture medium. In practise, a volume of fresh culture medium is added automatically at a rate proportional to the growth rate of the alga, while an equal volume of culture is removed.

Physical Parameters

Temperature

The temperature at which cultures are maintained should ideally be as close as possible to the temperature at which the organisms were collected; polar organisms (<10°C); temperate (10-25°C); tropical (>20°C). An intermediate value of 18-20°C is most often employed. Temperature controlled incubators usually use constant temperature (transfers to different temperatures should be conducted in steps of 2°C/week), although some models permit temperature cycling. In temperate regions ambient room temperature is generally acceptable for culturing purposes.

Light

Natural light is usually sufficient to maintain cultures in the laboratory. Cultures should never be exposed to direct sunlight (which may cause photopigment damage), and should therefore be placed next to a north-facing window (in the northern hemisphere).

Artificial lighting by fluorescent bulbs is often employed for culture maintainance and experimental purposes. Light intensity should range between 0.2-50% of full daylight (= 1660 $\mu E/s/m^2$), with 5-10% (c. 80-160$\mu E/s/m^2$) most often employed. Light quality (spectrum) depends on type of bulb used (see manufacturers technical data), the most common types being 'cool white' or 'daylight' bulbs. Light intensity and quality can be manipulated with filters. Many microalgal species do not grow well under constant illumination, and hence a light/dark (LD) cycle is used (maximum 16:8 LD, usually 14:10 or 12:12).

Mixing

Mixing of microalgal cultures may be necessary under certain circumstances: when cells must be kept in suspension in order to grow (particularly important for heterotrophic dinoflagellates); in concentrated cultures to prevent nutrient limitation effects due to stacking of cells and to increase gas diffusion. It should be noted that in the ocean cells seldom experience turbulence, and hence mixing should be gentle. The following methods may be used: bubbling with air (may damage cells); plankton wheel or roller table (about 1 rpm); gentle manual swirling. Most cultures do well without mixing, particularly when not too concentrated, but when possible gentle manual swirling (once each day) is recomended.

Types of Culture Vessel

Culture vessels should have the following properties: non toxic (chemically inert); reasonably transparent to light; easily cleaned and sterilized; provide a large surface to volume ratio (depending on organism).

Certain materials which could potentially be used for culture vessels may leach chemicals which have a deleterious effect on algal growth into the medium. The use of chemically inert materials is particularly important when culturing oceanic plankton and during isolation. A list of materials which should be safe, inhibitory, or toxic to algal cultues is given by Stein. Recommended materials for culture vessels and media preparation include:

- Teflon (very expensive, used for media preparation);
- Polycarbonate (expensive and becomes cloudy and cracks with repeated autoclaving);
- Polystyrene (cheaper alternative to Teflon and polycarbonate, not autoclavable);
- Borosilicate glass (has been shown to inhibit growth of some species).

Materials which should generally be avoided during microalgal culturing include all types of rubber, and PVC. The following types of culture vessel are most commonly used for microalgal culturing (but may not prove suitable for all species):

- Erlenmeyer flasks (glass or polycarbonate) with cotton, glass, polypropylene, or metal covers.
- tubes (glass or polycarbonate) with cotton, glass, polypropylene or metal caps.
- polystyrene tissue culture flasks, purchased as single-use sterile units (eg. Iwaki, Nunc, Corning).

Cleaning/Sterilization of Culture Materials

Cleaning procedure

- scrub (abrasive brushes not appropriate for most plastics) and soak with warm detergent (not domestic detergents, which leave a residual film on cultureware – use laboratory detergent such as phosphate-free Decon);
- rinse extensively with tap water;
- soak in 10% HCl for 1 day-1 week (not routinely necessary, but particularly important for new glass and polycarbonate material);
- rinse extensively with distilled and finally bidistilled water;
- leave inverted to dry in a clean, dust-free place.

Sterilization

Sterilization can be defined as a process which ensures total inactivation of microbial life (NB. not the same as desinfection, which is defined as an arbitrary reduction of bacterial numbers). The primary purpose of sterilization is to prevent contamination by unwanted organisms, but it may also serve to eliminate unwanted chemicals.

There are several sterilisation methods and the choice depends on the purpose and material used:

- Gas (ethylene oxide, often used for disposable plastic material);
- Moist heat - Autoclave or pressure cooker (1-2 Bar, 121°C, in pure saturated steam).
  - For sterility the steam must penetrate the material (wrap in material that allows access of steam - kraft paper or aluminium). Autoclave steam may introduce chemical contaminants; glass and polycarbonate vessels should be autoclaved containing a small amount of bidistilled water which is poured out (thus diluting contaminants) under sterile conditions immediately prior to use. Never close vessels (risk of implosion); use cotton wool bungs, or leave screw caps slightly open.
- Dry heat
  - Oven (at least 2 hours at 160°C). Cover neck of vessel with aluminium foil to maintain sterility on removal from oven.
  - Flame (heating to incandescence for metal)
  - Problems: few materials stand high temperatures (glass, teflon, silicone, metal, cotton).
- Radiation
  - X and g - radiation (industrial applications for disposable plastics that cannot stand more then 60°C, not always effective)
  - UV radiation 240-280 nm (not often used for culture materials).

Transfer protocol

The following procedures should always be used when preparing media or transferring cultures:

- work in clean place or preferably in a laminar flow cabinet (cabinet must be turned on at least 30 minutes before transfer; if equipped with UV lamps, leave on overnight prior to use).
- clean working surface with 70% alcohol (ethanol/isopropanol) prior to and after use.
- clean hands with disinfecting soap and rinse with 70% alcohol prior to all operations.
- when not using a laminar flow cabinet (and to be safe even when using a cabinet), sterilise (flame) the neck of vessel of origin before and after transfer (not possible with some plastic vessels, which must, therefore, be opened in a laminar flow cabinet).
- pipettes must be clean and sterile; use autoclavable tips for repeating pipettes (eg. Gilson), pre-wrapped sterile single use plastic/glass pipettes, or if using non-sterile glass pipettes (with cotton plugs), sterilise in the flame before use.

The frequency of culture transfer depends on species and culture conditions (it is advisable to follow growth for each unfamiliar species to get a feeling). Transfers should be conducted as population growth slows, preferably before population growth ceases (stationary phase). Culture transfers (usually 1-5 % of culture volume into fresh sterile medium) can be conducted directly (pouring, not used when vessel neck is flame sterilized), or more often with a pipette.

Culture Media

Media for the culture of marine phytoplankton consist of a seawater base (natural or artificial) which may be supplemented by various substances essential for microalgal growth, including nutrients, trace metals and chelators, vitamins, soil extract and buffer compounds.

Seawater base

The quality of water used in media preparation is very important.

- Natural seawater can be collected nearshore, but its salinity and quality is often quite variable and unpredictable, particularly in temperate and polar regions (due to anthropogenic pollution, toxic metabolites released by algal blooms in coastal waters). The quality of coastal water may be improved by ageing for a few months (allowing bacterial degradation of inhibitory substances), by autoclaving (heat may denature inhibitory substances), or by filtering through acid-washed charcoal (which absorbs toxic organic compounds). Most coastal waters contain significant quantities of inorganic and organic particulate matter, and therefore must be filtered before use.

- The low biomass and continual depletion of many trace elements from the surface waters of the open ocean by biogeochemical processes makes this water much cleaner, and therefore preferable for culturing purposes.
- Collect seawater from the front of the boat (or pump from subsurface) to avoid contamination. Seawater can be stored in polyethylene carboys, and should be stored in cool dark conditions.
- Artificial seawater, made by mixing various salts with deionized water, has the advantage of being entirely defined from the chemical point of view, but they are very laborious to prepare, and often do not support satisfactory algal growth. Trace contaminants in the salts used are at rather high concentrations in artificial seawater because so much salt must be added to achieve the salinity of full strength seawater.

Nutrients, Trace Metals, and Chelators

The term 'nutrient' is colloquially applied to a chemical required in relatively large quantities, but can be used for any element or compound necessary for algal growth.

Table 1: The average concentrations of constituents of potential biological importance found in typical seawater

Element	Average Molar concentration (range in brackets)
Group A	
Na^+	4.7×10^{-1}
K^+	1.02×10^{-2}
Mg^{2+}	5.3×10^{-2}
Ca^{2+}	1.03×10^{-2}
Cl^-	5.5×10^{-1}
SO_4^{2-}	2.8×10^{-2}
HCO_3^-	2.3×10^{-3}
BO_3^{3-}	4.2×10^{-4}
Group B	
Br^-	8.4×10^{-4}
F^-	6.8×10^{-5}
IO_3^-	4.4×10^{-7}
Li^+	2.5×10^{-5}
Rb^+	1.4×10^{-6}
Sr^{2+}	8.7×10^{-5}
Ba^{2+}	1×10^{-7}
MoO_4^{2-}	1.1×10^{-7}
VO_4^{3-}	2.3×10^{-8}
CrO_4^{2-}	4×10^{-9}
AsO_4^{3-}	2.3×10^{-8}
SeO_4^{2-}	1.7×10^{-9}
Group C	
NO_3^-	3×10^{-5} (10^{-8} to 4.5×10^{-5})
PO_4^{3-}	2.3×10^{-6} (10^{-7} to 3.5×10^{-6})
Fe^{3+}	1×10^{-9} (10^{-10} to 10^{-7})
Zn^{2+}	6×10^{-9} (5×10^{-11} to 10^{-7})
Mn^{2+}	5×10^{-10} (2×10^{-10} to 10^{-6})
Cu^{2+}	4×10^{-9} (5×10^{-10} to 6×10^{-9})
Co^{2+}	2×10^{-11} (10^{-11} to 10^{-10})
SiO_4^{4-}	1×10^{-4} (10^{-7} to 1.8×10^{-4})
Ni^{2+}	8×10^{-9} (2×10^{-9} to 1.2×10^{-8})

Group A: concentrations of these constituents exhibit essentially no variation in seawater, and high algal biomass cannot deplete them in culture media. These constituents do not, therefore, have to be added to culture media using natural seawater, but do need to be added to deionized water when making artificial seawater media.

Group B: also have quite constant concentrations in seawater, or vary by a factor less than 5. Because microalgal biomass cannot deplete their concentrations significantly, they also do not need to be added to natural seawater media. Standard artificial media (and some natural seawater media) add molybdenum (as molybdate), an essential nutrient for algae, selenium (as selenite), which has been demonstrated to be needed by some algae, as well as strontium, bromide and flouride, all of which occur at relatively high concentrations in seawater, but none of which have been shown to be essential for microalgal growth.

Group C: all known to be needed by microalgae (silicon is needed only by diatoms and some chrysophytes, and nickel is only known to be needed to form urease when algae are using urea as a nitrogen source). These nutrients are generally present at low concentrations in natural seawater, and since microalgae take up substantial amounts, concentrations vary widely (generally by a factor of 10 to 1000). All of these nutrients (except silicon and nickel in some circumstances) generally need to be added to culture media in order to generate significant microalgal biomass.

Nitrate is the N source most often used in culture media, but ammonium can also be used, and indeed is the preferential form for many algae since it does not have to be reduced prior to amino acid synthesis, the point of primary intracellular nitrogen assimilation into the organic linkage. Ammonium concentrations greater than 25μM are, however, often reported to be toxic to phytoplankton, so concentrations should be kept somewhat low.

Inorganic (ortho)phosphate, the P form preferentially used by microalgae, is most often added to culture media, but organic (glycero)phosphate is sometimes used, particularly when precipitation of phosphate is anticipated (when nutrients are autoclaved in the culture media rather than separately, for example). Most microalgae are capable of producing cell surface phosphatases which allow them to utilise this and other forms of organic phosphate as a phosphorus source.

The trace metals which are essential for microalgal growth are incorporated into essential organic molecules, particularly a variety of coenzyme factors which enter into photosynthetic reactions. Of these metals, the concentrations (or more accurately the biologically available concentrations) of Fe, Mn, Zn, Cu and Co (and sometimes Mo and Se) in natural waters may be limiting to algal growth. Little is known about the complex relationships between chemical speciation of metals and biological availability. It is thought that molecules which complex with metals (chelators) influence the availability of these elements. Chelators act as trace metal buffers, maintaining constant concentrations of free ionic metal. It is the free ionic metal, not the chelated metal, which influences microalgae, either as a nutrient or as a toxin. Without proper chelation some metals (such as Cu) are often present at toxic concentrations, and others (such as Fe) tend to precipitate and become unavailable to phytoplankton. In natural seawater, dissolved organic molecules (generally present at concentrations of 1-10mg l^{-1}) act as chelators. The most widely used chelator in culture media additions is EDTA (ethylenediaminetetraacetic acid), which must be present at high concentrations since most complexes with Ca and Mg, present in large amounts in seawater. EDTA may have an additional benefit of reducing precipitation during autoclaving. High concentrations have, however, occasionally been reported to be toxic to microalgae. As an alternative the organic chelator citrate is sometimes utilised, having the advantage of being less influenced by Ca and Mg. The ratio of chelator:metal in culture medium ranges from 1:1 in f/2 to 10:1 in K medium. High ratios may result in metal deficiencies for coastal phytoplankton (ie. too much metal is complexed), and many media therefore use intermediate ratios.

In today's aerobic ocean, iron is present in the oxidized form as various ferric hydroxides and thus is rather insoluble in seawater. While concentrations of nitrogen, phosphorus, zinc and manganese in deep water are similar to plankton elemental composition, there is proportionally 20 times less iron in deep water than is apparently needed, leading to the suggestion that iron may be the ultimate geochemically limiting nutrient to phytoplankton in the ocean (Brand, 1986). Very little is known about iron in seawater or phytoplankton uptake mechanisms due to the complex chemistry of the element. Iron availability for microalgal uptake seems to be largely dependent on levels of chelation. It is highly recommended that iron be added as the chemically prepared chelated iron salt of EDTA rather than as iron chloride or other iron salts; the formation of iron chelates is relatively slow, and iron hydoxides will form first in seawater, leading to precipitation of much of the iron in the culture medium.

Apparently as a result of the extreme scarcity of copper in anaerobic waters, copper did not begin to be utilised by organisms until the earth became aerobic and copper increased in abundance. Consequently copper does not seem to be an obligate requirement, algae either not needing it, or needing so little that free ionic copper concentrations in natural seawater are sufficient to maintain maximum growth rates. Copper may indeed be toxic, particularly to more primitive algae, and hence copper, if added to culture media at all, should be kept at low concentrations.

Vitamins

Roughly of all microalgal species tested have been shown to have a requirement for vitamin B12, which appears to be important in transferring methyl groups and methylating toxic elements such as arsenic, mercury, tin, thallium, platinum, gold, and tellurium (Brand, 1986), around 20% need thiamine, and less than 5% need biotin.

It is recommended that these vitamins are routinely added to seawater media. No other vitamins have ever been demonstrated to be required by any photosynthetic microalgae.

Soil extract

Prepared by heating, boiling, or autoclaving a 5 to 30% slurry of soil in fresh water or seawater and subsequently filtering out the soil. Soil extract has historically been an important component of culture media. The solution provides macronutrients, micronutrients, vitamins, and trace metal chelators in undefined quantities, each batch being different, and hence having unpredictable effects on microalgae. With increasing understanding of the importance of various constituents of culture media, soil extract is less frequently used. Soil extract should only be used on a non-experimental basis.

Buffers

The control of pH in culture media is important since certain algae grow only within narrowly defined pH ranges, and in order to prevent the formation of precipitates. Except under unusual conditions, the pH of natural seawater is around 8. Because of the large buffering capacity of natural seawater (due to a bicarbonate buffering system, HCO_3 being present at c. 2.2M) it is quite easy to maintain the pH of marine culture media. The buffer system is overwhelmed only during autoclaving, when high temperatures drive CO_2 out of solution and hence cause a shift in the bicarbonate buffer system and an increase in pH, or in very dense cultures of microalgae, when enough CO_2 is taken up to produce a similar effect. As culture medium cools after autoclaving, CO_2 reenters solution from the atmosphere, but certain measures must be taken if normal pH is not fully restored:

- The pH of seawater may be lowered prior to autoclaving (adjustment to pH 7-7.5 with 1M HCl) to compensate for subsequent increases.
- Certain media recipes include additions of extra buffer, either as bicarbonate, Tris (Tris-hydroxymethyl-aminomethane), or glycylglycine to supplement the natural buffering system. Tris may also act as a Cu buffer, but has occasionally been cited for its toxic properties to microalgae. Glycylglycine is rapidly metabolized by bacteria and hence can only be used with axenic cultures. These additions are generally not necessary if media are filter sterilized, unless very high cell densities are expected.
- The problem of CO2 depletion in dense cultures may be reduced by having a large surface area of media exposed to the atmosphere relative to the volume of the culture, or by bubbling with either air (CO2 concentration c.0.03%) or air with increased CO2 concentrations (0.5 to 5%). Unless there is a large amount of biomass taking up the CO2, the higher concentrations could actually cause a significant decline in pH. When bubbling is employed, the gas must first pass through an in line 0.2µm filter unit (eg. Millipore Millex GS) to maintain sterile conditions. For many microalgal species, aeration is not an option since the physical disturbance may inhibit growth or kill cells.

Media Preparation

The salinity of the seawater base should first be checked (30-35% for marine phytoplankton), and any necessary adjustments (addition of fresh water/evaporation) made before addition of enrichments.

Always use reagant grade chemicals and bidistilled (or purer) water to make stock solutions of enrichments. Gentle heating and/or magnetic stirring of stock solutions can be used to ensure complete dissolution. When preparing a stock solution containing a mixture of compounds, dissolve each individually in a minimal volume of water before mixing, then combine and dilute to volume.

Seawater, stock solutions of enrichments and the final media must be sterile in order to prevent (or more realistically minimize) biological contamination of unialgal cultures. Several methods are available for sterilization:

- autoclaving is the most widely used technique for sterilizing culture media, and is the ultimate guarantee of sterility (including the destruction of viruses). A commercial autoclave is best, but pressure cookers of various sizes are also suitable. Sterility requires 15 minutes at 1-2 Bar pressure and a temperature of 121°C in the entire volume of the liquid (ie. longer times for larger volumes of liquid; approximately 10 min for 100ml, 20 min for 2l, 35 min for 5 l). Bottles containing media should be no more than _ full, and should be left partially open or plugged with cotton wool or covered with aluminium foil. Ensure the heating elements are covered with distilled water, and the escape valve should not be closed until a steady stream of steam is observed. After autoclaving, the pressure release valve should not be opened until the temperature has cooled to below 80°C.
- Autoclaving is a process which has many effects on seawater and its constituents, potentially altering or destroying inhibitory organic compounds, as well as beneficial organic molecules. Because of the steam atmosphere in an autoclave, CO_2 is driven out of the seawater and the pH is raised to about 10, a level which can cause precipitation of the iron and phosphate added in the medium. Some of this precipitate may disappear upon re-equilibriation of CO_2 on cooling, but both the reduced iron and phosphate levels, and the direct physical effect of the precipitate may limit algal growth. The presence of EDTA and the use of organic phosphate may reduce precipitation effects. Addition of 5% or more of distilled water may also help to reduce precipitation (but may affect final salinity). The best solution, however, if media are autoclaved, is to sterilize iron and phosphate (or even all media additions) seperately and add them aseptically afterwards.
- Autoclave steam may contaminate the media (i.e. with trace metals from the autoclave tubing). Autoclaving also produces leaching of chemicals from the medium receptacle into the medium (silica from glass bottles, toxic chemicals from plastics). Autoclaving in well-used Teflon or polycarbonate vessels reduces leaching of trace contaminants.

- Autoclaving will cause evaporation of water, and hence an increase in salinity (usually of c. 1‰). Distilled water can be added prior to autoclaving to compensate for this increase.
- Pasteurization (heating to 90-95°C for 30 minutes) of media in Teflon or polycarbonate bottles is a potential alternative, reducing the problems of trace metal contamination and alteration of organic molecules inherent with autoclaving. Pasteurization does not, however, completely sterilize the seawater; it kills all eukaryotes and most bacteria, but some bacterial spores probably survive. Heating to 90-95°C for at least 30 minutes and cooling, repeated on two successive days ('tyndallisation') may improve sterilization efficiency; it is assumed that vegetative cells are killed by heat and heat resistant spores will germinate in the following cool periods and be killed by subsequent heating.
- Ultraviolet radiation can be used to sterilize seawater, but very high intensities are needed to kill everything in the seawater (1200 W lamp, 2-4h for culture media in quartz tubes). Such intense UV light necessarily alters and destroys the organic molecules in seawater and generates many long lived free radicals and other toxic reactive chemical species (Brand, 1986). Seawater exposed to intense UV light must, therefore, be stored for several days prior to use to allow the level of these highly reactive chemical species to decline.
- Sterilization by commercial microwave apparatus is another option. Microwave sterilization has not, as yet, been widely employed in culture media preparation due to uncertainties about sterilization efficiency. Trials should be conducted before use of this method to ensure sterility of seawater.
- Sterile filtration is probably the best method of sterilizing seawater without altering the chemistry of the seawater, as long as care is taken not to contaminate the seawater with dirty filter apparatus. Sterilization efficiency is, however, to some extent reduced compared with heat sterilization methods. Membrane filters of 0.2μm porosity are generally considered to yield water free of bacteria, but not viruses. 0.1μm filters can be used, but the time required for filtration of large volumes of culture media may be excessively long. The filtration unit must be sterile: for small volumes (<50ml) pre-sterilized single use filter units for syringe filtration (eg. Millipore Millex GS) can be used; for volumes up to 1 litre reusable autoclavable self-assembly filter units (glass or polycarbonate) with 47mm cellulose ester membrane filters (eg. Millipore HA) can be used with suction provided by a vacuum pump; for larger volumes an in-line system with peristaltic pump and cartridge filters may be the best option.
- Filter units (particularly disposable plastic systems), and the membrane filters themselves can also leak toxic compounds into the filtrate. The first portion of filtrate (eg. 5% of volume to be filtered) should be discarded to alleviate this problem.

Most stock solutions of culture medium additions can be sterilized separately by autoclaving, although vitamin stock solutions are routinely filtered through 0.2μm single use filter units (eg. Millipore Millex GS), since heat sterilization will denature these organic compounds. Filter sterilization of all additions may reduce uncertainties about stability of the chemical compounds and contamination from autoclave steam, but absolute sterilization is not guaranteed. Stock solutions can be stored in ultraclean sterile glass, polycarbonate or Teflon tubes/bottles. In order to minimize effects of any microbial contaminations, all stock solutions should be stored in a refrigerator at 4°C, except vitamin stocks which are stored frozen at –20°C and thawed immediately prior to use.

Pure, Axenic Cultures

If biological contaminants appear in a culture, the best remedy is to isolate a single cell from the culture with a micropipette, and try to establish a new, clean clonal culture. Alternatively the culture can be streaked on an agar plate in the hope of attaining a colony free of contaminants. Neither of these methods work well, however, for eliminating bacteria that attach firmly to the surface of microalgae. Placing a test tube of microalgal culture in a low-intensity 90 kilocycles/sec ultrasonic water bath for varying lengths of time (a few seconds to tens of minutes) can sometimes physically separate bacteria without killing the algae, making it easier to obtain an axenic culture by micropipette isolation.

Often, however, to achieve an axenic culture, antibiotics must be used. Best results appear to occur when an actively growing culture of algae is exposed to a mixture of penicillin, streptomycin and gentamycin for around 24 hours. This drastically reduces the growth of bacteria while allowing the microalgae to continue to grow, increasing the chances of obtaining an axenic cell when using micropippette or agar streaking isolation. Different algal species tolerate different concentrations of antibiotics, so a range of concentrations should be used (generally 50-500 mg/l). Other antibiotics that can be used include chloramphenicol, tetracycline, and bacitracin. Antibiotic solutions should be made with distilled water and filter-sterilized (0.2μm filter units) into sterile tubes, and should be stored frozen until use. Another approach is to add a range of antibiotic concentrations to a number of subcultures and then select the culture that has surviving algal cells but no surviving bacteria or other contaminants. Sterility of cultures should be checked by microscopic examination and by adding a small amount of sterile bacterial culture medium (eg. 0.1% peptone) to a microalgal culture and observing regularly for bacterial growth. Absence of bacterial growth does not, however, ensure that the microalgal culture is axenic, since the majority of bacteria do not respond to standard enrichments. In reality there is no way of demonstrating that a microalgal culture is completely axenic. In practise, therefore, axenic usually means 'without demonstrable unwanted prokaryotes or eukaryotes'. Some microalgal cultures may die when made axenic, probably due to the termination of obligate symbiotic relationships with bacteria.

CHAPTER 1

BIOMASS FROM MICROALGAE

Rathinam Raja, Hemaiswarya Shanmugam, Venkatesan Ganesan and Isabel S. Carvalho

Keywords: Microalgae, Biomass, Open raceway ponds, Photobioreactor, Nutrients

INTRODUCTION

Microalgae are most abundant primary unicellular producers found in all the aquatic systems such as, freshwater, seawater, hypersaline lakes and even in deserts and arctic ecosystems [1,2]. Subdivided into eukaryotic and prokaryotic algae, eukaryotic possessing defined cell organelles such as nuclei, chloroplasts, mitochondria, prokaryotes (cyanobacteria or blue-green) are primitive, possessing the simpler cellular structure of bacteria. They convert light energy and carbon dioxide (CO2) into biomass (e.g.) carbohydrates [3]. Their photosynthetic mechanism is similar to land-based plants, but due to a simple cellular structure, and submerged in an aqueous environment where they have efficient to access water, CO2 and other nutrients, they are more efficient in converting solar energy into biomass [4]. Microalgae can also grow in extreme environments; it could be produced on agricultural and non-agricultural lands. It could also make use of fresh, brackish, saline, wastewater, municipal sewage and industrial effluents [5]. Many microalgae species are able to switch from phototrophic to heterotrophic growth conditions. As heterotrophs, the algae rely on glucose or other carbon sources for carbon metabolism and energy and some algae can grow mixotrophically. Some of the important updates on the biomass of algae have been elaborated in the following sections. Microalgae are energy-rich feed stocks which have received so far more attention due to their facile adaptability to grow in the photobioreactors or open ponds for high yields and multiple applications. They have many inherent advantages, some of them are: higher productivity (biomass) in few days, easy adaptability to new environments and high-lipid content. Algae are cultivated not only for food source but also the cultivated biomass is used for feedstocks too [6] and scaling up algae farming could lead to yields of other commercially viable products [7,8]. Since two thirds of earth's surface is covered with ocean, algae would be the option with greatest potential for food including aquaculture industries, pharma products and global energy needs (Figure 1) [9].

SELECTION OF NEW STRAINS

Microalgae can be purified from a wide range of environments and they have to be screened for biofuel production, accumulation of desired storage compounds (high lipid strains), high-value products (unsaturated fatty acids), salt and temperature tolerance, resistance to predation and ease of harvesting. A variety of techniques can be used for isolation of microalgae, including physical extraction from crude water samples (micromanipulation), dilution to resolve individual cells, antibiotic selection and enrichment cultures using specific selection pressures (photoautotrophic conditions). Individual algal strains can be purified based on traditional method or high throughput FluorescenceActivated Cell Sorting (FACS) approaches before the resulting axenic cultures are cryopreserved for storage to prevent genetic drift. Screening of algae is often based on optimal growth or the production of specific metabolites and the use of response surface modeling and principle component analysis to identify these conditions and the key variables controlling those [10].

Citation: Raja R, Shanmugam H, Ganesan V, Carvalho IS (2014) Biomass from Microalgae. Oceanography 2: 118. doi:10.4172/2332-2632.1000118.

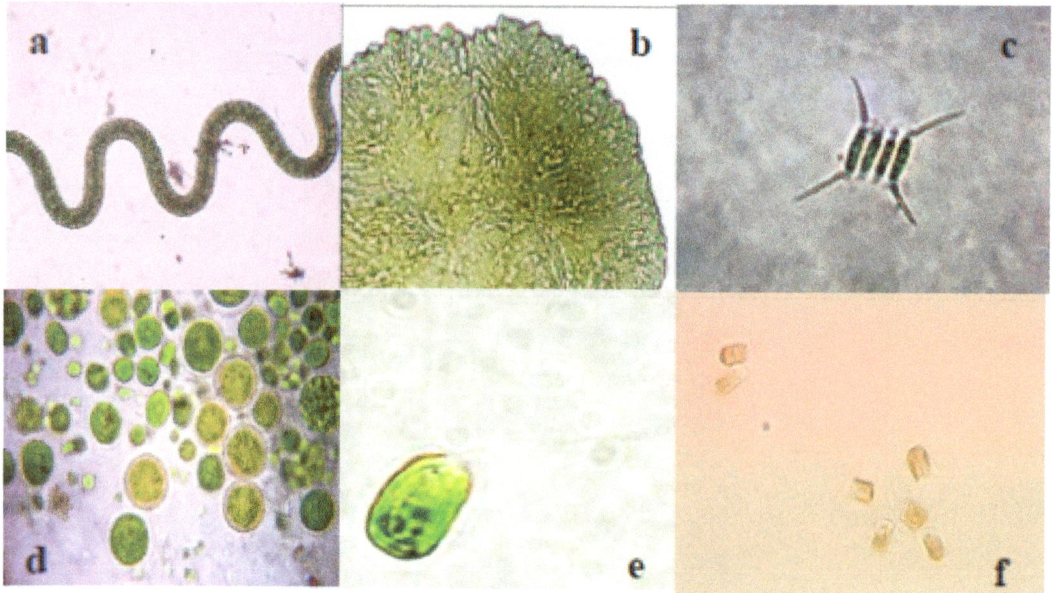

Figure 1: Few commercially important microalgal strains. (a) Arthrospira maxima (b) Botryococcus brauanii (c) Scenedesmus quadricauda (d) Chlorella vulgaris (e) Dunaliella salina (f) Chaetoceros muelleri.

It is important to select the strain that grows well in the particular geographical location and environment, as it easier to manipulate acclimatized strains. Most companies have recognized the importance of utilizing the highest performing algal strains, whereas some favour screening locally harvested natural strains others prefer to genetically modify strains [11]. The selected marine microalgae strains would be suitable for maximum biomass production, where the total biomass composition includes, total caloric value of the biomass, % lipids and lipid composition (for biodiesel), % starch and carbohydrate composition (for bioethanol and to identify higher value byproducts, % protein and protein composition (soluble/insoluble for food/feed purposes). There are lots of important characters ideally be (quantitatively) measurable. For these criteria one should do an outdoor cultivation to get maximum as well optimum biomass in large scale systems to commercialize the product [2]. Productivity of specific biomass would help to increase production of components such as fine chemicals, nutraceutical and lipids.

Strain development and nutrients

For autotrophic growth, microalgae require many elements, which fall under two categories macronutrients (required in g/l concentrations) and micronutrients (required in mg/l concentrations). In addition to macronutrients such as inorganic elements (P, N, and C), microalgae require sufficient light, and favorable temperatures to grow. Hydrogen (H) and oxygen (O) are also essential for algal growth. Micronutrients are elements are needed for algal growth in trace amounts such as calcium (Ca), magnesium (Mg), sodium (Na), potassium (K), iron (Fe), manganese (Mn), sulfur (S), zinc (Zn), copper (Cu), and cobalt (Co) [5]. It is also important to understand that different algal species do not have identical nutritional needs, and it also depends on their source of environment such as freshwater, marine, halotolerant/halophilic species. Screening of algae is often based on optimal growth or the production of specific metabolites and the use of response surface modelling and principle component analysis to identify these conditions and the key variables controlling the production of microalgal biomass [10]. Alternatively, since microalgae represent a much simpler system for genetic manipulations compared to higher plants, they could be exploited for strain improvement. Additionally microalgae have significant advantages such as they have life cycles of hours or days instead of seasonal cycles, their unicellular nature and absence of cell differentiation helps in miniaturization of breeding systems and thereby the cost; they can replicate both sexually and asexually making the versatile tools for generation of higher genetic diversity compared to prokaryotic organisms and finally selection, screening and UV and chemical mutagenesis can be more easily applied [10].

But both genetically and non-Genetically Modified Organisms (GMOs and non-GMOs) can be screened for efficient phenotypes. Genetic modifications in microalgae through a wide range of target mechanisms such as stress tolerance, product accumulation pathways, cellular chlorophyll contents, novel metabolic pathways, resistance to pathogens and competition, etc, make them better suitable for industrial applications. For instance, development of new genetic transformation tools for will provide metabolic engineering opportunities for overproduction of astaxanthin in microalgae. One possible application is to enhance astaxanthin production in H. pluvialis or C. zofingiensis by overexpressing PSY and CrtR-b genes, which previously have been shown to be possible rate-limiting steps for astaxanthin synthesis in H. pluvialis [12-14]. Additionally, an alternative strategy for enhancing astaxanthin production in microalgae is through random mutagenesis. Through chemical mutagenesis, several Haematococcus mutants with improved phenotypic traits have been generated, and one such example is Haematococcus astaxanthin-hyper-accumulating mutant strain MT 2877 [15].

There has been lot of advancements in algal genetic engineering, including genetic transformation, gene expression to use molecular genetics to advance algal biofuel production. An efficient transformation method was developed for Nannochloropsis sp., a fast-growing, unicellular alga capable of accumulating large amounts of oil by generated knockouts of the genes encoding nitrate reductase and nitrite reductase, resulting in strains that were unable to grow on nitrate and nitrate/ nitrite, respectively [16]. More recently, in a Chlamydomonas reinhardtii mutant with inactivated ADP-glucose pyrophosphorylase, an enzyme involved in starch synthesis, the triacylglycerol content exhibited a tenfold increase, from 2 to 20.5%, under nitrogen starvation [14]. Genetically modified microalgae can offer varied advantages, which is achieved by engineering the metabolic pathway to flux the metabolites towards the desired metabolite production or altering the regulatory factors such as transcription factors for over-expression of specific genes. On the whole collection, cultivation, selection and breeding of naturally occurring strains with high productivity or genetically modified microalgae with desired phenotype is required for both increase in biomass as well as high value product production.

Maintenance of axenic culture

Microlgae isolated from natural environments or genetically modified organisms are unlikely to be optimally adapted to the new environment, either in open pond cultivation or using photobioreactors. Contamination by predators and other fast growing heterotrophs have restricted the commercial production of algae. There are many factors that determine the species dominance such as resource-growth rate relationships for different algae, variable environmental conditions, inhibition of one organism by another through excrete dallelopathic substances, loss of growth due to predation or sinking. Free fatty acids, cell wall degradation products, exo-metabolites produced by algae and cyanobacteria, such as cyanobacterin [17] and fischerellin [18] could also have allelopathic potential. The effect of single or specific variables or combination variables on the growth of organisms in a laboratory scale can be made. The key issue is difficulty in simulation of the outdoor environment in the laboratory conditions or extrapolation of data from lab to mass culture condition. The problem of species dominance and competition in outdoor mass cultures can be controlled by variations of biota, light, temperature, pH, oxygen and nutrient supplies. For example, high ammonia or pH will inhibit most zooplankton infestations, or selecting a growth environment for specific microalgae species such as a very high alkalinity selects for Arthrospira and high salinity for Dunaliella, the two major species. These techniques are expensive and result in severe reduction of algal productivities. High density inoculations and semi-batch operations could reduce contamination in Chlorella production, but this often resulted in reduced overall productivity. Both Dunaliella and Chlorella are dominant in their optimal environments. Contamination can be better managed in closed photobioreactors, but upon continuous cultivation both open and closed systems become more susceptible. Careful strain isolation and characterization, cultivation parameters of individual microalgal candidates and the expected contaminants in the region can reduce the contamination issues [19].

Biomass production

Microalgae is the best choice for recombinant protein productions, for production of fine chemicals, pharma products, poultry feeds, feed stock and the source of future biofuel (biodiesel, hydrogen, methane and bioethanol) because they have simple and inexpensive growth requirements (free seawater, inexpensive nitrogen, phosphorus, and carbon sources), rapid growth rates with sufficient light. The high density and large-scale cultivation of microalgae is the preferred way for the development and production of these high-value products. This could be achieved by either growing algae in closed photobioreactors or open-pond systems.

Photobioreactor: Basically, Photobioreactors (PBRs) are closed cultivation systems to grow microalgae under photo-autotrophic conditions [20]. Ideal growth conditions for microalgal cultures are strain specific and the biomass productivity depends upon many factors. These include abiotic factors like temperature, pH, water quality, minerals, carbon dioxide, light cycle and intensity. Water, nutrients and CO2 are provided in a controlled way, while oxygen has to be removed. All parameters (nutrients, light regime, gas exchange) are maintained to realize optimal culture conditions. The contamination level is much lower compared to open systems.

Open raceway ponds: Open pond systems are shallow ponds (a maximum of 70 cm) in which algae are cultivated. Nutrients can be provided by mixing with few litres of water nearby paddle wheels. The water is typically kept in motion by paddle wheels and some mixing can be accomplished by appropriately designed guides. Raceway ponds are being in use in Israel, USA, China, India, Taiwan, Japan and other countries. Fertilizer is used for mass cultivation and the culture is agitated gently by paddle wheel (Figure 2d). The advantages of open ponds are low costs and that they are easy to operate. However, they are sensitive to contamination leading to introduction of unwanted fast growing organisms in the ponds. Especially heterotrophic organisms will graze on the autotrophic biomass and lead to loss of productivity. Therefore, the present commercial production of microalgae in open culture systems is restricted to only those organisms that can grow under extreme conditions (i.e.) high pH or salinity. Thus, a limited range of microalgae can be maintained as monoculture in open ponds in long-term operation. Currently there is lot of algae mass cultured and marketed commercially they are Skeletonema sp, Chaetoceros sp, Dunaliella sp, Arthrospira in high alkalinity and Chlorella in high levels of nutrients have been successfully cultivated.

Figure 2: (a) Culture scale-up for open raceway ponds (b) A typical lab-scale photobioreactor (c) French press (side view) (d) Open raceway pond (e) Flocculated culture shows algal clumping (f) CO2 cylinder (g) Culture storage tank (h) French press (front view) (i) Wet algal biomass collected from French press (j) Hot air dryer (Courtesy: Aquatic Energy LLC, Louisiana).

Biomass harvest

Biomass harvesting is a kind of technique and it accounts 15-20% of the production costs. The very small size of algae and their low concentration in the culture medium makes the cell recovery, harder one. The harvesting cannot be done by a single process because of the several species of algae with varying characteristics like shape, size and motility that influence to a big extent for their settling. Centrifugation is one of the most commonly used techniques to harvest microalgae in a lab scale as well in R&D laboratories. Algae are commonly used as a feed for aquaculture and they harvested by centrifuging to produce concentrates with longer shelf-life. Filter presses are used to recover fairly large microalgae like Arthrospira sp but are not suitable for smaller microalgae like Scenedesmus, Dunaliella, and Chlorella [19]. Using centrifuge in algal industries to harvest biomass is not a recommended and feasible one. Algae with autoflocculation characteristics simplify the harvesting step and minimize the energy and cost in the downstream processing of algae production [21]. The harvested algae biomass is dried so that the product can be stored without spoilage when algae are used as aquaculture feed (Figure 2).

Economic importance of algal biomass

Health supplements: Commercial use of microalgae as sources of specific chemicals began with D. salina for the production of β-carotene in the 1970's [1] followed by the use of Haematococcus pluvialis Flotow as a source of astaxanthin [22] and Crypthecodinium cohnii Seligo for long-chain Polyunsaturated Fatty Acid (PUFA) and Docosahexaenoic Acid (DHA) [23]. In the past few decades, microalgal biomass was directly utilized in the health food market with more than 75% of the annual microalgal biomass production being used for the manufacture of powders, tablets, capsules, or pastilles [2]. The two major species cultivated for this purpose are the unicellular green alga, Chlorella and more recently, filamentous blue-green alga (Cyanobacterium), Spirulina. A multitude of compounds are produced by microalgae which include Polyunsaturated Fatty Acids (PUFA), carotenoids, phycobiliproteins, polysaccharides and phycotoxins (Table 1).

Animal feed and aquaculture: Universally, more than 40 species of microalgae are used in aquaculture. Microalgae provide food for zooplankton they also help to stabilize and improve the quality of the culture medium. Indeed, for numerous freshwater and seawater animal species, the introduction of phytoplankton to rearing ponds leads to much better results in terms of survival, growth and transformation index [24]. The excreted biochemical compounds induce behavioural processes like initial prey catching and the regulation of bacterial population, probiotic effects and the stimulation of immunity [25,26]. Several factors can contribute to the nutritional value of a microalga (including its size and shape, digestibility, biochemical composition, enzymes, toxins and the requirements of animal feeding on the alga). Studies have attempted to correlate the nutritional value of microalgae with their biochemical profile [27]. However, results from feeding experiments that have tested microalgae differing in a specific nutrient are often difficult to interpret because of the confounding effects of other microalgal nutrients.

Microalgae grown to late logarithmic growth phase typically contain 30-40% protein, 10-20% lipid and 5-15% carbohydrate [28]. In the stationary phase, the proximate composition of microalgae can change significantly (e.g.) when nitrate is limiting, carbohydrate levels can double at the expense of protein [29]. There does not appear to be a strong correlation between the proximate composition of microalgae and nutritional value, though algal diets with high levels of carbohydrate are reported to produce the best growth for juvenile oysters, Ostrea edulis [30]. Larval scallops, Patinopecten yessoensis provided polyunsaturated fatty acids in adequate proportions. In contrast, high dietary protein provided best growth for juvenile mussels, Mytilus trossulus and Pacific oysters, Crassostrea gigas [31]. Microalgae are being in use for to improve the nutritional quality and gives coloration to the shrimps. Among these Chlorella grows well in nutrient-rich media, while Arthrospira sp requires a high pH of 9.5- 11 with appropriate concentration of bicarbonate. Similarly, Dunaliella salina grows at the very high salinity of 0.5-6 M [32].

Several algae such as Chaetoceros sp, Isochrysis sp, Skeletonema sp, Thalassiosira sp, Tetraselmis sp and Crypthecodinium cohnii are being useful in the aquaculture industry which does not have these selective advantages, must be grown in closed systems. Table 2 summarizes the commercial algal culture and their uses. Factors to be considered for production of microalgae include: the biology of the alga, cost of land, labor, energy, water, nutrients (climate if the culture is outdoors) and the type of final product. Microalgae are necessary from the second stage of larval development (zoea) and in combination with zooplankton from the third stage (myses). Naturally occurring microalgal blooms are encouraged in large ponds with low water exchange where the larvae are introduced. Sometimes fertilizers and bacteria are added to induce more favorable conditions. A sustainable and profitable biodiesel production from microalgae is possible. The biofuel can overcome the energy and environmental needs by integrating new technologies [33]. Large quantities of algal biomass needed for the production of biodiesel could be grown in photobioreactors combined with photonics and biotechnologies. However, more precise economic assessments of production are necessary to establish with petroleum derived fuels. The direct hydrothermal liquefaction is an energy-efficient technique for producing biodiesel from algae without the need to reduce the water content of the algal biomass [34].

Table 1: High value products from microalgae

Microalgae	Products	References
Pavlova, Nannochloropsis, Phaeodactylum	Eicosapentaenoic acid (EPA)	Borowitzka and Borowitzka [1], Chisti [37] Raja, et al. [2]
Crypthecodiuimu and *Schizochytrium*	Docohexaenoic acid (DHA)	
Spirulina	γ-linolenic acid (GLA)	
Porphyridium	Arachidonic acid (AA)	Borowitzka and Borowitzka [1], Chisti [37]
Spirulina platensis	Phycocyanin	
Porphyridium cruentum	Phycoerythrin Polysaccharides	
Dunaliella salina	β-carotene	
Haematococcus pluvialis	Astaxanthin	Borowitzka and Borowitzka [1], Chisti [37] Raja, et al. [2]
Aphanizomenon flosaquae	Mycosporine-like amino acids (MAA)	

Table 2: Commercial algal culture and its applications (Hemaiswarya et al.) [46]

Genus	Morphology	Purpose
Nannochloropsis sp	Small green algae	Growing rotifers and in fin fish hatcheries, used in reef tanks for feeding corals and other filter feeders, very high EPA level
Pavlova sp	Small golden-brown flagellate, very difficult to grow so it is not produced by many hatcheries	Used to increase the DHA/EPA levels in broodstock, oysters, clams, mussels and scallops, sterol composition so it is popular with cold water fish hatcheries (cod) for enriching rotifers
Isochrysis sp	Small golden-brown flagellate	Enrichment of zooplankton such as Artemia, used in shellfish hatcheries and used in some shrimp hatcheries, good size for feeding brine shrimp and copepods, oysters, clams, mussels, and scallops
Tetraselmis sp	Large green flagellate	Excellent feed for larval shrimps and contains natural amino acids that stimulate feeding in marine animals, used in conjunction with *Nannochloropsis* for producing rotifers, good size for feeding brine shrimp, standard feed for oysters, clams, mussels, and scallops, excellent feed for increasing growth rates and fighting zoea syndrome
Thalassiosira weissflogii	Large diatom	Used in the shrimp and shellfish larviculture, considered by several hatcheries to be the single best alga for larval shrimps, also good for feeding copepods and brine shrimps, post-set (200l and larger) oysters, clams, mussels, and scallops for brood stock conditioning
Dunaliella sp	Small green flagellate	Used to increase vitamin levels in some shrimp hatcheries and also for the coloration
Chaetoceros sp	Diatom	Used to increase vitamin levels in some shrimp hatcheries

Microalgae have high growth rates and photosynthetic efficiencies due to their simple structures. The efficiency is much higher (6- 8%) than that of terrestrial plants (it is, 1.8-2.2%). The idea of using microalgae is not new, but it is now being taken seriously in several countries because of the emerging concern about global warming that is closely associated with burning of fossil fuels [35]. Microalgal biomass contains approximately 50% carbon by dry weight; therefore, it is also used to produce methane by anaerobic digestion. The process is technically feasible, but it cannot compete with many other low-cost organic substrates that are available for anaerobic digestion [36]. Depending on species, microalgae produce different kinds of lipids, hydrocarbons, and other complex oils [37,38]. The extent of unsaturation of microalgal oil and its content of fatty acids with more than four double bonds can be reduced easily by partial catalytic hydrogenation of the oil. The challenge lies in harvesting algal biomass and the extraction of biodiesel [39,40]. The heterogeneity of algal species and growth parameters makes this bio-inspired option a technical challenge for scale-up consideration.

Greenhouse gas reductions and global warming

The combustion of fossil fuel generates carbon dioxide, a major greenhouse gas that is considered as a huge threat because of its potential to cause severe global warming. Microalgae are particularly considered for bio-fixation because of their ability to grow fast and fix greater amounts of carbon dioxide. The bio-mitigation of carbon dioxide and other flue gases by microalgae have significantly gained interest in reducing the emissions from coal-fired power plants. The algae biomass thus produced by capturing carbon can be used in generating valuable products such as fuel, animal feed and fertilizer. Power plants are the major sources of CO2 and release 5.7 giga tones of carbon dioxide per year [41]. CO2 is produced by both stationary and mobile sources. Microalgae offer a natural way to recycle carbon dioxide from the flue gas and thus help in reducing the effects of global warming and climate change. Microalgal biomass is composed of 45 to 50% carbon based on dry weight measurements [42]. The high carbon content of microalgae makes it suitable for storing carbon. CO2 present is flue gas can significantly raise the growth rates of microalgae. Microalgae can be engineered in open ponds or photobioreactors to maximize CO2 conversion to biomass thereby sequestering carbon and also producing a biofuel. The selection of microalgae is the most important factor in the bio-mitigation of carbon dioxide from flue gases generated by power plants. The algae should have high growth and CO2 utilization rates. The other suitable characteristics for carbon dioxide bio-fixation are their ability to tolerate SOx and NOx, and thrive in mass cultures without contamination.

Prominent issues in commercializing microalgae

In the current situation, commercially biodiesel is not produced from microalgae [36]. One of the biggest challenges is to reproduce the closed laboratory conditions on a large scale. It is understood that in the lab, it is easier to produce oil [43] and several companies are attempting to commercialize microalgal biodiesel [43,44]. Oil productivity is directly related with algal growth rate and the oil content of the biomass. Hence, it is well known that the microalgae with high oil productivities are desired for producing biodiesel. However, producing biomass is generally more expensive than growing crops. Temperature must remain within 20 to 30°C. To minimize the expense, biodiesel production must rely on freely available sunlight, despite daily and seasonal variations in light levels [38]. The crucial economic challenge for algae producers is to discover low cost oil extraction and harvesting methods [45]. Singh and Gu [45] suggested that the utilization of fatter algae with higher oil content (around 60%) in comparison to lower oil content algae can reduce up to half of the size and footprint of algae biofuels production systems and reduces the capital and operating costs. In addition, a cheaper and easier process can provided a better ground to commercialize. Using a non-lethal extraction called milking, can avoid a serious of processes such as harvesting, extraction etc. Carotenoids, high value lipids have also been selectively extracted from the green alga Chlorella sp. using decane [46]. Such methods have the capability to reduce the production cost significantly and simplify the biodiesel production process from algal biomass.

CONCLUSION

As far as the algal biomass production has concerned, currently there are lot of projects going on in many countries particularly in India, China, United States, European Union etc. and they are using these harvested biomass for multitude of applications. Algae also have the scope of growing in municipal sewages (or from effluents) containing heavy metals and unwanted chemicals which are bio-absorbed. This is the best biological method recommended by naturalists and scientists to protect the environment clean. At the same time, algae capturing atmospheric CO2 as well as contribute to reduce global warming considerably. The greatest problem in algae biomass production development of superior strains of algae is with hurdles in collection from natural habitats, isolation, selection, maintenance, and genetic improvement.

ACKNOWLEDGEMENTS

The authors would like to thank the Foundation for Science and TechnologyFCT, Portugal for their funding and also wish to express their sincere gratitude to Dr. Velusubramani, Senior Scientist, Beyond Petroleum, Naperville, Illinois, USA and Prof. R. Manivasakan, Indian Institute of Technology, Chennai, India for their critical review of the manuscript.

REFERENCES

1. Borowitzka MA, Borowitzka LJ (1988) Dunaliella. Microalgal biotechnology, Cambridge University Press, New York, USA.
2. Raja R, Hemaiswarya S, Ashok Kumar N, Sridhar S, Rengasamy R (2008) A perspective on the biotechnological potential of microalgae. Crit Rev Microbiol 34: 77-88.
3. Park J-H, Yoon J-J, Park H-D, Kim YJ, Lim DJ, et al. (2011) Feasibility of biohydrogen production from Gelidium amansii. Int J Hydrogen Energ 36: 13997-14003.
4. Karthikeyan S (2012) A critical review: microalgae as a renewable source for biofuel production. International Journal of Engineering Research and Technology 1: 1-6.
5. Raja R, Anbazhagan C, Ganesan V, Rengasamy R (2004) Efficacy of Dunaliella salina (Volvocales, Chlorophyta) in salt refinery effluent treatment. Asian J Chem 16: 1081-1088.
6. Williams PJB, Laurens LML (2010) Microalgae as biodiesel & biomass feedstocks: Review and analysis of the biochemistry, energetics and economics. Energy Environ Sci 3: 554-590.
7. Pienkos PT, Laurens L, Aden A (2011) Making Biofuel from Microalgae: So much potential coexists with so

many scientific, environmental and economic challenges. American Scientist.

8. Lam MK, Lee KT (2012) Microalgae biofuels: A critical review of issues, problems and the way forward. Biotechnol Adv 30: 673-690.

9. Hemaiswarya S, Raja R, Carvalho IS, Ravikumar R, Zambare V, et al. (2012) An Indian scenario on renewable and sustainable energy sources with emphasis on algae. Appl Microbiol Biotechnol 96: 1125-1135.

10. Larkum AW, Ross IL, Kruse O, Hankamer B (2012) Selection, breeding and engineering of microalgae for bioenergy and biofuel production. Trends Biotechnol 30: 198-205.

11. Lundquist TJ, Woertz IC, Quinn NWT, Benemann JR (2010) A Realistic Technology and Engineering Assessment of Algae Biofuel Production. Energy Biosciences Institute, University of California, Berkeley, California, USA.

12. Li Y, Sommerfeld M, Chen F, Hu Q (2008) Consumption of oxygen by astaxanthin biosynthesis: a protective mechanism against oxidative stress in Haematococcus pluvialis (Chlorophyceae). J Plant Physiol 165: 1783-1797.

13. Li Y, Han D, Hu G, Dauvillee D, Sommerfeld M, et al. (2010) Chlamydomonas starchless mutant defective in ADP-glucose pyrophosphorylase hyperaccumulates triacylglycerol. Metab Eng 12: 387-391.

14. Li Y, Sommerfeld M, Chen F, Hu Q (2010) Effect of photon flux densities on regulation of carotenogenesis and cell viability of Haematococcus pluvialis (Chlorophyceae). J Appl Phycol 22: 253-263.

15. Hu Z, Li Y, Sommerfeld M, Hu Q (2008) Enhanced protection against oxidative stress in an astaxanthin-overproduction Haematococcus mutant (Chlorophyceae). Eur J Phycol 43: 365-376.

16. Kilian O, Benemann CSE, Niyogi KK, Vick B (2011) High-efficiency homologous recombination in the oil-producing alga Nannochloropsis sp. P Natl Acad Sci USA 108: 21265-21269.

17. Gleason FK, Baxa CA (1986) Activity of the natural algicide, cyanobacterin, on eukaryotic microorganisms. FEMS Microbiol Lett 33: 85-88.

18. Hagmann L, Jüttner F (1996) Fischerellin A, a novel Photosystem-II-inhibiting allelochemical of the cyanobacterium, Fischerella muscicola with antifungal and herbicidal activity. Tetrahedron Lett 37: 6539-6542.

19. Hemaiswarya S, Raja R, Ravikumar R, Carvalho IS (2013) Algae taxonomy, genetics and breeding. Biofuel Crops: Production, Physiology and Genetics, CABI, GA, UK.

20. Richmond A (2004) Basic Culturing Techniques and Downstream processing of cellmass and products. Handbook of Microalgal culture-Biotechnology and Applied phycology. Blackwell publishing, Oxford, UK.

21. Brennan L, Owende P (2010) Biofuels from microalgae-A review of technologies for production, processing, and extractions of biofuels and co-products. Renewable and Sustainable Energy Reviews 14: 557-577.

22. Lorenz RT, Cysewski GR (2000) Commercial potential for Haematococcus microalgae as a natural source of astaxanthin. Trends Biotechnol 18: 160-167.

23. Kyle DJ, Reeb SE, Sicoiie VJ (1998) Dinoflagellate biomass, methods for its production and compositions containing the same. USA Patent Number-5711983.

24. Muller-Feuga A (2000) The role of microalgae in aquaculture: situation and trends. J Appl Phycol 12: 527-534.

25. Hong HA, Duc le H, Cutting SM (2005) The use of bacterial spore formers as probiotics. FEMS Microbiol Rev 29: 813-835.

26. Raja R, Hemaiswarya S (2010) Microalgae and immune potential. Dietary components and immune function - prevention and treatment of disease and cancer. Humana Press/Springer, USA.

27. Durmaz Y (2007) Vitamin E (a-tocopherol) production by the marine microalgae Nannochloropsis oculata (Eustigmatophyceae) in nitrogen limitation. Aquaculture 272: 717-722.

28. Fujii K, Nakashima H, Hashidzume Y, Uchiyama T, Mishiro K, et al. (2010) Potential use of the astaxanthin-producing microalga, Monoraphidium sp. GK12, as a functional aquafeed for prawns. J Appl Phycol 22: 363-369.

29. Liang H, Gong W-J, Chen Z-L, Tian J-Y, Qi L, et al. (2009) Effect of chemical preoxidation coupled with in-line coagulation as a pretreatment to ultrafiltration for algae fouling control. Desalination Water Treat 9: 241-245.

30. Ponis E, Probert I, Veron B, Mathieu M, Robert R (2006) New microalgae for the Pacific oyster, Crassostrea gigas larvae. Aquaculture 253: 618-627.

31. Knuckey RM, Brown MR, Barrett SM, Hallegraeff GM (2002) Isolation of new nanoplanktonic diatom strains and their evaluation as diets for the juvenile Pacific oyster (Crassostrea gigas). Aquaculture 211: 253-274.

32. Raja R, Hemaiswarya S, Balasubramanyam D, Rengasamy R (2007) Protective effect of Dunaliella salina (Volvocales, Chlorophyta) on experimentally induced fibrosarcoma on wistar rats. Microbiol Res 162: 177-184.

33. Schenk PM, Thomas-Hall SR, Stephens E, Marx UC, Mussgnug JH, et al. (2008) Second Generation Biofuels: High-Efficiency Microalgae for Biodiesel Production. Bioenergy Research 1: 20-43.

34. Patil V, Tran K-Q, Giselrød HR (2008) Towards sustainable production of biofuels from microalgae. Int J Mol Sci 9: 1188-1195.

35. Omer AM (2012) Renewable energy, emerging energy technologies and sustainable development: innovation in power, control and optimisation. ARPN J Sci Technol 2: 13-30.

36. Hussain K, Nawaz K, Majeed A, Feng L (2010) Economically effective potential of algae for biofuel production. World Appl Sci J 9: 1313-1323.

37. Guschina IA, Harwood JL (2006) Lipids and lipid metabolism in eukaryotic algae. Prog Lipid Res 45: 160-186.

38. Chisti Y (2007) Biodiesel from microalgae. Biotechnol Adv 25: 294-306.

39. Jang ES, Jung MY, Min DB (2005) Hydrogenation for low trans and high conjugated fatty acids. Compr Rev Food Sci Food Safety 4: 22-30.

40. Dijkstra AJ (2006) Revisiting the formation of trans-isomers during partial hydrogenation of triacylglycerol oils. Eur J Lipid Sci Tech 108: 249-264.

41. Kadam KL (2001) Microalgae Production from Power Plant Flue Gas: Environmental Implications on a Life Cycle Basis. National Renewable Energy Laboratory, technical report.

42. Schlesinger WH (1991) Biogeochemistry: An analysis of global change. Academic press, California, USA.

43. Kwok R (2009) Cellulosic ethanol hits roadblock. Nature 461: 582-583.

44. Mascarelli AL (2009) Gold rush for algae. Nature 461: 460-461.

45. Singh J, Gu S (2010) Commercialization potential of microalgae for biofuels production. Renewable and Sustainable Energy Reviews 14: 2596 2610.

46. Hejazi MA, Kleinegris D, Wijffels RH (2004) Mechanism of extraction of beta-carotene from microalga Dunaliellea salina in two-phase bioreactors. Biotechnol Bioeng 88: 593-600.

CHAPTER 2

DYNAMIC MODELING OF THE MICROALGAE CULTIVATION

Matteo Marsullo, Alberto Mian, Adriano Viana Ensinas, Giovanni Manente, Andrea Lazzaretto and François Marechal

Keywords: Microalgae cultivation phase, open raceway pond, flat panel photobioreactor, repeated batch cultivation, carbon dioxide

INTRODUCTION

Microalgae are single cell organisms, which can be found in colonies or individual cells. Their most interesting characteristic is the ability of realizing a photosynthetic reaction in a single cell. They are extremely resistant and may grow in many different environments, from fresh water to marine and hyper-saline water (Le et al., 2010; Mata et al., 2010). Microalgae can be seen as an interesting alternative to more typical biomass as a source for biofuels production. Studies on microalgae have been carried out since the 1980s, but in recent years their importance has grown fast: as Chisti (2007) claims, the reason is that microalgae appear to be the only source of biodiesel that has the potential to completely displace fossil diesel. Moreover, compared to other biofuels sources, such as traditional crops and wood, microalgae have several advantages: they grow extremely fast, reaching high areal and volumetric productivities; they do not require arable land; they ask for less freshwater than normal crops, being able to use wastewater; they can directly capture CO_2 released by industries and they overtake the food vs. fuel debate; and their ability to grow in a wide range of conditions, resisting to severe temperatures, pH, and salinity (NREL, 0000), makes them even more attractive.

In 1978, the National Renewable Energy Laboratory of the United States of America started a 20-year program (Aquatic Species Program) to develop renewable transportation fuels from algae. The researches explored both the genetic engineering for manipulating the metabolism of microalgae and the engineering of microalgae production systems (NREL, 0000). In the last decades of the twentieth century, many other academic studies have been carried out all over the world, focusing on different aspects: the biological studies try to identify the best strain for a specific purpose, the engineering work tries to define open or closed systems to optimize the growth process and the system analysis works aim at defining the possible impact of the microalgae as a source of biomass for different applications, adopting a holistic approach.

Microalgae can be cultivated with the main purpose of producing lipids that can substitute biodiesel (Rodolfi et al., 2009) or produce higher value products that can be used as base chemicals for the production of biobased chemical products (Hempel et al., 2012). Cultivation of microalgae can be realized in open ponds systems where hydrodynamics (Hadiyanto et al., 2013) and CO_2 supply (Yang, 2011) are key drivers of the growth efficiency. Pollution and contamination related to mass transfer between open systems and environment together with the use of mechanical methods to obtain a proper mixing in the reactor make it impossible for a large number of microalgae to grow: only the most resistant strains might be used in open systems (Borowitzka, 1999). An alternative is the use of closed systems named photobioreactors that have the potential of maximizing the growth rate by controlling the growth conditions (pH, nutrients) and the gas supply at the expense of a much higher (factor ten) energy consumption. Examples are flat panel reactors or tubular reactors (Cuaresma et al., 2011). Studies have been conducted regarding the geometric characterization of flat panel photobioreactors, being the vertical positioning and east–west orientation the most suitable for microalgae proliferation (Sierra et al., 2008). Laboratory scale prototypes have been tested and compared to evaluate the relation between light availability, composition and growth rate of microalgae finding that high light availability leads to a change in the microalgae composition more than in a growth rate increase (Münkel et al., 2013). Tubular reactors need a much higher energy consumption (factor ten) than flat panel photobioreactors to maintain a proper mixing and circulation of the solution in the reactor and to ensure the oxygen removal, since there might be a decrease of the growth rate if oxygen saturation is reached (Molina et al., 2001). Mathematical models for light distribution have shown that photo-inhibition is a key problem for outdoor closed photobioreactors, both for

Citation: Matteo Marsullo, Alberto Mian, Adriano Viana Ensinas, Giovanni Manente, Andrea Lazzaretto and François Marechal, Dynamic modeling of the microalgae cultivation, http://dx.doi.org/10.3389/fenrg.2015.0004.

flat panel technology and tubular reactors, being 100–200 W/m^2 the saturation light intensity for most of microalgae species (Fernández et al., 1997).

The land use and the solar efficiency of the microalgae system require adopting a holistic approach for the microalgae production. On the one hand, it is important to assess the overall conversion chain to define the substitution potential of the microalgae with respect of the extracted products. It is, therefore, important to adopt a holistic vision considering the complete life-cycle assessment for both the biobased products and the fossil one that are substituted (Azadi et al., 2014). Recent studies analyze the microalgae cultivation and transformation technologies evaluating their environmental impact in terms of GHG emissions and non-renewable energy consumption (Azadi et al., 2014) through LCA methodology, resulting that microalgae drying phase and water (Stephenson et al., 2010) and nutrient (Collet et al., 2011) reuse are the crucial aspects from an environmental point of view. From techno-economic analysis, irradiation conditions, mixing, medium, and CO_2costs, together with dewatering technology, are the most important cost factors (Norsker et al., 2011). The opportunity to use microalgae strains capable of sustaining high growth rates and high lipid content give to costs the potential for significant improvement in the future (Davis et al., 2011).

The goal of this paper is to develop a comprehensive modeling framework to assess and compare different microalgae cultivation technologies in a consistent and holistic manner. A dynamic model of microalgae cultivation is developed. It aims at representing the influence of the most significant parameters on the system productivity and consumptions. The model is able to evaluate the variation of the most significant parameters, which influence microalgae growth, and therefore to estimate the microalgae biomass productivity and the energy productivity and consumption. The cultivation technologies taken into account are the open raceway pond and the flat panel photobioreactor: they are considered the most suitable technologies for extensive microalgae cultivation for biofuel production (Jorquera et al., 2010). Microalgae growth rate depends on several factors, such as light accessibility and nutrient availability, temperature, pH and salinity of the water, and CO_2 and O_2 concentration in the water. As all these terms are not constant in time, the model has to consider the dynamic behavior of the system and integrate the possible control strategies to be adopted to optimize the system performance.

The paper aims at including the effect of all the physical and chemical parameters that mostly affect microalgae growth, giving a consistent and robust modeling framework in which the contributions of recent publications are improved to interact together. Repeated batch operating strategy byRadmann et al. (2007) has been applied to microalgae growth model and mass balances by Yang (2011) together with thermal balances by Slegers et al. (2013) and pH control strategy by Sills (2013); reactor geometry has been taken into account for both the flat panel photobioreactor and the open pond to define the effect of solar radiation to microalgae growth, using the Beer–Lambert equations, and geometric evaluations by Duffie and Beckman (2013), included in Slegers et al. (2011).

MATERIALS AND METHODS

Reactor Geometry

Two photo bioreactor technologies are considered in this work, which are likely to be the most promising technologies for extensive microalgae production.

Open raceway pond is the most used artificial system for microalgae cultivation. The geometry used in this work is a closed oval-shaped channel: the most common values for pond depth are between 0.2 and 0.5 m [0.3 m being the most used value (Chisti, 2007)]. Circulation around the oval ring and mixing are guaranteed with the use of a mechanical rotating device, usually a paddlewheel (Brennan and Owende, 2010). The paddlewheel is in continuous operation to prevent sedimentation, giving the water a speed between 0.15 and 0.25 m/s, the most commonly used velocity range for open pond microalgae cultivations (Doucha and Lívanský, 2006; Hadiyanto et al., 2013). To satisfy the microalgae CO_2 requirement, the open pond is equipped with submerged aerators. The main characteristics of the raceway pond used in the model have been taken from the literature (Borowitzka, 1999; James and Boriah, 2010; Norsker et al., 2011; Yang, 2011; Sompech et al., 2012) and presented in Table 1 and Figure 1. Open cultivation technologies have several problems, such as lower volumetric productivity compared to closed systems [0.1 compared to 1.5 kg m^{-3} day^{-1} (Chisti, 2007)], evaporation of the water through the open surface, contamination which causes a limitation in species sustainability and the need for large land area (Singh and Dhar, 2011).

Table 1. Open pond characteristics

Pond surface	1 ha
Length to width ratio	10
Pond depth	30 cm
Speed of the water	0.20 m/s
CO_2 absorption system efficiency	0.9

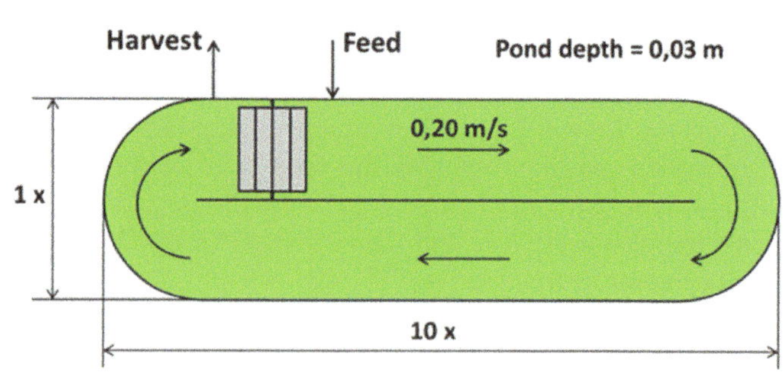

Figure 1. Raceway open pond.

Closed photobioreactors are designed to overcome the limitations of open systems. They have higher efficiency and biomass productivity, shorter harvesting times (2–3 days compared to 7–10 days for open pond), high surface-to-volume ratios, reduced contamination risks, and can be used to cultivate wider range of algal species than open systems (Bahadar and Bilal Khan, 2013). However, closed systems are more expensive to be constructed since they need high quality materials, and difficult to operate and scale up (Muñoz and Guieysse, 2006).

Flat panel photobioreactor is a flat, transparent vessel, made of glass, Plexiglas or plastic or other transparent material. The mixing of the water is carried out directly with air bubbling, which is introduced via a perforated tube at the bottom of the reactor. Flat panel PBRs are never thicker than 5–6 cm as the light entering the panel would not penetrate deeper in the culture. Only panels with both a height and width of <1 m have been studied (Janssen et al., 2003). The flat panel reactors are positioned vertically, closely spaced, to reach a higher photosynthetic efficiency through self-shading of the panels (Rawat et al., 2013). As for the open raceway pond, the main geometric characteristics have been taken from the literature (Janssen et al., 2003;Sierra et al., 2008; Pruvost et al., 2011; Ruiz et al., 2013; Kochem et al., 2014;Sugai-Guérios et al., 2014) and presented in Table 2 and Figure 2.

Table 2. Flat panel photobioreactor characteristics

Height of the panels	1.5 m
Thickness of the panels	5 cm
Distance between the panels	1 m
Reactor surface	1 ha
Operating temperature	Constant

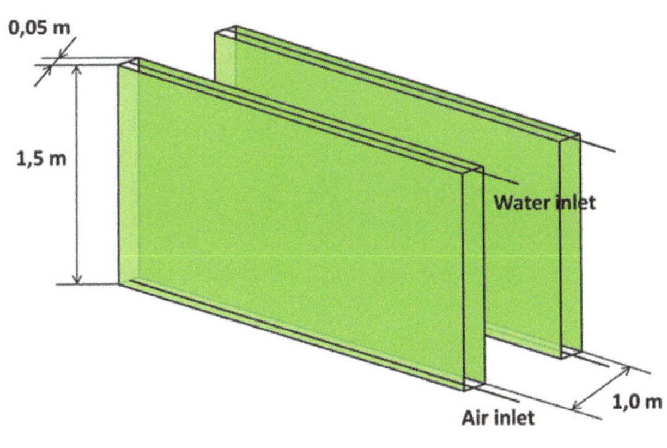

Figure 2. Flat panel photobioreactor.

Operating Strategy

The reactors can be operated either in continuous or in batch mode. The continuous operation requires a continuous extraction of the microalgae, while the batch operation extracts the produced microalgae after a certain time of production. An inoculum is, therefore, maintained in the reactor to serve as a basis for the new batch. The nutrient feeding strategy changes depending on the operating strategy: in continuous operation, two different reactors in series would be needed for lipid production, while in batch operation, a recipe for the batch feeding with nutrients has to be applied. As explained by Radmann et al. (2007), in the repeated batch cultivation after a certain period in the reactor, a specific culture volume is removed and replaced with an equal amount of fresh medium. Consequently, a part of cultivation medium is kept in the reactor as a starting inoculum. Repeated batch cultivation presents several operational advantages, the most important of which are the maintenance of a constant inoculum and high growth rates.

The operating strategy is also influenced by two considerations regarding the upstream and downstream processes. The downstream processing includes a settler tank where the extracted microalgae undergo a first concentration by sedimentation. The separated water is recycled while the products enter the downstream processing step. In this situation, the time required by the microalgae to grow can be decoupled from the time of the downstream technology. Second, in case of shortage of nutrients, the microalgae growth rate decreases as well as the production of new biomass, while the quantity of lipids in the biomass increases: this growth condition is particularly interesting if the downstream process aims at producing biodiesel. The models presented in this work do not consider the input condition of scarcity of nutrients, which instead are always supplied in excess.

MODEL

Cultivation Model Structure

The mass and energy produced by the cultivation systems in the form of biomass are calculated by the mathematical model through time-dependent mass and energy balances (Figure 3).

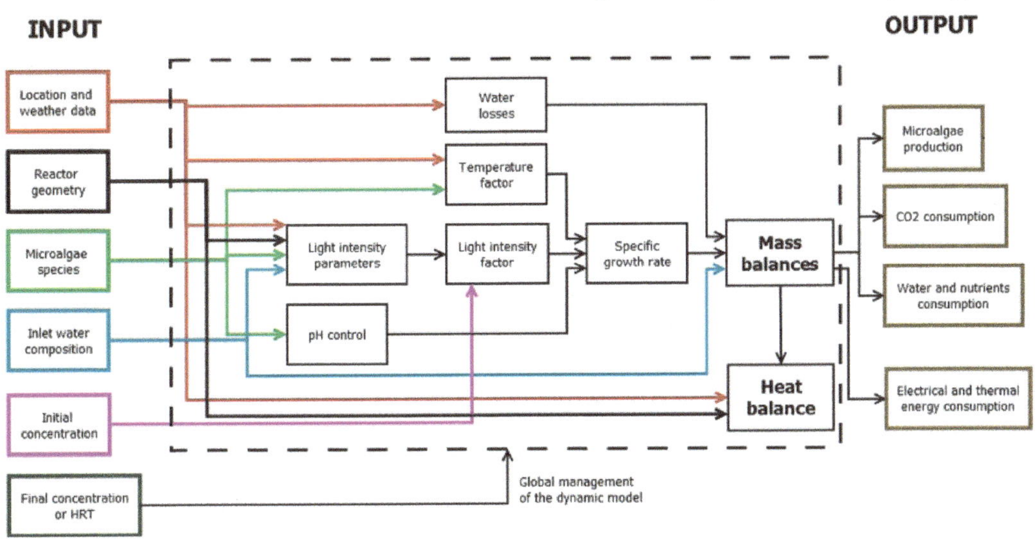

Figure 3. Model structure.

Before calculating the results of these balances, it is necessary to introduce some auxiliary equations. The auxiliary equations include the microalgae growth model, the pH control through CO_2 injection, and the equations to evaluate the radiation affecting the microalgae cells.

A set of differential equations is developed: it is solved by an explicit integration scheme with fixed time decomposition. To simplify the resolution of this system, the finite differences approximation was adopted. All the differential equations were treated as finite differences, and the backward difference formula was implemented

$$\frac{dx}{dt} = f(t) \rightarrow \frac{x(t) - x(t-1)}{\Delta t} = f(t) \tag{1}$$

This expression was preferred to the forward difference formula, being less vulnerable to instability.

If the backward difference formula is applied to matrices and vectors, the following expression is obtained:

$$\bar{x}(t) = F(t) * \Delta t + \bar{x}(t-1) \tag{2}$$

The model is a differential algebraic equations system that is converted into a set of algebraic equations by defining the proper solving sequence: some of the differential equations depend on one another, making impossible to solve them in a sequential logic: it is, therefore, necessary to solve them altogether as a system. Therefore, all the equations were manipulated to separate the known terms from the unknown variables and to obtain an easily solvable system of algebraic equations.

A dynamic approach to the resolution of energy and mass balances has been preferred to a steady state analysis because the laws describing the microalgae growth are time dependant; moreover, a dynamic model gives the opportunity of a deeper and more accurate evaluation of microalgae production and energy consumption over the time period.

Inputs to the Model

The following inputs are required by the model.

- Location data: latitude and longitude of the reactor are needed together with the hour difference between the location of the bioreactor and Greenwich.
- Weather data are also required by the model both to calculate the irradiation factor, which influences the microalgae growth and to evaluate the thermal balance. The open raceway pond model requires the global irradiation over a horizontal surface, whereas the flat panel model requires the direct and the diffuse radiation over a horizontal surface as two distinct data. Moreover, both the two models require other weather data, such as the atmospheric temperature, wind speed, and relative humidity. Hourly data are used in all models.
- Microalgae species characterization: many parameters of the model depend on this input, such as the maximum growth rate, the optimal growth temperature, and the saturation light irradiation. All the characteristics assume different values depending on the microalgae species. Table 3presents the model parameters considered for two different microalgae strains. The model assumes a constant pH that is kept at the optimal value for microalgae species using CO_2 injection: to evaluate the correct quantities of CO_2 to be injected, the pH optimal value is represented as an optimal value of alkalinity of the reactor. LHV is the lower heating value for wet microalgae biomass. I_S is the saturation light intensity, μ_{max} is the maximum microalgae growth rate, T_{let} is the lethal temperature, T_{opt} is the optimal temperature for microalgae growth (maximum growth rate), and β is a curve modulating constant for the temperature factor. The decay rate is a coefficient that is used to evaluate the microalgae mass losses. K_C and $K_{NA}$are the half-saturation constants, i.e., the compound concentration when the specific growth rate [day^{-1}] is μA=μ^A/2μA=μ^A/2. From Yang (2011), K_C is set to 0.001 molCO2/m30.001 molCO2/m3 while K_{NA} = 0.001 mol_{NA}/m^3. When nitrogen and carbon dioxide concentrations are next to the saturation values, these terms assume a value close to 1 and they do not affect the microalgae growth.
- The composition of the inlet water, which is used to fill the reactor at the beginning of each cultivation: since each cultivation cycle works as a batch reaction, nutrients must be supplied in sufficient quantities to feed the microalgae for the entire duration of the batch process. The model gives as input a surplus of nutrients in the water to guarantee they are not the limiting factor to microalgae growth. Then, as output, the model gives the exact quantity of nutrients consumed by the microalgae.
- The operating strategy, which is used to predict microalgae production. Two different strategies can be adopted: if the objective consists in searching the hydraulic retention time (HRT) required to reach a certain microalgae concentration in the reactor, the initial and the final microalgae concentrations are to be supplied as input information to the model. If the HRT is fixed along the whole time period, then the input data will be the initial concentration and the HRT, leaving the final concentration in the reactor free to change.

Table 3. Microalgae species characteristics

Typology	*P. tricornutum*	*T. pseudonana*
pH_opt	8.3	8.3
alk_opt [meq/l]	0.032	0.032
LHV [kJ/kg]	21,527	21,527
I_S [W/m²]	37.118	21.834
μ_{max} [1/day]	1.392	3.288
T_{let} [°C]	30	31
T_{opt} [°C]	21	24
β	1.57	1.83
Decay rate [1/day]	0.048	0.048
K_{NA} [mol_{NA}/m^3]	0.001	0.001
K_C [mol_C/m^3]	0.001	0.001

Outputs of the Model

The mass production of microalgae is the most significant output of the cultivation model. In addition to this parameter, it is possible to define two other global performance parameters: the volumetric productivity {expressed in [g/(l × day)] or in [kg/(m³ × day)]} and the areal productivity {expressed in [t/(ha × y)] or in [kg/m² × day)]}. These parameters refer to the volume of the reactor and to the area of the soil covered by the reactor. Moreover, the model is able to give as output the exact quantities of all the compounds consumed or produced in the time period considered by the simulation (usually 1 year). For example, the model gives the CO_2 that is needed to be bubbled in the reactor to maintain a constant pH level and to feed the microalgae. In this way, the model gives the possibility to analyze the carbon footprint of the technology or at least to understand the quantity of CO_2, which can be fixed by the biomass produced, being a part of the injected CO_2 lost to the atmosphere: this quantity can be relevant in the case of open raceway pond.

Fundamental outputs of the model are the total electrical and thermal energies required by the reactor: to evaluate these quantities, the energy balance is not sufficient, as it is necessary to include the electrical energy spent at the boundary of the system, to refill the reactor and to harvest the biomass from the reactor itself; the electrical energy also includes the quantities which are needed for mixing and for the air bubbling in the bioreactor. Finally, another output of the model is the quantity of water required by the reactor, without considering the possible water recirculation coming from the downstream process.

MODEL EQUATIONS

Growth model

Following Yang (2011), the growth rate is calculated by Eq. (3):

$$r_{gA} = \mu_A X_A \tag{3}$$

where X_A is the mass concentration of microalgae [g/m³] and μ_A is the specific growth rate [day⁻¹] that is calculated as:

$$\mu_A = \widehat{\mu}_A \left(\frac{CO_{2D}}{K_C + CO_{2D}}\right)\left(\frac{N_T}{K_{NA} + N_T}\right) f_I f_T \tag{4}$$

where the specific growth rate µ^Aµ^A depends on the microalgae species; CO_{2D},N_T are the quantities of dissolved nutrients and CO_2 [mol/m³] at time t. Nitrogen is the only nutrient that has been taken into account in the specific growth rate expression; other nutrients, such as phosphorus, are not explicitly considered, as suggested by Yang (2011), under the reasonable assumption that the metabolism of the microalgae is not limited or inhibited by these compounds. Although the growth rate may be affected by the presence of micro nutrients, they have not been introduced in the model to avoid the definition of a more complex reaction scheme with the corresponding experimental data to represent those influences. The specific growth rate can reach higher values thanks to the use of some micronutrients, such as iron and silicon (Chisti, 2007; Çelekli and Yavuzatmaca, 2009; Le et al., 2010; Ak, 2012; Bahadar and Bilal Khan, 2013). Nitrogen and CO_2 are considered limiting nutrients for microalgae growth and so microalgae growth rate dependence is expressed as saturation functions. K_C and K_{NA} are the half-saturation constants, as explained in the previous paragraphs. f_I is the light intensity factor representing the influence of the light on the microalgae production. Following Yang (Yang, 2011; Fernández et al., 2013), it is calculated by the following equation:

$$f_I = \frac{I_a}{I_s}\exp\left(1 - \frac{I_a}{I_s}\right) \tag{5}$$

where I_a is the average light intensity in the volume of the bioreactor at a given time t, while I_s is the saturation light intensity, which depends on the microalgae species being considered; saturation light intensity is usually in the range between 30 and 100 W/m² (Kumar et al., 2011). The average light intensity I_a depends on the weather, the turbidity caused by the microalgae and other substances in the water and on the reactor geometry. The model includes the hypothesis of perfect mixing: the whole quantity of microalgae in the water is affected by the same quantity of light radiation. f_T is the temperature factor. This term is equal to one and does not influence microalgae growth when the reactor temperature corresponds to the optimal growth temperature for the given microalgae species: this is what happens in the closed reactor, where a heat exchanger is used to keep a constant optimal temperature. Below and above the optimal growth temperature, the growth is negatively influenced by the water temperature. Above the optimal temperature, the value of f_T decreases fast and reaches 0 for a certain temperature that again depends on the algae species: this value is called lethal temperature (T_{let}). The expression of the temperature factor is shown in the following equation, taken from Slegers et al. (2013):

$$f_T = \left(\frac{T_{let} - T_w}{T_{let} - T_{opt}}\right)^{\beta_{algae}} \exp\left(-\beta_{algae}\left(\frac{T_{let} - T_w}{T_{let} - T_{opt}} - 1\right)\right) \tag{6}$$

β_{algae} depends on the algae species and it is a curve modulating constant.

Figure 4 shows the shape of the temperature factor f_T: it is possible to note that the equation used for the temperature factor suggested by Slegers et al. (2013) overestimates the growth for low temperature values, since the growth is not possible for low temperatures.

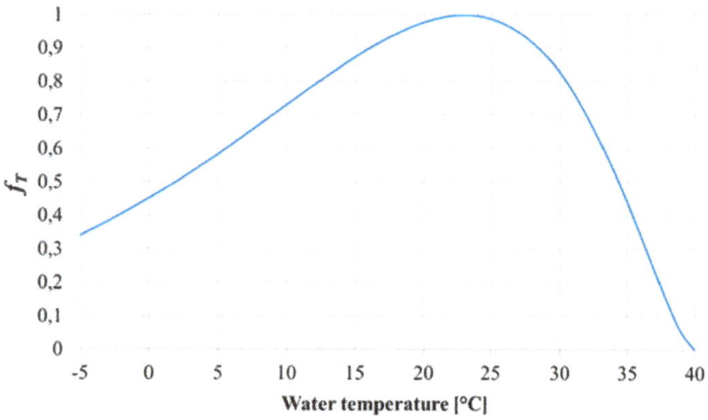

Figure 4. Temperature factor (Slegers et al., 2013).

pH control strategy

Each microalgae species has an optimal pH value: the model includes a control system to keep the optimal value of pH in the reactor through CO_2 injection. As shown in the following equations from Sills (2013), the pH can be controlled by keeping constant two parameters: dissolved CO_2 concentration and alkalinity. Since alkalinity does not change with time in the model, to keep pH constant, it is necessary to calculate the quantity of dissolved CO_2 which has to be maintained constant during time.

The total concentration (C_T) of carbonate species in solution is defined as

$$C_T = [H_2CO_3^*] + [HCO_3^-] + [CO_3^{2-}] \tag{7}$$

Dissolved CO_2 concentration is $[CO_2(ag)] \cong [H_2CO^*_3]$. The molar concentration of each carbonate species (as a fraction of C_T) depends on pH, according to the following equations:

$$[H_2CO_3^*] = \frac{C_T}{1 + \frac{k_1}{[H^+]} + \frac{k_1 k_2}{[H^+]^2}} = \alpha_0 C_T \tag{8}$$

$$[HCO_3^-] = \frac{C_T}{1 + \frac{[H^+]}{k_1} + \frac{k_2}{[H^+]}} = \alpha_1 C_T \tag{9}$$

From Eqs (8) and (9), α_0 and α_1 are calculated.

$$[CO_3^{2-}] = \frac{C_T}{1 + \frac{[H^+]}{k_2} + \frac{[H^+]^2}{k_1 k_2}} = \alpha_2 C_T \tag{10}$$

Finally:

$$C_T = \frac{\text{alk} - OH^- + H^+}{\alpha_1 + 2\alpha_2} \tag{11}$$

Since pH is the concentration of dissolved $[H^+]$ in the water and pH = 14 − pOH, all terms are known and the total concentration of carbonate species can be obtained by

$$C_T = \frac{\text{alk} - OH^- + H^+}{\alpha_1 + 2\alpha_2} \tag{12}$$

This term will be a part of the time-dependent CO_2 mass balance, which is used to calculate the CO_2 injection at each time t.

Mean light intensity for open pond

The light intensity factor in the specific growth rate expression contains I_a, that is the average light intensity in the bioreactor at a given time. Starting from global horizontal radiation, the model evaluates I_a using the Beer–Lambert's law, which assumes an exponential decay of the light intensity from the external surface of the cultivation system:

$$I_a(s) = I_0 \exp(-\sigma X_A s) \tag{13}$$

As it is explained by Béchet et al. (2013), $I_a(s)$ is the local light intensity, s is the distance from the external surface of the system to the position under consideration, I_0 is the incident light intensity, σ is the extinction coefficient, and X_A the cell concentration. The Beer–Lambert's law can be applied only if two conditions are verified by the culture medium: first, it must be isotropic (i.e., the optical properties of the broth are independent from light direction); this condition is often verified in well-mixed systems. Second, algae cells should not scatter light: this second condition is not always met, but the model considers both the two requirements always verified. The equations to calculate I_a strictly depend on the geometry of the bioreactor: for the open pond, as suggested by Yang (2011), an integration through the pond depth of the Beer–Lambert's law has been used:

$$I_a = \frac{1}{Z} \int_0^Z I_0 \exp(-K_e z)\, dz \tag{14}$$

where I_0 is obtained directly from the global horizontal radiation I_{GHR}: I_0 = PAR × I_{GHR}. PAR is the photosynthetic active radiation (PAR), i.e., the amount of solar radiation which is used by the microalgae for the photosynthesis and which corresponds to 45% of the total incoming light. K_e is related to algal concentration in the pond, and it is called extinction coefficient:

$$K_e = K_{e1} + K_{e2} X_A \tag{15}$$

Light intensity for flat panel

In case of flat panel, the complexity of the geometry requires the use of different equations from those applied for the open pond. The equations used to calculate the light intensity for flat panels are taken from Slegers et al. (2011). For this typology of reactors, the input data are the direct and diffuse radiation on the horizontal surface; moreover, the equations must take into account also the reflection of the radiation on the surface of the panel. As explained by Slegers et al. (2011), to evaluate the effect of direct light irradiation over a tilted surface (panels positioned vertically), it is necessary to introduce a geometrical parameter for direct radiation:

$$G_{\text{direct}}(t) = \frac{\cos\theta}{\cos\theta_z} \tag{16}$$

in which θ is the solar incidence angle, and θ_z is the solar zenith angle. The values assumed by these angles during each day of the year depend on the location, and have been calculated using the equations taken from Duffie and Beckman (2013).

For large scale cultivations, parallel positioned flat panels are used. Parallel placement causes shading, and consequently part of the panels no longer receive direct sky light. The shadow height on vertical reactor panels is given by

$$h_{\text{shadow}} = h - \frac{d\ \tan(90-\theta_z)}{\cos(\gamma)} \tag{17}$$

which is a function of the reactor height h [m], distance between the reactor panels d [m], solar elevation, which is equal to 90 − θ_z and angle between solar rays and the azimuth of the surface. If $h_{shadow} > 0$, the flat panel is divided into two parts. The upper part receives direct and diffuse radiation, the lower part only diffuse light. The separation between the upper and the lower part varies with the solar position. Parallel placement of the reactors also influences the penetration of diffuse sky light into the space between panels; the light intensity decreases from the top to the bottom. A similar situation can be seen with the penetration of light in urban street canyons (Robinson and Stone, 2004). For these reasons, the geometrical factor for diffuse light depends on the position over the surface of the panel.

At height $y < h_{shadow}$:

$$G_{\text{diffuse}} = \frac{1+\cos(\beta+u)}{2} \tag{18}$$

where u = atan(y/d), β [deg] is the slope of the reactor, i.e., the angle that the surface makes with the surface of the earth.

At height $y > h_{shadow}$:

$$G_{\text{diffuse}} = \frac{1+\cos(\beta)}{2} \tag{19}$$

All panels are treated similarly in the calculations. Moreover, ground reflection is low for parallel placed panels, and is therefore not taken into account.

The total amount of light falling on each reactor surface at a given height y, of a given time t is:

$$I_0(y,t) = G_{\text{direct}}(t)\, I_{\text{direct}}(t) + G_{\text{diffuse}}(y)\, I_{\text{diffuse}}(t) \tag{20}$$

At this point, it is fundamental to consider the reflected fraction of the irradiation, which does not enter into the reactor and so does not contributes to the microalgae growth. The amount of reflected light on each interface is related to the differences in refractive indices and the incidence angle (Sukhatme and Sukhatme, 1996).

The incidence angle for diffuse radiation which is considered to evaluate the light reflection is assumed to be 60°, as it is suggested by Duffie and Beckman (2013).

Light reflection by the flat panel walls follows Fresnel equations:

$$R_s = \left[\frac{\eta_i\cos(\theta_i) - \eta_t\sqrt{1-\left(\frac{\eta_i}{\eta_t}\sin(\theta_i)\right)^2}}{\eta_i\cos(\theta_i) + \eta_t\sqrt{1-\left(\frac{\eta_i}{\eta_t}\sin(\theta_i)\right)^2}}\right]^2 \tag{21}$$

$$R_p = \left[\frac{\eta_i\sqrt{1-\left(\frac{\eta_i}{\eta_t}\sin(\theta_i)\right)^2} - \eta_t\cos(\theta_i)}{\eta_i\sqrt{1-\left(\frac{\eta_i}{\eta_t}\sin(\theta_i)\right)^2} + \eta_t\cos(\theta_i)}\right]^2 \tag{22}$$

where θ_i [°] is the incidence angle, η_i [−] is the refractive index of the material before the interface, and η_t [−] is the refractive index for the material after the interface. Normal sunlight is non-polarized, therefore the overall reflection coefficient equals the average of the reflection coefficients for s-polarized andp-polarized light:

$$R' = \frac{R_s+R_p}{2} \tag{23}$$

The light reflected within the reactor wall is completely transmitted to the air, introducing the hypothesis of a non-absorbing material for the walls of the reactor. The light transmitted to the culture, which has to be calculated separately for direct and diffuse radiation, is

$$I_i(t) = I_0(t)(1 - R'_1 R'_2)T_m \tag{24}$$

T_m [−] is a factor which takes into account a possible low transparency of the material.

The calculation is performed for both the two sides of the reactor and, for parallel positioned panels, for each height. R_1' and R_2' are the reflection coefficients for the air–reactor wall interface and the reactor wall–culture volume interface, respectively. Two light intensity gradients exist in the culture volume. First, as a function of height due to shading and the penetration of diffuse light between parallel positioned panels. Second, in the liquid between the two reactor walls. The second gradient runs from the reactor wall to the center of the reactor and is caused by the absorption of light by the medium and the algae (Slegers et al., 2011). Only the PAR of the spectrum is absorbed by the algae. This accounts for about 45% of the total light. The Lambert–Beer's law is used for the overall light gradient in the culture volume, as it was done for the open pond:

$$I(y,z,t) = I_{\text{front}}\, e^{-(K_{e1}+K_{e2}X_A)z} + I_{\text{back}}\, e^{-(K_{e1}+K_{e2}X_A)(s-z)} \tag{25}$$

This equation gives the light intensity at location z [m] inside the reactor thickness [m], at a given height y [m] in the reactor at time t.

At this point, to simplify the model, the values of light intensity I(y,z,t) are integrated to find a mean value of irradiation for the whole culture inside the whole reactor at a given time t: with these integrations, a single value of radiation is obtained and used in the growth model, for the whole panel, at a given time t. The hypothesis of a perfect mixing inside the culture at each timet is necessary to integrate the equation in both height and depth.

Mass balances

Mass balances of nitrogen and oxygen inside the pond can be modeled in the same way, suggested by Yang (2011):

$$\frac{dM}{dt} = \mu_A X_A Y_{AM} - k_{Lg}\alpha\,(M - M^*) \tag{26}$$

where M is the concentration of the respective component in the water in the bioreactor, Y_{AM} is the mass of the respective component consumed or generated by the microalgae per unit mass of microalgae produced. The last term of the right-end side of Eq. (26) represents the mass transfer between the atmosphere and the pond, where $k_{Lg}\alpha$ is the mass transfer coefficient for a given element, M* is the saturation concentration of the associated dissolved element.

For total inorganic carbon, the mass balance assumes a different formulation since the CO_2 is injected continuously in the pond during the growth phase to keep a constant concentration of dissolved CO_2 in the reactor, balancing the losses of CO_2 to the atmosphere and the consumption of CO_2 by the microalgae.

$$\frac{dCO_2}{dt} = \mu_A X_A Y_{ACO_2} + f_{CO_2} - k_{Lg}\alpha\,(CO_2 - CO_2{}^*) \tag{27}$$

where fCO2fCO2 represents the flux of CO_2 introduced by the supply of gas flow into the system. Since the quantity of dissolved CO_2 is kept constant ($dCO_2/dt = 0$), the mass balance can be written again as

$$f_{CO_2} = -\mu_A X_A Y_{ACO_2} + k_{Lg}\alpha\,(CO_2 - CO_2{}^*) \tag{28}$$

Finally, the mass balance for microalgae species can be written as

$$\frac{dX_A}{dt} = \mu_A X_A - k_{dA} X_A \tag{29}$$

As it can be seen, all these equations are time dependent, and all strictly depend on one another: this means that they form altogether a system of differential equations. The strategy to solve the equations with the finite difference methodology is used to solve this system in Matlab, and therefore to solve the system of differential equations as a system of algebraic equations.

Thermal balance for open pond

The model includes the thermal balance as it is suggested by Slegers et al. (2013):

$$V_R cp_w \rho_w \frac{dT_w}{dt} = Q_{irr} - Q_{algae} - Q_{rad} - Q_{evap} - Q_{conv} - Q_{cond} \tag{30}$$

where V_R is the volume of the pond, cp_w the heat capacity of the growth medium, ρ_w the density of the growth medium, T_w the temperature in the pond, Q_{irr} the heat flow rate to the pond by the sunlight, Q_{algae} the light energy flow rate to algae during growth, Q_{rad} the heat flow rate by emission of long-wave radiation in the infrared region, Q_{evap} [W] the heat flow rate caused either by evaporation or condensation, Q_{conv} the heat flow rate by convection, and Q_{cond} the heat flow rate between pond and ground via conduction. The water in the pond is heated by sunlight that enters the culture volume. Solar energy that is not used by algae for growth is considered as thermal energy that is dissipated through the different thermal exchange mechanisms (Q_{rad},Q_{evap}, Q_{conv}, and Q_{cond}) or as a contribution to temperature growth in the pond. The total heat flow rate by the sunlight is given by

$$Q_{irr} = A_w I_{surface}(t) \tag{31}$$

where A_w [m^2] is the water surface area of the pond and $I_{surface}$ [W/m^2] is the total light arriving on the pond. Part of this light is absorbed by microalgae for growth:

$$Q_{algae} = h_{comb}\mu_A X_A V_R \tag{32}$$

which is a function of the lower heating value on a dry basis of algae biomassh_{comb} [J/kg], specific growth rate μ_A [s^{-1}] and the biomass concentration X_A[kg/m^3].

The water in the pond emits thermal energy by long-wave radiation. The overall long-wave radiation flow between water in the pond and sky is calculated using Duffie and Beckman (2013):

$$Q_{rad} = A_w \varepsilon_w \sigma_{SB}\left((T_w + 273.15)^4 - T_{sky}^4\right) \tag{33}$$

where ϵ_w [−] is the emissivity of the water in the infrared region, σ_{SB}[W/(m^{2}K^4)] the Stefan–Boltzmann constant, and T_{sky} [K] the equivalent sky temperature for clear sky days, which is expressed by Duffie and Beckman (2013) as

$$T_{\mathrm{sky}} = (T_{\mathrm{a}} + 273.15)(0.711 + 0.0056T_{\mathrm{dew}} + 0.000073T_{\mathrm{dew}}^2 + 0.013\cos(15t_{\mathrm{solar}}))^{0.25} \quad (34)$$

where T_a [°C] is the air temperature, T_{dew} [°C] the dew point temperature, and t_{solar} [–] the number of hours after solar midnight.

Evaporation has a large effect on the water temperature, especially in locations with low humidity and high wind velocities. The evaporation rate depends on the shape of the water area, wind velocity, and consequently also movement of the water. The evaporation flow is driven by the difference of water vapor pressures between ambient air the saturated water body. The evaporation energy flow is given by

$$Q_{\mathrm{evap}} = A_{\mathrm{w}} h_{\mathrm{evap}} (p'_{\mathrm{s}} - p'_{\mathrm{a}}) \quad (35)$$

The evaporation flow depends on the heat transfer coefficient for evaporationh_{evap} [W/(m^2 Pa)], the saturated water pressure p'_s [Pa] at water temperatureT_w, and the water pressure of air p'_a [Pa] at air temperature T_a. The evaporation rates have been calculated using the heat transfer coefficient h_{evap}found in Duffie and Beckman (2013):

$$h_{\mathrm{evap}} = 0.036 + 0.025v \quad (36)$$

where v [m/s] is the wind speed. The Antoine equation is applied to calculate the saturated water pressure p'_s [Pa] at water temperature T_w and the water pressure of the air p'_a [Pa] at air temperature T_a:

p′=RH 10(8.07131+log10(101325760))−1730.63233.46+Tap′=RH 10(8.07131+log10(101325760))−1730.

$$p' = \mathrm{RH}\, 10^{\left(8.07131 + \log_{10}\left(\frac{101325}{760}\right)\right) - \frac{1730.63}{233.46 + T_a}} \quad (37)$$

where RH [–] is the relative humidity and T_a [°C] the temperature.

Convection and evaporation are related processes, as it is shown in Eq. (37). The flow for passive and forced convection at the water surface mainly depends on the difference between water and air temperature. The convection flow is given by

$$Q_{\mathrm{conv}} = C_{\mathrm{Bowen}} \frac{p_{\mathrm{a}}(T_{\mathrm{w}} - T_{\mathrm{a}})}{p_{\mathrm{ref}}(p'_{\mathrm{s}} - p'_{\mathrm{a}})} Q_{\mathrm{evap}} \quad (38)$$

where C_{Bowen} is the Bowen constant [Pa/°C], p_a is the ambient pressure [Pa] and p_{ref} the reference pressure [Pa], and p'_s and p'_a are derived using equation

$$p' = \mathrm{RH}\, 10^{\left(8.07131 + \log_{10}\left(\frac{101325}{760}\right)\right) - \frac{1730.63}{233.46 + \mathrm{Ta}}} \quad (39)$$

Conductive heat transfer takes place between the open pond and the soil. The soil is assumed to be an infinite source for heat transfer. This heat transfer calculation is derived from Fourier's law:

$$Q_{\mathrm{cond}} = h_{\mathrm{soil}} A_{\mathrm{soil}} (T_{\mathrm{w}} - T_{\mathrm{soil}}) \quad (40)$$

where h_{soil} [W/(m^{2}°C)] is the heat transfer coefficient of the surrounding soil layer, A_{soil} [m^2] is the area of the pond that is embedded in the soil, and T_{soil}[°C] is the temperature of the soil surrounding the pond.

Thermal balance for flat panel

Due to the significant difference in the geometry of the reactor and the temperature strategy being considered, the thermal balance for the flat panel assumes a different form from that implemented for the open raceway pond:

$$V_{\mathrm{R}} \mathrm{cp_w} \rho_{\mathrm{w}} \frac{\mathrm{d}T_{\mathrm{w}}}{\mathrm{d}t} = Q_{\mathrm{irr}} - Q_{\mathrm{algae}} - Q_{\mathrm{exch}} - Q_{\mathrm{conv+cond}} \quad (41)$$

In a flat panel, a constant value of the water temperature T_w [°C] is desired, and obtained using a heat exchanger placed at the bottom of the reactor to remove or supply heat.

Thus, the thermal balance can be written as follows:

$$\frac{\mathrm{d}T_{\mathrm{w}}}{\mathrm{d}t} = 0 \rightarrow Q_{\mathrm{irr}} - Q_{\mathrm{algae}} - Q_{\mathrm{exch}} - Q_{\mathrm{conv+cond}} = 0 \quad (42)$$

Q_{irr} [W] is given by the following expression:

$$Q_{\mathrm{irr}} = 2\, A_{\mathrm{panel}}\, I_{\mathrm{mean}} \quad (43)$$

where A_{panel} is the surface of the reactor, which is multiplied by two, as it is necessary to consider both the front and the back surfaces; I_{mean} [W/m^2] is the radiation previously calculated taking into account both the direct and diffuse radiation and the reactor geometry that is the radiation which interacts with the water in the reactor and with the microalgae. As calculated for the open pond, a part of the incoming heat is used by the microalgae for growing:

$$Q_{\mathrm{algae}} = h_{\mathrm{comb}} \mu_{\mathrm{A}} X_{\mathrm{A}} V_{\mathrm{R}} \quad (44)$$

which is a function of the combustion energy of algae biomass h_{comb} [J/kg], specific growth rate μ_A [s^{-1}], biomass concentration X_a [kg/m^3], and reactor volume V_R [m^3].

The model takes into account the natural convection over the panel surface caused by the wind and the glass conductivity:

$$Q_{\mathrm{conv+cond}} = 2\, A_{\mathrm{panel}}\, U_{\mathrm{tot}}\, (T_{\mathrm{w}} - T_{\mathrm{atm}}) \quad (45)$$

where

$$U_{\mathrm{tot}} = U_{\mathrm{cond}} + U_{\mathrm{conv}} \quad (46)$$

where U_{cond} is the glass conductivity [W/(m^{2}K)] and U_{conv} is the conductivity of the natural convection that is obtained by Duffie and Beckman (2013). Using the equations above, the quantity of heat that has to be removed or supplied by the heat exchanger at each time t can be calculated. Heat exchange through radiation can be omitted from this balance, since the water temperature is always kept between 20 and 30°C, depending on microalgae growth optimal temperature.

Electrical energy consumption for harvesting, refilling, mixing and bubbling

For both the open pond and flat panel, the harvesting of the water from the reactor and its refilling are carried out in 8 h, during one night: 3.5 h are required for harvesting and 3.5 h for refilling. These operations are performed using a pump. The electrical consumption has been calculated as follows:

$$E_{\text{harv/refil}} = P_{\text{pump}} t_{\text{harv/refil}} \tag{47}$$

where the power of the pump is

$$P_{\text{pump}} = \frac{\rho g Q h}{\eta_{\text{pump}}} \tag{48}$$

where ρ is the water density [kg/m^3], g is the acceleration of gravity [m/s^2], η_{pump} [–] is the efficiency of the pump, set at 0.85, Q [m^3/s] is the volumetric flow rate which has to be pumped, and h [m] is the height difference between the two basins before and after the pump: thanks to the fact that the model includes also the design of a settler positioned after the bioreactor, it is possible to know the exact h for the harvesting, which has been increased to consider the losses in the pipes, whereas for the refilling data have been taken from the literature, considering h = 1 m for the open pond and 3 m for the flat panel: the difference between these two values is again associated with the energy losses in pipes.

The power required for mixing in the open pond is

$$P_{\text{mix}} = \frac{\rho g Q h}{\eta_{\text{paddlewheel}}} \tag{49}$$

where Q [m^3/s] is obtained from the water speed that is desired in the reactor (0.20 m/s) and from the cross-section of the open pond (which depends on the geometry), h is the given height difference before and after the paddle wheel, taken from literature (0.05 m). $\eta_{paddlewheel}$ is lower than the efficiency of a normal pump and is assumed to be equal to 0.25.

Both for the open pond and flat panel, a bubbling system has to be taken into account: for the flat panel, this system should be able both to supply the CO_2 necessary for the photosynthesis of the microalgae and to guarantee an adequate mixing inside the reactor. For this reason, the amount of air bubbled in the flat panel is higher than the air supplied to the open pond. These quantities are controlled by the CO_2 molar fraction inside the injected air which is 0.04 in the case of open pond and 0.02 for the flat panel. The design and the energy consumption of the bubbling system have been calculated through Belsim Vali modeling software, using a compressor. The result is that it is necessary to supply 4 kJ for each kg of air injected in the reactor.

RESULTS

Open Pond

Tables 4 and 5 show the input and output data in the open pond modeling:

Table 4. Input data for open pond simulations

HRT not fixed, X target fixed		HRT fixed, X target not fixed	
Location	Sevilla (SPA) Petrolina (BRA)	Location	Sevilla (SPA) Petrolina (BRA)
Typology	*P. tricurnutum* *T. pseudonana*	Typology	*P. tricurnutum* *T. pseudonana*
Xa_init [g/m^3]	100	Xa_init [g/m^3]	100
Xa_target [g/m^3]	490	HRT [day]	7
Tw_in [°C]	15	Tw_in [°C]	15
CO_2 rate [%]	0.04	CO_2 rate [%]	0.04
Z_pond [m]	0.3	Z_pond [m]	0.3
LW [–]	10	LW [–]	10

Table 5. Output from open pond simulations

	HRT not fixed X target fixed		HRT fixed X target not fixed	
	Petrolina	Sevilla	Petrolina	Sevilla
Number of harvesting in 1 year	45	50	–	–
HRT [day]	–	–	7	7
Mass microalgae [t/(ha y)]	57.29	64.68	54.04	67.88
CO_2 captured [t/(ha y)]	140	156	133	163
CO_2 injected [t/(ha y)]	243	260	235	268
CO_2 lost to atmosphere [t/(ha y)]	78.62	77.35	78.14	77.27
CO_2 ratio losses	0.324	0.298	0.33	0.29
CO_2 ratio algae	0.57	0.6	0.56	0.61
N absorbed [t/(ha y)]	5.8	6.55	5.57	6.8
Water injected [t/(ha y)]	109,210	121,640	120,220	119,600
Water evaporation [t/(ha y)]	5.8	5.01	5.83	5.24
Energy microalgae [kWh/(ha y)]	342,610	386,790	323,180	405,930
Electrical energy [kWh/(ha y)]	18,497	19,024	18,522	19,111
Thermal energy [kWh/(ha y)]	–	–	–	–
Volumetric productivity [kg/(m^3 x day)]	0.0523	0.0591	0.0494	0.062
Areal productivity [kg/(m^2 x day)]	0.0157	0.0177	0.0148	0.0186
NER	0.108	0.0603	0.1146	0.0942

The open pond productivity that is obtained by the model is consistent with the literature: Slegers et al. (2013) estimated an annual biomass production for a 1 ha surface in the Netherlands and Algeria equal to 41.5 and 63.7 t, respectively. Fernández et al. (2013) report a maximum microalgae areal productivity equal to 30 g/(m^2 day), higher than those resulting from the model calculations, but still of the same order of magnitude. Jiménez et al. (2003) obtained a volumetric productivity equal to 0.05 kg/(m^3 day) for a location in Southern Spain (Malaga), by cultivating the microalgae until a maximum concentration of 470 g/m^3: the cultivation conditions and the results from the dynamic model are consistent with the results of this work.

In this work, the Net Energy Ratio (NER)

$$\text{NER} = \text{Net Energy Ratio} = \frac{\text{total energy requirement for operation}}{\text{total energy production (biomass)}} \tag{50}$$

expresses the fraction of the energy produced in the cultivation system that is used by the system itself to generate the biomass. If this value is close to 1, the cultivation technology of microalgae is too energy intensive, requiring a big share of the energy produced. From Table 5, it appears that for both the locations analyzed by the model the NER is quite far from 1, showing that the energy demand for the operation of the cultivation system in Petrolina is 10.8% of the energy contained in the biomass produced, while for Sevilla is 6%. These values are quite promising for a potential production of microalgae in these locations, since the NER is far enough from 1: even if the energy content of open pond construction and materials are added to the total energy requirement, it seems reasonable to state that the plant would still be convenient from an energetic point of view. Another important output of the model from an environmental point of view is the quantity of CO_2 fixed by the microalgae during the growth process: the downstream process that transforms microalgae biomass into biofuel releases CO_2 to the atmosphere, with a negative environmental impact. Since the cultivation phase captures more CO_2 than the quantity released, using the data of CO_2 fixed in the algae it is possible to calculate a global CO_2 balance, for the entire biofuel production chain: this environmental analysis may lead to a comparison with other biofuel chain production and with traditional fossil fuels, to evaluate which product has a positive or negative environmental impact.

With the first operating strategy (meaning that the inputs are the initial and final concentration), one of the output would be the number of days that are needed to reach the final target concentration. The number of days is not the same along the year, being variable with weather conditions. During wintertime, the harvesting period in Sevilla is extremely long due to low irradiation and low temperatures, which make the microalgae growth in the pond slow and difficult. A possible strategy to overcome this limitation might be to interrupt the production during wintertime: in this case, the productivity during the whole year would decrease, since a part of the biomass would not be produced, but the NER would positively decrease, because of the absence of energy consumption during winter: the production lost is less than the energy saving obtained, and so the overall effect would be positive. For Petrolina, the HRT is more homogeneous along the year, but longer: the reason might be that weather conditions in Petrolina cause a strong increase of pond temperature that may cause a consequent reduction in microalgae productivity when pond temperature is higher than the optimal temperature for microalgae growth.

Results of dynamic simulation runs are reported in Figures 5–9, showing the effect of PAR and pond temperature over the specific growth rate and microalgae growth rate. Input parameters for simulation runs are shown in Table 6.

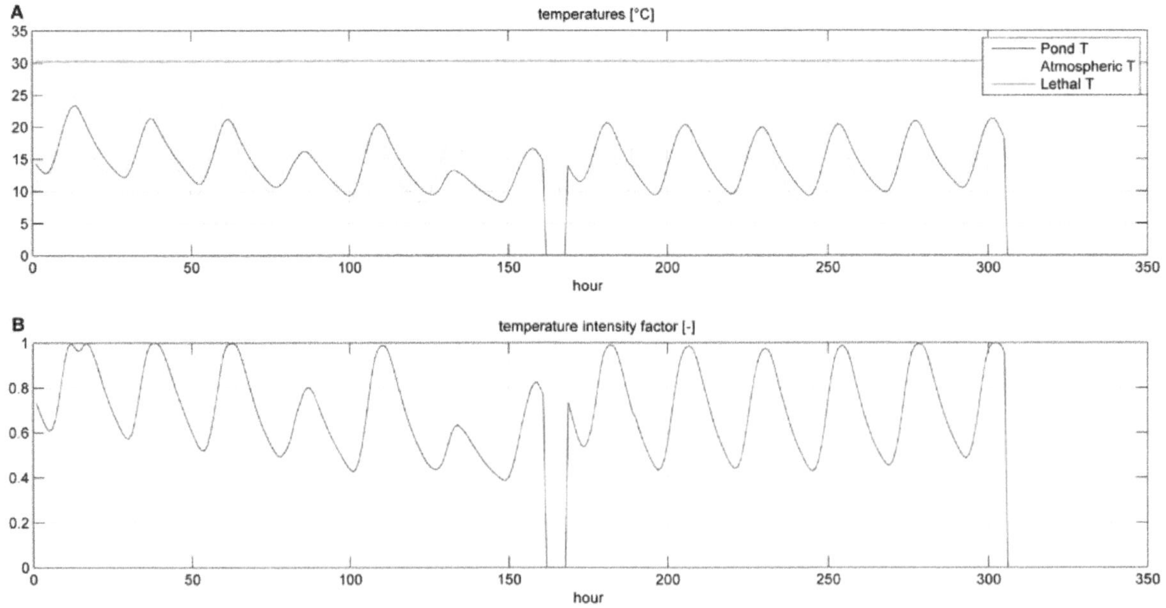

Figure 5. Temperatures (A) and temperature factor (B) for a simulation time equal to two HRTs.

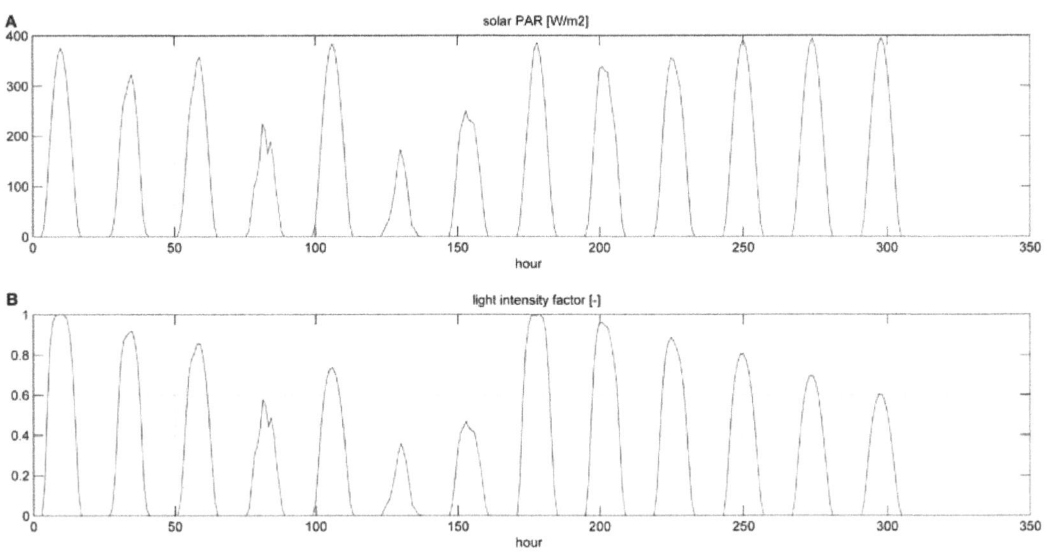

Figure 6. Photosynthetic active radiation (PAR) (A) and light intensity factor (B) for a simulation time equal to two HRTs.

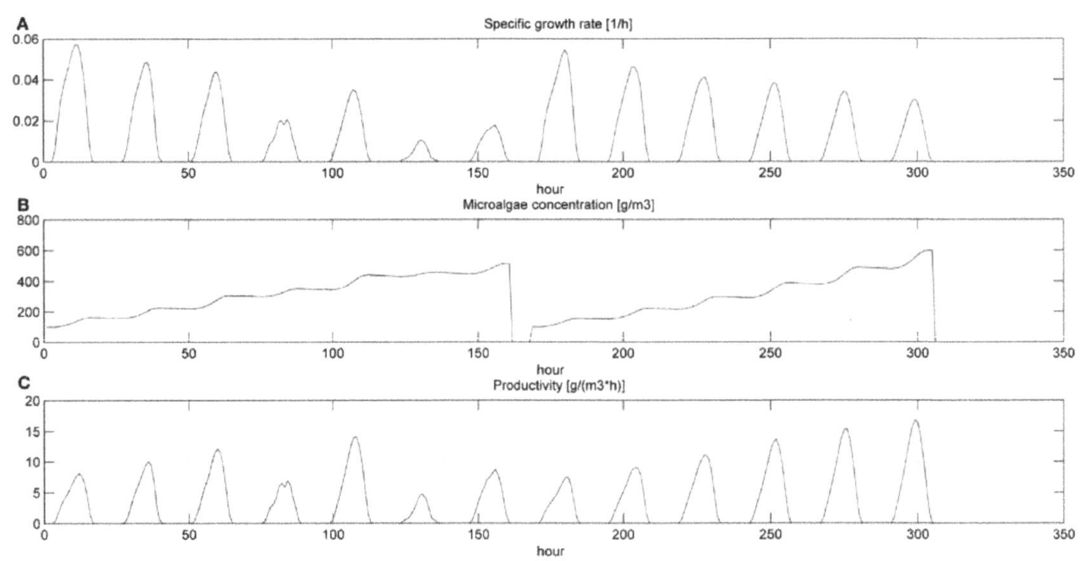

Figure 7. Specific growth rate (A), microalgae concentration (B) and microalgae volumetric productivity (C) in the open pond during two consequent batch cultivations (simulation time equal to two HRTs).

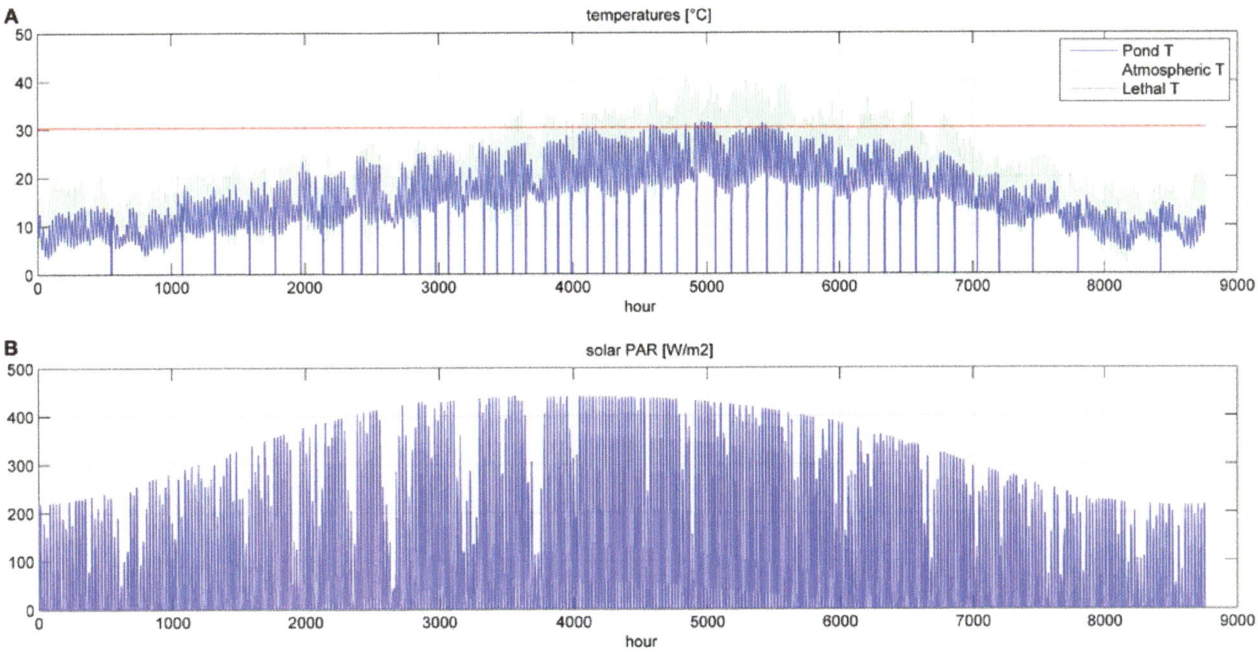

Figure 8. Temperatures (A) and photosynthetic active radiation (B) for a simulation time equal to 1 year.

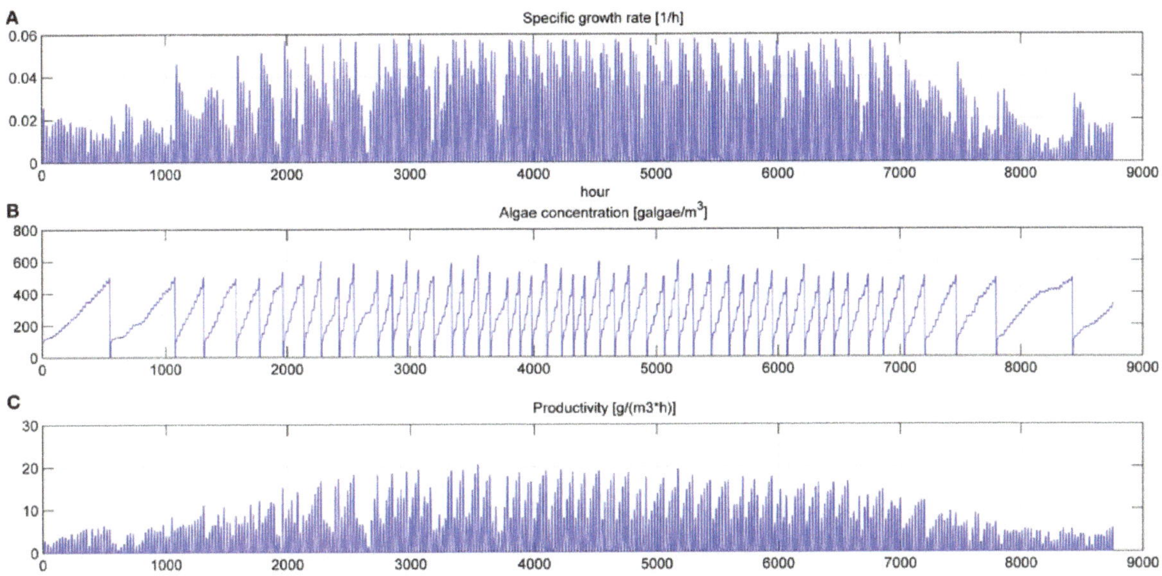

Figure 9. Specific growth rate (A), microalgae concentration in the pond (B) and microalgae growth rate (C) for a simulation time equal to 1 year.

Table 6. Input data for open pond simulation runs

HRT not fixed, X target fixed	
Location	Sevilla (SPA)
Typology	*P. tricurnutum*
Xa_init [g/m^3]	100
Xa_target [g/m^3]	490
Tw_in [°C]	15
CO_2 rate [%]	0.04
Z_pond [m]	0.3
LW [–]	10

Even if the simulation has been run for the entire year, as it is shown in Figures 8 and 9, Figures 5–7 report a small time range that includes two consequent batch cultivations, for an easier understanding of the light and temperature effect over the microalgae growth. Figure 5 includes the pond temperature values (A), calculated in each hour of the entire simulation period, and the consequent temperature factor (B). In Figure 5, the model does not evaluate the pond temperature during the hours when the pond is emptied. Figure 6 reports the PAR and the consequent light intensity factor (B). Figure 7 shows the specific growth rate (A) and the microalgae concentration (B) in the open pond during two consequent batch cultivations. Microalgae growth rate (C) is obtained, as explained in Eq. (3), by the product of the specific growth rate and the microalgae concentration. As shown in Figure 7B, each batch cultivation starts with the same initial concentration, while the final concentration depends on each batch: the model waits for the end of the solar radiation to start to empty the pond that takes almost the whole night.

In Figures 8 and 9, the time of the entire simulation run has been considered presenting input values (temperature in Figure 8A and PAR in Figure 8B) and output [specific growth rate in Figure 9A, microalgae concentration in the pond in Figure 9B and microalgae growth rate (productivity) in Figure 9C].

The time to reach the minimum microalgae target concentration to start the harvest in winter time in Sevilla is really long and it might be useful to interrupt the production for some months.

Flat Panel

Table 7 shows the input data for the flat panel model for both the two operating strategies.

Table 7. Input data for flat panel simulations

HRT not fixed, X target fixed		HRT fixed, X target not fixed	
Location	Sevilla Petrolina	Location	Sevilla Petrolina
Typology	*P. tricurnutum* *T. pseudonana*	Typology	*P. tricurnutum* *T. pseudonana*
Azimuth [°]	0 −90	Azimuth [°]	0 −90
Slope [°]	90	Slope [°]	90
Xa_init [g/m³]	3000	Xa_init [g/m³]	3000
Xa_target [g/m³]	6000	HRT [day]	4 or 5
CO_2 rate [%]	0.02	CO_2 rate [%]	0.02
h [m]	1.5	*h* [m]	1.5
s [m]	0.05	*s* [m]	0.05
d [m]	0.5	*d* [m]	0.5

The azimuth of the panel is an input, which may vary from 0°, when the panel faces south, to −90°, when the panel faces east. The model is not able to consider different slopes of the reactor, which may only be positioned vertically: this position is the most favorable for light distribution and dilution, leading to the highest values of productivity (Sierra et al., 2008). CO_2 molar concentration in the injected gases is lower than that of the open raceway pond, since the injected gases in the flat panel photobioreactor do not only have to supply the CO_2 needed by microalgae for photosynthesis, but also to generate the necessary mixing in the reactor: adequate mixing is essential to obtain good levels of productivity in every microalgae cultivation system. Due to the complex geometry of the flat panel photobioreactor, more geometrical data are needed as input if compared to the open raceway pond system. As explained by Fernández et al. (2013), heights <1.5 m and widths <0.10 m are preferred; following this indication and data from Slegers et al. (2011), the distance between the vertical panels has been set equal to 0.5 m, the height of each panel to 1.5 m and the thickness to 5 cm. Both height and distance between panels have been varied within a range of possible values in a further parametrical analysis. The initial and final concentrations have been suggested by Münkel et al. (2013); the values are higher than in the case of open raceway pond: the more sophisticate closed photobioreactor allows higher concentrations to be reached without compromising the productivity of the cultivation system. This is possible thanks to an optimal light distribution over the whole reactor for the entire operation time that is guaranteed also by an adequate mixing of the medium. The flat panel model, as well as the open raceway pond model, is able to produce the dynamic trend of all the time dependent physical quantities, which are included in the analysis. The global results for the flat panel reactor are shown in Tables 8and 9: Table 8 reports the results for the first operating strategy, i.e., the case in which both input and output microalgae concentrations are known, whereas Table 9 corresponds to the second operating strategy, in which the input parameters are the input concentration and the HRT.

Table 8. Results from flat panel simulations using the first operating strategy, meaning that input and output microalgae concentrations are known

	HRT not fixed South		HRT not fixed East	
	Petrolina	**Sevilla**	**Petrolina**	**Sevilla**
N harvesting	65	77	88	81
Mass microalgae [t]	289	345	393	366
CO_2 captured [t]	860	976	1083	1025
CO_2 injected [t]	956	1085	1204	1139
CO_2 ratio algae	0.9	0.9	0.9	0.9
N absorbed [t]	35.92	40.77	45.2	42.77
Water injected [t]	46,023	54,592	62,353	57,742
Energy microalgae [kWh]	1733,700	2063,700	2350,700	2191,100
Electrical energy [kWh]	35,634	40,476	44,906	42,473
Thermal energy [kWh]	1310,300	2124,000	1267,400	2093,900
Volumetric productivity [kg/(m^3 x day)]	0.5851	0.6965	0.7934	0.7395
Areal productivity [kg/(m^2 x day)]	0.0794	0.0946	0.1077	0.1004
NER	1.55	2.09	1.1165	1.95

Table 9. Results from flat panel simulation using the second strategy, meaning that input microalgae concentration and HRT are known

	HRT fixed South		HRT fixed East	
	Petrolina	**Sevilla**	**Petrolina**	**Sevilla**
HRT [day]	5	5	4	4
Mass microalgae [t]	284	367	390	380
CO_2 captured [t]	844	1041	1081	1052
CO_2 injected [t]	938	1157	1201	1168
CO_2 ratio algae	0.9	0.9	0.9	0.9
N absorbed [t]	35,023	43,044	45.11	43.88
Water injected [t]	46,691	52,620	62,422	57,201
Energy microalgae [kWh]	1701,300	2197,000	2336,600	2273,000
Electrical energy [kWh]	34,973	43,043	44,819	43,542
Thermal energy [kWh]	1292,400	2264,900	1262,700	2113,900
Volumetric productivity [kg/(m^3 x day)]	0.5742	0.7415	0.7886	0.7672
Areal productivity [kg/(m^2 x day)]	0.078	0.1007	0.1071	0.1041
NER	1.56	2.101	1.1192	1.8983
Xa final mean [kg/m^3]	5.87	6.707	6.016	6.077

For the two operating strategies, results are reported for both the locations (Sevilla and Petrolina) and for the two most significant orientations, that is when the flat panel is facing south (and north) and when it is facing east (and west). The east–west orientation appears to be preferable in both locations because it leads to a higher microalgae production: as explained by Sierra et al. (2008), if the orientation of the two faces of the reactor is east–west, the intercepted radiation is maximum during the first and last solar hours, because of the orientations toward sunrise and sunset. Thus, light availability during the daylight solar cycle is also more homogenous for this configuration. Areal and volumetric productivities are consistent with those found in literature: Chisti (2007) reports a volumetric productivity equal to 1.535 kg m^{-3} day^{-1}, which is higher than the values obtained from the model; if compared to some other works, the results of the dynamic model, in particular the areal productivity, seem to be optimistic: for example, the microalgae production coming from the model [~400 t/(ha × year)] is two times higher than the production obtained by Slegers et al. (2011) from a flat panel reactor located in Algeria which produces up to 200 t/(ha × year). Moreover, the volumetric productivity for a flat panel reported by Jorquera et al. (2010) is equal to 0.27 kg m^{-3} day^{-1}, whereas the model supplies values equal to 0.8 kg m^{-3} day^{-1}. From a recent work by Münkel et al. (2013), volumetric productivities equal to 1.25 kg m^{-3} day^{-1} have been reached in experimental analyzes. The difference from some values found in literature could be a consequence of a series of related factors: the microalgae species chosen may strongly influence the productivity of the reactor. Moreover, the model created in this work contains a temperature control, which fixes the temperature inside the reactor at the optimal level for microalgae growth: this means that the specific growth rate is not affected by the temperature factor (which is always equal to one), and consequently it is nearer to the maximum growth rate than in the real operating conditions. Furthermore, the locations chosen for the analysis present optimal values of irradiation, next to the saturation irradiation, where the light intensity factor affecting the growth rate is next to 1. From Tables 8 and 9 the high dependence on the orientation of the panel clearly appears in the productivity of Petrolina: if the two faces of the reactor are oriented toward east and west, the productivity is higher than in case of south and north orientation. A possible explanation might be related to radiation reflection by the panel: when the sun is high in the sky, the radiation hits the flat panel with an incidence angle close to 90°; if the incidence angle is too high, radiation could not enter the reactor, due to glass reflection. For this reason, if the panel is oriented toward south, the largest part of the radiation during summer is lost and does not contribute to microalgae growth: if the orientation if east–west, the radiation is collected during morning and afternoon with an incidence angle next to 0°. The same situation does not take place in Sevilla, because the sun does not reach high elevations during the whole year; the east–west orientation is preferable also in Sevilla, as suggested by Sierra et al. (2008).

In general, higher volumetric and areal productivity values have been reached both in Sevilla and in Petrolina, and the difference between the two locations is less remarkable than it was for the open pond; the flat panel photobioreactor growth model includes a temperature control, which maintains the temperature of the water at a constant level. In spite of the higher productivities, Sevilla appear to be less suitable for a flat panel photobioreactor than Petrolina, as shown by NER. The thermal energy requirement is extremely high in Sevilla, where winter time brings low atmospheric temperatures: the photobioreactor should operate only during summertime.

Microalgae mass production in flat panel reactor is much higher than in open raceway pond, CO_2 absorption is more efficient, but the thermal energy requirement implies a NER value >1. There are different possible strategies to solve this problem: keeping the water in the reactor within a range of suboptimal temperatures where the productivity of microalgae is still high and the thermal energy requirement is lower; leaving the temperature in the reactor without any control during the night: this might be an interesting solution also to limit the microalgae losses due to dark respiration, which are higher for optimal temperature. If water temperature is lower than the optimal value for growth, the metabolic energy required by the microalgae for their maintenance during the night is lower.

DISCUSSION

Unlike the most common existing models in the literature, which deal with a specific part of the overall cultivation process, the models presented in this paper include all physical and chemical quantities that mostly affect microalgae growth. These features allow the model to correctly predict the overall behavior of microalgae cultivation plants for energy production. All input parameters can be tuned and varied to obtain reliable predictions.

This paper aims to present a model for microalgae cultivation where geometric and physical parameters, presented in recent works by Slegers et al. (2011, 2013), Duffie and Beckman (2013), Béchet et al. (2013) and more, receive a robust analysis together with the chemical and biological aspects, as presented by Yang (2011), Chisti (2007), Le et al. (2010), and more.

A comparison with experimental data taken from the literature shows that the predictions are consistent, only slightly overestimating the productivity in case of closed photobioreactor. The reason for the overestimation might be found in the temperature control strategy of the reactor, which is kept constant in the simulation runs. This strategy is currently applied in laboratory scale plants but it is expensive both from an economic and energetic point of you for a large scale cultivation plant.

The model is a part of a wider activity on microalgae cultivation plants for energy production which consider the additional directions of work:

- Further analysis could be conducted at the boundaries of the system taken into consideration by the model, evaluating and modeling the downstream microalgae transformation process and the upstream technologies that generate the input streams entering the cultivation phase.
- The cultivation phase model might be modified to include other less important, but still valuable parameters that influence microalgae growth, such as some other nutrients like phosphorus.
- The microalgae growth rate equation could be compared with other models in the literature and with results coming from real pilot plants to make predictions more reliable.
- The flat panel model should be tested with different temperature control strategies, which might lead to a lower productivity but also to a lower thermal energy consumption, making the technology more interesting from an economic point of view.

Conflict of Interest Statement

The authors declare that the research was conducted in the absence of any commercial or financial relationships that could be construed as a potential conflict of interest.

REFERENCES

1. Ak, İ (2012). Effect of an organic fertilizer on growth of blue-green alga Spirulina platensis.Aquacult. Int. 20, 413–422. doi: 10.1007/s10499-011-9473-5
2. Azadi, P., Brownbridge, G., Mosbach, S., Smallbone, A., Bhave, A., Inderwildi, O., et al. (2014). The carbon footprint and non-renewable energy demand of algae-derived biodiesel. Appl. Energy 113, 1632–1644. doi:10.1016/j.apenergy.2013.09.027
3. Bahadar, A., and Bilal Khan, M. (2013). Progress in energy from microalgae: a review. Renew. Sustain. Energ. Rev. 27, 128–148. doi:10.1016/j.rser.2013.06.029
4. Béchet, Q., Shilton, A., and Guieysse, B. (2013). Modeling the effects of light and temperature on algae growth: state of the art and critical assessment for productivity prediction during outdoor cultivation. Biotechnol. Adv. 31, 1648–1663. doi:10.1016/j.biotechadv.2013.08.014
5. Borowitzka, M. A. (1999). Commercial production of microalgae: ponds, tanks, tubes and fermenters. J. Biotechnol. 70, 313–321. doi:10.1016/S0168-1656(99)00083-8
6. Brennan, L., and Owende, P. (2010). Biofuels from microalgae – a review of technologies for production, processing, and extractions of biofuels and co-products. Renew. Sustain. Energ. Rev.14, 557–577. doi:10.1016/j.rser.2009.10.009
7. Çelekli, A., and Yavuzatmaca, M. (2009). Predictive modeling of biomass production by Spirulina platensis as function of nitrate and NaCl concentrations. Bioresour. Technol. 100, 1847–1851. doi:10.1016/j.

biortech.2008.09.042

8. Chisti, Y. (2007). Biodiesel from microalgae. Biotechnol. Adv. 25, 294–306. doi:10.1016/j.biotechadv.2007.02.001
9. Collet, P., Hélias, A., Lardon, L., Ras, M., Goy, R.-A., and Steyer, J.-P. (2011). Life-cycle assessment of microalgae culture coupled to biogas production. Bioresour. Technol. 102, 207–214. doi:10.1016/j.biortech.2010.06.154
10. Cuaresma, M., Janssen, M., Vílchez, C., and Wijffels, R. H. (2011). Horizontal or vertical photobioreactors? How to improve microalgae photosynthetic efficiency. Bioresour. Technol. 102, 5129–5137. doi:10.1016/j.biortech.2011.01.078
11. Davis, R., Aden, A., and Pienkos, P. T. (2011). Techno-economic analysis of autotrophic microalgae for fuel production. Appl. Energy 88, 3524–3531. doi:10.1016/j.apenergy.2011.04.018
12. Doucha, J., and Lívanský, K. (2006). Productivity, CO2/O2 exchange and hydraulics in outdoor open high density microalgal (Chlorella sp.) photobioreactors operated in a Middle and Southern European climate. J. Appl. Phycol. 18, 811–826. doi:10.1007/s10811-006-9100-4
13. Duffie, J. A., and Beckman, W. A. (2013). Solar Engineering of Thermal Processes, 4th edn. New York: A Wiley-Interscience Publication, John Wiley and Sons, Inc.
14. Fernández, F. G. A., Camacho, F. G., Pérez, J. A. S., Sevilla, J. M. F., and Grima, E. M. (1997). A model for light distribution and average solar irradiance inside outdoor tubular photobioreactors for the microalgal mass culture. Biotechnol. Bioeng. 55, 701–714. doi:10.1002/(SICI)1097-0290(19970905)55:5
15. Fernández, F. G. A., Sevilla, J. M. F., and Grima, E. M. (2013). Photobioreactors for the production of microalgae. Rev. Environ. Sci. Biotechnol. 12, 131–151. doi:10.1007/s11157-012-9307-6
16. Hadiyanto, H., Elmore, S., Van Gerven, T., and Stankiewicz, A. (2013). Hydrodynamic evaluations in high rate algae pond (HRAP) design. Chem. Eng. J. 217, 231–239. doi:10.1016/j.cej.2012.12.015
17. Hempel, N., Petrick, I., and Behrendt, F. (2012). Biomass productivity and productivity of fatty acids and amino acids of microalgae strains as key characteristics of suitability for biodiesel production. J. Appl. Phycol. 24, 1407–1418. doi:10.1007/s10811-012-9795-3
18. James, S. C., and Boriah, V. (2010). Modeling algae growth in an open-channel raceway. J. Comput. Biol. 17, 895–906. doi:10.1089/cmb.2009.0078
19. Janssen, M., Tramper, J., Mur, L. R., and Wijffels, R. H. (2003). Enclosed outdoor photobioreactors: light regime, photosynthetic efficiency, scale-up, and future prospects. Biotechnol. Bioeng. 81, 193–210. doi:10.1002/bit.10468
20. Jiménez, C., Cossio, B. R., Labella, D., and Xavier Niell, F. (2003). The feasibility of industrial production of Spirulina (Arthrospira) in Southern Spain. Aquaculture 217, 179–190. doi:10.1016/S0044-8486(02)00118-7
21. Jorquera, O., Kiperstok, A., Sales, E. A., Embiruçu, M., and Ghirardi, M. L. (2010). Comparative energy life-cycle analyses of microalgal biomass production in open ponds and photobioreactors.Bioresour. Technol. 101, 1406–1413. doi:10.1016/j.biortech.2009.09.038
22. Kochem, L. H., Da Fré, N. C., Redaelli, C., Rech, R., and Marcílio, N. R. (2014). Characterization of a novel flat-panel airlift photobioreactor with an internal heat exchanger. Chem. Eng. Technol. 37, 59–64. doi:10.1002/ceat.201300420
23. Kumar, K., Dasgupta, C. N., Nayak, B., Lindblad, P., and Das, D. (2011). Development of suitable photobioreactors for CO2 sequestration addressing global warming using green algae and cyanobacteria. Bioresour. Technol. 102, 4945–4953. doi:10.1016/j.biortech.2011.01.054
24. Le, P. J., Williams, B., and Laurens, L. M. L. (2010). Microalgae as biodiesel & biomass feedstocks: review & analysis of the biochemistry, energetics & economics. Energy Environ. Sci. 3, 554–590. doi:10.1039/b924978h
25. Mata, T. M., Martins, A. A., and Caetano, N. S. (2010). Microalgae for biodiesel production and other applications: a review. Renew. Sustain. Energ. Rev. 14, 217–232. doi:10.1016/j.rser.2009.07.020
26. Molina, E., Fernández, J., Acién, F. G., and Chisti, Y. (2001). Tubular photobioreactor design for algal cultures. J. Biotechnol. 92, 113–131. doi:10.1016/S0168-1656(01)00353-4
27. Münkel, R., Schmid-Staiger, U., Werner, A., and Hirth, T. (2013). Optimization of outdoor cultivation in flat panel airlift reactors for lipid production by Chlorella vulgaris. Biotechnol. Bioeng. 110, 2882–2893. doi:10.1002/bit.24948
28. Muñoz, R., and Guieysse, B. (2006). Algal-bacterial processes for the treatment of hazardous contaminants: a review. Water Res. 40, 2799–2815. doi:10.1016/j.watres.2006.06.011
29. Norsker, N.-H., Barbosa, M. J., Vermuë, M. H., and Wijffels, R. H. (2011). Microalgal production – a close look at the economics. Biotechnol. Adv. 29, 24–27. doi:10.1016/j.biotechadv.2010.08.005
30. NREL, . (0000). Biomass Research – Publications. Available at:http://www.nrel.gov/biomass/publications.html?print
31. Pruvost, J., Cornet, J. F., Goetz, V., and Legrand, J. (2011). Modeling dynamic functioning of rectangular photobioreactors in solar conditions. AIChE J. 57, 1947–1960. doi:10.1002/aic.12389
32. Radmann, E. M., Reinehr, C. O., and Costa, J. A. V. (2007). Optimization of the repeated batch cultivation of microalga Spirulina platensis in open raceway ponds. Aquaculture 265, 118–126. doi:10.1016/j.aquaculture.2007.02.001
33. Rawat, I., Ranjith Kumar, R., Mutanda, T., and Bux, F. (2013). Biodiesel from microalgae: a critical evaluation from laboratory to large scale production. Appl. Energy 103, 444–467. doi:10.1016/j.apenergy.2012.10.004

34. Robinson, D., and Stone, A. (2004). Solar radiation modelling in the urban context. Solar Energy77, 295–309. doi:10.1016/j.solener.2004.05.010

35. Rodolfi, L., Chini Zittelli, G., Bassi, N., Padovani, G., Biondi, N., Bonini, G., et al. (2009). Microalgae for oil: strain selection, induction of lipid synthesis and outdoor mass cultivation in a low-cost photobioreactor. Biotechnol. Bioeng. 102, 100–112. doi:10.1002/bit.22033

36. Ruiz, J., Álvarez-Díaz, P. D., Arbib, Z., Garrido-Pérez, C., Barragán, J., and Perales, J. A. (2013). Performance of a flat panel reactor in the continuous culture of microalgae in urban wastewater: prediction from a batch experiment. Bioresour. Technol. 127, 456–463. doi:10.1016/j.biortech.2012.09.103

37. Sierra, E., Acién, F. G., Fernández, J. M., García, J. L., González, C., and Molina, E. (2008). Characterization of a flat plate photobioreactor for the production of microalgae. Chem. Eng. J. 138, 136–147. doi:10.1016/j.cej.2007.06.004

38. Sills, D. (2013). Modeling CO2 Requirements for Cultivation of Microalgae in Open Raceway Pond.

39. Singh, N. K., and Dhar, D. W. (2011). Microalgae as second generation biofuel. A review. Agron. Sustain. Dev. 31, 605–629. doi:10.1007/s13593-011-0018-0

40. Slegers, P. M., Lösing, M. B., Wijffels, R. H., van Straten, G., and van Boxtel, A. J. B. (2013). Scenario evaluation of open pond microalgae production. Algal. Res. 2, 358–368. doi:10.1016/j.biortech.2012.11.123

41. Slegers, P. M., Wijffels, R. H., van Straten, G., and van Boxtel, A. J. B. (2011). Design scenarios for flat panel photobioreactors. Appl. Energy 88, 3342–3353. doi:10.1016/j.apenergy.2010.12.037

42. Sompech, K., Chisti, Y., and Srinophakun, T. (2012). Design of raceway ponds for producing microalgae. Biofuels 3, 387–397. doi:10.4155/bfs.12.39

43. Stephenson, A. L., Kazamia, E., Dennis, J. S., Howe, C. J., Scott, S. A., and Smith, A. G. (2010). Life-cycle assessment of potential algal biodiesel production in the United Kingdom: a comparison of raceways and air-lift tubular bioreactors. Energy Fuels 24, 4062–4077. doi:10.1021/ef1003123

44. Sugai-Guérios, M. H., Mariano, A. B., Vargas, J. V. C., de Lima Luz, L. F., and Mitchell, D. A. (2014). Mathematical model of the CO2 solubilisation reaction rates developed for the study of photobioreactors. Can. J. Chem. Eng. 92, 787–795. doi:10.1002/cjce.21937

45. Sukhatme, K., and Sukhatme, S. P. (1996). Solar Energy: Principles of Thermal Collection and Storage. Noida, UP: Tata McGraw-Hill Education.

46. Yang, A. (2011). Modeling and evaluation of CO2 supply and utilization in algal ponds. Ind. Eng. Chem. Res. 50, 11181–11192. doi:10.1021/ie200723w

APPENDIX

Nomenclature

A_{soil}: area of the pond that is embedded in the soil [m²]
A_W: water surface area of the pond [m²]
C_{Bowen}: Bowen constant [Pa/°C]
CO_{2D}: dissolved CO_2 molar quantity in the bioreactor [mol/m³]
cp_W: heat capacity of the growth medium [J/(kg°C)]
C_T: total concentration of carbonate species in the bioreactor
d: distance between the panels [m]
$E_{harv/refil}$: electrical energy consumption for microalgae harvesting and reactor refilling [J]
$f_{CO\ 2}$: flux of CO_2 introduced by the supply of gas flow into the system [g/(m³ s)]
f_I: light intensity factor [–]
f_T: temperature factor [–]
g: gravitational acceleration [m/s²]
$G_{diffuse}$: geometrical factor for diffuse radiation [–]
G_{direct}: geometric factor for direct radiation [–]
h: height of the panels [m]
h_{comb}: lower heating value on a dry basis of algae biomass [J/kg]
h_{shadow}: height of the shadow on the panel behind [m]
h_{soil}: heat transfer coefficient of the surrounding soil layer [W/(m²°C)]
I_0: incident light intensity [W/m²]
I_a: average light intensity in the volume of the bioreactor at a given time t [W/m²]
I_{back}: radiation light intensity for the back surface [W/m²]
$I_{diffuse}$: diffuse radiation over a horizontal surface [W/m²]
I_{direct}: direct radiation over a horizontal surface [W/m²]
I_{front}: radiation light intensity for the front surface [W/m²]
I_{GHR}: global horizontal radiation [W/m²]
I_S: saturation light intensity for microalgae species [W/m²]
$I_{surface}$: total light arriving on the pond [J/(kg°C)]
K_C, K_{NA}: half-saturation constants [mol/m³]
K_e: light extinction coefficient [m⁻¹]
K_{e1}, K_{e2}: light extinction constants
$k_{Lg}\alpha$: mass transfer coefficient for a given element
k_1, k_2: carbonate species equilibrium constants

M: concentration of the respective component [g/m^3]
M^*: saturation concentration of the associated dissolved element [g/m^3]
N_T: dissolved nitrogen molar quantity in the bioreactor [mol/m^3]
PAR: photosynthetic active radiation [%]
p_a: ambient pressure [Pa]
P_{pump}: power of the installed pump for harvesting and refilling of the reactor [W]
p_{ref}: reference pressure [Pa]
p'_a: water pressure of the air [Pa]
p'_s: saturated water pressure [Pa]
Q: volumetric flow rate [m^3/s]
Q_{algae}: light energy flow to algae during growth [W]
Q_{cond}: heat flow between the pond and the ground via conduction [W]
Q_{conv}: heat flow by convection [W]
Q_{evap}: heat flow caused by either evaporation or condensation [W]
Q_{irr}: heat flow to the pond by sunlight [W]
Q_{rad}: heat flow by emission of long-wave radiation in the infrared region [W]
r_{gA}: microalgae growth rate [g/(m^3 s)]
RH: relative humidity [–]
R_p: reflection coefficient for p-polarized light [–]
R_s: reflection coefficient for s-polarized light [–]
R': overall reflection coefficient for each interface [–]
R'_1: overall reflection coefficient for the air – reactor interface [–]
R'_2: overall reflection coefficient for the reactor – culture interface [–]
s: position in bioreactor depth [m]
T_a: air temperature [°C]
T_{dew}: dew point temperature [°C]
$t_{harv/refil}$: time for microalgae harvesting and reactor refilling [s]
T_{let}: lethal temperature for microalgae species [K]
T_m: transparency of wall material [–]
T_{opt}: optimal temperature for microalgae species [K]
T_{sky}: equivalent sky temperature for clear sky days [K]
T_{soil}: temperature of the soil surrounding the pond [°C]
t_{solar}: number of hours after solar midnight [–]
T_w: water temperature in the bioreactor [K]
V_R: reactor volume [m^3]
X_A: microalgae concentration in the bioreactor [g/m^3]
y: position in reactor height [m]
Y_{AM}: mass of the respective component consumed or generated by the microalgae per unit mass of microalgae produced [$g_M/g_{microalgae}$]
z: pond depth [m] or photobioreactor thickness [m]
β: slope of the reactor [°]
β_{algae}: curve modulating constant for the temperature fac it depends on microalgae species
γ: azimuth angle [°]
ϵ_w: emissivity of the water in the infrared region [–]
η_i: refractive index of the material before the interface [–]
η_{pump}: efficiency of the pump [–]
η_t: refractive index of the material after the interface [–]
θ: solar incidence angle [°]
θ_i: incidence angle [°]
θ_z: solar zenith angle [°]
μ_A: microalgae specific growth rate [s^{-1}]
$\hat{\mu}_A$: maximum microalgae specific growth rate [s^{-1}]
ρ: water density [kg/m^3]
ρ_w: density of the growth medium [kg/m^3]
σ: light extinction coefficient [m]
σ_{SB}: Stefan–Boltzmann constant [W/(m^{2}K^4)]

CHAPTER 3 MICROALGAE GROWTH IN OUTDOOR POND CULTURES

M. Huesemann· B. Crowe, P. Waller, A. Chavis, S. Hobbs, S. Edmundson, M. Wigmosta

Keywords: Biomass growth model, Light absorption coefficient, Maximum specific growth rate, Dark respiration, Biomass loss rate in the dark, Raceway pond

INTRODUCTION

The use of microalgae for the production of carbon-neutral drop-in biodiesel and jet-biofuel has attracted increasing interest due to accelerating global climate change [1]. An economically feasible microalgae biofuels production process requires that strains be found that exhibit high annual biomass productivities (greater than 30 g/m^2-day) in outdoor culture systems [2]. Significant efforts have been undertaken in a variety of research organizations to find microalgae strains which, through either prospecting or genetic engineering, might be suitable for large-scale, economically viable biofuel production.

A major challenge is the identification and selection of strains that have the potential to achieve high annual biomass productivities in outdoor ponds. Even though a strain may grow well in laboratory cultures under particular incubation conditions (e.g., room temperature and relatively low light intensities), there is no guarantee that it will perform satisfactorily in outdoor pond cultures that are subjected to daily and seasonal water temperature and light fluctuations. Furthermore, unless experiments are specifically designed to generate data for input to a predictive growth model, it is impossible to extrapolate the findings from the laboratory to determine the operating conditions (i.e., culture depth, dilution rate) and the geographic pond location where biomass productivities (monthly or annual) for a given strain will be optimal. A possible solution, suggested earlier by Huesemann et al. [3], is the use of a biomass growth model to predict outdoor pond performance, using a limited number of strain-specific model input parameters that are measured in laboratory cultures in conjunction with measured or predicted, via the PNNL Biomass Assessment Tool (BAT) [4], [5], [6] and [7], sunlight intensity and pond water temperature data (or time series, termed “scripts”).

Since light and temperature are the primary determinants of biomass productivity in outdoor pond cultures operated under nutrient-replete (N, P, CO_2, etc.) and well-mixed conditions, it is necessary to know the selected strain’s specific growth rate as a function of these two variables. There is a large body of literature on the effects of increasing light intensity on a microalgae species’ maximum specific growth rate or photosynthetic oxygen evolution rate, resulting in useful knowledge about the quantum efficiency, the saturating light intensity, and potential photo-inhibition [8], [9] and [10]. There are also a few studies on the relationship between specific growth rate and temperature for multiple[11] and [12] or individual species [13], [14], [15] and [16]. However, this information is of limited use when faced with the task of identifying suitable microalgae for outdoor pond cultivation, given that each strain has its own unique light and temperature response curves. Consequently, for each selected strain, the maximum specific growth rate must be measured during the exponential growth phase in unshaded cultures as a function of light intensity and temperature.

Light and temperature also affect the rate of biomass loss overnight due to dark respiration. Biomass-specific dark respiration rates in microalgae are highly species-dependent, varying from 0.01 to 0.6 day^{-1}, and can comprise as much as 30% of the strain’s maximum specific growth rate [17]. Ryther and Guillard [18] determined the respiratory coefficient (g carbon respired per hour per gram chlorophyll) for five diatoms at 5, 10, 15, 20, and 25 °C, demonstrating a significant difference among these species. Grobbelaar and Soeder [19] measured respiratory oxygen uptake rates for two microalgae species in dark ponds and found these to be dependent on incubation temperature, time duration in the dark, and on the temperature and light intensity under which the microalgae had previously been cultivated. Torzillo et al. [20] cultivatedSpirulina platensis in outdoor tubular

---

Citation: M. Huesemann, B. Crowe, P. Waller, A. Chavis, S. Hobbs, S. Edmundson, M. Wigmosta, “Microalgae growth in outdoor pond cultures”, Algal Research, Volume 13, January 2016, Pages 195-206, ISSN 2211-9264, http://dx.doi.org/10.1016/j.algal.2015.11.008.

photobioreactors and reported that night biomass loss depended on the temperature and light intensity at which the cultures were grown. When culturing Chlorella pyrenoidosa, Ogbonna and Tanaka [21] observed that the magnitude of biomass loss during the dark period was positively correlated with the temperature during the dark period and negatively correlated with the temperature during the light period. Furthermore, cells in the light-sufficient exponential growth phase exhibited much larger biomass loss in the dark than those in the light-limited linear or stationary growth phase, possibly indicating that the average light intensity that cells receive during the light period affects the accumulation of photosynthetic energy storage products and thus the magnitude of biomass loss due to dark respiration of these compounds. Therefore, in order to model biomass loss in the dark, it will be necessary to know, for a given strain, how the rate of biomass loss during the dark period is affected by temperature and average light intensity during the light period.

Over a hundred mathematical and numerical models for predicting phytoplankton net primary productivity [e.g., 22] or microalgae biomass growth in photobioreactors and ponds have been published [23], [24], [25], [26], [27], [28], [29], [30] and [31]. Generally in these models, light attenuation within the culture is estimated and the rate of biomass growth is predicted as a function of either incident or absorbed light [3] and [31]. With few exceptions [32], [33], [34], [35] and [36], most models rely on Beer–Lambert›s law to determine the light intensity as a function of culture depth and biomass concentration. This, however, is problematic for high density cultures where light scattering can be significant [31]. In addition, models that attempt to incorporate biomass loss in the dark use assumed, rather than measured, dark respiration rates or maintenance energy requirements [23], [28], [30] and [31].

Unfortunately, current models are not useful for screening novel strains in terms of their potential to achieve high biomass productivities in outdoor cultures. This is due to the fact that most models are complex and require the input of large numbers of parameters, most of which are difficult, expensive or time-consuming to determine or, worse, have values which simply must be postulated. In addition, most models have not been validated or their validation has been questionable [3]. In a recent review of over 40 microalgae growth modeling approaches, Bechet et al. [31] state that "most of the models hitherto described in the literature have not been validated under conditions relevant to outdoor cultivation." As Chalup and Laws [37] wrote more than 20 years ago when reviewing existing microalgae growth models: "With a judicious choice of parameter values, the models almost always seem to give a good fit to available experimental data." If a model requires many input parameters, it is easy to make appropriate "judicious choices," convenient assumptions, or to select the "right" values from the extensive literature until a good model fit to the experimental culture data is obtained. This can lead to a phenomenon called "overfitting," resulting in a poor ability to predict because such a model will describe noise rather than actual trends [31]. While a reasonable model fit may show that the predictions are pointing in the right direction, it is questionable whether such a model, in the absence of rigorous validation, can be trusted to accurately forecast the performance of outdoor pond cultures.

Building on our earlier modeling work [3], our objective here is to develop and validate, for three different microalgae species, a biomass growth model based on (a) independently measured strain-specific input parameters (scatter-corrected biomass light absorption coefficient, specific biomass growth or loss rate as a function of light and temperature), (b) environmental parameters (incident light intensity, water temperature), and (c) key culture operational parameters (depth, dilution rate). This will enable the reliable prediction of biomass productivities in raceway ponds subjected to daily light and temperature fluctuations.

MICROALGAE BIOMASS GROWTH MODEL

The microalgae growth model presented here and validated for three independent strains has been designed to screen for those strains which exhibit significant potential for high biomass productivities in outdoor ponds under nutrient-replete conditions (Figure S1). It is assumed that light and temperature are the key and instantaneous determinants of microalgae growth and productivity, and that no other factors such as nutrients, CO_2, and mixing (i.e., mass-transfer) are limiting. Furthermore, it is assumed that the culture pH remains constant via feed-back controlled CO_2 addition and that there is no growth inhibition by photosynthetic oxygen or other compounds.

The pond water temperature (T) is easily measured or predicted using the mass and energy balance pond temperature model within the Biomass Assessment Tool. According to Beer–Lambert›s law, for a given biomass concentration (B), light (PAR, 400 to 700 nm) attenuates as a function of light penetration distance (z) as follows:

$$I(z) = I_0 \cdot e^{-k_a \cdot B z} \quad (1)$$

where Io is the incident light intensity at the culture surface, I(z) is the light intensity at depth (z), and ka is the biomass light absorption (extinction) coefficient for the entire PAR range (400–700 nm). Since the biomass concentration B increases in a growing culture, the light attenuation increases with time.

Beer-Lambert›s law states that the light absorbance ($A = \log_{10}(I0/I)$) is directly proportional to the compound's concentration and light path length. This relationship is linear but applicable only to monochromatic, collimated light in dilute light-absorbing solutions where scattering is negligible [38]. Although these conditions are generally not fulfilled in dense pond cultures illuminated by full spectrum sunlight, most microalgae biomass growth models employ Beer–Lambert's law to estimate the light intensity as a function of depth. In order to account for possible light scattering in dense microalgae cultures, we utilized an experimentally determined scatter-corrected biomass light attenuation coefficient (ksca), following a procedure adopted from Suh and Lee [39], for the prediction of light intensity as a function of light penetration distance z and biomass concentration B (see Material and Methods section):

$$k_{sca} = k_a \cdot \frac{K_B}{K_B + B} \cdot \frac{K_z}{K_z + z} \quad (2)$$

where KB and Kz are the light scattering coefficients related to biomass concentration (B) and light path length (z), respectively.

If the culture volume is discretized into n equal volume layers orthogonal to the incident light, the biomass concentration in each layer is assumed to increase exponentially fromB(t) to B(t + Δt) during time interval Δt as follows:

$$B(t+\Delta t)=B(t)\cdot e\mu\cdot\Delta t \quad (3)$$

where μ is the specific growth rate (day^{-1}) in the respective volume layer. The specific growth rate is affected by water temperature (T) and light intensity (I) in the respective volume layer. Assuming perfect mixing, the water temperature is the same in all volume layers and can be predicted as a function of time using the pond energy balance model which is part of the biomass assessment tool (BAT, see below). The light intensity in each volume layer can be predicted via Beer–Lambert's law (Eq. 1) using the experimentally determined scatter-corrected light attenuation coefficient (Eq.2).

Since the application of Eq. (3) requires knowledge of the specific growth rate at the particular temperature and light intensity within each discretized volume layer, it is necessary to know how μ varies with temperature and light intensity. In dilute cultures with minimum self-shading, the specific growth rate of microalgae can be experimentally determined as a function of temperature and light as follows:

$$\mu=f(TI) \quad (4)$$

where f(T,I) is the two dimensional array (or surface) of specific growth rates measured for different combinations of temperature and light intensity values. Since each microalgae strain has a unique response to light (i.e., compensation light intensity, saturating light intensity, photo-inhibition) and temperature (i.e., optimum temperature, temperature tolerance range), the function f(T,I) is strain-specific and must be experimentally determined prior to running the model. It is assumed that individual cells respond instantaneously to the new light conditions as they enter each successive volume layer and that they exhibit the corresponding experimentally determined specific growth rate for that particular light intensity. This response has been verified in our laboratory by measurements of conventional P–I curves which clearly indicate that changes in light intensity produce immediate changes in photosynthetic oxygen evolution [3] and [43]. Others have found the same [23], [40],[41] and [42].

Since biomass loss overnight due to dark respiration can have a significant, negative effect on biomass productivity [19], [20] and [21], it is necessary to know the rate of biomass loss (μ_{dark}) in the absence of light (I = 0) as a function of temperature (T) and the average light intensity (Iavg) to which the cells were exposed in the mixed pond culture during the preceding day:

$$\mu dark=f(TIavg). \quad (5)$$

Iavg is calculated by averaging the depth-integrated light attenuation profiles (Eqs.(1) and (2)) in the pond culture for each time step (Δt) over the entire day preceding the night in which biomass loss due to dark respiration occured. Biomass loss rates in the dark (μ_{dark}) as a function of temperature (T) and average light intensity (Iavg) were independently determined in laboratory experiments as described in Materials and Methods.

In summary, the modeling procedure involves the following steps:

- Select an initial biomass concentration, B(t = 0);
- Discretize the culture volume into n equal-sized parallel volume layers orthogonal to incident light;
- Calculate the light intensity at the midpoint of each of the n culture volume layers (Eqs. (1) and (2), Figure S1);
- For daytime conditions, determine the specific growth rate (μ) in each of the n culture volume layers using experimentally determined strain-specific growth rate data (Eq.4). Since cells in well mixed dense cultures during the daytime are exposed to high average light intensities at or near the surface of the pond while receiving little or no light during their short residence time in the bottom part for the pond and therefore may lose biomass due to dark respiration, the specific (negative) growth rate in the complete absence of light (I = 0 $\mu mol/m^2\cdot s^{-1}$) was assumed to be μ_{dark} for the experimentally determined case of high average light intensity during the late exponential growth phase (Table S1).
- For nighttime conditions, calculate the biomass loss rate in the dark (μ_{dark}) as a function of pond water temperature and the average light intensity during the preceding day, using experimentally determined biomass loss rate data (Eq. 5, Table S1);
- Calculate the increase (μ during day) or decrease (μ_{dark} during night) in biomass concentration in each of the n culture volume layers during time interval Δt (Eq. 3);
- Determine the new biomass concentration B(t + Δt) in the entire culture by numerically averaging the biomass concentrations of all n culture volume layers (B = ΣBi/n);
- Repeat steps 3 through 7 for each time step Δt until the desired final time.

This modeling routine was programmed in Visual Basic for Applications (VBA Microsoft Excel) and MATLAB® (for integration with the Biomass Assessment Tool), using n = 100 and Δt = 5 min.

MATERIALS AND METHODS

Microorganisms and media

Chlorella sorokiniana (DOE 1412) was isolated from surface water in Texas by Dr. Juergen Polle (Brooklyn College) and grown at pH 7 in freshwater BG-11 medium containing 17.6 mM NO_3 and 0.66 mM PO_4[3]. Nannochloropsis salina CCMP 1776 was obtained from the Bigelow culture collection and grown at pH 7.5 in f/2–Si artificial seawater medium prepared using Instant Ocean (added to achieve 35 ppt salinity) containing 4 mM NO_3 and 0.1 mM PO_4[44]. Picochlorum sp. was isolated from a mixed culture at Los Alamos National Laboratory and grown at pH 7 in artificial seawater (Instant Ocean) f/2–Si medium (salinity = 35 ppt) containing 8.8 mM NO_3 and

0.254 mM PO_4. The high concentration of N and P in all media and periodic monitoring ensured that no cultures were ever nutrient limited during the experiments.

Measurement of biomass concentration

Biomass concentration was measured as optical density at 750 nm (OD_{750}) and as ash-free dry weight (AFDW, g/L), as described in Van Wagenen et al. [45].

Measurement of the scatter-corrected biomass light absorption coefficient (ksca)

The respective microalgae culture was placed in a white translucent container (30 cm diameter, 42 cm height) mixed from below with a magnetic stirrer and illuminated from above with a multi-color LED panel simulating sunlight at ca. 2000 μmol/m^2·s^{-1}. The light intensity was measured with a leveled LI-COR underwater light sensor (LI-192SA) connected to a LI-COR light meter (model LI-250A) as a function of light penetration distance in 1 cm increments. Light attenuation profiles were measured twice (i.e., stepwise with increasing and decreasing depth) for at least 10 different biomass concentrations, ranging from an OD_{750} of ca. 0.1 to ca. 3.6. The biomass light absorption coefficient ka (Eqs. (1) and (2)) was separately determined in triplicate for dilute culture samples (OD_{750} < 0.3) in a 1 cm path-length cuvette by measuring light absorption relative to a medium blank with a quantum sensor (PAR, LI-190) as described in Huesemann et al. [3]. For each light attenuation profile, the Excel solver function was used to obtain the pair of KB and Kz values (Eq. 2) that gave the best fit via minimizing the sum of squared residuals. Since KB and Kz were functions of biomass concentration, interpolated values were used to predict the light attenuation profiles for the numerous incremental OD_{750} values during the modeling exercise.

Measurement of the specific growth rate (μ) as a function of incident light intensity and temperature

The maximum specific growth rate of each microalgae strain was determined as a function of temperature in dilute, light-sufficient, exponentially growing shaker flask batch cultures (OD_{750} < 0.3) placed on a thermal gradient incubator (8 temperatures, ranging from ca. 13 to 45 °C) as described in Van Wagenen et al. [45]. The culture pH was maintained at the desired set-point (7 or 7.5, see above) by sparging the respective media with CO_2-enriched air [3]. All cultures received continuous lighting at above saturating light intensity (≈ 300 μmol/m^2·s^{-1} or slightly higher), using a multi-colored LED panel simulating the sunlight spectrum. For the case of C. sorokiniana (DOE 1412), the maximum specific growth rate was also measured in dilute cultures as a function of light intensity ranging from 44 to 1950 μmol/m^2·s^{-1}[45].

Measurement of the photosynthetic oxygen evolution rate (P) as a function of incident light intensity and temperature

Temperature adapted samples were taken from the thermal gradient incubator (TGI) cultures, and the rate of photosynthetic oxygen evolution was measured as a function of light intensity (P–I curves) at each respective TGI temperature using a computer-controlled oxygraph Chlorolab system as described by Huesemann et al. [43].

Measurement of the biomass loss rate (μ_{dark}) in the dark as a function of temperature and average light intensity

The three microalgae strains were grown at ca. 24 °C in their respective media in well mixed 1 L Roux bottles sparged continuously with CO_2-enriched air to maintain the desired pH set-points (see Edmundson and Huesemann [46] for additional details). The cultures were illuminated with white LED lighting at an average surface light intensity of 500 μmol/m^2·s^{-1}. Following inoculation, the cultures were allowed to grow for 14 h under continuous illumination, split into four aliquots, which were then incubated in 125 mL shaker flasks for 10 h in the dark on a thermal gradient incubator (@70 rpm) at four temperatures (ranging from ca. 10 to 30 °C). During the dark incubation, cultures were sparged with humidified, sterile filtered CO_2-enriched air to avoid oxygen depletion and to maintain the medium pH. Evaporation due to sparging over the 10 h period was compensated by addition of sterile, deionized water. Biomass concentrations (duplicate AFDW) were measured at the beginning and end of the dark period and used to calculate specific dark loss rates (μ_{dark}). Experiments were carried out repeatedly ensuring reproducibility at three initial optical densities (i.e., OD_{750} of ca. 0.4, 1.0, and ca. 3.0) to simulate different growth stages (i.e., late exponential, linear, and late-linear, respectively) during the light period prior to the dark incubations. The biomass light absorption coefficient ka of culture samples was measured as described above and used to calculate the average light intensity (Iavg) "experienced" by the cells during the light period [47]:

$$I_{avg} = I_0 \cdot \frac{1 - e^{-k_a \, OD_{750} \, d}}{k_a \cdot OD_{750} \cdot d} \quad (6)$$

where d is the Roux bottle culture thickness (d = 0.045 m). The average light intensity was determined using the average OD_{750} during the light period for each of the three growth stages, i.e., average OD_{750} = 1/2 (final OD_{750} − initial OD_{750}).

Outdoor raceway pond culture experiments

N. salina was cultured in nutrient-sufficient f/2–Si medium in triplicate outdoor raceway ponds, maintained at pH 7.5 by periodic CO_2 addition via a feedback controlled solenoid valve, in Tucson, Arizona (USA), from January to April 2011, as described by Crowe et al.[44]. The raceway pond cultures (3 m^2 surface area, ca. 0.25 m depth, 800 L) were mixed with a paddlewheel to achieve an average channel flow velocity of about 0.28 m/s. The ponds were inoculated with exponentially growing photobioreactor seed cultures. Makeup water was added daily to compensate for evaporative losses and to maintain the culture depth at 25 cm. During the initial phases of the batch culture experiment until about March 9 or Day of Year (DOY) 69, the ponds were covered at night with insulation in an effort to keep water temperatures from decreasing below 10 °C (Figure S2).

Indoor raceway pond culture experiments

Two indoor 800 L temperature-controlled (via Labview™ software) fiberglass raceway ponds, of design similar to the outdoor ponds, each illuminated with a panel containing 4500 multi-colored computer-dimmable LEDs (Figure S3), were used to perform the culturing experiments for Chlorella and Picochlorum spp. (see Huesemann et al. [48] for more details on PNNL›s climate-simulation ponds).

Chlorella sp. had originally been cultured in batch and semi-continuous mode in triplicate outdoor raceway ponds in nutrient replete BG-11 medium, maintained at pH 7 via intermittent CO_2 sparging, in Rimrock, Arizona (USA), from June through August, 2012. However, as described in Huesemann et al. [48], there is evidence that these three outdoor pond cultures performed less optimally, due to various mishaps and possible environmental stressors, than two indoor pond cultures that were operated under conditions simulating the outdoor ponds, i.e., using light and temperature scripts (Figure S4) collected during the outdoor cultivation study in Arizona. Therefore, the biomass growth model for Chlorella sp. was validated using the more reliable data collected from the two indoor climate-simulation ponds. As in Arizona, the indoor ponds (culture depth = 25 cm) were operated first in batch culture mode (DOY 162 to 195) followed by semi-continuous culture mode at two different dilution rates (D = 0.45 day^{-1} from DOY 195 to 210, and 0.21 day^{-1} from DOY 210 to 221). During semi-continuous pond operation, a specified fraction of the culture volume (e.g., 45% if D = 0.45 day^{-1}) was removed each day and replaced by fresh medium until a steady-state, as indicated by more or less constant biomass concentrations, was achieved. The incident light intensity and pond water temperature data collected during the outdoor cultivation study (Figure S4) were loaded into the LED and temperature controllers to simulate the daily light and temperature fluctuations observed in the outdoor pond cultures at the Rimrock, Arizona, site.

For Picochlorum sp., two indoor climate-simulation ponds were filled with f/2–Si medium and operated in batch culture mode at pH 7, maintained via intermittent CO_2 sparging, for 28 days at the constant depth of 0.205 m. The biomass growth model was run, using the experimentally determined input parameters for Picochlorum sp. (Fig. 1 and Fig. 6, Table S1), in conjunction with the Biomass Assessment Tool (BAT) [4] to identify outdoor pond locations of maximum annual biomass productivity for this strain. After identifying the best outdoor pond location in Southern Florida, the BAT used 30 year meteorological data to generate light intensity and water temperature scripts for use in this cultivation study. The two indoor ponds were then operated using light and temperature scripts for May, the month of highest predicted biomass productivity at this location (Figure S5).

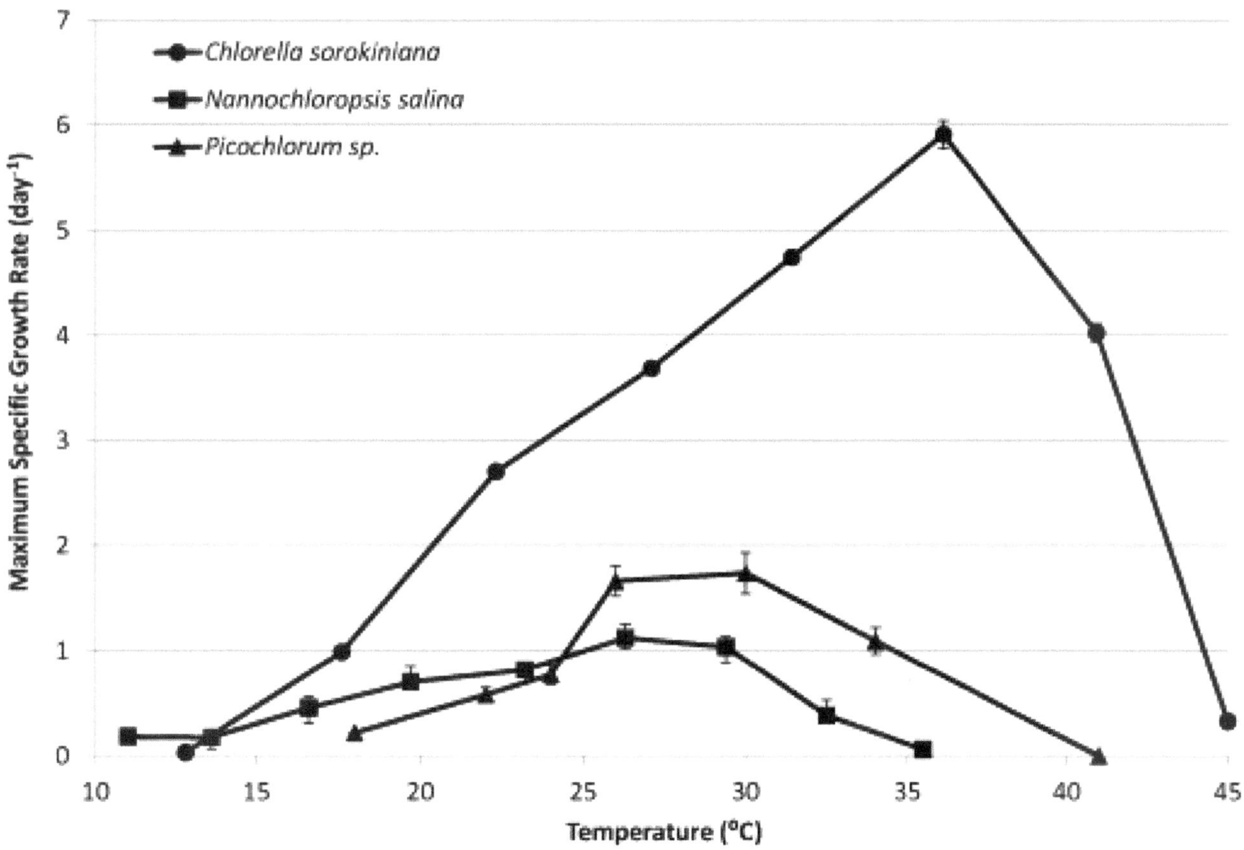

Figure 1. Maximum specific growth rate of Chlorella sorokiniana (DOE 1412), Nannochloropsis salina, andPicochlorum sp. measured in continuous light at above saturating light intensity (≈ 300 μmol/m^2·s^{-1} or slightly higher) as a function of temperature. Error bars are one standard deviation.

RESULTS AND DISCUSSION

C. sorokiniana: laboratory characterization and pond culture modeling

As shown in Fig. 1, C. sorokiniana exhibited a significantly higher maximum specific growth rate than N. salina or Picochlorum sp. over the entire temperature range tested. The optimum value of the maximum specific growth rate was 5.9 day^{-1} for Chlorella at 36.1 °C, and only 1.1 day^{-1} for N. salina at 26.3 °C and 1.8 day^{-1} for Picochlorum sp. at 30 °C, respectively. While all species failed to grow below 13 °C, Chlorella was significantly more thermo-tolerant, being capable of growing, although slowly, at 45 °C. By contrast, N. salina and Picochlorum sp. were unable to sustain growth above 36 and 41 °C, respectively.

Fig. 2 shows the maximum specific growth rate of C. sorokiniana as a function of both temperature and light intensity, data which serve as critical input parameters to the biomass growth model (Eq. 4). The maximum specific growth rate of Chlorella generally (i.e., for temperatures ≥ 22 °C) increased with light intensity until reaching a plateau at a saturating light intensity of about 250 $\mu mol/m^2 \cdot s^{-1}$, and declined to various degrees, depending on temperature, at higher light intensities, most likely due to photo-inhibition. As expected, the maximum specific growth rate at the saturating light intensity showed the same temperature dependency as in Fig. 1, with cultures at the optimum temperature of 36 °C growing the fastest at all light intensities.

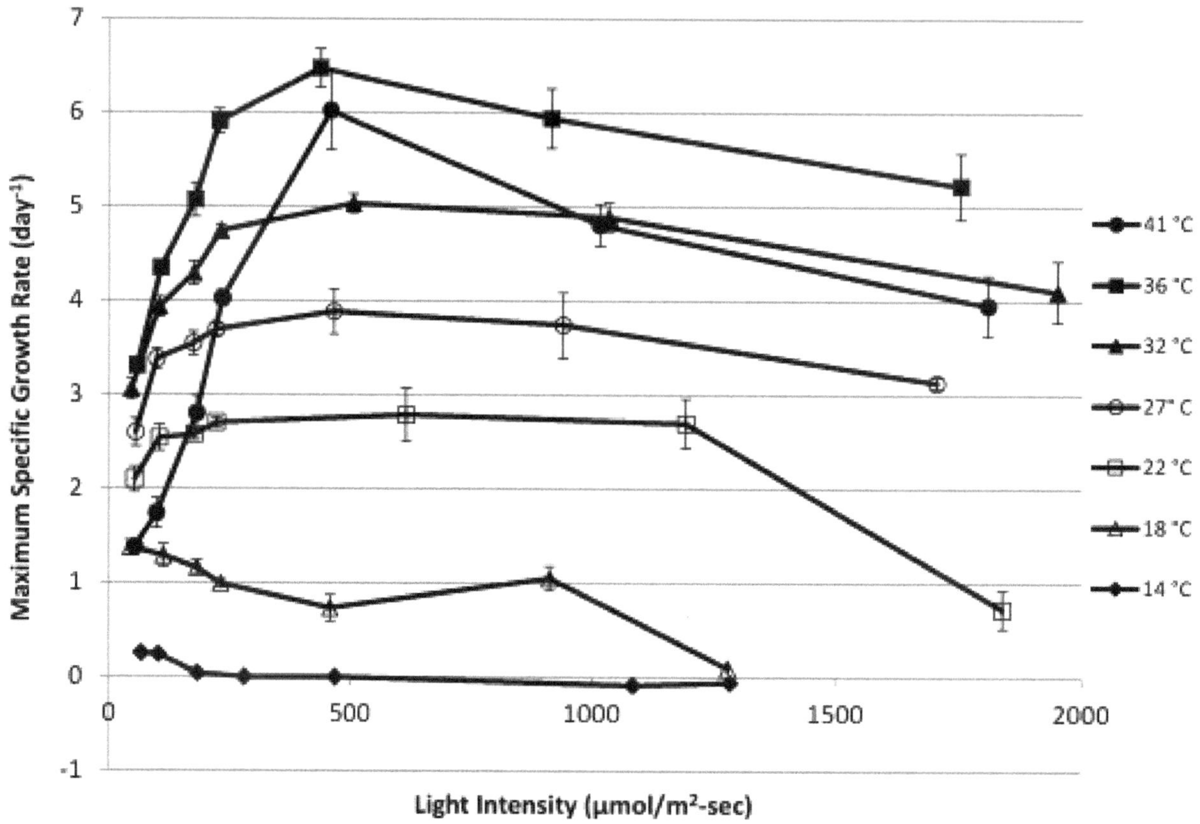

Figure 2. Maximum specific growth rate of Chlorella sorokiniana (DOE 1412) as a function of light intensity at seven temperatures (14, 18, 22, 27, 32, 36, and 41 °C). Error bars are one standard deviation.

Compared to the cultures maintained at 22 °C or higher, those grown at 14 and 18 °C did not exhibit the typical increase of maximum specific growth rate with increasing light intensity. By contrast, the maximum specific growth rates decreased with light intensity, resulting in complete or almost complete growth inhibition of the 14 °C and 18 °C cultures at 280 $\mu mol/m^2 \cdot s^{-1}$ and 1280 $\mu mol/m^2 \cdot s^{-1}$, respectively. The fact that cultures grown at lower temperatures (i.e., 14, 18, and 22 °C) are much more subject to photo-inhibition even at relatively low light intensities (i.e., the lower the temperature, the lower the light intensity at which photo-inhibition occurred) than cultures grown at higher temperature has also been observed by Jensen and Knutsen [49] and Vonshak et al. [50]. It has been postulated that the mechanisms for repairing the molecular damage of photo-inhibition are much more effective at higher temperature, thereby leading to reduced inhibition of photosynthesis under these conditions [49].

Another important input to the biomass growth model is the specific rate of biomass loss in the dark (μ_{dark}) as a function of average light intensity (Iavg) during the light period and culture temperature during the dark period (Eq. 5). The μ_{dark} data for all three strains used in this study have recently been published by Edmundson and Huesemann [46] and are given in Table S1 for reference. In general, the specific rate of biomass loss in the dark increases with increasing average light intensity during the light period and increasing temperature during the dark period. Finally, the light attenuation curves were measured for C. sorokiniana (DOE 1412) cultures at different cell densities (Figure S6-A) and used to calculate the scatter-corrected light attenuation coefficient (ksca) using non-linear regression (Eq. 2).

In order to determine whether the model developed above (Eqs. (1), (2), (3), (4) and (5)), together with the strain-specific input parameters given in Table S1, and Fig. 2 and S6-A, is able to predict the growth of C. sorokiniana (DOE 1412), we compared the model-predicted to the measured growth performance of this strain when cultured in two replicate indoor LED-lighted and temperature controlled ponds operated in batch and semi-continuous culture modes using light and temperature scripts that were recorded during a two month long outdoor pond culture experiment during the summer of 2012 at the University of Arizona Research Station (Figure S4 and Materials & Methods). As mentioned above in Step 4 of the modeling procedure, it was assumed that Chlorellacells, in well mixed ponds during the daytime, repeatedly lose biomass in the lower aphotic zone following exposure to light at the culture surface. Since the cells are exposed to relatively high light intensity in the upper photic zone, particularly in the early growth stages when the culture is still relatively thin, μ_{dark} values for the case of high average light intensity during the late-exponential growth phase were chosen as model inputs (Table S1).

As shown in Fig. 3A, the model-predicted and measured biomass concentrations compared reasonably well during the exponential and the early linear batch growth phase, i.e., for the first 10 days following inoculation when the OD_{750} was below 2. However, the model significantly under-predicted the measured biomass concentration in the later part of the batch culture phase when cell densities reached an OD_{750} as high as 6. The sawtooth pattern of the model-predicted concentration curve reflects the periodic increase of biomass during the day, followed by biomass loss at night due to dark respiration. Since the outdoor ponds were sampled each morning around 10 am, after a period of biomass loss overnight followed by very little growth in the early morning hours due to low water temperatures, the measured biomass concentrations should be compared to the predicted early morning low biomass concentrations following the dark period (i.e., the diurnal low points of the curve predicted by the model).

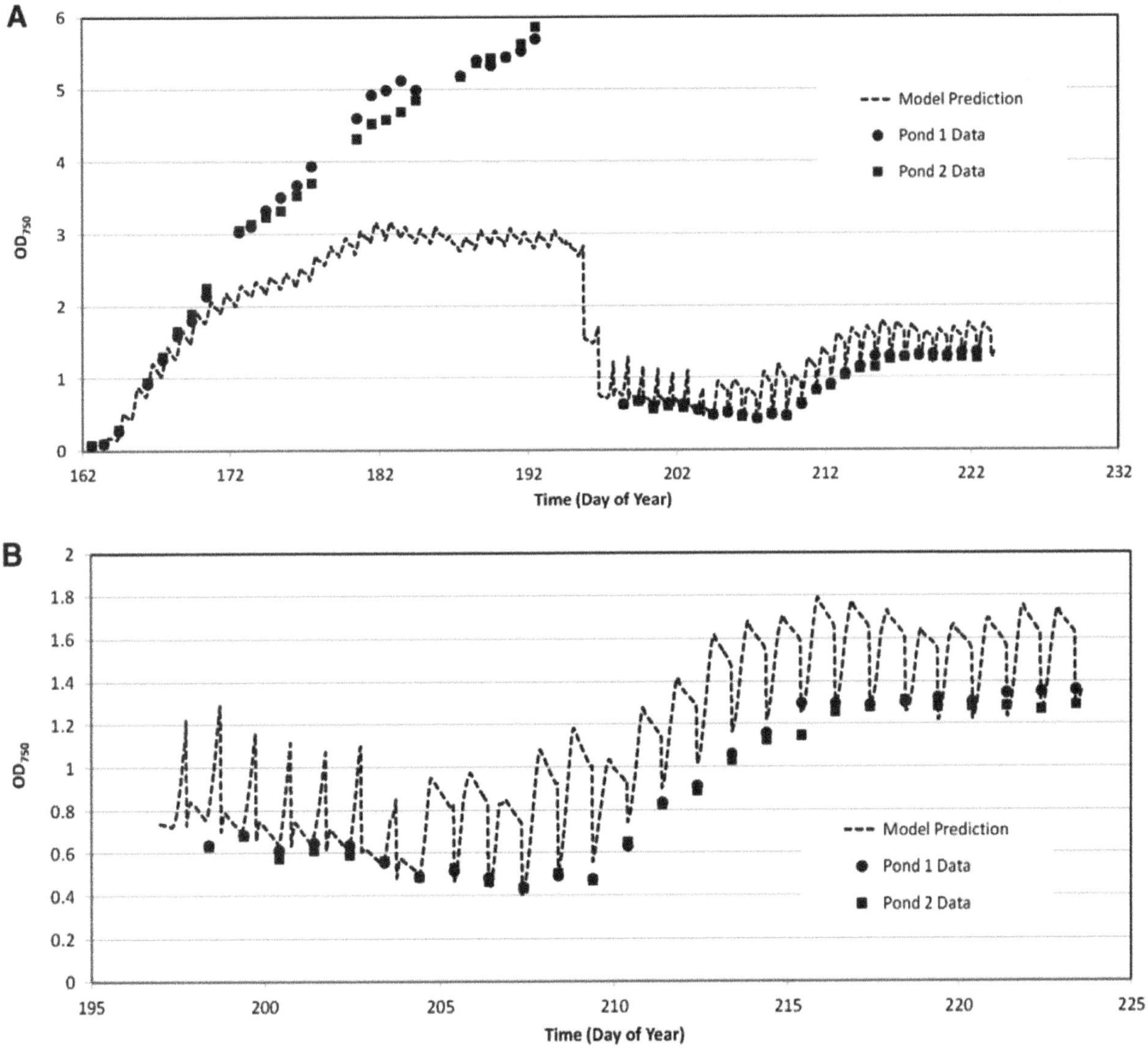

Figure 3. Predicted and measured biomass concentrations (OD_{750}) as a function of time in two indoor LED-lighted and temperature-controlled ponds simulating outdoor Arizona raceways during the batch and subsequent two semi-continuous culture phases. (A) Entire batch and semi-continuous culture experiment; and (B) Semi-continuous culture experiment only, operated at two dilution rates as described in Material and Methods. The AFDW-based areal biomass productivity during the first 12 days of linear batch culture growth, and during the first and second semi-continuous culture steady state was ca. 12.7 g/m^2·day^{-1}, 11.5 g/m^2·day^{-1}, and 12.1 g/m^2·day^{-1}, respectively.

The model was particularly successful in predicting the steady-state biomass concentrations that were measured during the subsequent semi-continuous culturing mode at two different dilution rates (Fig. 3B). As was the case for the analysis of the batch culture data, the measured biomass concentrations must be compared to the diurnal low points of the concentrations predicted by the model. It should be noted that the model correctly predicts the two different declines in biomass concentration that occur each day, i.e., the loss of biomass overnight due to dark respiration and the reduction of biomass due to daily dilution with fresh medium at the specified dilution rate. Until DOY 205, the ponds were diluted each evening at a dilution rate of 0.45 day^{-1}; this explains the large predicted decrease in biomass concentration followed by a smaller decrease overnight. After DOY 205, the ponds were diluted each morning; this explains the small predicted decrease in biomass concentration overnight followed by the larger decrease due to daily dilutions. As expected, the predicted decrease in biomass concentrations due to daily dilutions becomes smaller when the dilution rate was reduced to 0.21 day^{-1}at DOY 210. The model not only predicted the steady-state biomass concentrations at each dilution rate but also the transition from the first to second steady-state. In conclusion, the biomass growth model was successful in predicting the performance of both batch and semi-continuous C. sorokiniana cultures as long as the OD_{750} was less than 2, or equivalently the AFDW was less than ca. 400 mg/L. Given that most industrial scale microalgae ponds are expected to be operated at concentrations below this threshold, i.e., AFDW < 500 mg/L [51], this biomass growth model should be valuable in predicting the biomass productivities under different climatic conditions and pond operational schemes (depth, dilution rate).

It should be noted, however, that the model considerably underpredicted the biomass concentrations during the later phase of the batch culture experiment. A possible hypothesis is that the model over-predicted the biomass losses in the aphotic zone of dense pond cultures during the daytime because the chosen input parameters for μ_{dark}were too negative due to the above mentioned assumption that cells are exposed to high average light intensity near the surface of the pond and thus exhibit relatively large negative biomass loss rates in the darker, lower part of the pond. However, as the culture density increases with time, the average light intensity that cells are exposed to at or near the surface declines, resulting in less negative μ_{dark} values (Table S1). The use of such less negative μ_{dark} values, experimentally determined for cells grown under lower light intensities, would result in model predictions even more closely simulating the biomass growth measurements. However, since no information is presently available in the peer-reviewed literature regarding the effects of light exposure on short-term (seconds to minutes) biomass losses due to dark respiration in the aphotic zone of the pond, we were unable to incorporate this mechanism in the modeling algorithm. Clearly, more research is needed to understand how the rapid cycling of microalgae cells through the light and dark zones of a well-mixed dense pond culture affects the rate of biomass loss in the aphotic zone.

N. salina: Laboratory characterization and pond culture modeling

Since it was extremely time-consuming to measure the maximum specific growth rate ofC. sorokiniana (DOE 1412) as a function of light intensity at seven temperatures (Fig. 2), we evaluated whether the much more rapid measurements of photosynthetic oxygen evolution rate as a function of light intensity (P–I curves) could serve as a proxy for the relationship between maximum specific growth rate and light intensity (μ–I curves). It was found that with the exception of temperatures below 20 °C, the shape of the P–I curves for C. sorokiniana (DOE 1412) were similar to the shape of the μ–I curves (Figure S7 for 27 °C). Assuming that the observed similarity in shape between P–I and μ–I curves is not limited to C. sorokiniana, the P–I curves were measured for N. salina at seven different temperatures (Fig. 4) and the relative shapes of the P–I curves were used in conjunction with the μ–T curve measured at saturating light intensity (Fig. 1) to generate the two dimensional μ(T,I) array that is needed as model input (Eq. 4).

As in the case of Chlorella, the specific rate of biomass loss in the dark (μ_{dark}) was measured for N. salina as a function of average light intensity (Iavg) during the light period and culture temperature during the dark period (Table S1) and entered as model input parameter (Eq. 5). The scatter-corrected light attenuation coefficient (ksca) was calculated via linear regression (Eq. 2) of N. salina culture light attenuation data shown in Figure S6-B.

In order to determine whether the model developed above (Eqs. (1), (2), (3), (4) and (5)), together with the strain-specific input parameters given in Table S1, and Fig. 1 and Fig. 4and S6-B, is able to predict the growth of N. salina, we compared the model-predicted to the measured growth performance of this strain when cultured in three replicate outdoor raceways operated in batch culture mode for three months during the winter and spring of 2011 at the University of Arizona in Tucson, AZ. As shown in Figure S2, the recorded pond water temperatures were rather low during the first month of operation, reaching − 1.8 °C on February 3, 2011 (DOY 34), resulting in the formation of floating ice clumps.

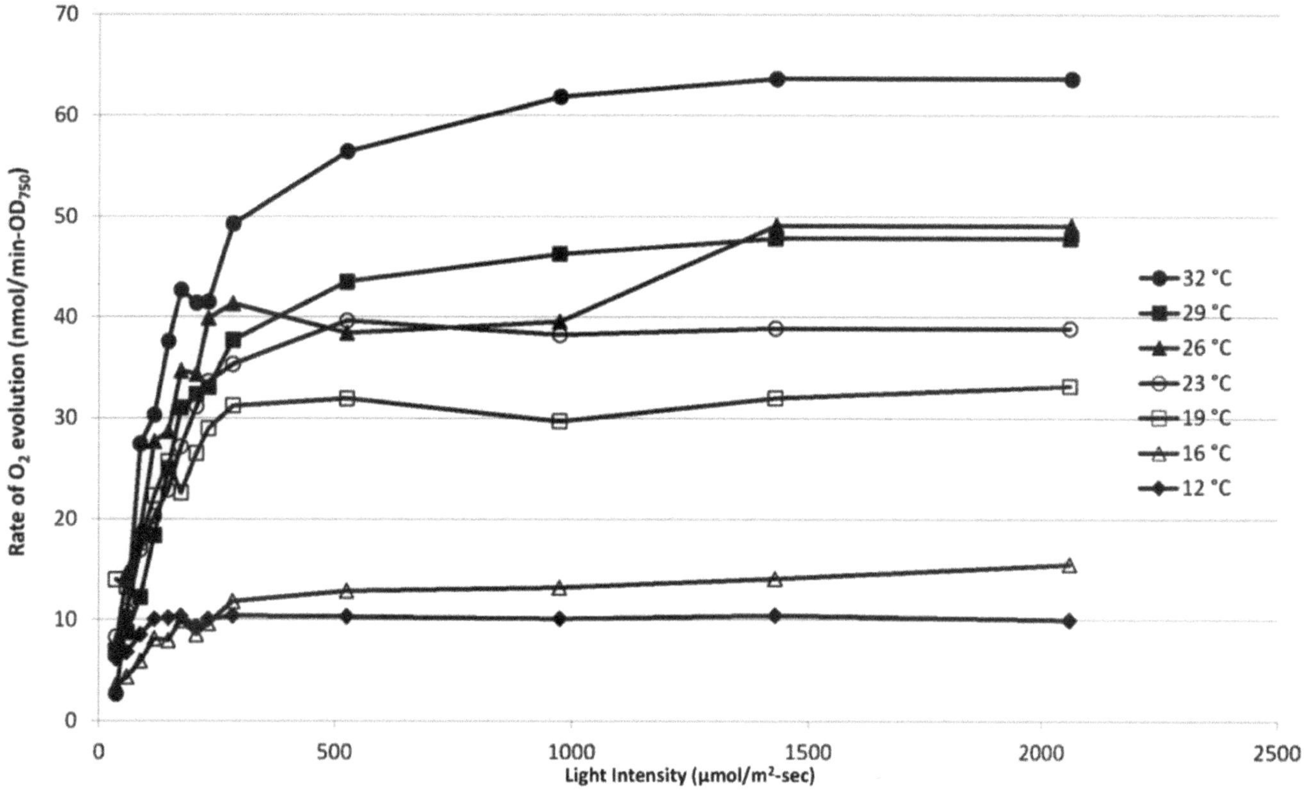

Figure 4. Rate of photosynthetic oxygen evolution of Nannochloropsis salina as a function of light intensity at seven temperatures (12, 16, 19, 23, 26, 29, and 32 °C).

When running the model for N. salina assuming that biomass losses due to dark respiration occur not only overnight but also in the aphotic zone of the pond during the daytime, as was done with success for C. sorokiniana (see above), the predicted biomass concentrations and productivity during the linear growth phase were significantly lower than measured in the outdoor ponds (data not shown). However, when assuming that biomass losses in the dark occur only overnight and not during the daytime in the aphotic zone, the overall shape of the predicted and measured biomass concentration versus time curves look remarkably similar (Fig. 5A). While the model closely predicted the measured biomass concentrations until DOY 45, it significantly underpredicted the biomass productivity (i.e., biomass concentration versus time slope) between DOY 45 and 60. After DOY 60, the model-predicted and measured concentration versus time slopes were very similar, indicating that the model accurately predicts the biomass productivity during the linear growth phase.

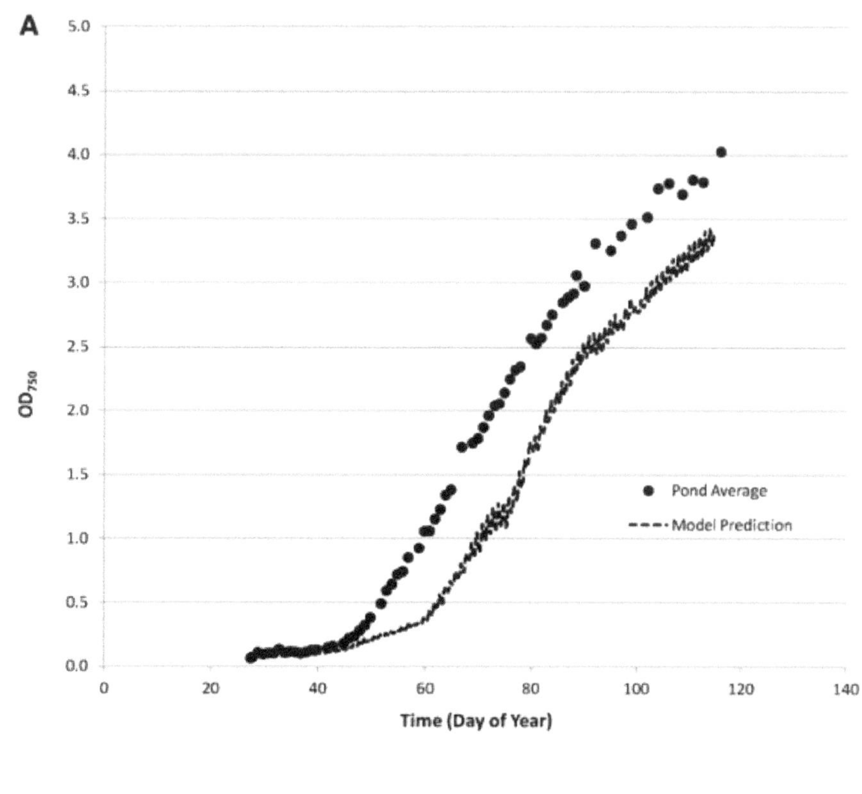

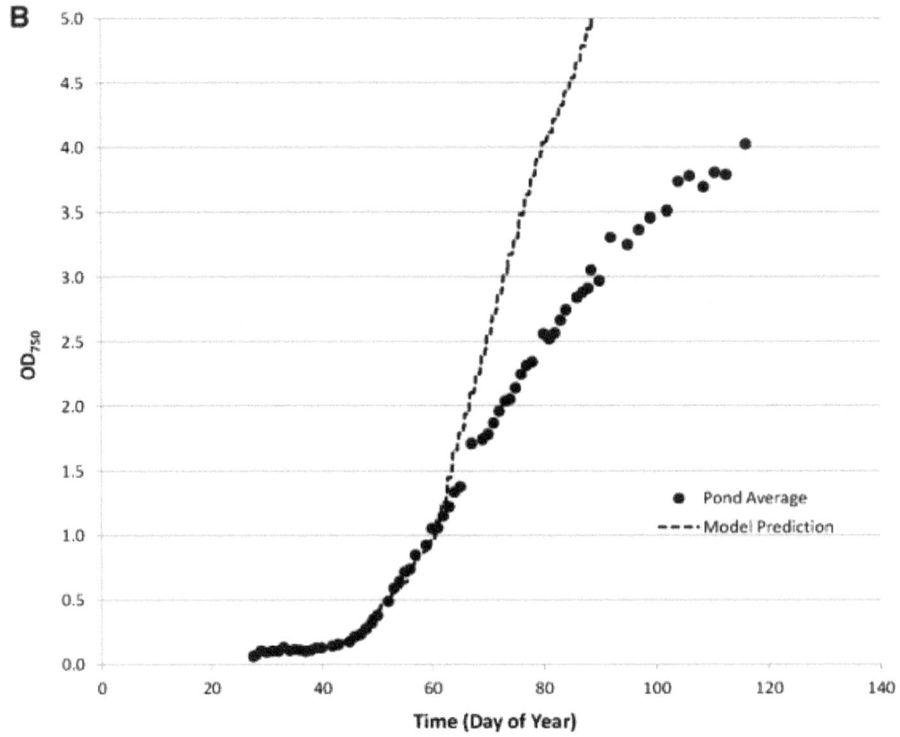

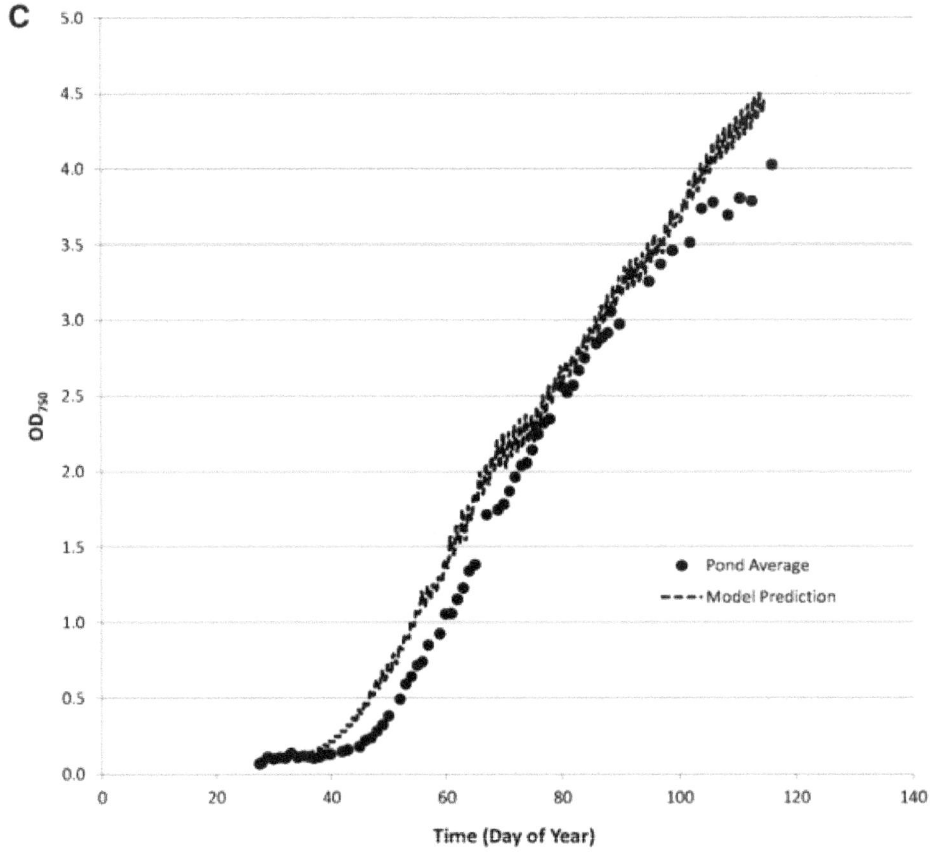

Figure 5. Predicted and measured average biomass concentrations (OD_{750}) of Nannochloropsis salina as a function of time in three outdoor Arizona ponds operated in batch culture mode. (A) Model run using original maximum specific growth rate data (Fig. 1) measured in the presence of continuous light and assuming overnight biomass losses due to dark respiration; (B) Same as (A) but assuming no overnight biomass losses due to dark respiration; (C) Same as (A) but using maximum specific growth rate data at 5 and 10 °C that were measured in the presence of diurnal 14 h:10 h light:dark cycles. The measured AFDW-based areal biomass productivity during the linear growth phase was ca. 3.1 $g/m^2 \cdot day^{-1}$.

There are several potential hypotheses as to why there was a time lag between model predicted and measured biomass concentrations. One possibility is that the freezing of the outdoor pond culture in early February (DOY = 34) imparted unknown stresses on the cells, possibly resulting in smaller biomass loss rates during the recovery period than were used as model inputs using laboratory measurements (Table S1) conducted under controlled and non-freezing conditions. Running the model with larger biomass loss rates measured in the laboratory rather than potentially smaller biomass loss rates of cells recovering from freezing would explain the observed model under-prediction until DOY 60. Another possibility is that because the outdoor ponds were tightly covered overnight until about DOY 69, there was insufficient oxygen transfer from the air into the culture during the night to enable biomass losses due to dark respiration [21]. Indeed, when the model was run assuming no biomass loss due to dark respiration overnight, the predicted and measured biomass concentrations were almost exactly the same until about DOY 60 (Fig. 5B), providing theoretical support for this hypothesis. Unfortunately, the dissolved oxygen concentrations were not recorded during the outdoor pond culture experiment. Therefore, it is impossible to confirm lack of dissolved oxygen as a cause for the potential absence of biomass loss due to dark respiration overnight. Finally, a third possibility is that the model input parameters for the specific growth rate values at low temperatures were unrealistically small because they were measured in the laboratory using continuous light at above saturating light intensity. As was shown in Fig. 2 for C. sorokiniana (DOE 1412), cells fail to grow at low temperature when exposed to continuous light at high intensity as a result of photo-inhibition. When the maximum specific growth rates of N. salina were measured in continuous light at above saturating light intensity (> 300 μmol/m²·s⁻¹), growth was still observed at 11 °C (Fig. 1) but not at 5 °C (data not shown).

In order to determine whether the absence of growth at 5 °C was possibly due to light stress causing photo-inhibition, the specific growth rate measurements were repeated using cultures subjected to alternating 14 h:10 h light/dark cycles. Under these conditions, significant growth was observed at 5 °C (0.111 ± 0.074 day^{-1}, n = 3) and the maximum specific growth rate at 10 °C was much higher (0.588 ± 0.056 day^{-1}, n = 3) than measured previously in continuous light at 11 °C (0.185 ± 0.023 day^{-1}, n = 2). Apparently, as has been observed earlier [52] and [53], when cells are given a chance to recover from light stress during extended dark periods, they exhibit significantly higher growth rates (and photosynthetic oxygen evolution rates) than in continuous light.

When re-running the model using specific growth rate values at 5 and 10 °C measured under more realistic conditions of alternating day–night light cycles, the previous model-predicted lag period disappeared completely (Fig. 5C). Although the model now predicts a slightly earlier initiation of growth than was observed in the outdoor ponds, perhaps because the specific growth rates measured in the laboratory were maximal (and larger than in the actual pond) due to the long adaptation period to a constant low temperature, the overall shape and slope (i.e., volumetric and areal productivity) of the model predicted biomass concentration versus time curve is surprisingly close to the measured data. Therefore, it can be concluded that the main reason for the model-predicted lag during the cold period (Fig. 5A) was the use of specific growth rate values (Fig. 1) that were too small because they had been measured under conditions causing light stress, i.e., in continuous light at low temperatures.

Picochlorum sp.: laboratory characterization and pond culture modeling

As in the case of N. salina, the P–I curves were measured for Picochlorum sp. at different temperatures (Fig. 6) and the relative shape of the P-I curves was used in conjunction with the μ–T curve measured at saturating light intensity (Fig. 1) to generate the two dimensional μ(T,I) array that is needed as model input (Eq. 4). As before, the μ_{dark} data for Picochlorum sp. (Table S1) and the scatter-corrected light attenuation coefficient (ksca) obtained from measured light attenuation curves of this strain (Figure S6-C) were used as model inputs. The model was then run using light and temperature scripts generated by the Biomass Assessment Tool (BAT) for 20.5 cm deep ponds in Southern Florida for the month of May (Figure S5).

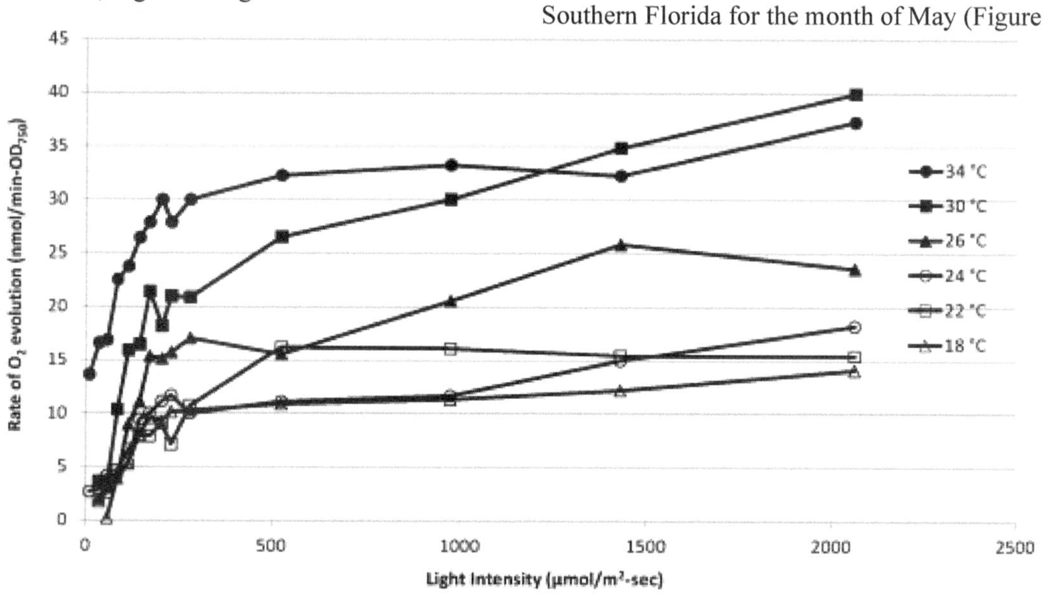

Figure 6. Rate of photosynthetic oxygen evolution of Picochlorum sp. as a function of light intensity at six temperatures (18, 22, 24, 26, 30, and 34 °C).

As was the case for N. salina, the model significantly under-predicted the growth performance of Picochlorum sp. during both the exponential and linear growth phase if biomass losses in the dark were assumed to occur not only overnight but also in the aphotic zone of the pond during the daytime (data not shown). However, when assuming that biomass losses in the dark occur only overnight, the model-predicted rate of increase in biomass concentrations over the four week culturing period was reasonably close to the measured values (Fig. 7).

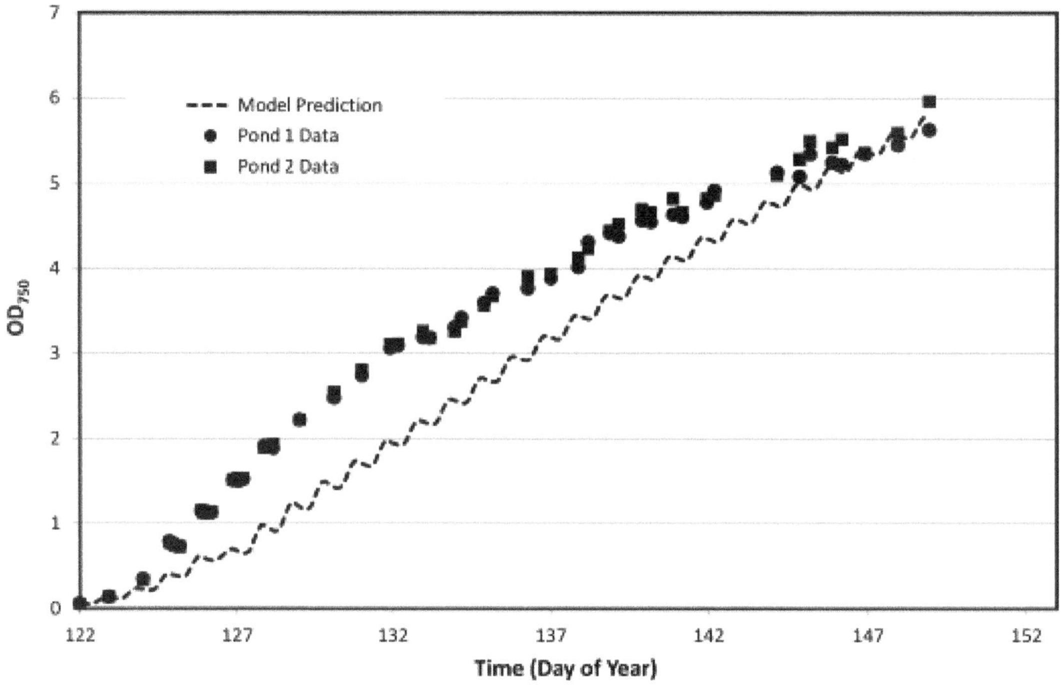

Figure 7. Predicted and measured biomass concentrations (OD_{750}) of Picochlorum sp. as a function of time in two indoor LED-lighted and temperature-controlled ponds operated in batch culture mode using a BAT-generated light and temperature script for 20.5 cm deep ponds in Southern Florida, for the month of May. The measured AFDW-based areal biomass productivity during the first 10 days of linear batch culture growth was ca. 15.3 $g/m^2 \cdot day^{-1}$.

It is not clear why the model did not predict the rapid growth observed in the two ponds during the exponential growth phase. One possibility is that the model did not account of changes in pigmentation due to light adaption and thus a constant ka value of 88.3 (1/m) was used throughout the entire run. However, when performing the pond culture experiments, it was observed that immediately following inoculation, the ka value of cells was only 69 (1/m) but subsequently increased to ca. 88.3 (1/m), possibly due to adaption of the cells to light limited conditions. A smaller k_a value during the exponential growth phase will translate into a greater model-predicted biomass growth rate (see Sensitivity Analysis below).

Sensitivity analysis

Since strain-specific input parameters (i.e., μ(T) P–I curves, μ_{dark}(T, Iavg), ka) are determined experimentally, a certain degree of variability is unavoidable. Therefore, a sensitivity analysis was performed to assess how a 10% increase in the various model input parameter values will change the predicted steady-state biomass concentrations (OD_{750}) for the two semi-continuous C. sorokiniana pond cultures described above (Fig. 3B).

When maximum specific growth rates (μ) were erroneously over-estimated by 10%, the average steady-state biomass concentrations were over-predicted by about 11% (i.e., 13% and 9% for the first and second steady-state, respectively). Conversely, average steady-state biomass concentrations were under-estimated by about 4 and 12% due to a 10% increase, relative to the base case values, in the dark biomass loss rate (μ_{dark}) and the biomass light absorption coefficient (ka), respectively. Errors such as these may combine to produce either larger errors, or they may tend to cancel and reduce the final error, resulting in a predicted steady-state biomass concentration closely approximating the base case scenario with the original input parameters [3].

Errors in final modeling results may also enter through variability in measurements relating to culture depth (d), incident light intensity (Io), and water temperature (T), the latter two being functions of time. Average steady-state biomass concentrations were over-estimated by 2 and 16% due to a 10% positive error in the Io or T values, respectively. Conversely, biomass concentrations are under-estimated by 9% due to a 10% positive error in the d parameter. Errors in the above three physical parameters can be easily controlled, preventing them from contributing significantly to overall model error. By contrast, in the more involved experimental measurements of strain-specific biological parameters, variability cannot be so easily controlled.

Modeling knowledge gaps and research needs

Although we were successful in validating a truly predictive biomass growth model for outdoor ponds subjected to light and temperature fluctuations for three different microalgae strains without "adjusting" parameters or curve-fitting, a number of major uncertainties still remain. In particular, it is not clear how vertical mixing induced light:dark cycles, assumed to have durations in the range of ca. 33 to 100 s in raceway ponds [54], affect two critical processes, i.e., recovery from photo-inhibition and biomass loss in the dark zone of the pond during the daytime.

Regarding photo-inhibition, it is well known that the level of inhibition not only depends on the instantaneous light intensity but also to some degree on the light history. For example, algae cells pre-exposed to high light intensity have lower productivities (or photosynthetic oxygen evolution rates) than cells pre-exposed to low light intensity, a phenomenon called the hysteresis effect [52]. Similarly, Merchuk et al. [53] demonstrated that short dark periods in dense turbulent air-lift photobioreactor microalgae cultures facilitated better recovery from damage by photo-inhibition. Since cells in well-mixed and relatively dense microalgae raceway pond cultures are exposed to random short-term (seconds to minutes) light–dark cycles [54], they also have the opportunity to recover from light stress during short dark periods and thus the effect of photo-inhibition on biomass productivity is likely to be minimal. A possible exception are optically thin pond cultures (i.e., without dark zones) exposed to high sunlight intensity in the morning when water temperatures are still low, as was observed by Vonshak et al. [50].

To incorporate the potential effects of photo-inhibition into the biomass growth model, it will be necessary to better understand how short-term light:dark cycles affect maximum specific growth rates, a critical model input parameter, particularly at low temperatures. Clearly, quantifying maximum specific growth rates at low temperature in continuous light, as was done in this study, is likely to result in unrealistically small values since cells have no opportunity to recover from photo-inhibition, i.e., they are continuously light stressed. Using these unrealistically small maximum specific growth rate values will then result in a significant under-prediction of biomass productivities in outdoor ponds, as was observed for N. salina during cold season cultivation (Fig. 5). Therefore, it is suggested that future maximum specific growth rate measurements should be conducted at a minimum with diurnal light:dark cycles, and perhaps even rapid short-term light–dark cycles to simulate the light history of cells in well-mixed dense pond cultures.

Regarding biomass loss rates in the dark zone of the pond during the daytime, there are, to the best of our knowledge, no conclusive published studies on biomass losses during the short duration (seconds to minutes) dark periods that are encountered by microalgae cells in well mixed dense cultures. Lee and Pirt [55] found that the specific growth rate ofChlorella sp. increased linearly with the fraction of light provided during light–dark cycle experiments and that cells maintained their maximum growth rate, as measured in continuous light, for up to 9.2 s in subsequent dark periods, presumably by metabolizing stored energy compounds accumulated during the light period. Similarly, Merchuk et al.[53] observed a linear increase of the specific growth rate with light fraction for the red microalgae Porphyridium sp., and found that cells could remain for up to 6 s in the dark without reduction in the maximum growth rate observed in continuous light. These findings indicate that short dark periods (a few seconds) have no detrimental effect on the specific growth rate measured in continuous light and that longer dark periods (i.e., smaller light fractions) only decrease the specific growth rate in a linear fashion, suggesting the absence of biomass losses during short dark periods [56], [57], [58],[59] and [60].

By contrast, preliminary results from our laboratory indicate that C. sorokiniana (DOE 1412) cultures exposed to alternating light and dark periods of 1 and 10 min duration exhibit measurable biomass losses during the 10 min dark periods. However, if the dark periods are only 2 min long, no biomass losses could be detected (data not shown). Based on the modeling results presented above, it appears that the presence of biomass losses during short-term duration dark periods is highly strain-specific since the modeling of Chlorella pond cultures (Fig. 3) was only successful when assuming biomass losses in the aphotic zone during the daytime while the modeling of Nannochloropsis andPicochlorum pond cultures (Fig. 5 and Fig. 7) was only successful when assuming the opposite. Clearly, in order to improve the predictive power of any microalgae biomass growth model, more research is needed to quantify not only biomass losses during the long dark periods overnight [31] but also during the much shorter dark periods to which the cells are exposed when circulating in and out of the aphotic zone of the pond during the daytime.

In addition to photo-inhibition and biomass loss in the dark, there are several other knowledge gaps that need to be addressed to improve our ability to model microalgae cultures subjected to fluctuating light and temperature. Although the biomass light absorption coefficient (ka) remained more or less constant (with random variations) throughout the entire duration of the pond culture experiments that were carried out withChlorella, Nannochloropsis, and Picochlorum spp., it is well known that cellular pigment composition and content change in response to variations in light intensity, temperature, nutrient content, dilution rate and the resulting steady-state biomass concentrations [61],[62] and [63]. If there are significant changes in pigmentation during the culturing periods, particularly as a result of nutrient limited conditions, it is no longer correct to assume a constant ka value as model input parameter. Furthermore, a change in pigmentation also affects the correlation between optical density and ash-free dry weight, and thereby increases the uncertainty of biomass productivity predictions. Consequently, neglecting light acclimation may negatively affect the accuracy of biomass productivity predictions for outdoor cultivation, necessitating additional research in this area [31].

Finally, partial inhibition of biomass growth by photosynthetically produced oxygen can significantly affect the accuracy of model predictions [64] and [65]. In our strain characterization experiments, the maximum specific growth rates were measured in flask cultures continuously sparged with CO_2-enriched air, which ensures that photosynthetically oxygen is continually stripped from the medium, thereby avoiding growth inhibition by oxygen. However, outdoor pond cultures can experience large diurnal fluctuations in dissolved oxygen concentrations (from ca. 5 mg/L in the early morning following dark respiration overnight to ca. 40 mg/L in the early afternoon following intense photosynthetic activity during midday, data not shown) in response to changes in incident light intensity. Although there appeared to be no evidence of significant inhibition by oxygen in any of our pond cultures, given the relatively close model predictions of biomass growth for all three strains, inhibition by photosynthetic oxygen could potentially occur in other microalgae strains, thus warranting additional research.

In conclusion, the biomass growth model presented herein can serve as a useful tool for screening promising microalgae strains for their ability to exhibit high biomass productivities in outdoor ponds or photobioreactors operated in batch or semi-continuous culture mode, requiring knowledge only of the strain-specific parameters (μ, μ_{dark}, k_{sca}) and the sunlight and water temperature conditions. Obtaining these parameters for each strain is presently a laborious undertaking, but with greater adoption of this model, it will be possible to miniaturize and automate much of this work, allowing for rapid throughput of promising strains [66]. This will greatly reduce the requirement for outdoor pond tests, which presently are the bottleneck in the development of microalgae biomass and biofuels production. Finally, the biomass growth model, when integrated with PNNL's Biomass Assessment Tool [4], allows for the generation of strain-specific biomass productivity maps for different seasons and outdoor pond operational schemes, a topic of currently ongoing research.

ACKNOWLEDGMENTS

The authors would like to acknowledge funding of this work by the US Department of Energy under contract DE-EE0003046 awarded to the National Alliance for Advanced Biofuels and Bioproducts. Additional support by the Department of Energy Science Undergraduate Laboratory Internship Program was provided to Boris Chubukov and V.J. Tocco, whose early work on method development and validation is greatly appreciated. The authors have declared no conflict of interest.

SUPPLEMENTARY TABLES AND FIGURES

Table S1: Average specific biomass loss rates (μ_{dark}) during a 10 hour dark period calculated based on change in ash-free dry weight (AFDW) over time at four different temperatures and three different growth phases (or equivalent average light intensities, I_{avg}) of 14 hour duration. Values in parentheses are the standard error of the mean. Data are from Edmundson and Huesemann (2015)

μ_{dark} (day^{-1})						
			Temperature (°C)			
	Growth Phase	I_{avg} (μmoles/m^2-sec)	**15.1**	**19.3**	**24.7**	**28.7**
Pico-chlorum sp. (WT-LANL)	Late-Ex-ponential	145.2	-0.181 (0.047)	-0.204 (0.018)	-0.232 (0.035)	-0.204 (0.024)
	Linear	72.7	-0.045 (0.023)	-0.099 (0.011)	-0.087 (0.022)	-0.149 (0.018)
	Late-Linear	34.9	-0.135 (0.021)	-0.055 (0.014)	-0.043 (0.014)	-0.120 (0.016)
		I_{avg} (μmoles/m^2-sec)	**11.0**	**17.3**	**24.4**	**30.0**
Chlorella soroki-niana (DOE 1412)	Late-Ex-ponential	130.8	-0.093 (0.024)	-0.311 (0.037)	-0.409 (0.029)	-0.367 (0.015)
	Linear	71.6	-0.111 (0.023)	-0.176 (0.015)	-0.163 (0.029)	-0.167 (0.039)
	Late-Linear	40.3	-0.083 (0.023)	-0.043 (0.025)	-0.053 (0.006)	-0.123 (0.025)
		I_{avg} (μmoles/m^2-sec)	**9.8**	**15.3**	**20.0**	**23.7**
Nanno-chlorop-sis salina (CCMP 1776)	Late-Ex-ponential	143.3	-0.040 (0.020)	-0.212 (0.018)	-0.447 (0.010)	-0.524 (0.019)
	Linear	91.5	-0.048 (0.005)	-0.127 (0.016)	-0.272 (0.018)	-0.337 (0.004)
	Late-Linear	49.3	-0.049 (0.012)	-0.102 (0.009)	-0.140 (0.007)	-0.231 (0.007)

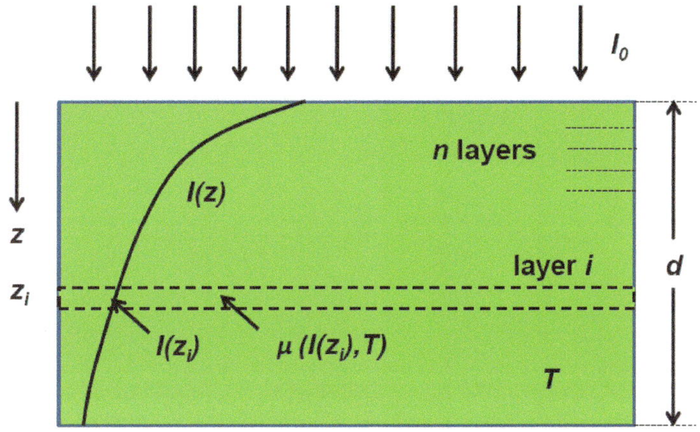

Figure S1: Conceptual schematic of the biomass growth model. All symbols are defined in the text.

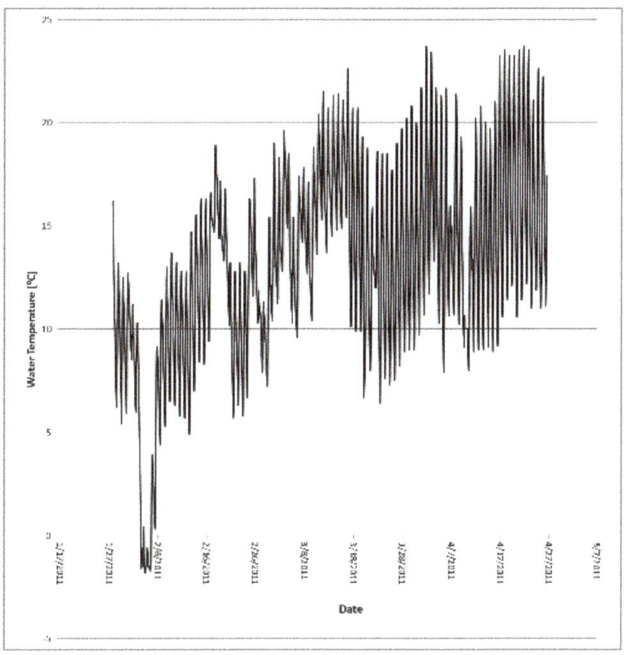

Figure S2: Average water temperature in triplicate Nannochloropsis salina raceway pond cultures during the winter and spring of 2011 at The University of Arizona in Tucson, AZ.

A

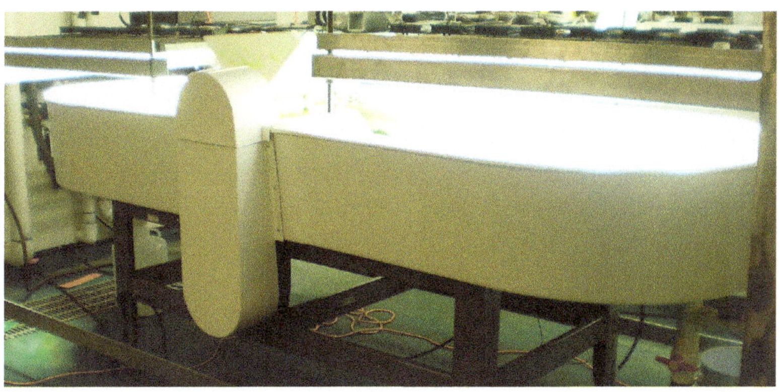

B

Figure S3: PNNL indoor LED-lighted and temperature controlled raceway pond. **A.** Side view. **B**. View of the LED panels and actively growing a Chlorella sorokiniana (DOE1412) culture.

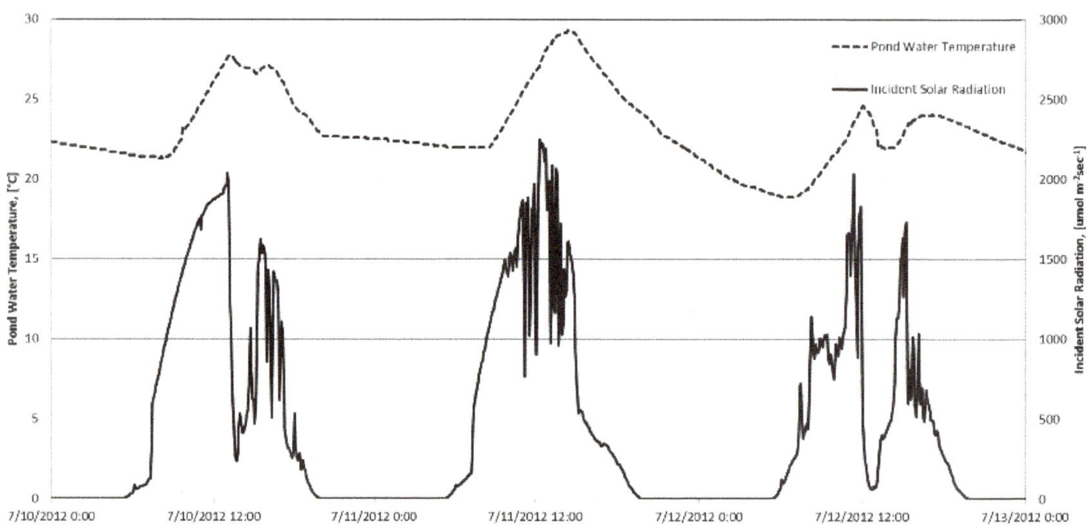

Figure S4: Pond water temperature (·····) and incident solar radiation (—) during three consecutive days in July 2012 at The University of Arizona Research Station in Rimrock, AZ. Note that temporary declines in solar radiation during midday due to extensive cloud cover resulted in concomitant temporary declines in pond water temperature.

A

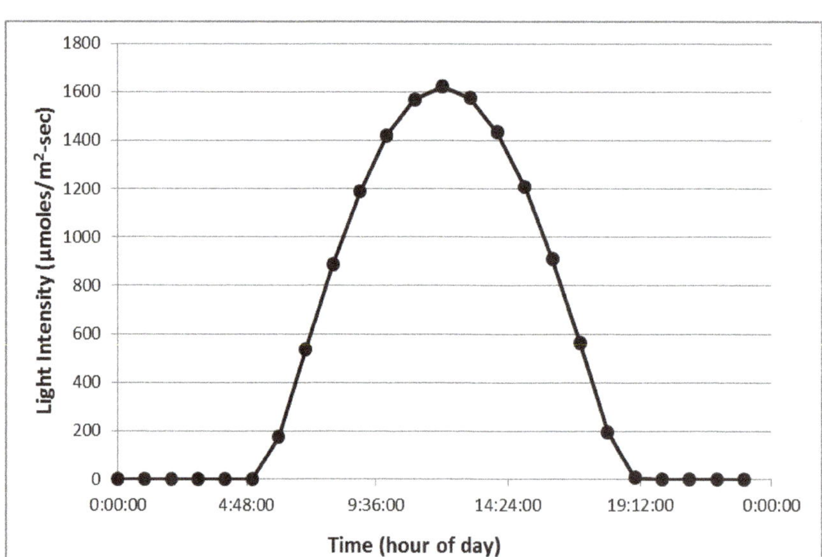

B

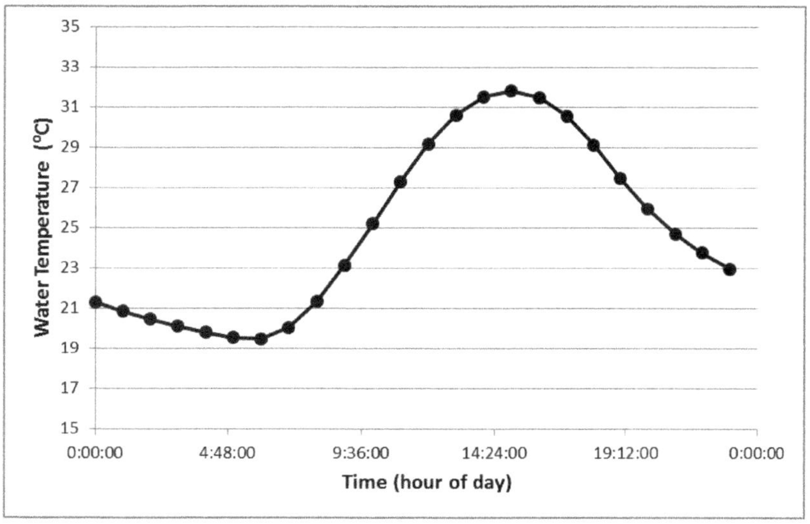

Figure S5: Incident solar radiation (A) and pond water temperature (B) predicted by the Biomass Assessment Tool for 20.5 cm deep ponds in Southern Florida during the first day of May. Light and temperature scripts for the entire month of May were used as input to the biomass growth model to predict the pond culture performance of Picochlorum sp.

A

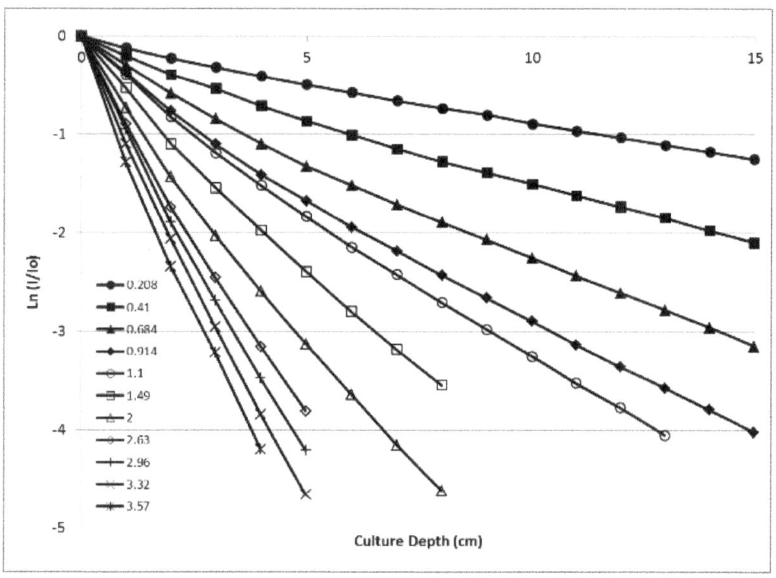

B

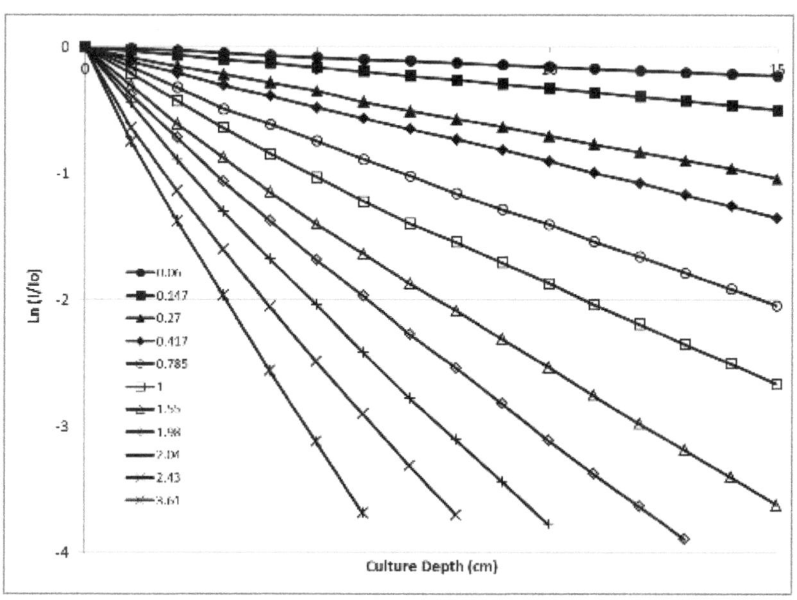

C

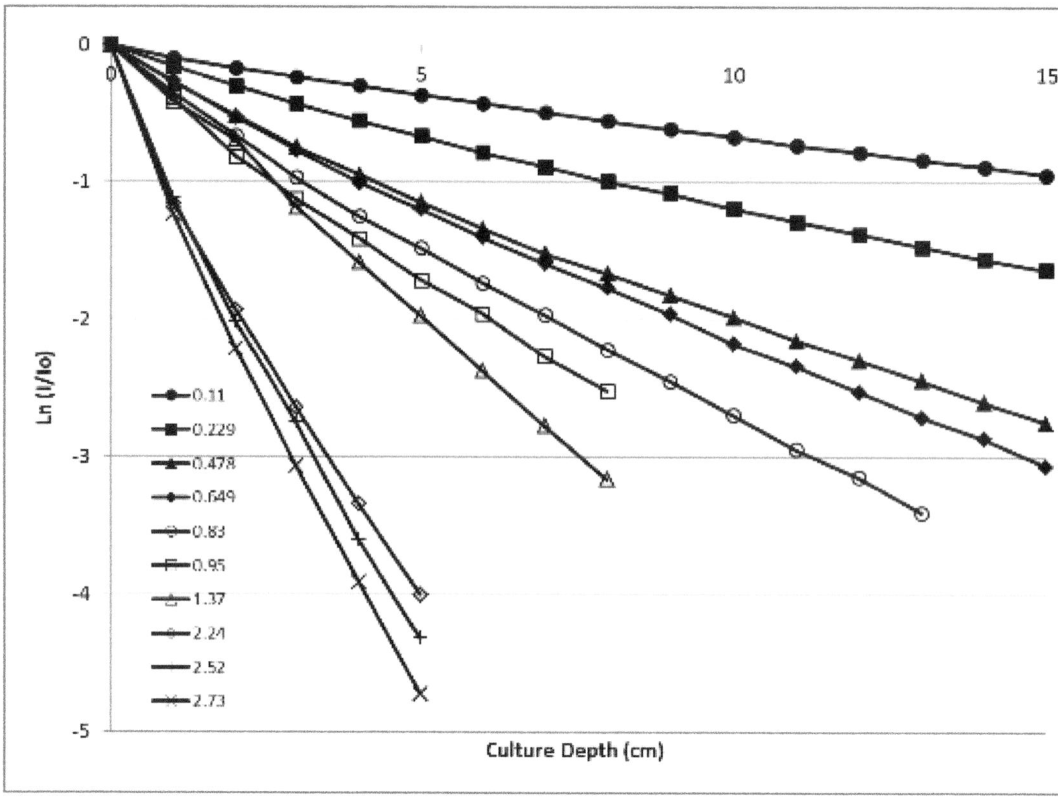

Figure S6: Natural logarithm of the ratio of light intensity (I) over incident light intensity (I_o) as a function of culture depth (z) for different biomass concentrations (OD_{750}) of (a) Chlorella sorokiniana (DOE 1412), (b) Nannochloropsis salina, and (c) Picochlorum sp. The biomass light absorption coefficients (k_a) for Chlorella, Nannochloropsis, and Picochlorum were 49.5, 40.4, and 88.3 (1/m), respectively.

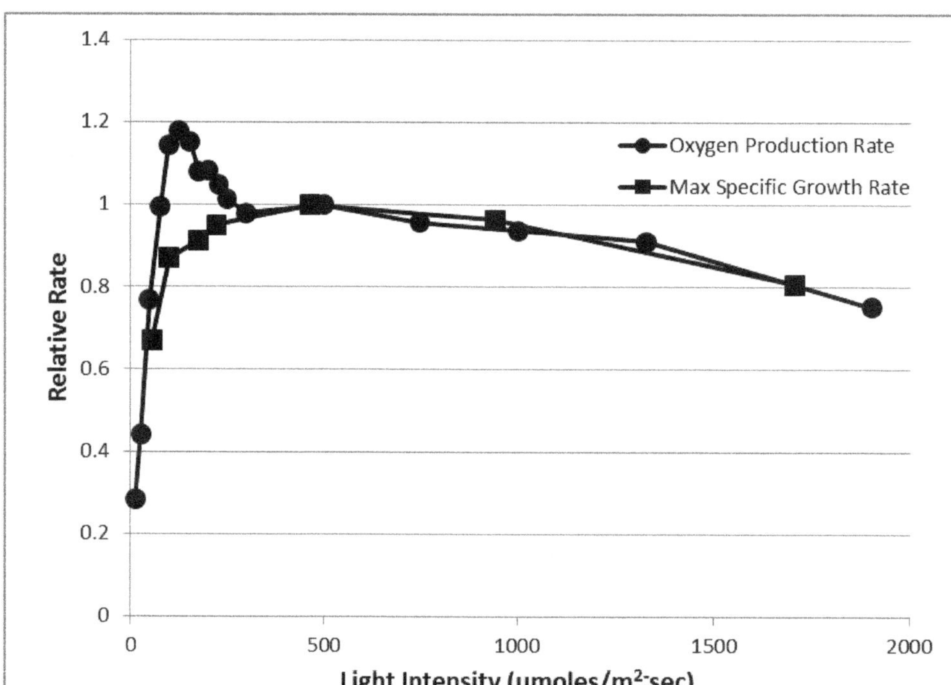

Figure S7: Normalized* photosynthetic oxygen evolution rate and maximum specific growth rate of Chlorella sorokiniana (DOE 1412) as a function of light intensity at 27 °C. It is not clear why oxygen evolution rates exhibited a peak around 100 µmol/m^2·sec. (*Normalized to values at 464 µmoles/m^2-sec).

REFERENCES

1. U.S. Department of Energy, National Algal Biofuels Technology Roadmap, College Park, MD, 2010, http://www.energy.gov/sites/prod/files/2014/03/f14/algal_biofuels_roadmap.pdf.
2. U.S. Department of Energy, Multi-year Program Plan, 2014, http://www.energy.gov/ sites/prod/files/2014/07/f17/mypp_july_2014.pdf.
3. M.H. Huesemann, J. Van Wagenen, T. Miller, A. Chavis, S. Hobbs, B. Crowe, A screening model to predict microalgae biomass growth in photobioreactors and raceway ponds, Biotechnol. Bioeng. 110 (2013) 1583–1594, http://dx.doi.org/10.1002/bit. 24814.
4. M.S. Wigmosta, A.M. Coleman, R.J. Skaggs, M.H. Huesemann, L.J. Lane, National microalgae biofuel production potential and resource demand, Water Resour. Res. 47 (2011), http://dx.doi.org/10.1029/2010WR009966 n/a–n/a.
5. E.R. Venteris, R.L. Skaggs, A.M. Coleman, M.S. Wigmosta, A GIS cost model to assess the availability of freshwater, seawater, and saline groundwater for algal biofuel production in the United States, Environ. Sci. Technol. 47 (2013) 4840–4849, http://dx.doi.org/10.1021/es304135b.
6. E.R. Venteris, R.L. Skaggs, M.S. Wigmosta, A.M. Coleman, A national-scale comparison of resource and nutrient demands for algae-based biofuel production by lipid extraction and hydrothermal liquefaction, Biomass Bioenergy 64 (2014) 276–290, http://dx.doi.org/10.1016/j.biombioe.2014.02.001.
7. A.M. Coleman, J.M. Abodeely, R.L. Skaggs, W.A. Moeglein, D.T. Newby, E.R. Venteris, et al., An integrated assessment of location-dependent scaling for microalgae biofuel production facilities, Algal Res. 5 (2014) 79–94, http://dx.doi.org/10.1016/j.algal. 2014.05.008.
8. P.G. Falkowski, Z. Dubinsky, K. Wyman, Growth–irradiance relationships in phytoplankton, Limnol. Oceanogr. 30 (1985) 311–321, http://dx.doi.org/10.4319/lo. 1985.30.2.0311.
9. R.J. Geider, B.A. Osborne, J.A. Raven, Light dependence of growth and photosynthesis in Phaeodactyium tricornutum (Bacillariophyceae)1, J. Phycol. 21 (2004) 609–619, http://dx.doi.org/10.1111/j.0022-3646.1985.00609.x.
10. T. Platt, A.D. Jassby, The relationship between photosynthesis and light for natural assemblages of coastal marine phytoplankton, J. Phycol. 12 (1976) 421–430, http://dx.doi.org/10.1111/j.1529-8817.1976.tb02866.x.
11. R.W. Eppley, Temperature and phytoplankton growth in the sea, Fish. Bull. 70 (1972) 1063–1085http://fishbull.noaa.gov/70-4/eppley.pdf.
12. J.C. Goldman, E.J. Carpenter, A kinetic approach to the effect of temperature on algal growth, Limnol. Oceanogr. 19 (1974) 756–766, http://dx.doi.org/10.4319/lo.1974.19.5.0756.
13. J.A. Yoder, Effect of temperature on light-limited growth and chemical composition of Skeletonema costatum (Bacillariophyceae), J. Phycol. 15 (1979) 362–370, http:// dx.doi.org/10.1111/j.1529-8817.1979.tb00706.x.
14. J.F. Coles, R.C. Jones, Effect of temperature on photosynthesis-light response and growth of four phytoplankton species isolated from a tidal freshwater river, J. Phycol. 36 (2000) 7–16, http://dx.doi.org/10.1046/j.1529-8817.2000. 98219.x.
15. D.J.S. Montagnes, M. Franklin, Effect of temperature on diatom volume, growth rate, and carbon and nitrogen content: reconsidering some paradigms, Limnol. Oceanogr. 46 (2001) 2008–2018, http://dx.doi.org/10.4319/lo.2001.46.8.2008.
16. P.W. Boyd, T.A. Rynearson, E.A. Armstrong, F. Fu, K. Hayashi, Z. Hu, et al., Marine phytoplankton temperature versus growth responses from polar to tropical waters — outcome of a scientific community-wide study, PLoS ONE 8 (2013), e63091, http:// dx.doi.org/10.1371/journal.pone.0063091.
17. R.J. Geider, B.A. Osborne, Respiration and microalgal growth: a review of the quantitative relationship between dark respiration and growth, New Phytol. 112 (1989) 327–341, http://dx.doi.org/10.1111/j.1469-8137.1989.tb00321.x.
18. J.H. Ryther, R.R.L. Guillard, Studies of marine planktonic diatoms: III. Some effects of temperature on respiration of five species, Can. J. Microbiol. 8 (1962) 447–453, http://dx.doi.org/10.1139/m62-058.
19. J.U. Grobbelaar, C.J. Soeder, Respiration losses in planktonic green algae cultivated in raceway ponds, J. Plankton Res. 7 (1985) 497–506, http://dx.doi.org/10.1093/ plankt/7.4.497.
20. G. Torzillo, A. Sacchi, R. Materassi, A. Richmond, Effect of temperature on yield and night biomass loss in Spirulina platensis grown outdoors in tubular photobioreactors, J. Appl. Phycol. 3 (1991) 103–109, http://dx.doi.org/10.1007/BF00003691.
21. J.C. Ogbonna, H. Tanaka, Night biomass loss and changes in biochemical composition of cells during light/dark cyclic culture of Chlorella pyrenoidosa, J. Ferment. Bioeng. 82 (1996) 558–564, http://dx.doi.org/10.1016/S0922-338X(97)81252-4.
22. M. Behrenfeld, P.G. Falkowski, A consumer's guide to phytoplankton primary productivity models, Limnol. Oceanogr. 42 (1997) 1479–1491, http://dx.doi.org/10.4319/lo.1997.42.7.1479.
23. J.C. Goldman, Outdoor algal mass cultures—II. Photosynthetic yield limitations, Water Res. 13 (1979) 119–136, http://dx.doi.org/10.1016/0043-1354(79)90083-6.
24. B.M.A. Kroon, H.A.M. Ketelaars, H.J. Fallowfield, L.R. Mur, Modelling microalgal productivity in a high rate algal pond based on wavelength dependent optical properties, J. Appl. Phycol. 1 (1989) 247–256, http://dx.doi.org/10.1007/ BF00003650.
25. R.S. Sukenik, Y. Levy, P.G. Levy, Z. Falkowski, Dubinsky, optimizing algal biomass production in an outdoor pond: a simulation model, J. Appl. Phycol. 3 (1991) 191–201, http://dx.doi.org/10.1007/BF00003577.

26. O. Bernard, Hurdles and challenges for modelling and control of microalgae for CO2 mitigation and biofuel production, J. Process Control 21 (2011) 1378–1389, http:// dx.doi.org/10.1016/j.jprocont.2011.07.012.
27. Y. Packer, T. Li, Q. Andersen, Y. Hu, M. Kuang, Sommerfeld, growth and neutral lipid synthesis in green microalgae: a mathematical model, Bioresour. Technol. 102 (2011) 111–117, http://dx.doi.org/10.1016/j.biortech.2010.06.029.
28. J. Quinn, L. de Winter, T. Bradley, Microalgae bulk growth model with application to industrial scale systems, Bioresour. Technol. 102 (2011) 5083–5092, http://dx.doi. org/10.1016/j.biortech.2011.01.019.
29. O. Bernard, B. Rémond, Validation of a simple model accounting for light and temperature effect on microalgal growth, Bioresour. Technol. 123 (2012) 520–527, http:// dx.doi.org/10.1016/j.biortech.2012.07.022.
30. S.C. James, V. Janardhanam, D.T. Hanson, Simulating pH effects in an algal-growth hydrodynamics model 1, J. Phycol. 49 (2013) 608–615, http://dx.doi.org/10.1111/jpy.12071.
31. Q. Béchet, A. Shilton, B. Guieysse, Modeling the effects of light and temperature on algae growth: state of the art and critical assessment for productivity prediction during outdoor cultivation, Biotechnol. Adv. 31 (2013) 1648–1663, http://dx.doi. org/10.1016/j.biotechadv.2013.08.014.
32. J.F. Cornet, C.G. Dussap, G. Dubertret, A structured model for simulation of cultures of the cyanobacterium Spirulina platensis in photobioreactors: I. Coupling between light transfer and growth kinetics, Biotechnol. Bioeng. 40 (1992) 817–825, http:// dx.doi.org/10.1002/bit.260400709.
33. J.F. Cornet, C.G. Dussap, A simple and reliable formula for assessment of maximum volumetric productivities in photobioreactors, Biotechnol. Prog. 25 (2009) 424–435, http://dx.doi.org/10.1002/btpr.138.
34. F.G.A. Fernández, F.G. Camacho, J.A.S. Pérez, J.M.F. Sevilla, E.M. Grima, A model for light distribution and average solar irradiance inside outdoor tubular photobioreactors for the microalgal mass culture, Biotechnol. Bioeng. 55 (1997) 701–714, http://dx.doi. org/10.1002/(SICI)1097-0290(19970905)55:5b701::AIDBIT1N3.0.CO;2-F.
35. L. Pottier, J. Pruvost, J. Deremetz, J.-F. Cornet, J. Legrand, C.G. Dussap, A fully predictive model for one-dimensional light attenuation by Chlamydomonas reinhardtii in a torus photobioreactor, Biotechnol. Bioeng. 91 (2005) 569–582, http://dx.doi.org/10. 1002/bit.20475.
36. J. Pruvost, G. Van Vooren, B. Le Gouic, A. Couzinet-Mossion, J. Legrand, Systematic investigation of biomass and lipid productivity by microalgae in photobioreactors for biodiesel application, Bioresour. Technol. 102 (2011) 150–158, http://dx.doi.org/10. 1016/j.biortech.2010.06.153.
37. M.S. Chalup, E.A. Laws, A test of the assumptions and predictions of recent microalgal growth models with the marine phytoplankter Pavlova lutheri, Limnol. Oceanogr. 35 (1990) 583–596, http://dx.doi.org/10.4319/lo.1990.35.3.0583.
38. J.D. Ingle, S.R. Crouch, Spectrochemical Analysis, First, Prentice Hall, New Jersey, 1988.
39. I.S. Suh, S.B. Lee, A light distribution model for an internally radiating photobioreactor, Biotechnol. Bioeng. 82 (2003) 180–189, http://dx.doi.org/10. 1002/bit.10558.
40. R. Bosma, E. van Zessen, J.H. Reith, J. Tramper, R.H. Wijffels, Prediction of volumetric productivity of an outdoor photobioreactor, Biotechnol. Bioeng. 97 (2007) 1108–1120, http://dx.doi.org/10.1002/bit.21319.
41. E.M. Grima, J.A.S. Pérez, F.G. Camacho, J.M.F. Sevilla, F.G.A. Fernández, Productivity analysis of outdoor chemostat culture in tubular air-lift photobioreactors, J. Appl. Phycol. 8 (1996) 369–380, http://dx.doi. org/10.1007/BF02178580.
42. Y.S. Yun, J.M. Park, Kinetic modeling of the light-dependent photosynthetic activity of the green microalga Chlorella vulgaris, Biotechnol. Bioeng. 83 (2003) 303–311, http://dx.doi.org/10.1002/bit.10669.
43. M.H. Huesemann, T.S. Hausmann, R. Bartha, M. Aksoy, J.C. Weissman, J.R. Benemann, Biomass productivities in wild type and pigment mutant of Cyclotella sp. (Diatom), Appl. Biochem. Biotechnol. 157 (2009) 507–526, http://dx.doi.org/10.1007/ s12010-008-8298-9.
44. B. Crowe, S. Attalah, S. Agrawal, P. Waller, R. Ryan, J. Van Wagenen, et al., A comparison of Nannochloropsis salina growth performance in two outdoor pond designs: conventional raceways versus the ARID pond with superior temperature management, Int. J. Chem. Eng. 2012 (2012) 1–9, http://dx.doi.org/10.1155/2012/920608.
45. J. Van Wagenen, T.W. Miller, S. Hobbs, P. Hook, B. Crowe, M. Huesemann, Effects of light and temperature on fatty acid production in Nannochloropsis salina, Energies 5 (2012) 731–740, http://dx.doi.org/10.3390/en5030731.
46. S.E. Edmundson, M.H. Huesemann, The dark side of algae cultivation: characterizing night biomass loss in three photosynthetic algae, Chlorella sorokiniana, Nannochloropsis salina, and Picochlorum sp, Algal Res. (2015) 470–476, http://dx. doi.org/10.1016/j.algal.2015.10.012.
47. B.C. Benson, K.A. Rusch, Investigation of the light dynamics and their impact on algal growth rate in a hydraulically integrated serial turbidostat algal reactor (HISTAR), Aquac. Eng. 35 (2006) 122–134, http:// dx.doi.org/10.1016/j.aquaeng.2005.09.005.
48. M.H. Huesemann, T. Dale, A. Chavis, B. Crowe, S. Twary, A. Barry, et al., Simulation of outdoor pond cultures using indoor LED-lighted and temperature-controlled raceway ponds and Phenometrics™ photobioreactors, Algal Res. (2015) (To be submitted).
49. S. Jensen, G. Knutsen, Influence of light and temperature on photoinhibition of photosynthesis in spirulina platensis, J. Appl. Phycol. 5 (1993) 495–504, http:// dx.doi. org/10.1007/BF02182508.
50. G. Vonshak, J. Torzillo, S. Masojidek, Boussiba, sub-optimal morning temperature induces photoinhibition

in dense outdoor cultures of the alga Monodus subterraneus (Eustigmatophyta), Plant Cell Environ. 24 (2001) 1113–1118, http://dx.doi.org/10. 1046/j.0016-8025.2001.00759.x.

51. R.E. Davis, D.B. Fishman, E.D. Frank, M.C. Johnson, S.B. Jones, C.M. Kinchin, et al., Integrated evaluation of cost, emissions, and resource potential for algal biofuels at the national scale, Environ. Sci. Technol. 48 (2014) 6035–6042, http://dx.doi. org/10.1021/es4055719.

52. P.G. Falkowski, T.G. Owens, Effects of light intensity on photosynthesis and dark respiration in six species of marine phytoplankton, Mar. Biol. 45 (1978) 289–295, http://dx.doi.org/10.1007/BF00391815.

53. J.C. Merchuk, M. Ronen, S. Giris, S. (Malis) Arad, Light/dark cycles in the growth of the red microalga Porphyridium sp. Biotechnol. Bioeng. 59 (1998) 705–713, http:// dx.doi.org/10.1002/(SICI)1097-0290(19980920)59:6b705::AID-BIT7N3.0.CO;2-J.

54. J.L. Mendoza, M.R. Granados, I. de Godos, F.G. Acién, E. Molina, C. Banks, et al., Fluiddynamic characterization of real-scale raceway reactors for microalgae production, Biomass Bioenergy 54 (2013) 267–275, http://dx.doi. org/10.1016/j.biombioe.2013. 03.017.

55. Y.-K. Lee, S.J. Pirt, Energetics of photosynthetic algal growth: influence of intermittent illumination in short (40 s) cycles, Microbiology 124 (1981) 43–52, http://dx. doi.org/10.1099/00221287-124-1-43.

56. J.U. Grobbelaar, Do light/dark cycles of medium frequency enhance phytoplankton productivity? J. Appl. Phycol. 1 (1989) 333–340, http://dx.doi.org/10.1007/ BF00003470.

57. J.U. Grobbelaar, Turbulence in mass algal cultures and the role of light/dark fluctuations, J. Appl. Phycol. 6 (1994) 331–335, http://dx.doi.org/10.1007/BF02181947.

58. J.U. Grobbelaar, B.M.A. Kroon, T. Burger-Wiersma, L.R. Mur, Influence of medium frequency light/dark cycles of equal duration on the photosynthesis and respiration of Chlorella pyrenoidosa, Hydrobiologia 238 (1992) 53–62, http://dx.doi.org/10.1007/ BF00048773.

59. M. Janssen, T.C. Kuijpers, B. Veldhoen, M.B. Ternbach, J. Tramper, L.R. Mur, et al., Specific growth rate of Chlamydomonas reinhardtii and Chlorella sorokiniana under medium duration light/dark cycles: 13–87 s, J. Biotechnol. 70 (1999) 323–333, http://dx.doi. org/10.1016/S0168-1656(99)00084-X.

60. M. Janssen, M. Janssen, M. De Winter, J. Tramper, L.R. Mur, J. Snel, et al., Efficiency of light utilization of Chlamydomonas reinhardtii under medium-duration light/ dark cycles, J. Biotechnol. 78 (2000) 123–137, http://dx.doi.org/10.1016/S0168- 1656(99)00233-3.

61. R.J. Geider, H.L. MacIntyre, T.M. Kana, A dynamic model of photoadaptation in phytoplankton, Limnol. Oceanogr. 41 (1996) 1–15, http://dx.doi.org/10.4319/ lo.1996. 41.1.0001.

62. R.J. Geider, H.L. MacIntyre, T.M. Kana, A dynamic regulatory model of phytoplanktonic acclimation to light, nutrients, and temperature, Limnol. Oceanogr. 43 (1998) 679–694, http://dx.doi.org/10.4319/lo.1998.43.4.0679.

63. J. Van Wagenen, M.L. Pape, I. Angelidaki, Characterization of nutrient removal and microalgal biomass production on an industsrial waste-stream by application of the decelerationstat technique, Water Res. 75 (2015) 301–311, http://dx.doi.org/ 10.1016/j. watres.2015.02.022.

64. P.C. Keymer, S. Pratt, P.A. Lant, Development of a novel electrochemical system for oxygen control (ESOC) to examine dissolved oxygen inhibition on algal activity, Biotechnol. Bioeng. 110 (2013) 2405–2411, http:// dx.doi.org/10.1002/bit.24905.

65. C. Sousa, A. Compadre, M.H. Vermuë, R.H. Wijffels, Effect of oxygen at low and high light intensities on the growth of Neochloris oleoabundans, Algal Res. 2 (2013) 122–126, http://dx.doi.org/10.1016/j.algal.2013.01.007.

66. J. Van Wagenen, S.L. Holdt, D. De Francisci, B. Valverde-Pérez, B.G. Plósz, I. Angelidaki, Microplate-based method for high-throughput screening of microalgae growth potential, Bioresour. Technol. 169 (2014) 566–572, http://dx.doi.org/10. 1016/j.biortech.2014.06.096.

CHAPTER 4

INFLUENCE OF SILICATE ON ENRICHMENT OF HIGHLY PRODUCTIVE MICROALGAE FROM A MIXED CULTURE

Peter R. Mooij, Lisanne D. de Jongh, Mark C. M. van Loosdrecht, Robbert Kleerebezem

Keywords: Mixed cultures, Selective environment, Green algae, Diatoms, Storage compounds

INTRODUCTION

Microalgae are efficient producers of triacylglycerides (Hu et al. 2008) and starch (Markou et al.2012) and have been proposed to supply these as feedstock for food, fuels, and chemicals to a biobased economy (Wijffels and Barbosa 2010). Nowadays, microalgal research mainly focuses on pure culture applications. Production of bulk compounds, such as triacylglycerides, at an industrial scale is however troublesome in an axenic microalgal culture (Kazamia et al. 2012; Shurin et al. 2013; McBride et al. 2014). Previously, we described an ecology-based enrichment and cultivation method which allows for stable storage compound production under nonaxenic conditions (Mooij et al. 2013, 2015). The basis of this approach is uncoupling of carbon fixation in the light and nitrogen uptake in the dark by limiting the presence of an essential growth nutrient (such as nitrate) to the dark phase. Limiting the presence of nitrate to the dark period provides a competitive advantage for storage compound-producing microalgae. Microalgae that efficiently convert CO_2 into storage compounds in the light phase can take up the nitrate in the dark phase for biomass production, resulting in an enrichment culture consisting of efficient storage compound-producing microalgae (Mooij et al. 2013).

Despite their overwhelming abundance in nature, diatoms are largely overlooked in microalgal biofuel and biomass production research. Diatoms possess however various characteristics which make them interesting candidates for large-scale storage compound production (Hildebrand et al. 2012). These include the capacity to accumulate large amount of lipids when exposed to silicate limitation, good resistance to predators, and good harvestability. The presence of diatoms in nature is to a large extent regulated by the amount of available silicate (Werner 1977). At nonlimiting silicate levels, diatoms are effective competitors for limiting nutrients and are able to effectively utilize nutrient pulses (Litchman 2007). At external silicate concentrations exceeding 2 mM, diatoms typically represent more than 70 % of the phytoplankton community (Egge and Aksnes 1992). With increasing silicate to phosphorus ratios, diatom abundance increases in competition experiments between diatoms and nonsiliceous algae (Sommer 1985).

The first objective of this research was to investigate if selective enrichment of storage compound-producing diatoms can be established in an open system with a surplus of silicate in the medium. We expect the enriched microalgae community to differ in a system with and without silicate supply. The functionality of storage compound production is nonetheless expected to be present under both conditions, as it is a consequence of the uncoupling of carbon fixation and nitrogen uptake.

MATERIALS AND METHODS

Operating conditions

Two 1.5-L bioreactors (Applikon, The Netherlands) with a diameter of 11 cm and height of 17 cm were operated in a sequenced batch mode with the following operational parameters (Table 1). The system was operated under nonaxenic conditions. The medium composition for both reactors is described in Table 2. Na_2SiO_3 (11.42 mM) was supplied to the silicate-excess reactor and withheld from the silicate-depleted reactor. Nitrate was supplied as the sole nitrogen source at the beginning of the dark period and was designed to be the limiting factor for algal growth. A solid retention time of 41 h and a cycle length of 24 h (Table 1) imply that every cycle 59 % (700 mL) of the culture was harvested and replaced with medium as in Table 2. NO_3-N (14 mg) was therefore dosed to the system at the start of every dark period. A mixture of several freshwater samples collected from the upper layer of canals, ponds, and lakes in the vicinity of Delft, the Netherlands, was

Citation: Peter R. Mooij, Lisanne D. de Jongh, Mark C. M. van Loosdrecht, Robbert Kleerebezem, Influence of silicate on enrichment of highly productive microalgae from a mixed culture, DOI: 10.1007/s10811-015-0678-2.

used as inoculum. Figure 1 describes the operational cycle for both reactors. A Bio Controller ADI 1030 (Applikon, The Netherlands) controlled Masterflex pumps (Cole-Parmer, USA) and mass flow controllers (Brooks Instruments, The Netherlands). The Bio Controller itself was controlled by a PC with MFCS_win software (Sartorius Stedim Systems, Germany).

Table 1. Operational parameters

Parameter	Unit	Value
Light period	h	16
Dark period	h	8
Cycle length	h	24
Solid retention time	h	41
Liquid volume	L	1.2
Temperature	°C	28
Stirrer speed	rpm	200
Gas flow reactor in and out	NmL min^{-1}	50
Gas recycle over reactor	NmL min^{-1}	1200
Gas composition	% CO_2 in air	5
Average light intensity at inner reactor surface	µmol m^{-2} s^{-1}	650
Light source	–	HPS lamps
pH setpoint	–	7.5
Base	–	1.0 M NaOH
Acid	–	0.5 M HCl

Table 2. Medium composition

Compound	Unit	Value
$NaNO_3^-$	mM	1.43
H_3BO_3	mM	1.66
$CaCl_2 \cdot 2H_2O$	mM	1.07
$MgSO_4 \cdot 6H_2O$	mM	0.64
K_2HPO_4	mM	0.23
$FeCl_3 \cdot 6H_2O$	mM	0.11
$Na_2EDTA \cdot 2H_2O$	mM	0.11
$MnCl_2 \cdot 4H_2O$	µM	7.71
$CuSO_4 \cdot 5H_2O$	µM	0.03
$ZnSO_4 \cdot 7H_2O$	µM	0.69
$CoCl_2 \cdot 6H_2O$	µM	0.43
$Na_2MoO_4 \cdot 2H_2O$	µM	0.77
$NaHSeO_3$	µM	0.10
$NaVO_3$	µM	0.01
Allylthiourea	mg L^{-1}	10

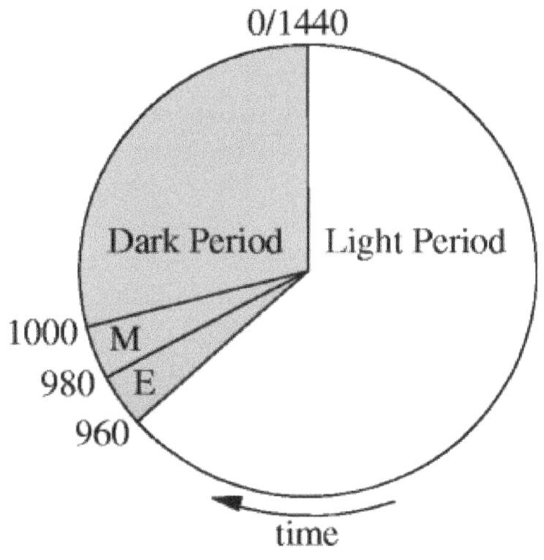

Figure 1 Operational cycle for reactors with effluent removal (E) and medium supply (M) at the start of the dark period. Numbers indicate the cumulative time in minutes from the start of the cycle

Analytical methods

Samples were taken at the transition from dark to light and from light to dark. NO_3^- was determined spectrophotometrically using Dr. Lange LCK 339 NO_3^- cuvette tests (Hach Lange, Germany). Silicate was determined spectrophotometrically using Dr. Lange LCW 028 SiO_2cuvette tests (Hach Lange, Germany). Measurements of organic dry weight, lipids, glucose-polymers, and analysis of the microalgal community structure were done as described by Mooij et al. (2014) with the following modification. Glucose polymers were heated with 0.9 M HCl instead of 0.6 M HCl. Species succession was quantified by taking pictures using a Leica DM500B light microscope (Leica Microsystems, Germany) at × 200 magnification. These pictures were used to count and sort at least 300 microalgal cells per sample.

RESULTS AND DISCUSSION

Microalgal community structure

The first aim of this study was to investigate the influence of silicate presence on the selection of storage compound producing microalgae. In the absence of silicate, the operational conditions imposed resulted in the enrichment of the green alga Chlamydomonas sp. (Lobochlamys segnis, Table 3). With silicate in excess a coculture of the diatom Nitzschia sp. (accounting for 60 % of the population) and the green alga Chlamydomonas was established (Fig. 2). Apparently, two microalgae could coexist under the given conditions, although only one resource, nitrate, was designed to be limited. A possible explanation for the occurrence of a coculture could be a different metabolic response to the pulse-wise addition of nitrate at the start of the dark period. Diatoms are experts in nutrient uptake and storage (Litchman 2007). Nitrate storage up to an intracellular concentration of 273 mM has been reported (Kamp et al.2011). This would allow diatoms to divide during periods without external nitrate, such as during the light period in this experiment. Nitrate storage in green algae is less documented, possibly limiting the period suitable for cell division to the dark period for green algae. This metabolic difference could possibly explain the observed coculture.

Table 3. Identity of species according to microscopic observation and PCR-DGGE analysis

Reactor	Species determined by microscope	Species determined by PCR-DGGE	RNA gene used	Identity (%)
Silicate excess	Chlamydomonas sp.	Lobochlamys segnisKMMCC 1045	18S	100
	Chlorella sp.	Chlorella luteoviridisCCAP 211/5B	18S	96
		Chlorella sorokinianachloroplast	16S	99
	Nitzschia sp.	Nitzschia frustulumchloroplast	16S	100
Silicate deplete	Chlamydomonas sp.	Lobochlamys segnisKMMCC 1045	18S	99
	Chlorella sp.	Chlorella luteoviridisCCAP 211/5B	18S	98
		Chlorella sorokinianachloroplast	16S	99
	Nitzschia sp.	Nitzschia frustulumchloroplast	16S	99

Only the main species present are depicted

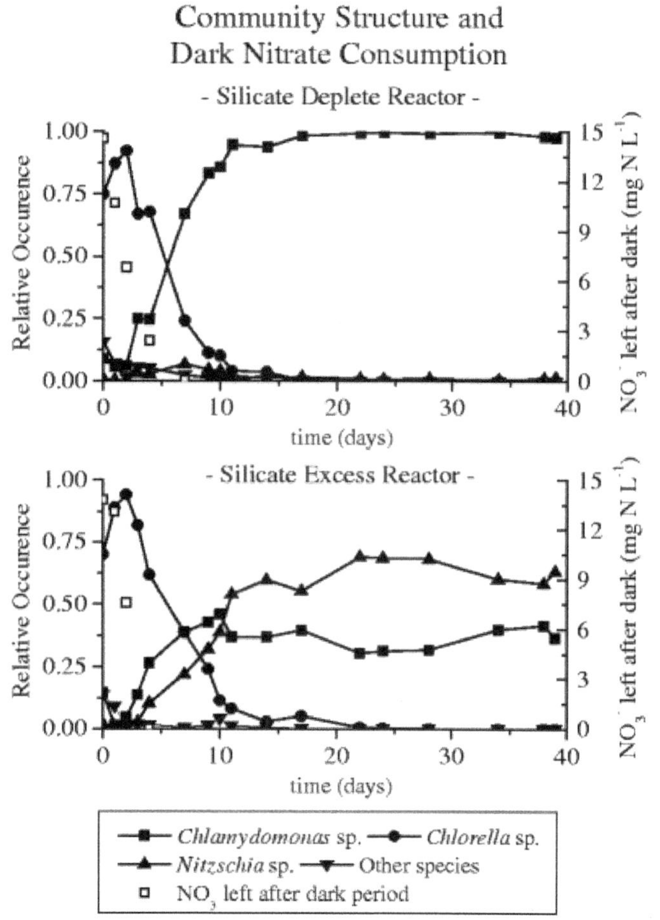

Figure 2 Community structure (black lines, left y-axis) and nitrate left over after the dark phase (black open squares, right y-axis) in time for the silicate-depleted (top) and silicate-excess (bottom) reactors

The dominance of Chlorella luteoviridis (Table 3) increased in both reactors during the first days of the experiment (Fig. 2). After dominating both systems for around 95 % on day 3,Chlorella numbers steeply declined in the next days. A possible explanation lies in the different conditions before and after day 5 of operation. Nutrients were dosed at the start of every dark phase in both reactors. During the first days of the experiment, the limiting nutrient nitrate was not fully consumed in the dark phase (Fig. 2). The presence of nitrate in the light phase favored nutrient uptake and cell division over storage compound production, and this apparently facilitated Chlorella enrichment during this transition period. From day 5 onward, all nitrate dosed at the start of the dark phase was consumed in the dark phase; nitrate was therefore limited during the entire light phase. From this moment on, algae were not able to directly divide in the light phase. The decline of Chlorella sp. in the enrichment culture coincided with the onset of nitrogen limitation during the light period. Apparently, Chlamydomonas andNitzschia outcompete Chlorella if carbon fixation and nitrate uptake are uncoupled. In previous work, Chlorella luteoviridis dominated the system throughout the experiment (Mooij et al.2013). Operational differences between the previous and current work, such as the nitrogen source used and the solid retention time, could explain the disappearance of C. luteoviridisunder the conditions applied in this experiment.

Storage compound productivity in the cycle

The second aim of this study concerned the functional characteristics of both systems. In both reactors, large amounts of glucose polymers were produced, increasing from 17 ± 4 to 53 ± 5 % and from 14 ± 4 to 43 ± 4 % on an organic matter basis during the light period for the silicate-depleted and silicate-excess reactors, respectively (Fig. 3). Diatoms are known to produce chrysolaminarin, a β-1,3-D-glucan, under nitrogen limitation (Granum and Myklestad 2001). Green algae produce starch under nitrogen-limited conditions (Markou et al. 2012). Both of these storage compounds will be measured as glucose monomers using our analytical methods. Lipid levels showed the same trends but were always significantly lower. The lipid fraction increased from 6 ± 1 to 7 ± 2 % and from 8 ± 2 to 10 ± 2 % on an organic matter basis during the light period for the silicate-depleted and silicate-excess reactors, respectively (data not shown). Comparing productivity values with literature values is difficult in the microalgae field. Studies differ in light input, reactor design, reactor operation, type of limitation, and other operational parameters. For pure cultures, glucose content values ranging from 41 to 62.1 % of total dry solids are reported (Dragone et al. 2011; Pirt and Pirt 1977; Brányiková et al. 2011). The values obtained for our enrichment cultures are in the same range but are reported on an organic matter basis. Due to their siliceous cell wall, the ash content of diatoms accounts for up to 50 % of the total dry solids. Comparing productivity between algae classes on a total dry solid basis should therefore be done with caution (Hildebrand et al. 2012).

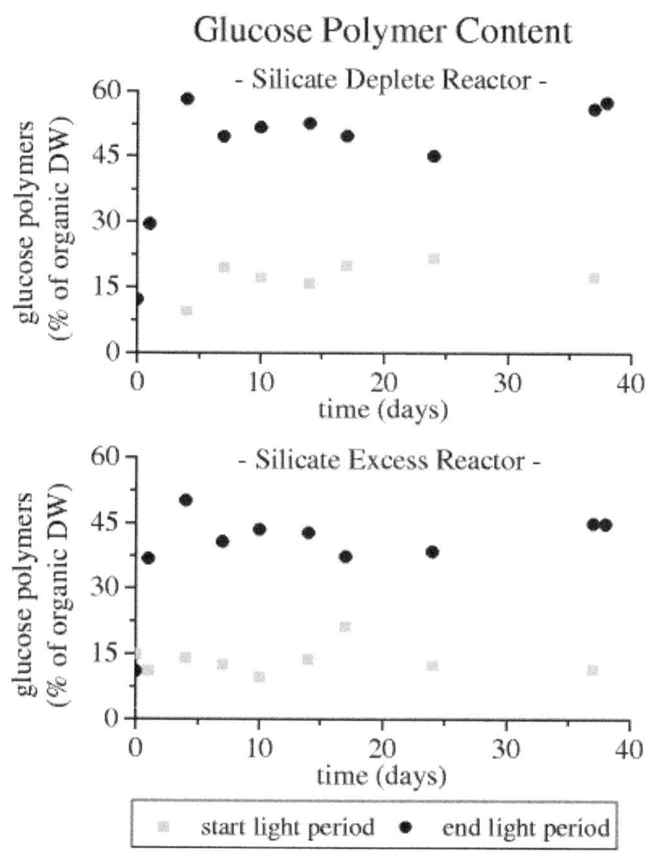

Fig. 3 Fraction of glucose polymers on organic dry weight basis at the start (gray squares) and at the end (black circles) of the light period in time for the silicate-depleted (top) and silicate-excess (bottom) reactors

Stability

From day 10 onward, storage compound production and the microalgal population were stable in both reactors for the remaining 27 days. With around 60 % of the reactor volume harvested every day, these 27 days correspond to 30+ generations of microalgae. Gas and liquid flows leaving and entering the reactors were not sterilized, and the reactors were manually cleaned, opening the reactors fully, every 3–4 days. Despite these disturbances, the systems were very stable both in terms of functional performance and microbial community structure, emphasizing the robustness of the approach used.

CONCLUSION

In the work presented here, we show that uncoupling of carbon fixation in the light and nitrogen uptake in the dark under silicate-excess conditions enriched a diatom-dominated, glucose-polymer-producing microalgae community. In the absence of silicate, a green-algae-dominated community was obtained. Both communities showed the same characteristic of producing high amounts of storage compounds in the light period and stable community structure over time. These results indicate that the proposed method will enrich in any environment a storage compound-producing algae which thrives in this specific environment. As a consequence, it allows stable storage compound production in open, and therefore cheap, cultivation systems.

ACKNOWLEDGMENTS

This research was funded by the Dutch Foundation for Technical Sciences (STW) as project 11610 in the Waste 2 Resource (W2R) program. The authors are grateful to Ben Abbas and Mitchell Geleijnse for the PCR-DGGE work.

Conflict of interest

The authors declare that there is no conflict of interest.

REFERENCES

1. Brányiková I, Maršálková B, Doucha J, Brányik T, Bišová K, Zachleder V, Vítová M (2011) Microalgae—novel highly efficient starch producers. Biotechol Bioeng 108:766–776
2. Dragone G, Fernandes BD, Abreu AP, Vicente AA, Teixeira JA (2011) Nutrient limitation as a strategy for increasing starch accumulation in microalgae. Appl Energy 88:3331–3335
3. Egge JK, Aksnes DL (1992) Silicate as regulating nutrient in phytoplankton competition. Mar Ecol Prog Ser 83:281–289
4. Granum E, Myklestad SM (2001) Mobilization of β-1,3-glucan and biosynthesis of amino acids induced by NH_4^+ addition to N-limited cells of the marine diatom Skeletonema costatum(Bacillariophyceae). J Phycol 5:772–782
5. Hildebrand M, Davis AK, Smith SR, Traller JC, Abbriano R (2012) The place of diatoms in the biofuels industry. Biofuels 3:221–240
6. Hu Q, Sommerfeld M, Jarvis E, Ghirardi M, Posewitz M, Seibert M, Darzins A (2008) Microalgal triacylglycerols as feedstocks for biofuel production: perspectives and advances. Plant J 54:621–639.
7. Kamp A, de Beer D, Nitsch JL, Lavik G, Stief P (2011) Diatoms respire nitrate to survive dark and anoxic conditions. Proc Natl Acad Sci U S A 108:5649–5654.
8. Kazamia E, Aldridge DC, Smith AG (2012) Synthetic ecology—a way forward for sustainable algal biofuel production? J Biotechnol 1:163–169
9. Litchman E (2007) Resource competition and the ecological success of phytoplankton. In: Falkowski, Knoll AH (eds) Evolution of primary producers in the sea. Elsevier, Amsterdam, pp 351–375
10. Markou G, Angelidaki I, Georgakakis D (2012) Microalgal carbohydrates: an overview of the factors influencing carbohydrates production, and of main bioconversion technologies for production of biofuels. Appl Microbiol Biotechnol 96:631–645.
11. McBride RC, Lopez S, Meenach C, Burnett M, Lee PA, Nohilly F, Behnke C (2014) Contamination management in low cost open algae ponds for biofuels production. Ind Biotechnol 10:221–227
12. Mooij PR, Stouten GR, Tamis J, van Loosdrecht MCM, Kleerebezem R (2013) Survival of the fattest. Energy Environ Sci 6:3404–3406
13. Mooij PR, de Graaff DR, van Loosdrecht MCM, Kleerebezem R (2014) Starch productivity in cyclically operated photobioreactors with marine microalgae—effect of ammonium addition regime and volume exchange ratio. J Appl Phycol 27:1–6
14. Mooij PR, Stouten GR, van Loosdrecht MCM, Kleerebezem R (2015) Ecology-based selective environments as solution to contamination in microalgal cultivation. Curr Opin Biotechnol 33:46–51.
15. Pirt MW, Pirt SJ (1977) Photosynthetic production of biomass and starch by Chlorella in chemostat culture. J Chem Technol Biotechnol 27:643–650
16. Shurin JB, Abbott RL, Deal MS, Kwan GT, Litchman E, McBride RC, Mandal S, Smith VH (2013) Industrial-strength ecology: trade-offs and opportunities in algal biofuel production. Ecol Lett 16:1393–1404.
17. Sommer U (1985) Comparison between steady state and non-steady state competition: experiments with natural phytoplankton. Limnol Oceanogr 30:335–346
18. Werner D (1977) The biology of diatoms. Blackwell Scientific Publishing, Oxford
19. Wijffels RH, Barbosa MJ (2010) An outlook on microalgal biofuels. Science 329:796–799.

CHAPTER 5

CULTIVATION OF SHEAR STRESS SENSITIVE AND TOLERANT MICROALGAL SPECIES IN A TUBULAR PHOTOBIOREACTOR EQUIPPED WITH A CENTRIFUGAL PUMP

Michiel H. A. Michels, Atze Jan van der Goot, Marian H. Vermuë, René H. Wijffels

Keywords: Shear stress, Sensitivity, Microalgae, Recirculation pump, Tubular photobioreactor

INTRODUCTION

Skeletonema, Chaetoceros, Phaeodactylum, Isochrysis, Pavlova, and Tetraselmis are frequently used microalgae in feed applications for shellfish in hatcheries (Muller-Feuga 2013). These species would be candidates for large scale production given the increasing demand for microalgal feed of constant and high quality for aquaculture. An increased level of microalgae (components) in fish feed is interesting for the development of fish feed with reduced environmental impact (Draganovic et al. 2013).

For controlled large-scale cultivation of microalgae, closed photobioreactors (PBRs) are often advocated (Borowitzka 1999; Acién et al. 2012). Closed PBRs were combine with low contamination risk and almost no CO_2 losses leading to a high productivity. Furthermore, the cultivation can be fully automated and culture conditions can be highly controlled (Pulz 2001). However, hydrodynamic forces evoked by turbulent mixing in closed PBRs may limit the application of these closed systems. These forces must not exceed the level that would lead to detrimental effects (Chisti 2010; Tredici 2010), suggesting the importance of limiting the shear stress in such PBR systems to levels that can be tolerated by the microalgae. This is especially important when considering the fact that digestibility of microalgael species is positively related to the fragility of the cells (Gladue and Maxey 1994) and therefore to the way many microalgal species used in aquaculture can be cultivated.

Table 1 summarizes the culture systems being employed for the cultivation of the most frequently used microalgal species for aquaculture. All these species can successfully be grown in bubble columns, bags, and carboys; systems that use aeration for mixing. Other types of PBRs, like horizontal tubular PBRs with a shorter light path to produce microalgae at a higher density, are used by aquaculture facilities more often lately (Zmora et al. 2013). Mixing in these systems is induced through recirculation of the microalgal culture in the tubular PBR using pumps, of which centrifugal pumps and airlift pumps are the most common ones (Alías et al.2004; Molina et al. 2001). Centrifugal pumps are efficient in gas–liquid mass transfer (Fadavi and Chisti 2005) and energy use (Norsker et al. 2011). However, microalgal productivity can be influenced negatively by using centrifugal pumps, due to cell damage occurring inside the pump (Carvalho et al. 2006). This could be an explanation for the very few records of microalgae species used in aquaculture, which have been successfully cultivated in tubular PBRs with centrifugal pumps (Table 1). It is still not clear if shear stress sensitivity of certain microalgal species enables successful cultivation in closed PBRs. Although shear stress sensitive species are being cultivated in tubular PBRs with airlift pumps, shear stress levels occurring in the tubes of PBRs could already be too high resulting in loss of productivity. Therefore, there is a need to study the relationship between shear stress sensitivity of microalgal strains and shear stress levels encountered in closed PBRs.

Citation: Michiel H. A. Michels, Atze Jan van der Goot, Marian H. Vermuë, René H. Wijffels, Cultivation of shear stress sensitive and tolerant microalgal species in a tubular photobioreactor equipped with a centrifugal pump, DOI: 10.1007/s10811-015-0559-8.

Table 1 Most frequently used microalgal species as feed for larvae of mollusks, shrimp, and live prey for fish larvae and their culture systems

Class	Species	Culture system	Mixing	References
Bacillariophyceae	Skeletonema costatum	Raceway	Air	Hussenot et al. (1998)
		Polyethylene bags	Air	Pronker et al. (2015)
		Airlift PBR, bubble column	Airlift, air	Monkonsit et al. (2011)
	Chaetoceros muelleri, Chaetoceros gracilis, Chaetoceros calcitrans	Polycarbonate carboys	Air	Camus and Zeng (2012)
		Polyethylene bags	Air	Kaspar et al. (2014); Pronker et al. (2015)
		Bubble column	Air	Lee et al. (2011)
		Airlift PBR, bubble column	Airlift, air	Krichnavaruk et al. (2005)
	Phaeodactylum tricornutum	Polyethylene bags	Air	Pronker et al. (2015)
		Bubble column	Air	Lee et al. (2011)
		Tubular PBR	Airlift	Acién Fernández et al. (2000)
		Tubular PBR	Centrifugal pump	Silva Benavides et al. (2013)
Prymnesiophyceae	Isochrysis galbana	Polycarbonate carboys	Air	Camus and Zeng (2012)
		Polyethylene bags	Air	Dunstan et al. (1993); Kaspar et al. (2014); Pronker et al. (2015)
		Bubble column	Air	Lee et al. (2011)
		Airlift PBR	Airlift	Loubière et al. (2009)
		Tubular PBR	Airlift	Molina Grima et al. (1994); Van Bergeijk et al. (2010)
		Tubular PBR	Centrifugal pump	Van Bergeijk et al. (2010)
	Pavlova lutheri, Pavlova salina	Polycarbonate carboys	Air	Camus and Zeng (2012)
		Polyethylene bags	Air	Dunstan et al. (1993); Pronker et al. (2015)
Prasinophyceae	Tetraselmis suecica, Tetraselmis chuii	Polycarbonate carboys	Air	Camus and Zeng (2012)
		Polyethylene bags	Air	Moheimani (2013); Pronker et al. (2015)
		Bubble column	Air	Lee et al. (2011)
		Annular column	Air	Chini Zittelli et al. (2006)
		Green wall panel reactor	Air	Bondioli et al. (2012)
		Tubular PBR	Centrifugal pump	Michels et al. (2014)

The aim of the study is to determine the tolerance to shear stress of four different microalgal species. Shear cylinders were used to quantify the threshold values of shear stress for the different microalgal species. These microalgae species were all tested in a tubular PBR with a variable-frequency-drive centrifugal pump to determine the capability of growth. Growth or lack of growth of microalgae will be related to estimated shear stress levels inside the reactor and shear tolerance levels determined for all four species.

MATERIAL AND METHODS

Organisms and medium

Tetraselmis suecica, Isochrysis galbana, and Skeletonema costatum were obtained from Seasalter Shellfish (Whitsable) Ltd. (Kent, UK), and Chaetoceros muelleri (CCMP 1316) was provided by NIOZ (Royal Netherlands Institute for Sea Research). Walne medium modified from Laing (1991) was used for the cultivation of the microalgal species. The modified Walne medium consists of solution A (macro- and micronutrients), C (vitamins), and D (silicate, which is only needed for diatoms). Solution A contains 0.8 g $FeCl_3$, 0.4 g $MnCl_2 \cdot 4H_2O$, 33.6 g H_3BO_3, 45.0 g EDTA, 20.0 g $NaH_2PO_4 \cdot 2H_2O$, 100.0 g $NaNO_3$, 21 mg $ZnCl_2$, 20 mg $CoCl_2 \cdot 6H_2O$, 9.0 mg $(NH_4)_6Mo_7O_{24} \cdot 4H_2O$, and 20 mg $CuSO_4 \cdot 5H_2O$ in 1 L of distilled water, with the pH adjusted to 4.0 with concentrated HCl. Solution C consists of 1.0 g vitamin B_1 and 0.05 g vitamin B_{12} in 1 L of distilled water. Solution D contains 40.0 g $Na_2SiO_3 \cdot 5H_2O$ in 1 L of distilled water. Medium for maintaining the cultures in Erlenmeyer flasks was made by adding 1 mL solution A, 0.1 mL solution C, and 2 mL solution D per liter of filtered and deironized saline groundwater (30 g L^{-1}). A double dose of solutions A, C, and D was supplied to the culture in the tubular photobioreactor during turbidostat operation to avoid nutrient depletion. Solution D was only added for the cultivation of the diatoms S. costatum and C. muelleri.

Generation of shear stress levels tolerance

Batch cultures of T. suecica, I. galbana, and S. costatum from 3-L Erlenmeyer flasks containing 2 L medium after 1 week of growth at 20 °C under white fluorescent light (150 μmol photons m^{-2} s^{-1}; PAR) were used for shear stress experiments. The three microalgal strains were exposed to different shear stress levels in shear cylinders as Couette devices (Van Riemsdijk et al. 2010). Shear stress exposed to the microalgae can be determined with:

$$\tau = \dot{\gamma} \cdot \eta$$

where τ is the shear stress (Pa), γ˙γ˙ is the shear rate (s^{-1}), and η is the apparent viscosity (Pa s). Shear stress levels can be varied by increasing the shear rate or by increasing the apparent viscosity of the medium. The shear rate applied in the shear cylinders is related to the rotational speed with a conversion factor of rotational speed (rpm) to shear rate (s^{-1}) of 2.157 (Michels et al. 2010). Locust bean gum (LBG), which does not affect the viability directly (Michels et al.2010), was used as a thickener to increase the apparent viscosity.

S. costatum and I. galbana with 0.3 % LBG in the medium were exposed to different shear rates of 0, 8.6, 43, 216, and 1079 s^{-1}. T. suecica also was exposed to the same shear rates, but with two LBG concentrations: 0.5 and 0.75 % LBG, to obtain higher viscosity. All exposures were done in triplicate at 4 °C with an exposure time of 1 h. Since LBG solutions can be described as non-Newtonian fluids, the apparent viscosity was measured at the different shear rates in order to calculate the applied shear stress levels. The shear stress levels applied in the shear cylinders were determined with a rheometer (type Physica MCR 301, Anton Paar) at 4 °C according to the method described in Michels et al. (2010). The power-law functions that describe the relation between shear rate and shear stress in 0.3, 0.5, and 0.75 % LBG are τ=0.034γ˙0.943τ=0.034γ˙0.943 with $R^2 = 0.9998$, τ=0.227γ˙0.763τ=0.227γ˙0.763 with $R^2 = 0.993$ and τ=1.175γ˙0.618τ=1.175γ˙0.618 with $R^2 = 0.987$, respectively. These functions were used to calculate the shear stress levels applied at different rotational speed levels and LBG concentrations. Table 2 presents the calculated shear stress levels related to rotational speed, shear rate, and LBG concentration.

Table 2 Shear stress applied in relation with rotational speed, shear rate, and LBG concentration

Rotational speed (rpm)	Shear rate (s^{-1})	Shear stress 0.3 % LBG (Pa)	Shear stress 0.5 % LBG (Pa)	Shear stress 0.75 % LBG (Pa)
0	0	0	0	0
4	8.6	0.26	1.2	4.5
20	43	1.2	4.0	12
100	216	5.4	14	33
500	1079	25	47	88

Measurement of effect of shear stress on viability

The effect of applying shear stress on the viability of T. suecica, S. costatum, and I. galbana was measured by using fluorescein diacetate (FDA). Viable cells contain esterases that convert FDA into fluorescein and diacetate. Viable cells show fluorescence caused by fluorescein (Altman et al. 1993; Rotman and Papermaster 1966). After exposure in the shear cylinders, 1 mL of algae was incubated with 10 μL FDA solution (11 mM) for 20 min. The total cell concentration and viable cell concentration were determined with a hemocytometer (DHC-B02-5 Büker Türk) using a fluorescence microscope. The viability of the sheared algae was calculated as the percentage of fluorescing algae and compared to the shear stress tolerance of C. muelleri(Michels et al. 2010).

Because S. costatum can form chains up to 8–10 cells per chain, the effect of shear stress on the distribution in the number of cells per chain was also measured. The number of cells per chain was counted for at least 50 chains per sample. The Mann–Whitney U test was done to determine any statistical differences in cells per chain distribution per applied shear stress.

Growth of microalgae in a tubular photobioreactor

Growth tests with T. suecica, I. galbana, S. costatum, and C. muelleri were carried out in a tubular PBR with a total volume of 40 L. The tubular PBR consists of 20-m-long loop connected to a degasser. The loop is made of Plexiglas tubes with an external diameter of 50 mm and an internal diameter 43.6 mm. This culture system is equipped with a variable-frequency-drive centrifugal pump (SealPro KR-32-95, ARBO) with an impeller diameter of 95 mm and a pressure side diameter of 32 mm, to circulate the microalgal culture. The location and main operating mode of the tubular PBR are described in Michels et al. (2014).

The four microalgal species were separately used for inoculation of the tubular PBR with a minimal starting cell concentration of 200,000 cells mL^{-1}. Temperature was controlled at 20 ± 0.5 °C. The pH is measured at the end of the tube before the degasser and controlled at pH 8.40 via CO_2 supply. The initial pumping speed at which the microalgae were recirculated was 0.37 m s^{-1}, with a rotational speed of the impeller of 10 s^{-1}.

The increase in cell concentration was followed by taking daily samples for 7 days. The run in the PBR was terminated when the cell concentration of an algae species did not increase during this period. The PBR was cleaned thoroughly and inoculated with the same species for a second time. The daily samples were inspected with a microscope to check the shape of the microalgae cells, the motility of the flagellates and potential bacterial occurrence.

The runs in which the algae concentration increased during the first 7 days were further used to study the effect of pumping speed on the net volumetric productivity at turbidostat conditions, with the biomass concentration set at about 0.5 g L^{-1}. The pumping speed of the recirculation pump was 2.0, 2.4, 2.8, 3.2, and 3.6 m^3 h^{-1}, respectively. Those trials were done for a period of 14 days.

Calculation of shear stress levels in the tubular photobioreactor

The corresponding flow velocities and Reynolds numbers in the tubes and at the pressure side of the pump were calculated, with a known culture density of 1024 kg m^{-3} and an apparent viscosity of $1.8 \cdot 10^{-3}$ Pa s (Michels et al. 2010). Those conditions were used to estimate the average shear stress levels in the tubes. The Blasius equation was used to calculate the shear stress at the wall of the tube where flow will be laminar (Durst et al. 1996). The equation is:

$$\tau = C_f \frac{1}{2} \rho \bar{u}^2$$

with $C_f = 0.0791 Re^{-1/4}$

where τ is the average shear stress (Pa), C_f is the Fanning friction factor (dimensionless), ρ is the density (kg m^{-3}), $\bar{u}$ is the average flow velocity (m s^{-1}), and Re is the Reynolds number. The Reynolds number is expressed by:

$$Re = \frac{\rho \cdot \bar{u} \cdot D}{\eta}$$

where D is the internal diameter (m).

The shear stress in the centrifugal pump depends on the apparent viscosity, rotational speed of the impeller, and the Reynolds number. The local shear stress adjacent to the rotating impeller of the centrifugal pump can be estimated with:

$$\tau = 6.30 \eta N Re_L^{0.5}$$

where N is the rotational speed of the impeller and the local Reynolds number at the local diameter is defined as:

$$Re_L = \frac{N d_L^2 \rho}{\eta}$$

where d_L is the local diameter of the impeller (Chisti 2010; Wichterle et al. 1996).

Determination of the net volumetric productivity

The net volumetric productivity (P_V, g L^{-1} day^{-1}) was determined as the product of the net-specific growth rate (µ, d^{-1}) and the biomass concentration (C_X, g L^{-1}). At turbidostat conditions, the net-specific growth rate is equal to the dilution rate (D, d^{-1}), defined as the daily harvest volume divided by the volume of the PBR (Michels et al. 2014).

The daily photon fluxes were derived from the measured photosynthetically active radiation (PAR) as an average daily incident photon flux density (PFD, mol photons m^{-2} day^{-1}) (Michels et al. 2014). The PFD was measured with a LiCor LI190 PAR sensor.

The average daily photon flux density during culture ranged from 0.69 to 13.3 mol photons m^{-2}day^{-1}.

RESULTS AND DISCUSSION

Shear stress tolerance

Figure 1 shows the effect of shear stress on the viability of T. suecica, I. galbana, S. costatum, and C. muelleri. An adverse effect was found for I. galbana, where viability decreased suddenly to 74.9 % evoked by a shear stress level of 5.4 Pa. The viability of I. galbana was not reduced any further when the shear stress increased to 25 Pa. I. galbana is a flagellate and a member of the non-calcified coccolithophytes with only a plasma membrane covering, which makes the naked cell fragile (Graham et al. 2009b; Zhu and Lee 1997).

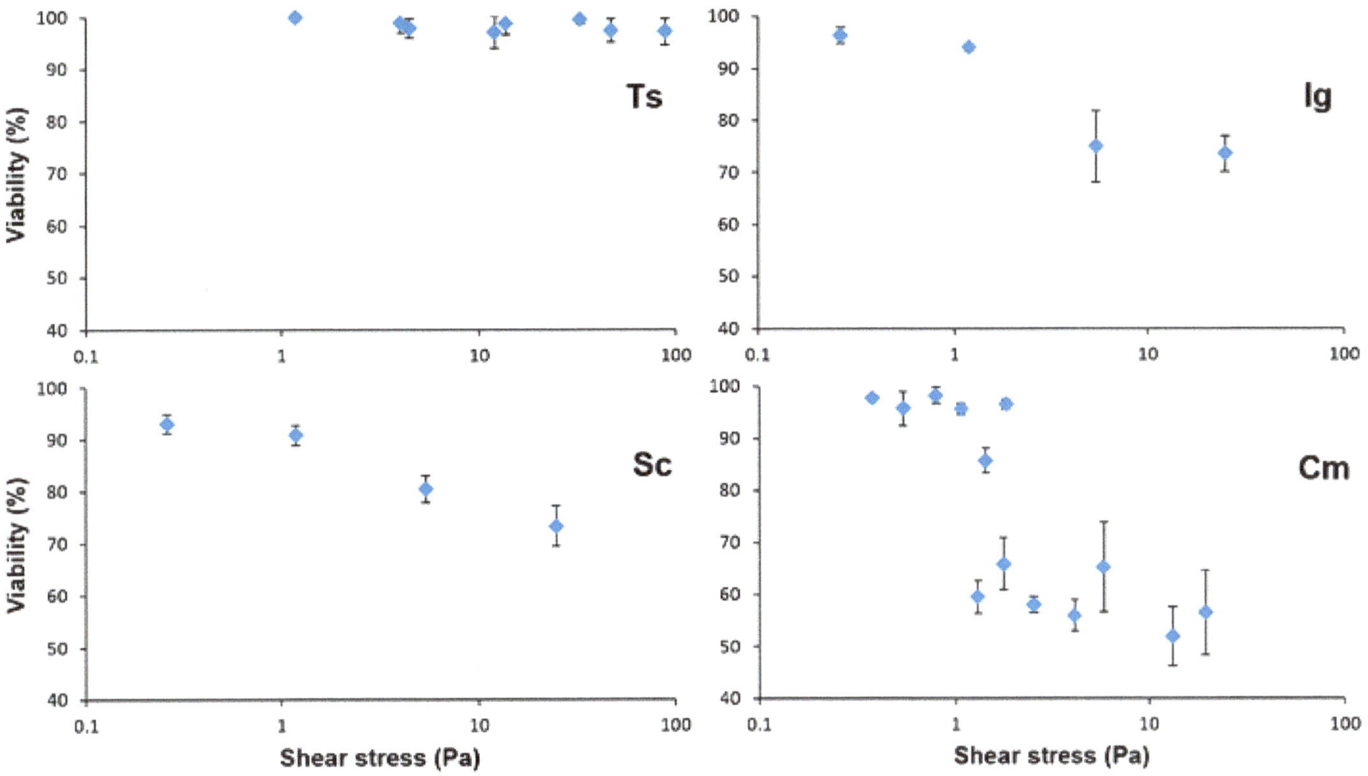

Figure 1 Effect of shear stress on Tetraselmis suecica (Ts), Isochrysis galbana (Ig), Skeletonema costatum (Sc), and Chaetoceros muelleri (Cm). Data from effect of shear stress onChaetoceros muelleri were obtained from Michels et al. (2010)

The effect of shear stress on S. costatum was a bit different. The viability of S. costatumdecreased further from 80.5 to 73.4 %, when the shear stress increased from 5.4 to 25 Pa. This observation can be ascribed to a change in morphology of S. costatum due to the increased shear stress. Since this diatom forms chains, breakages of the chains were expected at higher shear stress levels (Sauriau and Baud 1994). The effect of shear stress on the distribution of cells per chain is shown in Fig. 2. The distribution of cells per chain of unexposed S. costatumand cells exposed to shear stress of 0.26 Pa did not differ, while significant differences in distribution of cells per chain were found between all the exposed shear stress levels. Shear stress levels higher than 0.26 Pa caused a significant reduction in the average chain length with a progressive increase of chains with 1–3 cells. The average chain length reduced linearly on a semi-logarithmic scale from 3.46 ± 0.18 to 2.49 ± 0.04 cells per chain with corresponding shear stresses of 0.26 and 25 Pa. Although a shear stress of 1.2 Pa already caused a reduction of the average number of cells per chain, the viability did not decrease significantly (Fig. 1). Shear stress probably has first an impact on the intercellular junctions causing chain breakage, which is then followed by other cell structures like the siliceous frustules being damaged causing mortality (Sauriau and Baud 1994).

Data of C. muelleri were used as reference and were obtained from Michels et al. (2010). C. muelleri was found to be shear stress sensitive with a threshold value of shear stress between 1 and 1.3 Pa. Higher shear stress levels caused a decrease in viability with a certain fraction of the cells being sensitive to shear stress (Michels et al. 2010).

T. suecica exposed to shear stress levels up to 88 Pa did not show any adverse effects on the viability (Fig. 1). The tolerance to high shear stress of T. suecica is probably caused by the rigid cell wall composed of layers of scales attached to the cell membrane (Graham et al. 2009a).

For the algae species that were susceptible to shear stress, the viability was decreased with 20 to 40 %. The fact that not all cells were inactivated could be explained by the fact that cells are not equally susceptible to shear damage during the full growing cycle. It has been reported that microalgae are more vulnerable to shear stress during cell division (García Camacho et al. 2007; Stoecker et al. 2006). This will cause a loss of viability of only a certain percentage of the shear stress sensitive species I. galbana, S. costatum, and C. muelleri (Fig. 1). A plausible explanation why the flagellate T. suecica is not shear stress sensitive during cell division, is that the flagella are shed during the division prior to mitosis, and cytokinesis takes place within the rigid parental wall (Graham et al. 2009a).

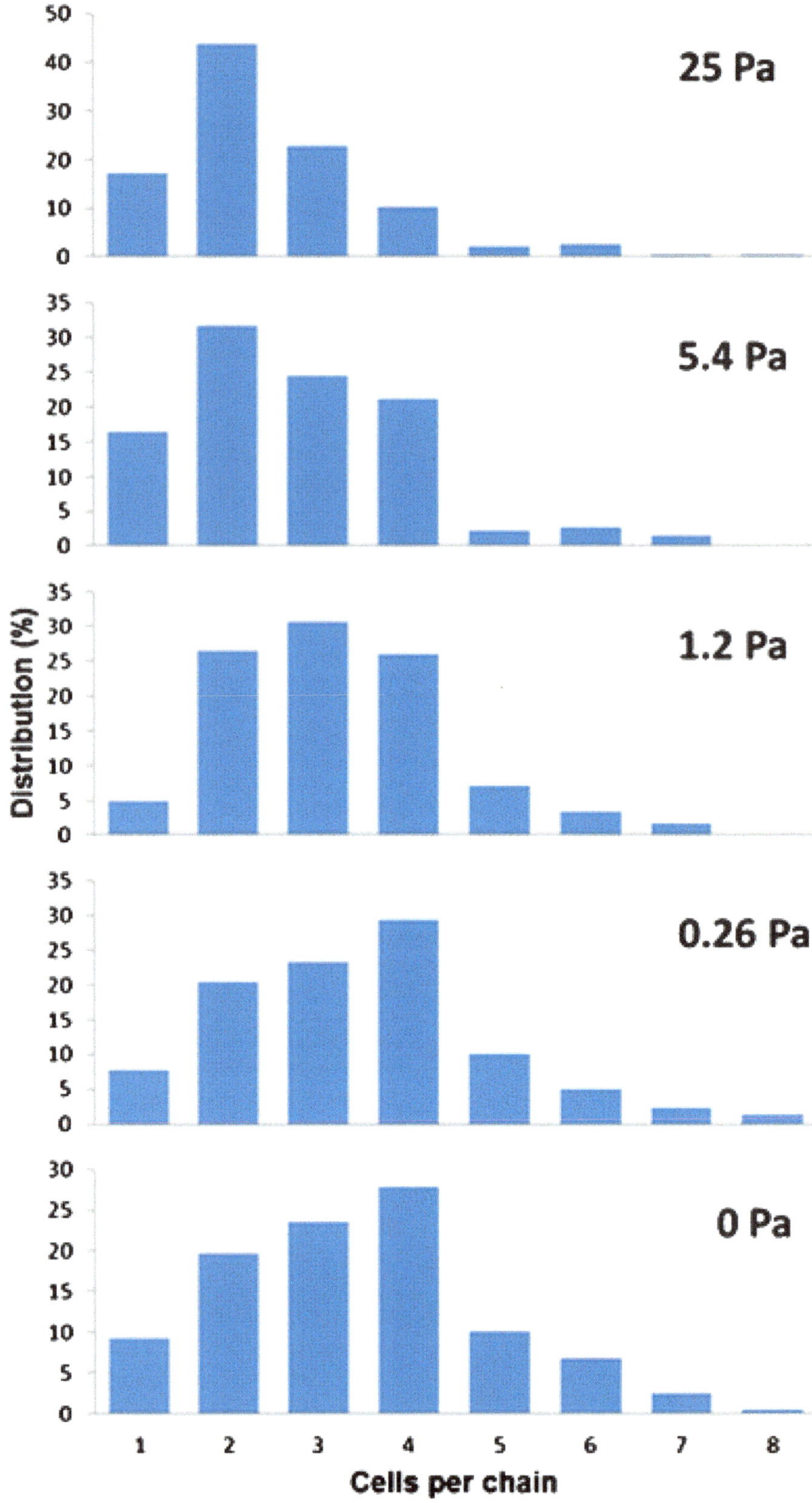

Figure 2 Effect of shear stress on the distribution of cells per chain of Skeletonema costatum

Growth tests with shear stress sensitive and tolerant microalgal species

T. suecica, I. galbana, S. costatum, and C. muelleri were all tested for their capability to grow in a tubular PBR in which the culture is recirculated using a centrifugal pump. T. suecica was the only species tested that grew well in the tubular PBR. The net volumetric productivity of T. suecica cultivated at different pumping speeds is shown in Fig. 3. The biomass concentration was kept constant at 0.52 ± 0.05 g L^{-1} applying turbidostat conditions. The net volumetric productivity linearly increased with the daily photon flux received. During the entire experiment, the microalgae were exposed to relatively low light intensity. At these low light intensities, no light saturation or light inhibition occurs. A linear increase of productivity with light input is therefore to be expected (Geel et al. 1997).

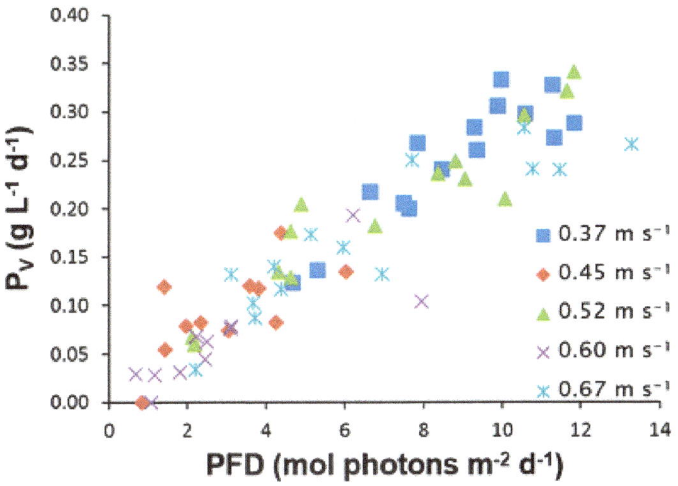

Figure 3 Net volumetric productivity of Tetraselmis suecica versus the daily photon flux density at different pumping speeds

No differences in net volumetric productivity of T. suecica receiving similar light were found between the cultivation periods at different pumping speeds. Higher pumping speed evokes increased shear stress and could affect the net productivity negatively by its potential detrimental effects on the microalgae. On the other hand, it can also have a positive effect on the net volumetric productivity due to a better mass and gas transfer and shorter light/dark cycles that are produced when mixing is increased (Contreras et al. 1998; Leupold et al. 2013; Vejrazka et al. 2012). Neither negative nor positive effects of pumping speed on the productivity of T. suecica, however, were found during this study.

The average net volumetric productivities with corresponding average daily photon flux densities for the five runs with a different pumping speed are given in Table 3. During the experimental period at limiting light conditions during the fall, the maximum net volumetric productivity of the microalgae receiving a daily photon flux density of 11.8 mol photons m^{-2} day^{-1} was 0.34 g L^{-1} day^{-1}, while daily light inputs lower than 1 mol photons m^{-2} day^{-1} resulted in a zero or close to zero productivity (Fig. 3). The daily photon flux density of 1 mol photons m^{-2} day^{-1} was obviously the compensation point of photosynthesis, at which the rate of photosynthesis equals the respiration rate. The compensation point of photosynthesis of T. suecica was similar to the value of Chlamydomonas reinhardtii (Takache et al. 2010).

Table 3 Average net volumetric productivity and average daily photon flux density at different runs

Pumping speed (m^3 h^{-1})	Net volumetric productivity (g L^{-1} day^{-1})	Daily photon flux density (mol m^{-2} day^{-1})
2.0	0.25 ± 0.06	8.76 ± 2.17
2.4	0.09 ± 0.05	3.02 ± 1.58
2.8	0.20 ± 0.09	7.14 ± 3.35
3.2	0.07 ± 0.05	2.94 ± 2.23
3.6	0.17 ± 0.08	6.66 ± 3.56

Biofouling at high pumping speed

Although the highest pumping speed did not lead to a lower net volumetric productivity, the culture of T. suecica was affected negatively. Biofouling started to occur in the tubes 1 day after the pumping speed was set at 3.6 m^3 h^{-1}, at which the average shear stress in the tubes and at the pressure side of the pump were 1.60 and 5.10 Pa, respectively (Table 4). Due to the presence of most of the biofouling at the bottom of the tubes, the cells were still able to receive light, which did not result in a lower net volumetric productivity. Biofouling in the tubes increased enormously during the 2 weeks T. suecica was cultivated at this high pumping speed. Therefore,T. suecica should be cultured at lower pumping speeds to ensure cells will not be damaged. An additional benefit of applying a lower pumping speed is the lower energy costs.

Table 4 Flow velocities, Reynolds numbers, and average shear stress levels in tubes and at pressure side of the pump at different pumping speeds

	Tubes			Pressure side of pump		
Pumping speed (m^3 h^{-1})	Flow velocity (m s^{-1})	Reynolds number	Average shear stress (Pa)	Flow velocity (m s^{-1})	Reynolds number	Average shear stress (Pa)
2.0	0.37	$9.2 \cdot 10^3$	0.57	0.69	$1.3 \cdot 10^4$	1.82
2.4	0.45	$1.1 \cdot 10^4$	0.79	0.83	$1.5 \cdot 10^4$	2.51
2.8	0.52	$1.3 \cdot 10^4$	1.03	0.97	$1.8 \cdot 10^4$	3.29
3.2	0.60	$1.5 \cdot 10^4$	1.30	1.11	$2.0 \cdot 10^4$	4.15
3.6	0.67	$1.7 \cdot 10^4$	1.60	1.24	$2.3 \cdot 10^4$	5.10

Growth of recirculated I. galbana, S. costatum, and C. muelleri in the tubular PBR was not observed, not even at the lowest pumping speed of 2.0 $m^3\ h^{-1}$. The cell concentration of these three microalgae species did not increase during the 7 days tested. General observations in chronological order were a decrease of cell concentration within 1 or 2 days, an increase of number of bacteria, followed by foam formation and finally biofouling. In the case of I. galbana, the remaining viable cells did not lose their motility and no alteration of the shape could be observed. Regarding S. costatum, longer chains disappeared after 1 day with more short chains and single cells as a result. Broken and disintegrated cells of S. costatum were seen after a few days. The shape of the remaining cells of C. muelleri did not change immediately, but after only a few days the cells became spherical and lost their spines. Shear stress most likely caused the incapability of I. galbana, S. costatum, and C. muelleri to grow in the tubular PBR.

Relative high shear-stress levels encountered in the photobioreactor

Lack of growth of these microalgae strains could be correlated to the shear stress sensitivity of the cells investigated. Unfortunately, exact shear stress values are difficult to predict in turbulent flow, but the stress at the wall can be calculated easily, due to the fact that flow is laminar at the wall. The shear stress at the wall is 0.57 Pa at a pumping speed of 2.0 $m^3\ h^{-1}$. A similar calculation leads to a wall shear stress of 1.82 Pa in the pressure side of the pump (Table 4). It can be expected that even higher shear stress values can occur in other parts of the reactor. For example, maximum shear stress levels of about 2 Pa were reported in bends of tubes with similar diameters (5 cm) and at similar velocity (0.35 $m\ s^{-1}$) in the tubes based on computational fluid dynamics (Ramírez-Duque and Ramos-Lucumi 2011). Even higher shear stress levels will occur in the cavity of the centrifugal pump, where the shear forces are not equally distributed. With an impeller of 95 mm and rotational speed of 10 s^{-1}, the maximum shear stress occurring in the centrifugal pump could be calculated. The maximum shear stress at the tip of the impeller in the centrifugal pump at a pumping speed of 2.0 $m^3\ h^{-1}$ is 26 Pa. Comparing this shear stress level with the shear stress that evokes loss of viability of the cells (Fig. 1), it is obvious that local shear stress in the pump must have been detrimental to the shear stress sensitive microalgae species. However, a large part of the reactor might still have favorable conditions for cell cultivation.

Simulation of the possible damage caused by high circulation rates

Unfortunately, the high circulation rate of the culture will inevitably lead to the situation that all cells will pass a high shear zone and most of them will pass the zone several times. This can be demonstrated by a simple simulation as shown below. In this simulation, it is assumed that a shear sensitive cell will be damaged if the cell enters a high shear stress zone inside a certain part of the reactor (e.g., in the cavities of the circulation pump), where the shear stress level is beyond the threshold of the algae. Computational fluid dynamics (CFD) analyses on shear stress in centrifugal pumps show that the probability of cells entering regions inside the pump with shear stress levels that damage the cells depends on the residence time of the cells inside the pump, the rotational speed of the impeller, and the shape of the impeller (Song et al. 2003; Takiura et al. 1998; Zhou et al. 2003). The magnitude of the damage is therefore related to the proportion of the cells, , that pass this high shear stress zone per passage and the amount of passages (n) over time:

$$\frac{N_r}{N_0} = (1-\varphi)^n$$

where N_r is the cell concentration of the sensitive algae remaining viable and N_0 is the cell concentration of sensitive algae before exposure. Figure 4 shows the percentage of intact cells remaining over time assuming that 1, 5, or 10 % of the cells are being damaged per passage.

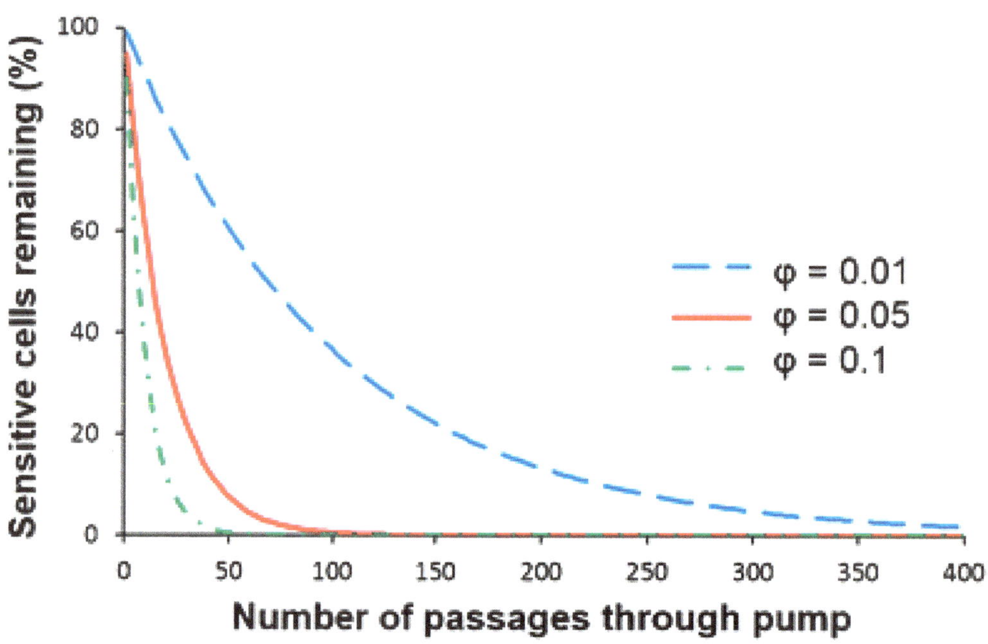

Figure 4 Percentage of remaining cells as an effect of passages through a pump and proportion of cells being damaged per passage

The number of passages through the pump depends on the length of the tubular PBR and the flow velocity. At a pumping speed of 2 $m^3 h^{-1}$, the culture is recirculated 50 times per hour in the PBR. This means that the number of passes per day is about 1200. Even when only 1 % of the cells pass a high shear zone, it is clear that all sensitive cells would be inactivated within 1 day. However, not all cells might be shear stress sensitive. Most likely, the cells which are in the dividing stage are especially vulnerable to shear stress. Assuming that at a given time about 30 % of the microalgal cells are in the dividing stage (Coats and Heinbokel 1982) and that the generation time is about 1 day, the time that a cell is vulnerable to shear stress is about 7.2 h per day. In those 7.2 h, the number of passages is 360, which leads to an inactivation of 97 % of the sensitive cells. Only 3 % will remain for further growth, which is clearly too low to maintain or even increase the number of microalgae cells. This means that in the tubular PBR, almost all cells are expected to be inactivated within one or a few days, which is in line with the experimental observations.

Design of closed photobioreactor systems for shear sensitive microalgae

Theoretically, two possible routes are possible to avoid cell damage due to shear stress. The first one is reducing or avoiding high shear stress zones in the reactor. This option, however, might be quite difficult to achieve. Lowering the flow rate too much will lead to other problems, because high turbulence is needed for keeping the microalgae suspended and the use of light and nutrients is enhanced by turbulent mixing (Richmond 2013). Another option is to reduce the number of passages. Many commercial tubular PBRs consist of longer tubes with a length of 100 m (Pulz et al. 2013) and with flow velocities applied in PBRs between 0.3 and 0.5 m s^{-1}(Norsker et al. 2011). This reduces the number of passages of the shear stress sensitive cells, which are in the dividing stage, to about 90 per day. In the case of 1 % of the cells being affected per pump passage in a commercial PBR, 60 % of the sensitive cells would still be damaged. It is obvious that more cells will be damaged, if a higher percentage of the cells is affected per passage through the pump (Fig. 4). Considering that the time needed for cell proliferation is about 1 day, the time is too short to overcome the damage done by high shear stress, even if the high shear stress region is small.

The tolerance to shear stress of various strains seems to be selective to the choice of recirculation pumps. Microalgae with rigid cell walls are shear stress tolerant, while species lacking a cell wall, coccolithophores with calcium carbonate containing coccospheres and diatoms with fragile siliceous frustules are sensitive to shear stress (Leupold et al. 2013; Moheimani et al. 2011; Vandanjon et al. 1999). Phaeodactylum tricornutum is an exception among the diatoms. Its cell wall structure does not contain siliceous valves but rather the cells hace a more rigid polysaccharide cell wall (Borowitzka and Volcani 1978; Tesson et al. 2009). It therefore can be successfully cultivated in tubular PBRs with centrifugal pumps, although some reduction of productivity has been reported at higher flow rates causing high shear zones in the reactor (Alías et al. 2004; Silva Benavides et al. 2013).

The fact that I. galbana, S. costatum, and a species of the genus Chaetoceros were reported to be successfully cultivated in tubular PBRs recirculated with airlift pumps (Molina Grima et al.1994; Krichnavaruk et al. 2005; Loubière et al. 2009; Monkonsit et al. 2011; Van Bergeijk et al.2010) suggests that hydrodynamic forces exerted in the tubes of PBRs were probably not high enough to negatively affect the growth. Airlift pumps, which are causing lower shear stress than centrifugal pumps (Carvalho et al. 2006), seem to be the best option for the recirculation of shear stress sensitive microalgae. Peristaltic pumps, eccentric rotor pumps, and diaphragm pumps may be interesting alternatives as well. These pumps have been shown to be more gentle than centrifugal or positive displacement rotary vane pumps (Jaouen et al. 1999). However, further research on the actual shear stress levels occurring in these pumps is recommended in order to find whether shear stress levels are sufficiently low to allow successful algae cultivation in PBRs.

CONCLUSION

Four different microalgae species used as feed for shellfish in hatcheries were tested for their shear stress sensitivity. T. suecica was found to be the only shear stress tolerant species tested, which viability was not negatively affected by a maximum applied shear stress level of 88 Pa. T. suecica was also successfully grown in a tubular PBR driven by a centrifugal pump. I. galbana,S. costatum, and C. muelleri were not able to grow in the tubular PBR recirculated by a centrifugal pump at its lowest speed. Shear stress levels between 1.2 and 5.4 Pa caused a reduction in viability of the shear stress sensitive species I. galbana, S. costatum, and C. muelleri. In order to increase the feasibility of the production of microalgae for aquaculture in fully-automated PBRs, high shear stress zones in the reactor (including the pump) should be avoided when designing a culture system for shear stress sensitive microalgal species.

ACKNOWLEDGMENTS

This study is supported by the Zeeuwse Tong project (Zeeland Sole project), co-funded by the European Fisheries Fund. The authors would like to thank Julius Frölich, Solveig Höfer, Zanina Ilieva, Scott Quainoo, Anna Jeworrek, Lennet Leuchter, Jolien Scholten and Jens Schouwenaars for their assistance working with the photobioreactor.

REFERENCES

1. Acién FG, Fernández JM, Magán JJ, Molina E (2012) Production cost of a real microalgae production plant and strategies to reduce it. Biotechnol Adv 30:1344–1353
2. Acién Fernández FG, Sánchez Pérez JA, Fernández Sevilla JM, García Camacho F, Molina Grima E (2000) Modeling of eicosapentaenoic acid (EPA) production from Phaeodactylum tricornutum cultures in tubular photobioreactors. Effects of dilution rate, tube diameter, and solar irradiance. Biotechnol Bioeng 68:173–183
3. Alías CB, García-Malea López MC, Acién Fernández FG, Fernández Sevilla JM, García Sánchez JL, Molina Grima E (2004) Influence of power supply in the feasibility of Phaeodactylum tricornutum cultures. Biotechnol Bioeng 87:723–733
4. Altman SA, Renders L, Rao G (1993) Comparison of

trypan blue dye exclusion and fluorometric assays for mammalian cell viability determinations. Biotechnol Prog 9:671–674

5. Bondioli P, Della Bella L, Rivolta G, Chini Zittelli G, Bassi N, Rodolfi L, Casini D, Prussi M, Chiaramonti D, Tredici MR (2012) Oil production by the marine microalgae Nannochloropsissp. F&M-M24 and Tetraselmis suecica F&M-M33. Bioresour Technol 114:567–572
6. Borowitzka MA (1999) Commercial production of microalgae: ponds, tanks, tubes and fermenters. J Biotechnol 70:313–321
7. Borowitzka MA, Volcani BE (1978) The polymorphic diatom Phaeodactylum tricornutum: Ultrastructure of its morphotypes. J Phycol 14:10–21
8. Camus T, Zeng C (2012) Reproductive performance, survival and development of nauplii and copepodites, sex ratio and adult life expectancy of the harpacticoid copepod, Euterpina acutifrons, fed different microalgal diets. Aquac Res 43:1159–1169
9. Carvalho AP, Meireles LA, Malcata FX (2006) Microalgal reactors: a review of enclosed system designs and performances. Biotechnol Prog 22:1490–1506
10. Chini Zittelli G, Rodolfi L, Biondi N, Tredici MR (2006) Productivity and photosynthetic efficiency of outdoor cultures of Tetraselmis suecica in annular columns. Aquaculture 261:932–943
11. Chisti Y (2010) Shear sensitivity. In: Flickinger MC (ed) Encyclopedia of industrial biotechnology, bioprocess, bioseparation and cell technology, vol 7. Wiley, New York, pp 4360–4398
12. Coats DW, Heinbokel J (1982) A study of reproduction and other life cycle phenomena in planktonic protists using an acridine orange fluorescence technique. Mar Biol 67:71–79
13. Contreras A, García F, Molina E, Merchuk J (1998) Interaction between CO_2-mass transfer, light availability, and hydrodynamic stress in the growth of Phaeodactylum tricornutum in a concentric tube airlift photobioreactor. Biotechnol Bioeng 60:317–325
14. Draganovic V, Jørgensen SE, Boom R, Jonkers J, Riesen G, Van Der Goot AJ (2013) Sustainability assessment of salmonid feed using energy, classical exergy and eco-exergy analysis. Ecol Indic 34:277–289
15. Dunstan GA, Volkman JK, Barrett SM, Garland CD (1993) Changes in the lipid composition and maximisation of the polyunsaturated fatty acid content of three microalgae grown in mass culture. J Appl Phycol 5:71–83
16. Durst F, Kikura H, Lekakis I, Jovanović J, Ye Q (1996) Wall shear stress determination from near-wall mean velocity data in turbulent pipe and channel flows. Exp Fluids 20:417–428
17. Fadavi A, Chisti Y (2005) Gas–liquid mass transfer in a novel forced circulation loop reactor. Chem Eng J 112:73–80
18. García Camacho F, Gallardo Rodríguez JJ, Sánchez Mirón A, Cerón García MC, Belarbi EH, Molina Grima E (2007) Determination of shear stress thresholds in toxic dinoflagellates cultured in shaken flasks. Implications in bioprocess engineering. Process Biochem 42:1506–1515
19. Geel C, Versluis W, Snel JFH (1997) Estimation of oxygen evolution by marine phytoplankton from measurement of the efficiency of Photosystem II electron flow. Photosynth Res 51:61–70
20. Gladue RM, Maxey JE (1994) Microalgal feeds for aquaculture. J Appl Phycol 6:131–141
21. Graham LE, Wilcox LW, Graham J (2009a) Green algae I - Introduction and prasinophyceans. In: Algae, 2nd edn. Pearson, San Francisco pp 353–372
22. Graham LE, Wilcox LW, Graham J (2009b) Haptophytes. In: Algae, 2nd edn. Pearson, San Francisco pp 170–185
23. Hussenot J, Lefebvre S, Brossard N (1998) Open-air treatment of wastewater from land-based marine fish farms in extensive and intensive systems: current technology and future perspectives. Aquat Living Resour 11:297–304
24. Jaouen P, Vandanjon L, Quéméneur F (1999) Stress of microalgal cell suspensions (Tetraselmis suecica) in tangential flow filtration systems: the role of pumps. Bioresour Technol 68:149–154
25. Kaspar HF, Keys EF, King N, Smith KF, Kesarcodi-Watson A, Miller MR (2014) Continuous production of Chaetoceros calcitrans in a system suitable for commercial hatcheries. Aquaculture 420–421:1–9
26. Krichnavaruk S, Loataweesup W, Powtongsook S, Pavasant P (2005) Optimal growth conditions and the cultivation of Chaetoceros calcitrans in airlift photobioreactor. Chem Eng J 105:91–98
27. Laing I (1991) Cultivation of marine unicellular algae. MAFF Lab Leafl Num 67:31
28. Lee SJ, Go S, Jeong GT, Kim SK (2011) Oil production from five marine microalgae for the production of biodiesel. Biotechnol Bioprocess Eng 16:561–566
29. Leupold M, Hindersin S, Gust G, Kerner M, Hanelt D (2013) Influence of mixing and shear stress on Chlorella vulgaris, Scenedesmus obliquus, and Chlamydomonas reinhardtii. J Appl Phycol 25:485–495
30. Loubière K, Olivo E, Bougaran G, Pruvost J, Robert R, Legrand J (2009) A new photobioreactor for continuous microalgal production in hatcheries based on external-loop airlift and swirling flow. Biotechnol Bioeng 102:132–147
31. Michels MHA, Slegers PM, Vermuë MH, Wijffels RH (2014) Effect of biomass concentration on the productivity of Tetraselmis suecica in a pilot-scale tubular photobioreactor using natural sunlight. Algal Res 4:12–18
32. Michels MHA, Van Der Goot AJ, Norsker NH, Wijffels RH (2010) Effects of shear stress on the microalgae Chaetoceros muelleri. Bioprocess Biosyst Eng 33:921–927
33. Moheimani NR (2013) Long-term outdoor growth and lipid productivity of Tetraselmis suecica,Dunaliella tertiolecta and Chlorella sp (Chlorophyta) in bag photobioreactors. J Appl Phycol 25:167–176

34. Moheimani NR, Isdepsky A, Lisec J, Raes E, Borowitzka MA (2011) Coccolithophorid algae culture in closed photobioreactors. Biotechnol Bioeng 108:2078–2087

35. Molina E, Fernández J, Acién FG, Chisti Y (2001) Tubular photobioreactor design for algal cultures. J Biotechnol 92:113–131

36. Molina Grima E, Sánchez Pérez JA, García Camacho F, García Sánchez JL, Acién Fernández FG, Lopéz Alonso D (1994) Outdoor culture of Isochrysis galbana ALII-4 in a closed tubular photobioreactor. J Biotechnol 37:159–166

37. Monkonsit S, Powtongsook S, Pavasant P (2011) Comparison between airlift photobioreactor and bubble column for Skeletonema costatum cultivation. Eng J 15:53–64

38. Muller-Feuga A (2013) Microalgae for aquaculture: the current global situation and future trends. In: Richmond A, Hu Q (eds) Handbook of microalgal culture. Wiley, Chichester, pp 613–627

39. Norsker NH, Barbosa MJ, Vermuë MH, Wijffels RH (2011) Microalgal production - a close look at the economics. Biotechnol Adv 29:24–27

40. Pronker AE, Peene F, Donner S, Wijnhoven S, Geijsen P, Bossier P, Nevejan NM (2015) Hatchery cultivation of the common cockle (Cerastoderma edule L.): From conditioning to grow-out. Aquacult Res 46:302–312

41. Pulz O (2001) Photobioreactors: production systems for phototrophic microorganisms. Appl Microbiol Biotechnol 57:287–293

42. Pulz O, Broneske J, Waldeck P (2013) IGV GmbH experience report, industrial production of microalgae under controlled conditions: innovative prospects. In: Richmond A, Hu Q (eds) Handbook of microalgal culture. Wiley, Chichester, pp 445–460

43. Ramírez-Duque JL, Ramos-Lucumi MA (2011) Hydrodynamic computational evaluation in solar tubular photobioreactors bends with different cross sections. Ciencia. Tecnologia y Futuro 4:59–72

44. Richmond A (2013) Biological principles of mass cultivation of photoautotrophic microalgae. In: Richmond A, Hu Q (eds) Handbook of microalgal culture. Wiley, Chichester, pp 169–204

45. Rotman B, Papermaster BW (1966) Membrane properties of living mammalian cells as studied by enzymatic hydrolysis of fluorogenic esters. Proc Natl Acad Sci USA 55:134–141

46. Sauriau PG, Baud JP (1994) Artificial filament breakage of the diatom Skeletonema costatumintended for mollusc aquaculture. Aquaculture 123:69–81

47. Silva Benavides AM, Torzillo G, Kopecký J, Masojídek J (2013) Productivity and biochemical composition of Phaeodactylum tricornutum (Bacillariophyceae) cultures grown outdoors in tubular photobioreactors and open ponds. Biomass Bioenerg 54:115–122

48. Song X, Throckmorton AL, Wood HG, Antaki JF, Olsen DB (2003) Computational fluid dynamics prediction of blood damage in a centrifugal pump. Artif Organs 27:938–941

49. Stoecker DK, Long A, Suttles SE, Sanford LP (2006) Effect of small-scale shear on grazing and growth of the dinoflagellate Pfiesteria piscicida. Harmful Algae 5:407–418

50. Takache H, Christophe G, Cornet JF, Pruvost J (2010) Experimental and theoretical assessment of maximum productivities for the microalgae Chlamydomonas reinhardtii in two different geometries of photobioreactors. Biotechnol Prog 26:431–440

51. Takiura K, Masuzawa T, Endo S, Wakisaka Y, Tatsumi E, Taenaka Y, Takano H, Yamane T, Nishida M, Asztalos B (1998) Development of design methods of a centrifugal blood pump with in vitro tests, flow visualization, and computational fluid dynamics: Results in hemolysis tests. Artif Organs 22:393–398

52. Tesson B, Genet MJ, Fernandez V, Degand S, Rouxhet PG, Martin-Jézéquel V (2009) Surface chemical composition of diatoms. ChemBioChem 10:2011–2024

53. Tredici MR (2010) Photobiology of microalgae mass cultures: Understanding the tools for the next green revolution. Biofuels 1:143–162

54. Van Bergeijk SA, Salas-Leiton E, Cañavate JP (2010) Low and variable productivity and low efficiency of mass cultures of the haptophyte Isochrysis aff. galbana (T-iso) in outdoor tubular photobioreactors. Aquac Eng 43:14–23

55. Van Riemsdijk LE, Snoeren JPM, Van Der Goot AJ, Boom RM, Hamer RJ (2010) Particle size effects in colloidal gelatin particle suspensions. J Food Eng 101:394–401

56. Vandanjon L, Rossignol N, Jaouen P, Robert JM, Quéméneur F (1999) Effects of shear on two microalgae species. Contribution of pumps and valves in tangential flow filtration systems. Biotechnol Bioeng 63:1–9

57. Vejrazka C, Janssen M, Streefland M, Wijffels RH (2012) Photosynthetic efficiency ofChlamydomonas reinhardtii in attenuated, flashing light. Biotechnol Bioeng 109:2567–2574

58. Wichterle K, Sobolík V, Lutz M, Denk V (1996) Shear rate on centrifugal pump impeller. Chem Eng Sci 51:5227–5228

59. Zhou W, Zhao Z, Lee T, Winoto S (2003) Investigation of flow through centrifugal pump impellers using computational fluid dynamics. Int J Rotating Mach 9:49–61

60. Zhu CJ, Lee YK (1997) Determination of biomass dry weight of marine microalgae. J Appl Phycol 9:189–194

61. Zmora O, Grosse DJ, Zou N, Samocha TM (2013) Microalgae for aquaculture: practical implications. In: Richmond A, Hu Q (eds) Handbook of microalgal culture. Wiley, Chichester, pp 628–652

CHAPTER 6

MICROALGAL CULTURE FOR CHLORELLA SP. USING A HOLLOW FIBER MEMBRANE MODULE

Yoshihiko Sano, Akihiko Horibe, Naoto Haruki and Yugo Okino

Keywords: Microalgal culture, Membrane transport, Hollow fiber membrane, Microalgal growth rate

INTRODUCTION

Several attempts have been successfully made to produce the biomass fuel from the microalgae [1,2]. Some microalgae are capable of producing the biofuel which is carbon neutral to the environment, absorbing light, nutrition and carbon dioxide. Microalgae have a high oil production capacity as compared with common food sources, such as corn and soybean. There is no competition with food industry since microalgae are not a common food source. A report published by Chisti [1] indicates that the production efficiency of microalgae is about ten times as high as palm oil. Moreover, microalgae do not require extensive arable fields to cultivate them and they can be cultivated irrespectively of the season. Various microalgal reactors have been developed in order to cultivate microalgae [2]. Especially, open ponds are the most widely applied in industrial processes [3], which are composed of long single or multiple loop channels placed outdoors. Open ponds are possible to produce microalgae at low cost as compared with other systems by taking account of building and maintenance costs. On the other hand, closed photobioreactors have drawn attention in views of controlling temperature and light intensity at favorable conditions for microalgal species. Moreover, photobioreactors allow us to maintain a high biomass production and reduce the risk of contamination [4]. Various photobioreactors illuminated naturally or artificially lit were proposed such as inside open ponds [5,6], flat-plate [7], airlift [8,9] and tubular [10].

Citation: Sano Y, Horibe A, Haruki N, Okino Y (2016) Microalgal Culture for Chlorella sp. using a Hollow Fiber Membrane Module. J Membra Sci Technol 6: 147. doi:10.4172/2155-9589.1000147.

A number of researchers investigated the effect of important factors on microalgal culture, such as light, minerals, temperature and CO_2 concentration. Several papers [11-13] indicated that light intensity is one of the most important limiting factors since microalgae carry out photosynthesis. Thus, microalgae culture systems are illuminated by artificial light or solar light. On the other hand, the optimal temperature for microalgae cultures is generally between 20 and 25°C, although this may vary with the composition of the culture medium and the microalgal species [14]. Moreover, it is also well known that the carbon dioxide also plays an important role in microalgal culture through investigations for the effect of CO_2 concentration on the microalgal growth rate in photobioreactors [15-17]. Some microalgal production facilities use combustion gases of fossil fuels so as to increase productivity of the microalgae. Chiu et al. [16] introduced the mixed air controlled at high CO_2 concentration 10-15%v/v into microalgal culture of Chlorella sp. Furthermore, Hirata et al. [17] revealed that Chlorella sp. can grow in up to 40% CO_2 concentration at 30°C.

In most conventional microalgae culture, an air bubbling has been adopted to supply the carbon dioxide. However, in the medical field, especially in most blood oxygenators, hollow fiber membranes are widely used to supply gases into the blood. According to Carvalho et al. [18], the membrane makes it possible to diffuse the carbon dioxide into the microalgae culture in lower gas pressures, as no need to counterbalance hydrostatic heads. Photobioreactors using membranes have been reported for culture of mammalian cells [19],Chlamydomonas reinhardtii [20], P. tricornutum in seawater [21] and Anabaena Variabilis [22].

In this study, a hollow fiber culture system has been proposed for supplying the carbon dioxide to the microalgae for Chlorella sp. In order to examine the usefulness of hollow fiber membranes for the microalgal culture, we shall measure the microalgal growth rate for chlorella sp. and the effective mass transfer coefficient of the carbon dioxide through the hollow fiber Silicon tubes exploiting the proposed photobioreactor. Moreover, the microalgal growth rate using hollow fiber membranes are compared against that of a non-membrane photobioreactor. Subsequently, we shall evaluate the effect of the volume flow rate and CO_2 concentration of the feed air through the hollowfibers on the microalgal growth rate.

MATERIALS

Culture of Chlorella sp.

The green alga Chlorella sp., in which the diameter is about 2-10 μm spherical, has been employed in this study. Chlorella sp. has been focused on a potential source of supplements and energy source because they have high photosynthetic efficiency. It is well known that Chlorella sp. can grow in high CO_2 concentration in excess of 10% [16,17]. In this study, the culture medium was adopted Hyponex medium widely used for the plant growth so as to take account of industrial applications of microalgal production.

Previously, Chlorella sp. was cultivated in 100 mL glass beakers illuminated by the white light emitting diode (LED), under the several temperature 20-30°C, the dilution rate of Hyponex medium, the light intensity 600~6000lx at the top of microalgal culture in beakers. It was found that the microalgal growth rate of Chlorella sp. is the highest under the condition at temperature 26°C and the light intensity 2000lx, conducting a series of preliminary experiments in beakers. Moreover, the culture medium concentrations were composed (in mgL-1): TN=24, NH_4=11.6, NO_3=4.2, P_2O_5=40, K_2O=20, MgO=0.02, MnO=0.004, B_2O_3=0.02. Thus, these conditions were applied to cultivate Chlorella sp. in the photobioreactor using the hollow fiber membranes in this study. The microalgal concentration was determined through its correlation to absorbance of the dry weight for chlorella sp. at 690 nm by using an absorption spectrometer (AS ONE, ASV11D).

Hollow fiber devices

Figure 1 schematically shows a hollow fiber membrane device for the microalgal culture, in which thousands of small silicon tubes are arranged so as to allow to the carbon dioxide diffuse into the microalgal culture through the inside of the hollow fibers. Two types of hollow fiber devices, M60-3000, M40-3000 produced by Nagayanagi Kogyo (Japan), were applied for the photobioreactor of chlorella sp. Geometric characteristics of the membrane module are provided by the manufacturer in Table 1. All of hollow fiber membranes are made of Silicon, the thickness of membranes varies 40-60 μm in M40-3000 and M60-3000. Silicon membranes employed in this study can be classified into a non-porous membranes. The diffusion process of CO_2 through this silicon membrane is that the CO_2 molecules are adsorbed on the membrane interface, and then diffuse through the membrane, finally dissolved to the liquid side. Permeability is closely related to the partial pressure. The pressure inside of the hollow fiber can be considered as the atmospheric pressure since the outlet of the hollow fiber membrane module is opened in this study. Therefore, the partial pressure would depend on CO_2 concentration.

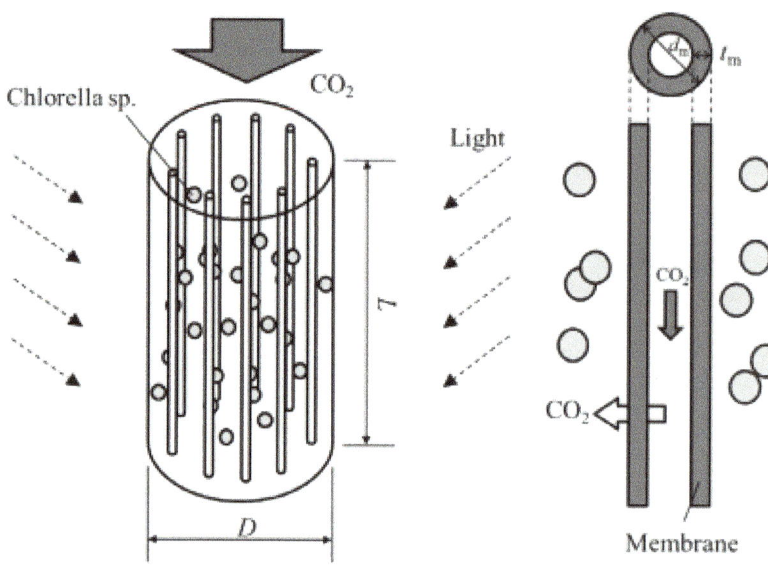

Figure 1: Photobioreactor using a hollow fiber membrane device.

Table 1: Geometric characteristics of hollow fiber memane devices.

	N [-]	d_m [mm]	t_m [μm]	D [mm]	L [mm]
M40-3000	3000	0.25	0.04	31	135
M60-3000	3000	0.32	0.06	31	135

METHODS

Microalgae culture using a hollow fiber membrane device

The experimental set-up used for the hollow fiber culture system is schematically illustrated in Figure 2. The hollow fiber membrane device was placed in the tunnel illuminated by eight LED lights. The light intensity was adjusted to 2000lx at the surface of the cylinder for photobioreactor using a hollow fiber membrane device. The temperature of the mixed air supplied by an air pump was adjusted to 26°C as it passes through a heat exchanger in the isothermal bath. CO_2 concentration of the mixed air was controlled by varying flow rate of air and CO_2 gas from CO_2 tank. The mixed air, which is controlled to the arbitrary temperature and concentration, is fed into the hollow fibers for photobioreactor, and then the carbon dioxide penetrates through the hollow fiber membrane and diffuses into microalgal culture filled outside of hollow fiber membranes. An air from one end of a hollow fiber membrane device can be fed back to this device the using a threeway valve located posteriorly in a hollow fiber membrane device. However, in this fundamental study, we focused on microalgal culture when an exhausted air was not fed back to a hollow fiber membrane device again.

In order to examine the usefulness of the proposed photobioreactor filled with hollow fiber membranes, we shall conduct an experiment on microalgal culture to measure the microalgal concentration of chlorella sp. Then, the microalgal growth rate of the proposed photobioreactor using hollow fiber membranes are compared against that of a nonmembrane photobioreactor, which is a cylindrical acrylic tank as large as hollow fiber membrane devices. Subsequently, the effects of the volume flow rate and CO_2 concentration of the feed mixed air through the hollow fibers on the microalgal culture were investigated to verify the usefulness of the microalgal growth rate in the proposed photobioreactor.

Diffusion of CO_2 through hollow fiber membranes

The overall mass transfer coefficient of the carbon dioxide from the hollow fibers to the pure water filled outside the hollow fibers was estimated by exploiting the hollow fiber culture system illustrated in Figure 2. Pure water was filled in the liquid phase of a hollow fiber membrane device (namely shell side of the membrane), and then pure CO_2 gas is fed into the inner side of hollow fibers. Temperature of liquid and gas phases is controlled 26ºC the same condition as for microalgal culture in this study. Total carbon concentration was measured using a multi-water quality meter (MM60, ToaDKK) with a carbonic acid electrode (CE-2041, ToaDKK).

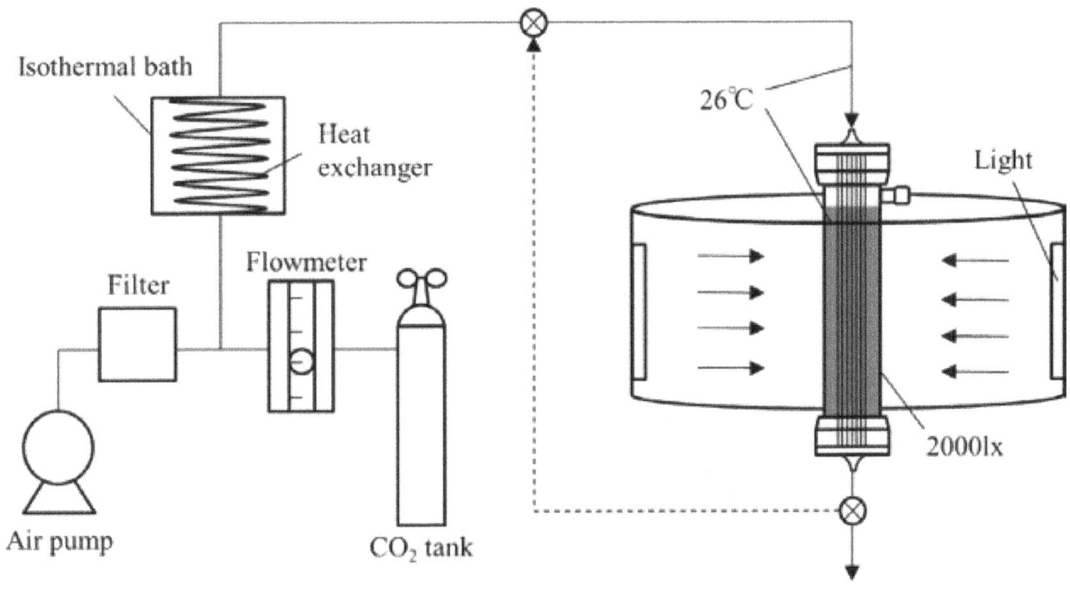

Figure 2: Hollow fiber culture system.

The volumetric mass transfer rate of the carbon dioxide through the membranes may be given in terms of the effective mass transfer coefficient h_{eff} as follows:

$$\frac{1}{V}\int_{A_{lint}} D_l \frac{\partial c}{\partial x_j} n_{lj} dA = a_l h_{eff}\left(\overline{c}^{\,*} - \overline{c}\right) \quad (1)$$

where $n_l j$ is the unit vector pointing outward from the liquid side to membrane side, while A_{lint} and al are the interfacial area and the specific surface area between the liquid and membrane phases. D_l is the carbon dioxide diffusion coefficient in the liquid phase. The bulk average of a certain local variable φ in the liquid phase is defined as such that c is the intrinsic average of the total CO_2 concentration in the liquid phase, while c * is its equilibrium concentration, namely, the intrinsic average of the total CO_2 concentration in the liquid phase when the liquid phase is in equilibrium with the gas phase. V_l is the volume space which the liquid phase occupies, thus the porosity $\varepsilon_l \equiv V_l/V$ is the volume fraction of the liquid space.

$$\overline{\varnothing} \equiv \frac{1}{V_l}\int_{V_l} \varnothing dV \quad (2)$$

Further simplification is possible as one notes the difference in the mass transfer resistances among the gas phases, membrane phases, liquid phases and the surface of the cells for the microalgal culture. According to Goldman et al. [23] and Becker [24], the resistance at the gas-liquid interface is much larger than at the boundary layer of the cells. Thus, the total mass transfer resistance between the liquid and gas phases is defined as,

$$\frac{1}{h_{eff}} = \left(\frac{d_m}{d_m - t_m}\right)\frac{1}{h_g} + \frac{d_m}{D_m} ln\left(\frac{d_m}{d_m - t_m}\right) + \frac{1}{h_l} \quad (3)$$

where h_g and h_l are the interstitial mass transfer coefficient for the gas-membrane interface and the membrane-liquid interface, D_m is the carbon dioxide diffusion coefficient in the membrane phase. We shall consider an unsteady mass balance equation for the carbon dioxide passing through membrane as follows:

$$V_l \frac{d\bar{c}}{dt} = A_l h_{eff} (\bar{c}* - \bar{c}) \quad (4)$$

which, with the initial condition, namely, $\bar{c} = \bar{c}|_{t=0}$ already measured, gives us an exponential time dependent of the total CO_2concentration for the liquid phase:

$$\frac{\bar{c} - \bar{c}|_{t=0}}{\bar{c}* - \bar{c}|_{t=0}} = 1 - exp\left(-\frac{A_l h_{eff}}{V_1} t\right) = 1 - exp\left(-\frac{a_l h_{eff}}{\varepsilon_l} t\right) \quad (5)$$

The effective mass transfer coefficient is adjusted by fitting the measured temporal development of the total CO_2 concentration in the liquid phase to the foregoing equation.

RESULTS

Diffusion of CO_2 through hollow fiber membranes

All measured values of the total CO_2 concentration in the liquid phase using the hollow fiber membrane devices are plotted against the elapse time in Figure 3 for the case of T=26°C, V_l=50 ml, using pure water and pure CO_2 gas. As shown in this figure, the total CO_2concentration in the liquid phase suddenly rises after the feed of CO_2 gas to the hollow fibers, and then achieves the critical value for the equal amount of pure CO_2 gas. The volumetric effective mass transfer coefficient $a_l h_{eff}/\varepsilon_l$ and the effective mass transfer coefficient h_{eff} are tabulated in Table 2, estimated by fitting the measured total CO_2 concentration developments to the CO_2 concentration profile obtained from an equation 5. In the same table, the experimental data reported by Carvolho et al. [18] are tabulated for reference. As can be seen from Table 2, the values of the volumetric effective mass transfer coefficient in present devices are higher than those of HIFO and HIFI provided by Carvolho et al. [18] since specific surface area of present devices is larger than that of HIFO and HIFI. Therefore, present hollow fiber devices can easily be adapted to control the CO_2 concentration in microalgal culture with varying the pressure of inner side of membrane or the number of hollow fibers.

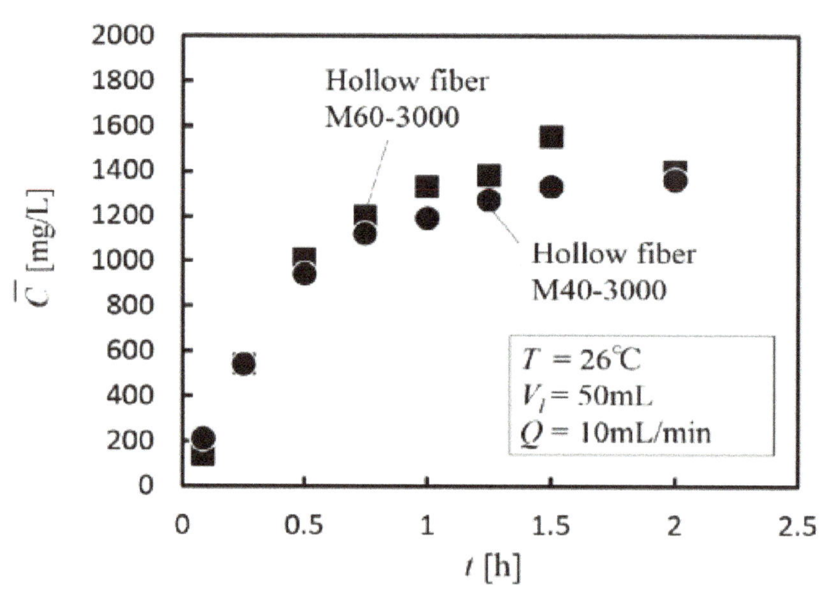

Figure 3: Development of total CO_2 concentration.

Table 2: Effective mass transfer coefficient of the carbon dioxide

Device	$a_l h_{eff}/\varepsilon_l$ [min^{-1}]	A [m^2]	a_l [m^{-1}]	h_{eff} [m/s]	Reference
M40-3000	3.17×10^{-2}	0.20	3160	1.34×10^{-7}	Present study
M60-3000	3.41×10^{-2}	0.30	4050	9.50×10^{-2}	Present study
HIFO	1.48×10^{-2}	1.70	1700	1.45×10^{-2}	Carvolho et al. [18]
HIFI	1.33×10^{-2}	0.14	140	1.59×10^{-2}	Carvolho et al. [18]

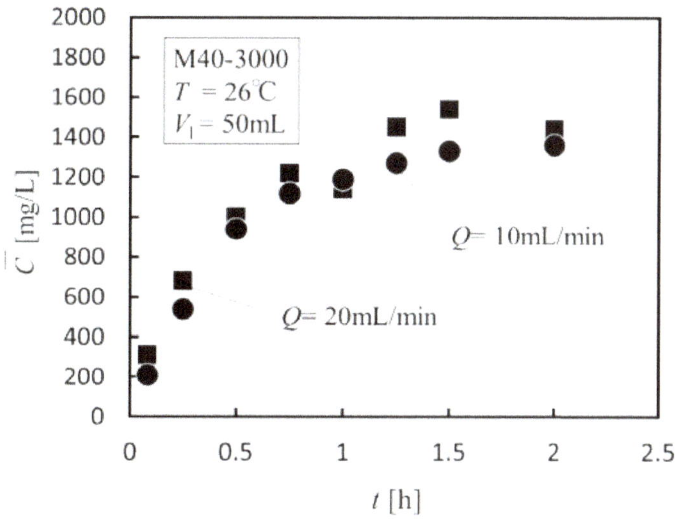

Figure 4: Effect of the volume flow rate of the feed gas on the carbon diffusion through hollow fiber membranes

Effect of the volume flow rate of the feed gas on the carbon diffusion is illustrated in Figure 4. The difference in the two sets of plots is indiscernible, which indicates that the volume flow rate of the feed gas is sufficiently larger than carbon diffusion rate through hollow fiber membranes in this study. Therefore, in order to efficiently supply CO_2 to microalgae culture, a feedback system of an excess gas may be adopted in an industrial application.

Microalgae culture using a hollow fiber membrane device

The picture of the cultured microalgae in the hollow fiber membrane device (M40-3000) after 120 hours is shown in Figure 5, in whichChlorella sp. is cultured in the space formed by the hollow fibers. In Figure 6, the development of the cell concentrations M using hollow fiber membrane device M40-3000 are compared against that using a non-membrane photobioreactor in which mixed air is fed by air bubbling at the bottom. CO_2 concentration of mixed air was set to 20 %v/v. Experimental period was set to 5 days in which cell concentrations may increase linearly, although it is known that the microbial growth can be expressed an exponential function. As can be seen in Figure 6, the microalgal growth rate using hollow fiber membrane device is found to be three times greater than that of a non-membrane photobioreactor. This high cultivation rate of microalgae using hollow fiber membranes would be due to reduction of shear stress associated with supplying gas, since carbon dioxide concentration in the culture medium reached in equilibrium regardless of both supply method after two hours. If adopting the feedback of mixed air, it is possible to significantly reduce the amount of carbon dioxide. These data indicate the usefulness of the hollow fiber module in microalgal culture in terms of cultivation rate and amount of carbon dioxide.

Figure 5: Picture of the cultured microalgae in the hollow fiber membrane device (M40-3000) (T=26°C, Vl=50 mL, Lx=2000lx, Q=25 mL/min, after t=120 h).

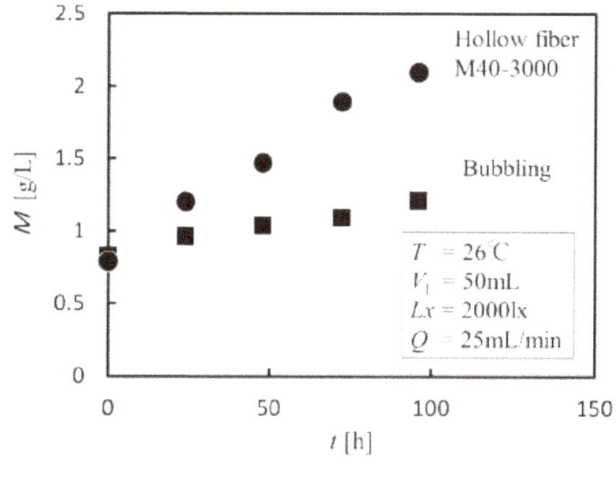

Figure 6: Development of cell concentration for comparison between the hollow fiber device and air bubbling.

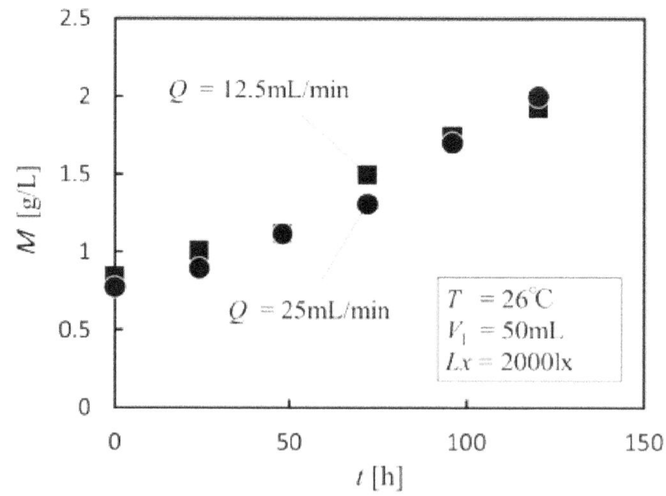

Figure 7: Microalgal growth rate obtained from M40-3000 and M60-3000

The values of the cell concentrations obtained from M40-3000 and M60-3000 are plotted against the elapse time in Figure 7. The microalgal growth rate of M40-3000 is larger than that of M60-3000, despite that the difference in dissolution behavior of carbon dioxide were indiscernible. Arrangement of hollow fibers appears to be important factors for microalgal culture in terms of the light attenuation through the bundle of hollow fibers. Actually, the microbial growth was not observed in the module with 6000 hollow fibers in preliminary experiments.

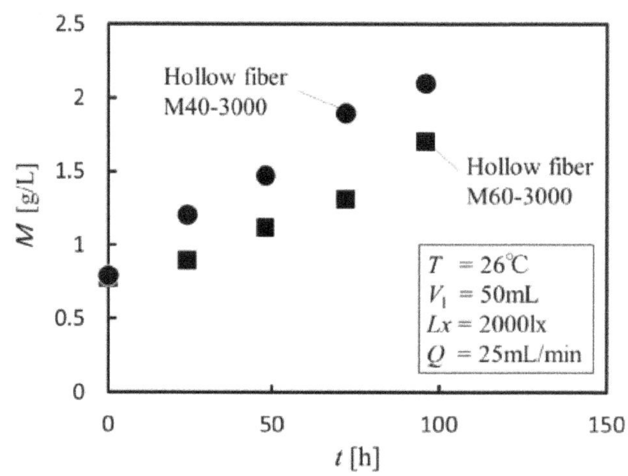

Figure 8: Effect of the volume flow rate of the feed mixed air on the microalgal growth rate.

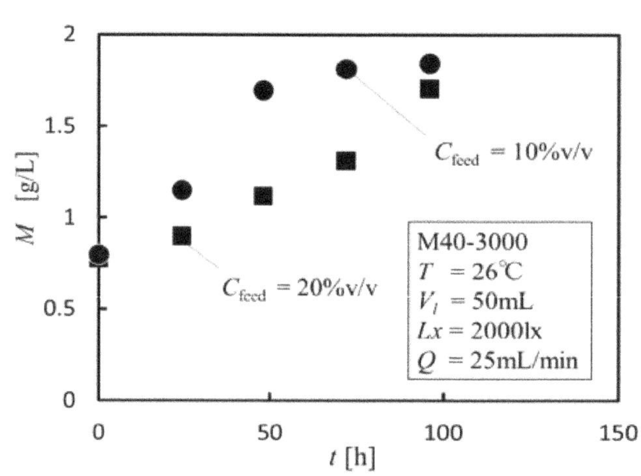

Figure 9: Effect of CO2 concentration of mixed air on the microalgal growth rate.

Effect of the volume flow rate of the feed mixed gas on the microalgal growth rate is illustrated in Figure 8. It is found that the variation of the flow rate does not affect the microalgal growth rate. Moreover, the effect of CO2 concentration of mixed air on the microalgal growth rate is illustrated in Figure 9. There exists an optimum carbon dioxide concentration for each microalgae, and then the appropriate conditions will lead to the further cultivation rate for the microalgal culture using the hollow fiber membrane device. Thus, the hollow fiber membrane is usefulness for the microalgae culture in terms of the microalgal growth rate and the dissolution of the carbon dioxide.

CONCLUSIONS

The hollow fiber culture system was proposed for supplying the carbon dioxide to the microalgae for Chlorella sp. The overall mass transfer coefficient of the carbon dioxide between pure CO_2 gas inside the hollow fibers and pure water filled outside the hollow fibers was estimated by exploiting the hollow fiber culture system. The microalgal growth rate using hollow fiber membranes is found to be three times greater than that of a non-membrane photobioreactor. Moreover, the appropriate conditions of CO_2 concentration and the number of hollow fibers will lead to the further growth rate for the microalgal culture using the hollow fiber membrane device. The present study clearly indicates that the hollow fiber membrane is usefulness for the microalgae culture in terms of the microalgal growth rate and the dissolution of the carbon dioxide.

ACKNOWLEDGEMENTS

The work has been partially supported by Okayama Foundation for Science and Technology.

REFERENCES

1. Chisti Y (2007) Biodiesel from microalgae. BiotechnolAdv 25: 294-306.
2. Carvalho AP, Meireles LA, Malcata FX (2006) Microalgal reactors: a review of enclosed system designs and performances. BiotechnolProg 22: 1490-1506.
3. Chaumont D (1993) Biotechnology of algal biomass production: a review of systems for outdoor mass culture. J ApplPhycol 5: 593-604.
4. Ugwu CU, Aoyagi H, Uchiyama H (2008) Photobioreactors for mass cultivation of algae. BioresourTechnol 99: 4021-4028.
5. Hase R, Oikawa H, Sasao C, Morita M, Watanabe Y (2000) Photosynthetic production of microalgal biomass in a raceway system under greenhouse conditions in Sendai city. J BiosciBioeng 89: 157-163.
6. Tredici MR, Materassi R (1992) From open ponds to vertical alveolar panels: the Italian experience in the development of reactors for the mass cultivation of photoautotrophic microorganisms. J ApplPhycol 4: 221-231.
7. Hu Q, Guterman H, Richmond A (1996) A flat inclined modular photobioreactor for outdoor mass cultivation of photoautotrophs. BiotechnolBioeng 51: 51-60.
8. Rubio FC, Ferna´ndezFGA, Sa´nchezPe´rez JA, Garci ´a Camacho F, Rubio FCM, et al. (1999) Prediction of dissolved oxygen and carbon dioxide concentration profiles in tubular photobioreactors for microalgal culture BiotechnolBioeng 62: 71-86.
9. Harker M, Tsavalos AJ, Young AJ (1996) Autotrophic growth and carotenoid production of Haematococcuspluvialisina30literairliftphotobioreactor. J Ferment Bioeng 82: 113-118.
10. Ugwu CU, Ogbonna JC, Tanaka H (2002) Improvement of mass transfercharacteristics and productivities of inclined tubular photobioreactors by installation of internal static mixers. ApplMicrobiolBiotechnol 58: 600-607.
11. Kcmond A, Lichtenberg E, Stahl B, Vonshak A (1990) Quantitative assessment of the major limitations on the productivity of Spialinaplatensis in open raceways. J ApplPsychol2: 195-206.
12. Oncel SS,Akpolat O (2006)An integrated photobioreactor system for the production of Spirulinaplatensis. Biotechnology 5: 365-372.
13. Kommareddy AR, Anderson GA(2003) Study of Light as a Parameter in the Growth of Algae in a Photo-Bio Reactor (PBR), ASAE Paper No. 034057, ASAE, St. Joseph, Michigan.
14. BitogJP, Lee IB, Lee CG, Kim KS, Hwang HS, et al. (2011) Application of computational fluid dynamics for modeling and designing photobioreactors for microalge production: A reviw. J Comp ElectAgri 76:131-147.
15. Rahaman MSA, Cheng LH, Xu XH, Zhang L, Chen HL (2011) A review of carbon dioxide capture and utilization by membrane integrated microalgal cultivation processes. J Renew Sustain EnerRev 15: 4002-4012
16. Chiu SY, Kao CY, Chen CH, Kuan TC, Ong SC, et al. (2008) Reduction of CO2 by a high-density culture of Chlorella sp. in a semicontinuousphotobioreactor. BioresourTechnol 99: 3389-3396.
17. Hirata S, Hayashitani M, Taya M, Setsuji T(1996) Carbon dioxide fixation in batch culture of Chlorella sp. using a photobioreactor with a sunlight-collection device. J Ferment Bioeng81:470-2.
18. Carvalho AP, Malcata FX (2001) Transfer of carbon dioxide within cultures of microalgae: plain bubbling versus hollow-fiber modules. BiotechnolProg 17: 265-272.
19. Gallagher SL, Tharakan JT, Chau PC (1987) Anintercalatedspiral wound hollow fiber bioreactor for the culture of mammalian cells. Biotechnol Tech 1: 91-96.
20. Chen F, Johns MR (1995)A strategy for high cell density culture of heterotrophic microalgae with inhibitory substrates. J ApplPhycol7: 43-46.
21. Marsot P,Cembella AD,Mouhri K (1992)Croissance de la biomasse azote´e du Phaeodactylumtricornutum (Bacillariophyceae) en culture discontinue dialysanteet non-dialysante. Can JMicrobiol 38: 945-952.
22. Markov SA,Bazin MJ, Hall DO (1995) Hydrogen photoproduction and carbon dioxide uptake by

immobilized Anabaena Variabilis in a hollow-fiber photobioreactor. Enzyme MicrobTechnol17: 306-310.

23. Goldman JC, Dennett MR, Riley CB (1981) Inorganic Carbon Sources and Biomass Regulation in Intensive Microalgal Cultures. BiotechnolBioeng 23: 995-1014.

24. Becker E W (1994) Large-Scale CultivationIn Microalgae: Biotechnology and Microbiology, Becker E W, Cambridge University Press: UK pp: 63-171.

CHAPTER 7

IMPACT OF LIGHT COLOR ON PHOTOBIOREACTOR PRODUCTIVITY

Tim de Mooij, Guus de Vries, Christos Latsos, René H. Wijffels, Marcel Janssen

Keywords: Chlamydomonas reinhardtii, Areal biomass productivity, Photosynthetic efficiency, Color of light, Spectral light composition

INTRODUCTION

Microalgae are an attractive source for biofuels and bulk chemicals due to their high photosynthetic efficiency (PE). At low light intensities, microalgae can achieve values up to 80% of the theoretical maximum PE of 0.125 mol CO_2 fixed per mol photons absorbed[1], [2], [3] and [4]. However, maximum PE values, as measured under low light conditions, will never be realized in microalgae mass cultures exposed to direct sunlight. The reason is the inherent nature of light. Unlike most chemical substances, light energy cannot be dissolved in the culture medium. Therefore there will always be a steep light gradient proceeding from a high level of sunlight to virtual darkness. Because of the high incident light intensity it is practically impossible to obtain the maximum light use efficiency in microalgae mass cultures. In a high density microalgae culture, most sunlight energy is absorbed in a small volume fraction of the photobioreactor on the light-exposed side. In this volume fraction, cells are coerced to absorb more light energy than the amount that can be converted to biochemical energy by their photosynthetic machinery. This leads to oversaturation and, consequently, waste of sunlight energy through heat dissipation [5]. The result is a PE that is dramatically lower than that which can be obtained under low light conditions [6].

As the photosynthetic machinery is easily oversaturated, the key to optimization is to reduce the amount of light energy absorbed per photosynthetic unit. This can be achieved by proper reactor design using the light dilution principle [7], [8] and [9]. However, high material costs limit its application. Considering efficient light utilization is a bottleneck of biological nature, modifications to the light harvesting complex of microalgae would possibly be more effective [10].

In our previous study [11], we evaluated the areal biomass productivity of four different antenna size mutants [12], [13], [14], [15] and [16] under simulated mass culture conditions. These mutants were expected to show improved productivity because of their lower pigment content compared to the wild-type thereby assuring less light absorption per cell. However, none of the studied mutants performed better than the wild-type, possibly due to impaired photo protection mechanisms induced by the antenna complex alterations. Another explanation is the inadvertent side effects caused by the actual process of genetic engineering resulting in reduced fitness of the strains. These genetic side effects will have to be eliminated to fully benefit from the potential of antenna size reduction by genetic engineering.

In order to demonstrate the potential of antenna size reduction on an experimental scale, light absorption can also be minimized by shifting the wavelength of the emitted light to the weakly absorbed green region. When supplying narrow-beam LED light (small peak width) at high light intensities, it is the wavelength specific absorption capacity of the algae that determines the extent of photosystem saturation and, consequently, the light use efficiency. Although there is a strong and prevalent agreement that red and blue light are optimal for algal cultivation because of the corresponding peaks in the algal absorption spectrum [17], [18], [19], [20], [21], [22] and [23], the opposite could possibly be true for high density cultures. In dilute cultures, not all incoming light energy is absorbed and, therefore, light absorption is the limiting factor for maximizing

Citation: Tim de Mooij, Guus de Vries, Christos Latsos, René H. Wijffels, Marcel Janssen, Impact of light color on photobioreactor productivity, Algal Research, Volume 15, April 2016, Pages 32-42, ISSN 2211-9264, http://dx.doi.org/10.1016/j.algal.2016.01.015.

productivity. On the contrary, high density cultures are characterized by the fact that all incoming light is absorbed anyway by direct or indirect control of biomass concentration via chemostat or turbidostat operation [11] and [24]. Therefore, since total light absorption is already guaranteed in mass cultures by applying a high biomass density, productivity is limited by the efficiency at which the absorbed light is converted to biochemical energy, and not by the efficiency of light absorption. We hypothesize that in high density mass cultures the utilization of weakly absorbed light (green-yellow) maximizes productivity while strongly absorbed light (blue and red) causes more oversaturation and is suboptimal for mass culture cultivation. Indeed, the action spectra of microalgal photosynthesis as determined by Emerson and Lewis [2] and by Tanada [25] indicate that green-yellow light (500–600 nm) is used at high efficiency once it is absorbed.

A microalgal growth model [26] was employed to estimate photobioreactor productivity as a function of light intensity and the spectral composition of light. The model takes into account the change of the spectral composition with increasing reactor depth because of preferential light absorption by microalgae. For example, white light becomes greener. The model allows calculation of the optimal biomass concentration (Cx) leading to maximal productivity. For each color of light, as well as sunlight, the areal biomass productivity, the biomass specific growth rate, and the optimal biomass concentration were computed. Next to overall reactor productivity, this model provides insight into the light use efficiency at different positions in the reactor and how this depends on light color.

In this study, we aim to deliver a proof of concept that the biomass specific light absorption rate determines the volumetric biomass productivity in microalgae mass cultures. We do not consider microalgae cultivation using artificial light as a viable process for producing bulk chemicals as the associated energy costs are high whereas sunlight is at no cost and abundantly available [27]. In this study, we employ artificial light only as a tool to generate different specific light absorption rates by spectral tuning. We measured the areal biomass productivity of cultures exposed to warm white, orange-red (peak 642 nm), deep-red (peak 661 nm), blue (peak 458 nm), and yellow light (peak 596 nm). The area reflects the illuminated surface area of the photobioreactor. Cultivation took place in continuously operated bench-scale flat plate photobioreactors. For each color of light, the applied light intensity was 1500 µmol photons m^{-2} s^{-1}. The biomass concentration was controlled at a fixed level that was high enough to absorb all incoming light energy. By comparing the biomass specific light absorption rate with the measured productivity of cultures exposed to different colored lights, insight was obtained into the importance of minimizing light absorption per cell to maximize productivity.

MATERIALS AND METHODS

Organisms and medium

Chlamydomonas reinhardtii CC-1690 was obtained from the Chlamydomonas Resource Center (University of Minnesota). The algae were cultivated in a filter sterilized (pore size 0.2 µm) medium (Sueoka high salt, HS) with the following composition (in g L^{-1}): urea, 0.99; KH_2PO_4, 0.706; K_2HPO_4, 1.465; $MgSO_4 \cdot 7H_2O$, 0.560; $CaCl_2 \cdot 2H_2O$, 0.114 and 20 mL L^{-1} of a 100 times concentrated Hutner›s trace elements solution [28]. The cultures were pre-cultivated in 250 mL shake flasks containing 100 mL of medium at pH 6.7 and at 25 °C at a light intensity of 200–300 µmol photons m^{-2} s^{-1}.

Photobioreactor setup and operation

The microalgae were continuously cultivated in flat-panel air-lift photobioreactors (Algaemist, Technical Development Studio, Wageningen University, the Netherlands) with a working volume of 0.4 L, an optical depth of 14 mm, and an illuminated area of 0.028 m^2 (See Fig. 1 for a schematic overview). The reactors were equipped with a black cover on the backside to prevent exposure to ambient light. Warm white light was provided by Bridgelux LED lamps (BXRA W1200, Bridgelux, USA) which are integrated in the Algaemist system. Other colors of lights were provided using 20 × 20 cm, SL 3500 LED panels of Photon Systems Instruments (PSI, Czech Republic). The following colors were used: blue (peak 458 nm, spectral half-width 20 nm); orange-red (642 nm; 20 nm); deep red (661 nm; 20 nm); and yellow (596 nm; 60 nm). The yellow light source was equipped with an optical low-pass filter (630 nm) to cut of red light. Unless explicitly stated otherwise, all cultures grown in yellow light described in this paper were supplemented with ± 50 µmol photons m^{-2} s^{-1} of blue light. The rationale behind this procedure is clarified in the results section of this paper. In Fig. 2 the emission spectra of all light sources are shown and these are supplemented with the solar light spectrum and the wavelength specific absorption cross section of C.reinhardtii. Please refer to Table S2–S6 of the supplementary material for a light intensity distribution across the illuminated reactor surface, which is provided for each light source. Reactor temperature was maintained at 25 °C, and the pH was kept at 6.7 (± 0.1) by means of on-demand CO_2supply. The medium that was fed to the reactor had a pH of 7.0 and to maintain the setpoint of 6.7 in the reactor, CO_2 supply rate was such that both CO_2 and HCO_3^- were present at concentrations of at least a magnitude higher than the saturation constant of Rubisco for CO_2 and HCO_3^-. The reactors were operated in turbidostat mode to ensure a constant light regime; a light sensor measures the transmission through the reactor and if light transmission was below the setpoint, the culture was automatically diluted with fresh medium employing a peristaltic pump. Further details of the photobioreactor setup and its operation are provided in de Mooij et al. [11], with the exception that the gas stream of di-nitrogen was, at all times, 200 mL min^{-1} (± 20).

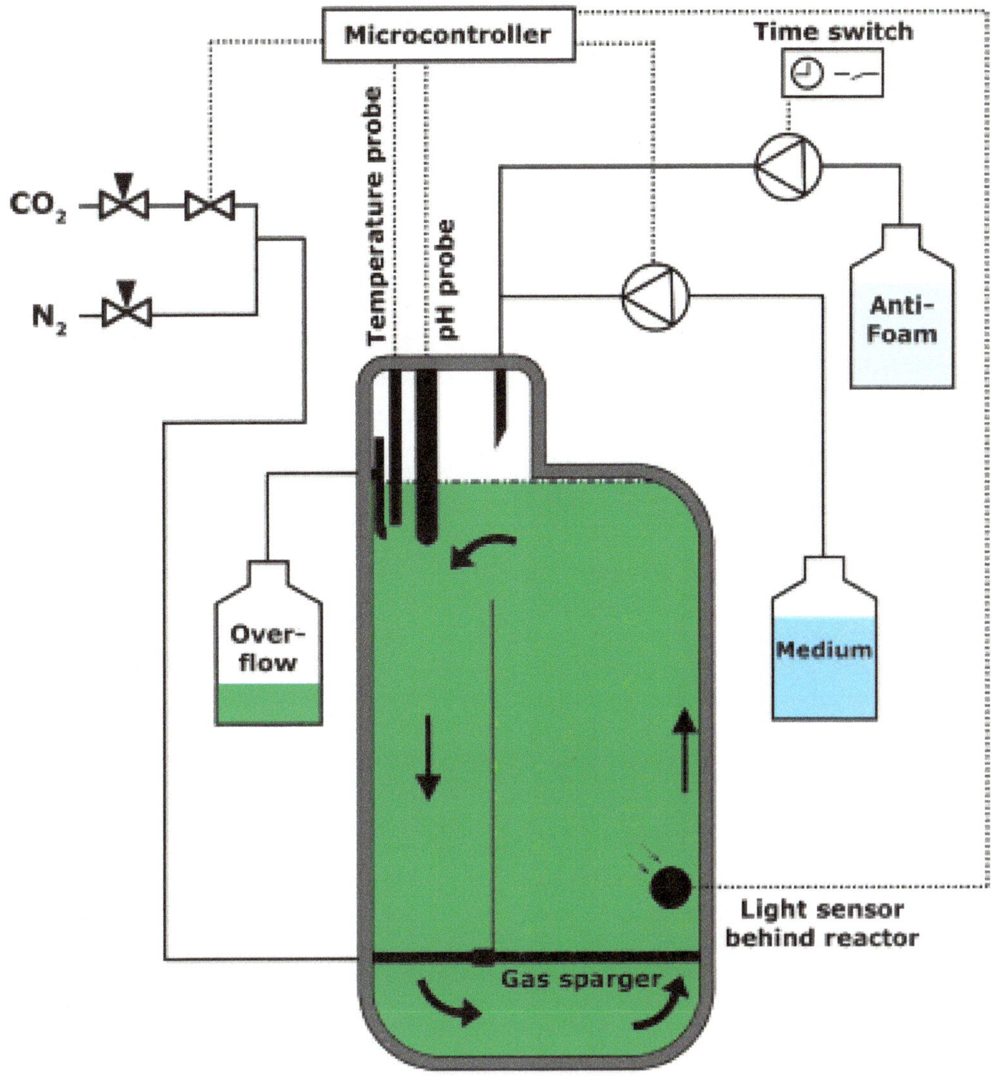

Figure 1. Schematic overview of the experimental setup.

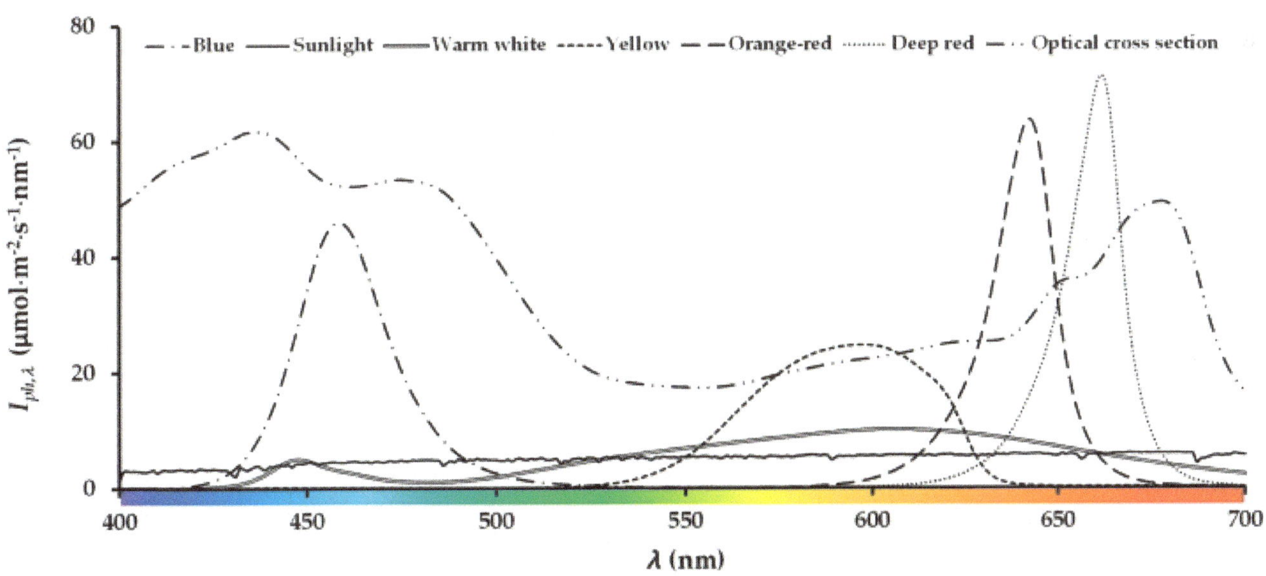

Figure 2. Emission spectra of the PSI LED panels (blue, orange-red, deep red and yellow) and the Algaemist Bridgelux LED panel (warm white). The curves illustrate a photon flux density of 1500 µmol photons m^{-2} s^{-1} when integrated across the entire PAR range (400–700 nm). This is the light intensity employed in the experiments. In addition, the solar light spectrum and the absorption cross section ($a_{x,\lambda}$) ofChlamydomonas reinhardtii (unitless) are plotted. Numerical values for $a_{x,\lambda}$ and all emission spectra can be found in Tables S7–S13 of the supplementary material. The ASTM G173-03 standard solar spectrum was obtained from the website of the National Renewable Energy Laboratory (NREL, USA) [29].

Analyses

Biomass dry weight concentration

To determine the biomass dry weight content, the culture broth was passed through glass fiber filters as described by Kliphuis et al. [30] and, subsequently, the mass difference between the dried empty filters and the dried filters with microalgae was recorded. All measurements on an individual sample were performed in triplicate.

Absorption cross section

Light absorption was measured in a double beam spectrophotometer (UV-2600, Shimadzu, Japan) equipped with an integrating sphere (ISR-2600). A reactor sample was transferred to a cuvette (100.099-OS, Hellma, Germany) with a short light path of 2 mm. The same reactor sample was analyzed for its dry weight content. This allowed normalization of the absorption cross section, resulting in a biomass specific absorption cross section. Additional details of this protocol have been described by de Mooij et al.[11]. Only samples from the cultures grown under yellow and warm white light were diluted with medium (1:1) because of the higher biomass density. All other samples were not diluted.

Maximum photosystem II quantum yield

The maximum photosystem II (PSII) quantum yield (Fv/Fm) was measured from samples withdrawn directly from the reactor. Samples were diluted to obtain a biomass density of approximately 25 mg L^{-1}. Following dark adaptation for ten minutes, fluorescence of the microalgae samples was measured with a technique based on pulse amplitude modulated (PAM) fluorometry (AquaPen-C AP-C 100, PSI, Czech Republic). According to the manufacturer's manual, the actinic light intensity was 270 μmol photons m^{-2} s^{-1}; the saturating light intensity was 2430 μmol photons m^{-2} s^{-1}; and the measuring light intensity was 0.03 μmol photons m^{-2} s^{-1}. The excitation light wavelength was 620 nm.Fv/Fm is the ratio of variable fluorescence to maximal fluorescence (Eq. (1)). In this equation, Fo is the dark-adapted minimum level of fluorescence. Fm is the maximum level of fluorescence, measured after a short light pulse of high intensity. Fv (variable fluorescence) is defined as the difference between Fm and Fo, and represents the difference between fluorescence intensities with closed and open reaction centers.

$$F_v/F_m = (F_m - F_0)/F_m \quad (1)$$

RESULTS

Model estimation of biomass productivity and biomass yield on light for different colors of light

Photobioreactor productivity was estimated employing a microalgae growth model. The model predicts photosynthetic rates at every location in the reactor based on the local light intensity. The light intensity is calculated for each wavelength at every point in the reactor to account for preferential light absorption by microalgae and the resulting change in spectrum composition. A description of the model and a list of the model parameters used are located in appendix A. The following model calculations are based on parameters used in the experiments performed: an ingoing light intensity of 1500 μmol photons m^{-2} s^{-1}, a reactor depth of 14 mm, and the absorption cross section of a continuous mass culture grown under warm white light in turbidostat mode (see Fig. 2). This mass culture was characterized by complete absorption of the incident light and with an outgoing light intensity of 10 μmol photons m^{-2} s^{-1}. These conditions were chosen to maximize reactor productivity as for C. reinhardtii, 10 μmol photons m^{-2} $s^{-1}$was found to be the photosynthetic compensation point [6] and [31], where the net photosynthesis rate is equal to zero. At higher biomass concentrations, dark zones are created where cell maintenance is a dominant process, which reduces the reactor productivity. At lower biomass concentrations, light passes the culture without being absorbed and without contributing to the overall productivity. Therefore, in this study light color was studied at biomass concentrations that were optimized for each light color. Running the cultivations at the same fixed biomass concentration would not be a fair comparison as it does not allow us to use the full potential of each color of light. Biomass concentration optimization is essential in a mass culture setup to maximize biomass productivity.

Using the model, we estimated the optimal biomass concentration (Cx) for each color of light to maximize the areal productivity, assuming the bioreactor is operated at a constant biomass density (turbidostat) and constant light intensity. As can be seen in Fig. 3, strongly absorbed light in colors such as blue and deep red result in low biomass concentrations (1.0 g L^{-1} and 1.3 g L^{-1}, respectively) while a weakly absorbed light in a color such as yellow gives a biomass concentration of 2.8 g L^{-1}. By combining the local light absorption rate with the estimated biomass concentrations for all light sources, we calculated the biomass specific light absorption (qph, mol_{ph} mol_x^{-1} s^{-1}) at each position inside the photobioreactor (Fig. 4A). Under the described conditions, blue light results in the highest qph while the cultures grown under yellow light absorb the least light energy per unit of biomass. Fig. 4B illustrates the local specific growth rate (μ) as a function of the local light intensity in the reactor. Each light intensity corresponds to a certain location in the reactor. A culture exposed to blue light grows at μmax if the light intensity is higher than approximately 100 μmol photon s m^{-2} s^{-1} while a culture exposed to yellow light requires about 500 μmol photons m^{-2} s^{-1} to support maximum growth. Although it shows the high sensitivity for blue light, this does not imply that the reactor productivity of a mass culture will be higher under blue light. The reason is that when grown under blue light, only low biomass concentrations can be supported and that the light use efficiency is low which limits the volumetric productivity. The spatially averaged μ values can be found inFig. 3. A table with all the estimated model values can be found in appendix B.

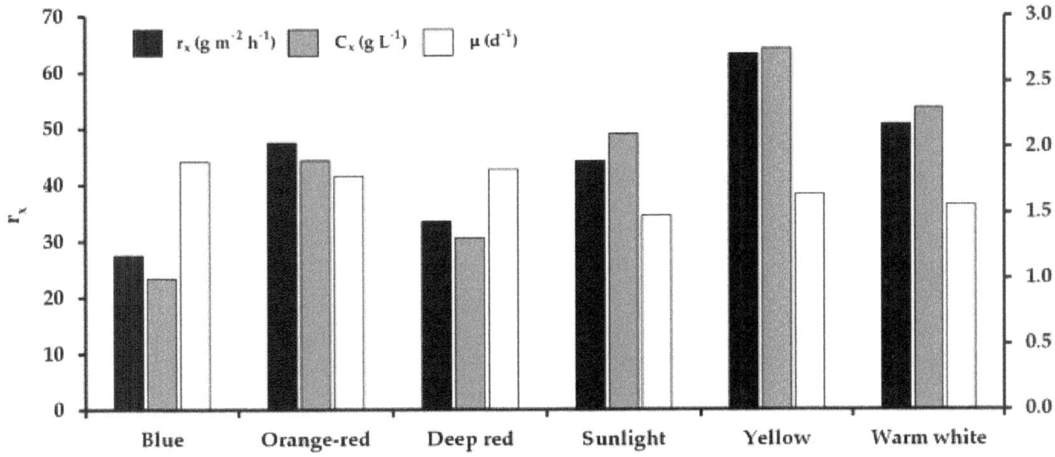

Figure 3. Model estimation of areal biomass productivity (rx), biomass concentration (Cx), and biomass specific growth rate (μ) at $I_{ph,in}$ = 150 0 μmol photons m^{-2} s^{-1} for different light colors.

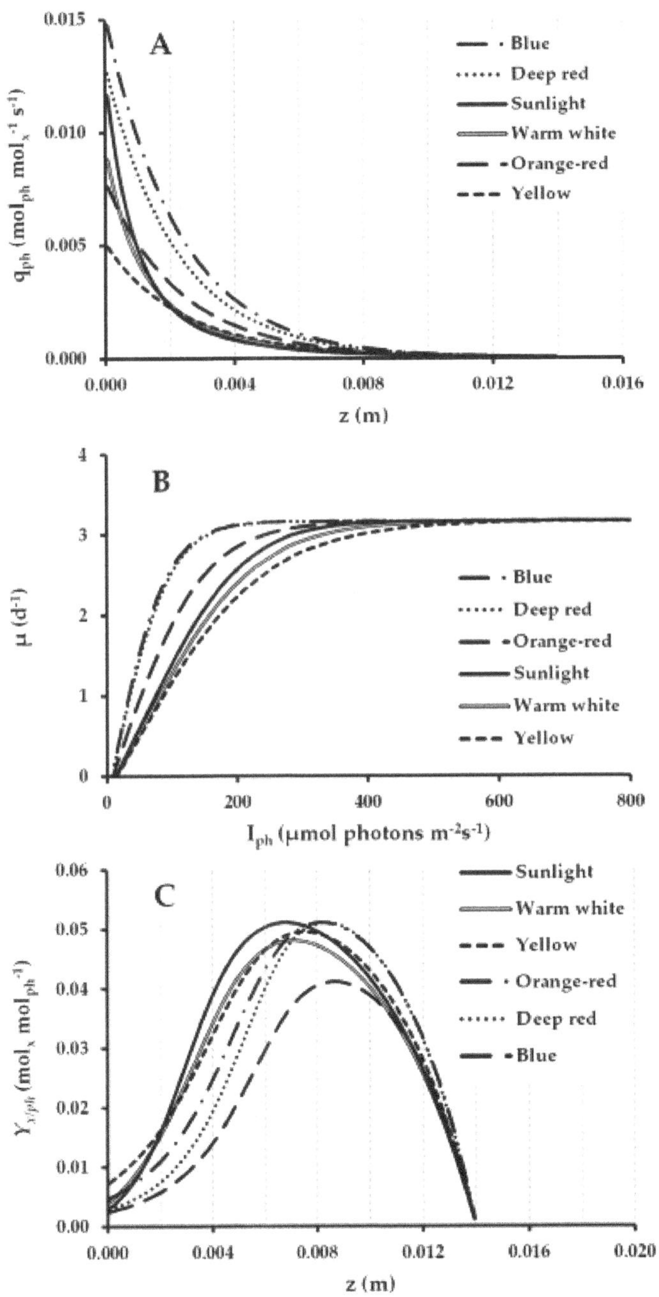

Figure 4. (A) Model estimation of the biomass specific light absorption rate (qph) as a function of location z in the reactor. (B) Estimation of the local biomass specific growth rate (μ) as a function of the local light intensity. The spatially averaged μ can be found in Fig. 3. At light intensities exceeding ± 500 μmol photons m^{-2} s^{-1}, all colors of light generate photosystem saturation, resulting in maximum growth rate. Blue and deep red lights begin to saturate at ± 100 μmol photons m^{-2} s^{-1}. (C) Biomass yield on light energy (Yx/ph) as a function of location z in the reactor. Of major interest is the reactor volume near the light exposed reactor surface (depth: 0–0.002m) as most light energy is absorbed in this region.

Fig. 4C depicts the local biomass yield on light energy (mol$_x$ mol$_{ph}^{-1}$) as a function of reactor depth. In general, it can be observed that, in a mass culture, weakly absorbed light results in higher yields than strongly absorbed light. In the deeper, darker part of the reactor, the biomass yield on light energy (Yx/ph) decreases for all colors of light as cell maintenance becomes a significant factor relative to the photosynthetic activity. To maximize productivity, the biomass concentration was chosen in such a way that at the back of the reactor the local biomass yield on light is zero. This is at the photosynthetic compensation point. Stated differently, at every position in the reactor there is a positive contribution to the reactor productivity.

The biomass productivity expressed per unit of illuminated surface area (rx) is presented in Fig. 3. The highest productivity (63 g m^{-2} d^{-1}) is predicted for cultures exposed to yellow light while the strongly absorbed blue light is expected to result in a productivity of 27 g m^{-2} d^{-1}. Warm white light, whose spectrum contains a significant fraction of weakly absorbed light (e.g., green and yellow), results in productivity as high as 51 g m^{-2} d^{-1}. Deep red light is estimated to result in lower productivity than orange-red light. This is explained by the fact that the deep red light spectrum is overlapping the chlorophyll a absorption peak (see Fig. 2) while the orange-red peak is located in a less absorbing region of the algae absorption spectrum.

The light spectrum changes with increasing reactor depth because of the preferential absorption of blue and red light by green microalgae. The light becomes greener as the red and blue fractions are rapidly absorbed. As a consequence, warm white LED light and sunlight are quickly converted into green light with increasing culture depth. As it is evident from Fig. 4A and C, the color of light influences the local qph and Yx/ph primarily in the first 2 mm of the culture. In high light conditions at the surface of the reactor, the highest Yx,ph is observed employing yellow light. However, at a depth ≥ 2 mm, higher yields can be obtained with warm white light and sunlight. Considering that 53% of the incoming light energy is absorbed within the first 2 mm, the photosynthetic efficiency in this surface layer has a dominant effect on reactor productivity.

Experimental areal biomass productivity for different colors of light

Based on the model predictions, a large difference in productivity can be expected between weakly and strongly absorbed light colors. Except for sunlight, we performed reactor experiments with all of the colors of lights mentioned. Areal biomass productivity was measured at an ingoing light intensity (Iph in) of 1500 μmol photons m^{-2} s^{-1}. The outgoing light intensity (Iph out) was maintained at 10 μmol photons m^{-2} s^{-1} by turbidostat control. The cultures exposed to yellow light were subjected to 1450 μmol photons m^{-2} s^{-1} and supplemented with 50 μmol photons m^{-2} s^{-1} of blue light, as will be discussed in detail later. In Fig. 5, the areal biomass productivity, biomass concentration, and the dilution rate are presented. The highest productivity was obtained employing yellow light (54 g m^{-2} d^{-1} ± 5). A slightly lower value was found for warm white light (50 g m^{-2} d^{-1} ± 2). Cultures exposed to blue, orange-red, and deep red all yielded a productivity of approximately 29 g m^{-2} d^{-1}. For the exact values of light intensity and obtained experimental data of each experiment, please refer to Table S1 of the supplementary material.

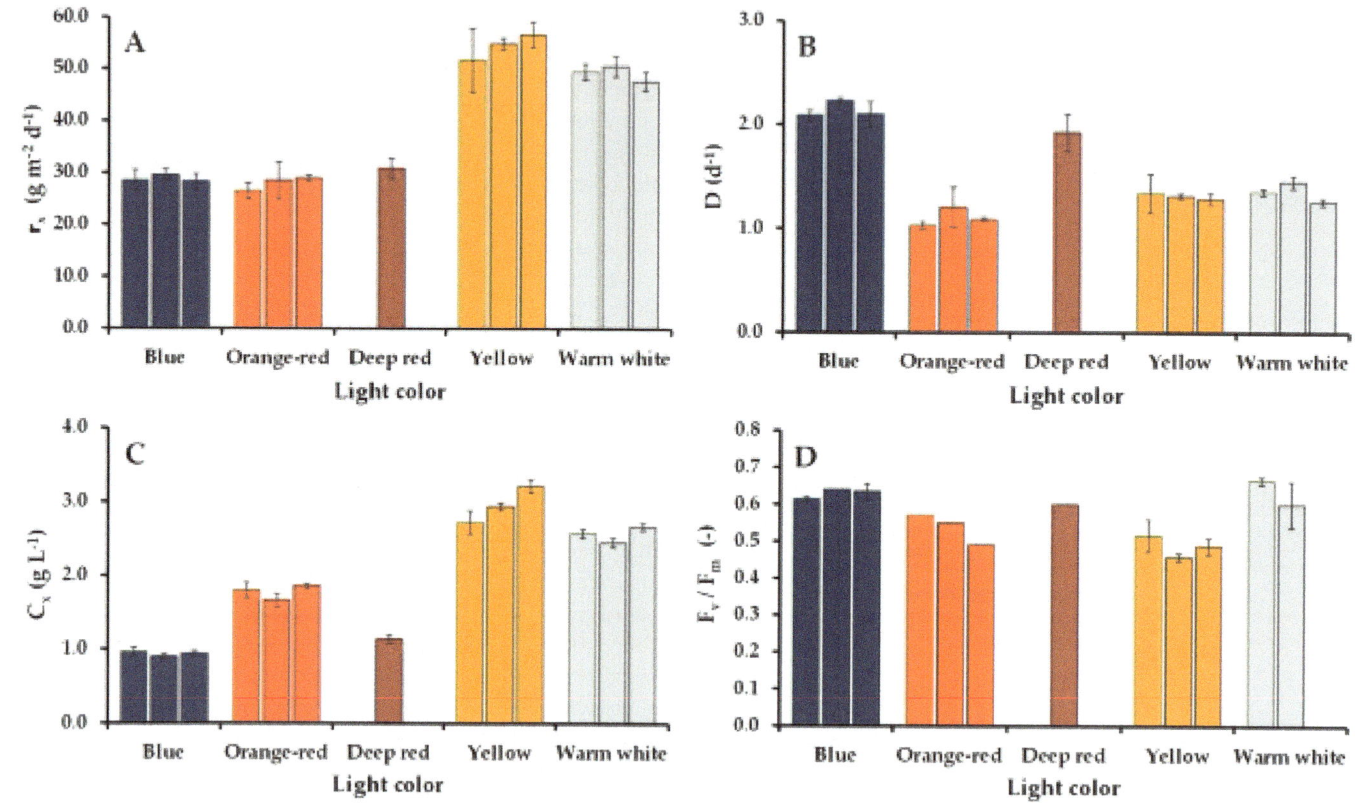

Figure 5. (A) Areal biomass productivity (rx). (B) Dilution rate (D = μ). (C) Biomass concentration (Cx). (D) Dark adapted maximum Fv/Fm. The cultures were continuously operated in turbidostat mode atIph in = 1500 μmol photons m^{-2} s^{-1}, Iph ou_t = 10 μmol photons m^{-2} s^{-1}. Each bar in (A), (B), and (C) represents the average ± SD of n ≥ 4 data points that were measured on a daily basis within one reactor experiment.

The highest biomass concentration was measured for cultures exposed to yellow light (2.96 g L^{-1} ± 0.12) and the lowest for cultures grown under blue light (0.92 g L^{-1} ± 0.05). Since all cultures were turbidostat controlled and were as such, forced to absorb 1490 μmol photons m^{-2} s^{-1}, the biomass concentration presented in Fig. 5B inherently demonstrates the ability of the algal biomass to absorb light of different colors. A low biomass concentration corresponds to a relatively high biomass specific light absorption rate which was accompanied by a high rate of energy dissipation.

In our experiments, the specific growth rate μ equals the reactor dilution rate D, as can be deduced from the biomass balance over the photobioreactor [11]. As expected, the low biomass concentration in cultures grown under blue light is accompanied by a high dilution rate (2.2 d^{-1} ± 0.04) because cells cultivated under blue light will be light saturated at relatively low light intensities. Compared to other light colors, the light intensity will be high enough to saturate the cells in a larger volume fraction of the reactor. The result is a higher spatially averaged biomass specific growth rate. However, since volumetric productivity is the product of biomass concentration and dilution rate, the low biomass concentration limits productivity. It is remarkable that, even though the culture exposed to orange-red exhibited a lower biomass concentration than cultures grown under warm white, which indicated a higher biomass specific light absorption, this lower biomass concentration was accompanied by a lower dilution rate.

The maximum Fv/Fm value of dark adapted samples (see Fig. 5D) withdrawn from the reactor represents photosystem

II quantum efficiency and is an indicator of photoinhibition or down-regulation of photosystem II activity [32], [33] and [34]. The highest values were obtained for the cultures exposed to blue (0.63 ± 0.02) and white (0.64 ± 0.05) light. The lowest Fv/Fm value was obtained for the cultures exposed to yellow light (0.50 ± 0.04). Cultures exposed to orange-red light also demonstrated reduced values (0.53 ± 0.06), indicating that photosystems did not function at full capacity.

Cultivation under both orange-red and deep-red light was difficult. Several experiments at an Iph in of 1500 μmol photons m^{-2} s^{-1} failed as no stable growth could be obtained. In most cases, there was biomass growth for a few days, after which growth suddenly ceased completely and was accompanied by cell agglomeration. In some cultivations, productivity fluctuated considerably from day to day. Applying orange-red illumination, three out of six experiments were successful, which means that stable, day to day productivity values were obtained for at least six days. Applying deep red light, only one experiment out of five was successful. Assuming that the high light intensity did not allow unconstrained growth of the algae when applying deep red light, two additional experiments were performed at an incident light intensity of 850 μmol photons m^{-2} $s^{-1}$and an outgoing light intensity of 10 μmol photons m^{-2} s^{-1}. As depicted in Fig. 6, cultures grown under deep red had a lower biomass concentration and, therefore, a higher biomass specific light absorption rate compared to those grown under warm white light. The dilution rate, however, was not higher compared to cultures grown under warm white light and, therefore, the productivity was also lower (18 vs 36 g m^{-2} d^{-1} for warm white). Otherwise stated, at 850 μmol photons m^{-2} s^{-1}, the light use efficiency of deep-red light was also lower than for warm white light. Maximum Fv/Fm values were low (0.50 ± 0.03) for the deep red culture. The culture grown under white light exhibited a Fv/Fm value of 0.61.

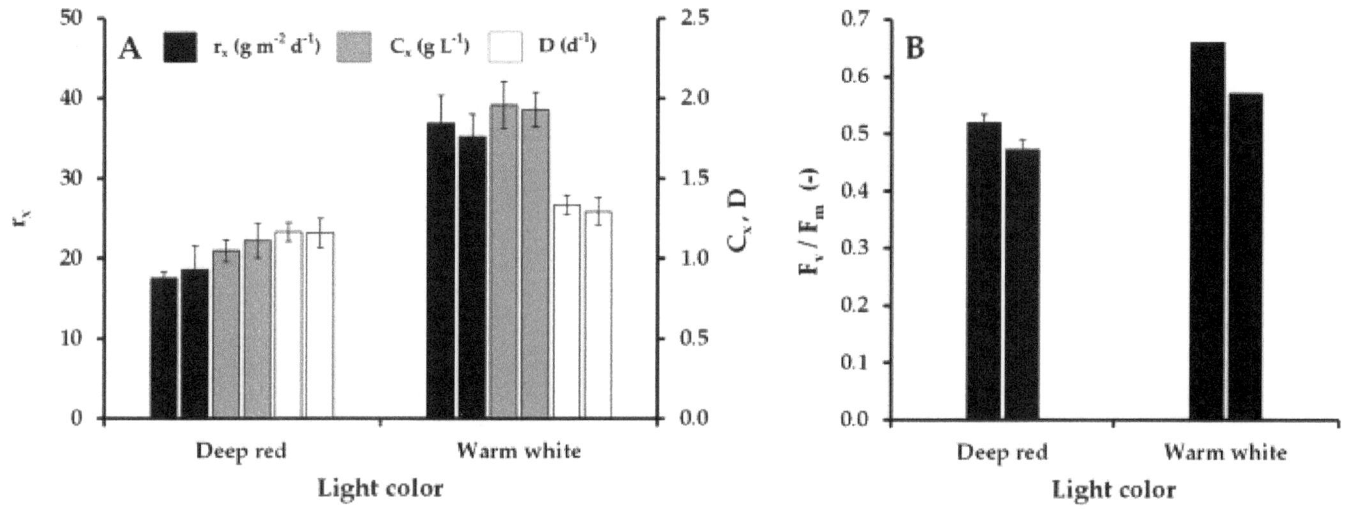

Figure 6. (A) Areal biomass productivity (rx), biomass concentration (Cx), dilution rate (D = μ). (B) Dark adapted maximum F_v/F_m. Reactors were operated at Iph in = 850 μmol photons m^{-2} s^{-1}, Iph out = 10 μmol photons m^{-2} s^{-1}. Each bar in (A) represents the average ± SD of n ≥ 5 data points that were measured on a daily basis within one reactor experiment. The white light data was obtained from de Mooij et al. [11].

Absorption cross section under different light colors

Fig. 7 shows the measured light absorption spectra of cultures grown under different light colors. In the continuously operated turbidostat cultures with ingoing light intensities as high as 1500 μmol photons m^{-2} s^{-1}, the absorption cross section of the microalgae did not markedly change as a function of light color. Up to seven measurements were performed per culture and the experimental variation within these measurements was higher than the variation between the different cultures.

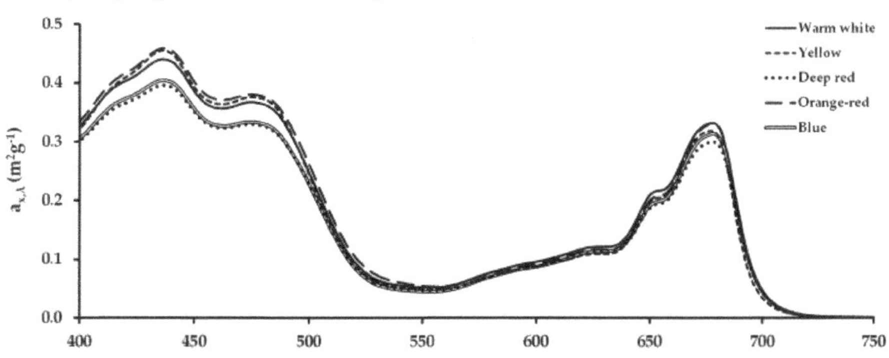

Figure 7. Absorption cross section (ax λ) of Chlamydomonas reinhardtii exposed to different colors of light. The optical cross section has been normalized to the measured biomass concentration in the reactor. Lines represent the average of at least seven measurements from three reactor experiments. For the deep red culture, the line represents the average of three measurements within a single reactor experiment.

The effect of blue light supplementation to monochromatic yellow light

We began our experiments exploiting a single yellow light source. Productivity was far below what was estimated. The cultures were unstable as productivity and biomass concentration fluctuated from day to day. In addition, maximum Fv/Fm values were low, indicating a low PSII quantum efficiency. Pigment content was also considerably lower than measured for all other light colors. For this reason, the yellow light was supplemented with a moderate quantity (± 50 μmol photons m^{-2} s^{-1}) of blue light. The total light intensity thus was 1500 μmol photons m^{-2} s^{-1} Unless explicitly stated otherwise (see Fig. 8), all cultures grown in yellow light described in this paper were supplemented with ± 50 μmol photons m^{-2} s^{-1} of blue light.

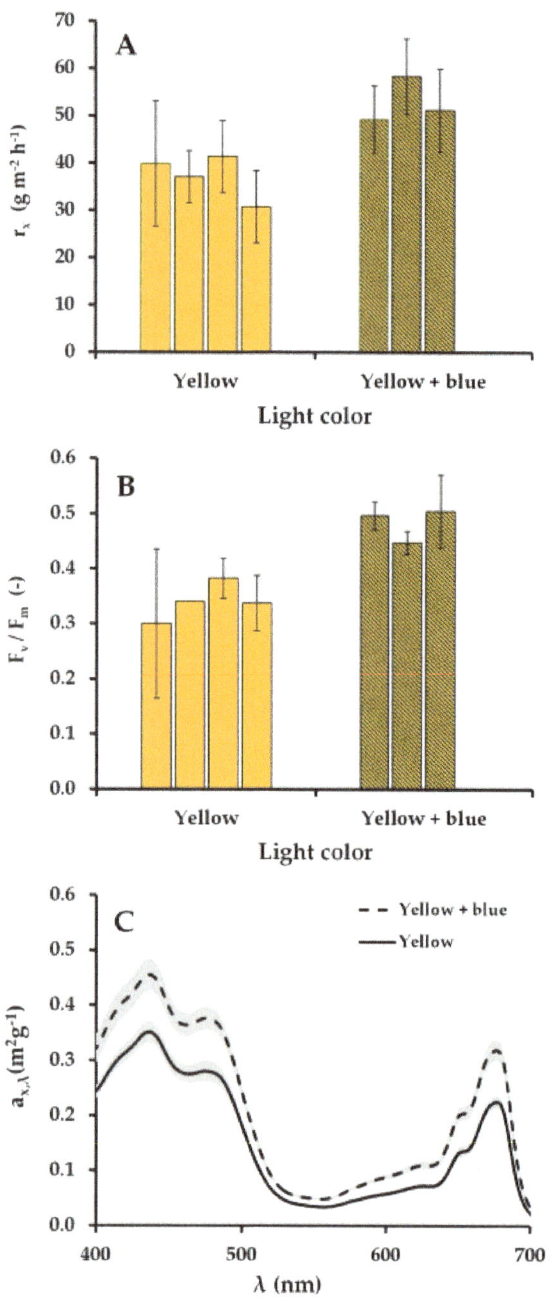

Figure 8. (A) Areal biomass productivity (rx). (B) Dark adapted maximum Fv/Fm. (C) Absorption cross section (ax λ). The cultures were exposed to yellow light and yellow light supplemented with ± 50 μmol photons m^{-2} s^{-1} of blue light. Each bar in (A) represents the average ± SD of $n \geq 4$ data points that were measured on a daily basis within one reactor experiment. The lines in (C) represent the average ± SD of at least ten measurements within three reactor experiments.

By applying blue light supplementation, the volumetric productivity increased from 37 g m^{-2} d^{-1} ± 11 to 52 g m^{-2} d^{-1} ± 8 (Fig. 8A), and cultivation was more stable. Furthermore, the maximum Fv/Fm value was clearly higher (0.49 ± 0.05 vs 0.34 ± 0.06), indicating improved functioning of photosystem II (Fig. 8B). The absorption cross section, depicted in Fig. 8C, was demonstrated to be higher in the case of blue light supplementation. The ratio between absorption by carotenoids and chlorophyll a was comparable for both situations.

DISCUSSION

Microalgae photosynthesis is inefficient at high light intensities. Not considering photobioreactor design, two approaches can be distinguished to increase the photosynthetic efficiency: genetic engineering of the microalgae or spectrally tailoring the light source via light engineering. Our previous study showed that the current generation of Chlamydomonas antenna size mutants is not able to outperform the productivity of the wild-type strain under mass culture conditions [11]. To provide a more solid foundation for the hypothesis that biomass productivity is a function of the amount of light absorbed per cell, in this work, we shifted the emission of artificial illumination to both the low and high absorption region of the spectrum by selecting four different colors of light.

Our model successfully predicted the biomass productivity for different colors of light. The biomass concentration could be accurately estimated as in a turbidostat controlled culture this is a function of the incident light intensity, the outgoing light intensity, and the absorption cross section (ax λ) of the cells. Calculation of the dilution rate, and the areal productivity is more challenging since there are many factors that influence the light use efficiency. The model assumes that good mixing prevents severe photodamage and therefore, photoinhibition is not considered. At very high light intensities, this assumption might not be valid, rendering the model prediction overly optimistic. Our model assumes that the microalgae suspension is exposed to a homogeneous light intensity distribution. In reality there can be substantial differences between for example the middle of the light exposed surface and the relatively dark corners. Depending on the distribution, this may lead to under- or overestimation of biomass productivity. Please refer to Table S2–S6 of the supplementary material to see the light intensity distribution of the light sources that were employed in our experiments.

In accordance with our model predictions, cultures exposed to yellow light resulted in the highest areal productivity (54 g m^{-2} d^{-1}) closely followed by cultures grown under warm white light (50 g m^{-2} d^{-1}). The three strongly absorbed colors (blue, orange-red, and deep red) resulted in areal productivities that were almost half of the areal productivity measured for yellow light. Cultures were difficult to grow under red light and this affected the productivity. However, our substantial number of successful reactor experiments with different colors of light confirms our model-based expectation that, under mass culture conditions, productivity is inversely correlated with biomass specific light absorption.

Analogous to our results, Kubin et al. [35] also showed that maximal productivity withChlorella vulgaris was obtained exploiting weakly absorbed green light. They also measured productivity values for blue light as being half of that for green and white light. Mattos et al. performed short term oxygen evolution experiments and conclude that weakly absorbed colors of light such as green results in a higher photosynthetic efficiency for high density cultures [36]. In these experiments the cells were not allowed to acclimate to the different colors of light and the applied light regime during the measurements and therefore these conditions do not simulate mass culture conditions. Instead of replacing blue and red light by green light, they suggest that green light should be supplemented to strongly absorbed colors of light [35].

The amount of nitrogen source present in the cultivation medium supports biomass concentrations up to 4.5 g L^{-1}. To ensure that nitrogen limitation did not occur, we increased the urea content for the cultures exposed to yellow light. There was no measurable effect of the urea supplementation and, therefore, we conclude that the medium was indeed sufficient for unconstrained growth.

No substantial difference in absorption cross section of the microalgae was observed after cultivating them under different light colors. Apparently, under mass culture conditions and irrespective of the color of light, the light regime in which the algae rapidly alternate between 10 and 1500 μmol photons $m^{-2} s^{-1}$ leads to the same level of pigmentation. The microalgal pigment content is highly dependent on the perceived light intensity. In the process of photoacclimation, the pigment content deceases with increasing light intensity, which reaches a plateau at high light intensities [37]. It could have been expected that pigment content correlates to the biomass specific light absorption rate [38]. If this was the case in our experiments, blue light should have resulted in a lower pigment content to compensate for the higher absorption capacity for blue light. Likewise, yellow light should have resulted in an increased pigment content to harvest more of the weakly absorbed yellow light. As the mechanism behind pigment acclimation in response to light quality has not yet been unravelled [39] and the importance of other (wavelength specific) light acclimation responses has not yet been studied in detail, our observation is difficult to explain. In literature, statements regarding pigment accumulation under different colors of light are contradictory. This is most likely due to the fact that it is difficult to distinguish between the effect of light intensity and light quality, as the color of light determines the ease of absorption and therefore the biomass specific light absorption rate (qph). For a fair comparison, the pigmentation should be compared for cultures exposed to different colors of light, but with the same biomass specific light absorption rate, which can be challenging to achieve in photobioreactors with steep light gradients.

Remarkable is the fact that the lowest Fv/Fm value was obtained for cultures exposed to yellow light (0.49 ± 0.04) while this culture yielded the highest volumetric productivity. The areal productivity for yellow light was almost double compared to blue light, where aFv/Fm value of 0.63 was measured. This suggests, therefore, that part of the photosystems (PSII) became inactive which reduced the biochemical conversion capacity, however, yellow light could still be used at a higher efficiency than, for example, blue light. However, Fv/Fm values should preferably be measured with the same color of light as the color of the cultivation light as, for higher plants, it was observed that this is required to measure maximum quantum yield values [40]. The rationale behind this statement is that the PSI/PSII stoichiometry is optimized for the light the plant is exposed to and when there is a sudden change in light spectrum, there might be an imbalanced excitation of the two photosystems [41]. This could have affected our results as we applied red light (620 nm) for our measurements.

The hypothesis of this study is that the degree of photosystem saturation dictates the photosynthetic efficiency of the microalgae culture and that photosystem saturation can be controlled by applying different colors of light. The rationale has been previously discussed in literature [35], [36], [41] and [42] and applies to both microalgae mass cultures and to the canopy of horticulture crops. In both situations, weakly absorbed light (green or yellow) is expected to increase the photosynthetic efficiency as less energy is dissipated in the surface layer of the photobioreactor or the outer zone of the canopy. Indeed, several experimental studies demonstrated that green light supplementation led to increased productivity of crops [43], [44] and [45]. Sforza et al. used a spectral converter filter to convert the green and yellow light to red light with the intention to maximize the portion of useful light for photosynthesis [46]. No significant improvement was found. According to our hypothesis this approach would actually decrease the productivity under high light conditions as the culture will become even more oversaturated. To maximize productivity in such a setup, the red and blue light should be shifted to the green range.

It remains ambiguous whether yellow light suffices for optimal growth. To our knowledge, blue light supplementation to yellow or green light has not been studied previously. Yellow light could possibly be more difficult for cultivation than green light as the green emission spectrum of some light sources partly overlaps with the blue region. In our experiments, cultures that were supplemented with a moderate amount of blue light gave a higher productivity, had more stable cultivation (i.e. less day-to-day variation in productivity), and had enhanced cell fitness as indicated by a higher Fv/Fm value. The improvement in performance cannot be attributed to the energy content of the additional 50 μmol photons $m^{-2} s^{-1}$ of blue light, as this is only a 3.5 % increase in total light intensity. This finding posits the following tentative hypothesis: blue light acts as a trigger for metabolic regulatory mechanisms that are essential for stable cultivation under the described mass culture conditions.

Higher plants were ascertained to exhibit photoprotection mechanisms that are solely activated by blue light [47]. Authors of the same paper also observed that blue light is exploited by plants as an indicator of over-excitation and the need to switch to a state enhancing thermal energy dissipation.

In addition, for the diatom Phaeodactylum tricornutum, blue light was determined to be essential for the activation of photoprotection under high light as an increased NPQ capacity and a larger pool of xanthophyll cycle pigments could only be observed in cultures grown under blue light [48]. In another study, it was hypothesized that, inChlorella, blue light produces the same effects that are normally observed for strong white light [17]. Blue light is also known to affect several metabolic pathways and induce gene expression in both algae and plants via blue light receptors [21]. In horticulture, the beneficial effects of blue light supplementation have been demonstrated in several studies. Blue light supplementation was found to double the photosynthetic capacity and prevent abnormal growth in cucumber plants [49]. In spinach, blue light was discovered to enhance the acclimation responses to high light conditions and to increase the chlorophyll content [50]. Other greenhouse plants were found to have increased biomass accumulation, increased vegetative growth, and expanded leafs under blue light supplementation [45]. To conclude, blue light seems to play a key role in the survival and development of photosynthetic organisms. Also our experiments with Chlamydomonasindicate that exposure to blue light is essential for optimal growth under high light conditions, probably caused by wavelength-dependent activation of photoprotection and dissipation mechanisms.

Maintaining a stable culture under red light was difficult. Under deep red light at 1500 μmol photons m^{-2} s^{-1}, only one experiment out of eight was successful. Productivity was slightly lower than was estimated by our model based on the light emission spectrum of the deep red light source. Possibly, 1500 μmol photons m^{-2} s^{-1} of deep red light was too intense for the photosystems. On the one hand, this is striking since the biomass specific light absorption rate is lower than that of blue light. On the other hand, the regulatory mechanisms triggered by the color of light seems to be more complex than initially expected. Therefore, it cannot be excluded that, under high light conditions, a balanced mix of wavelengths is required for optimal growth. At 850 μmol photons m^{-2} s^{-1} of deep red light two experiments were successful and reproducible. At this light intensity, severe damage to the photosystems is less probable. As expected, based on our theory that strongly absorbed light decreases light use efficiency, the biomass productivity was lower than for the culture in warm white light. Also, under orange-red light at 1500 μmol photons m^{-2} s^{-1}, productivity was lower than our model predicted. The use of orange-red light for microalgae cultivation is common and generally without complications. Kliphuis et al. [30], for example, used the same light source as we did, but worked with light intensities below 100 μmol photons m^{-2} s^{-1}. The high intensities of red and yellow light in this study have not been previously reported forChlamydomonas. Therefore, we suggest that the high light intensity must have been the explanation for the poor performance. For cultures exposed to yellow light, blue light supplementation was found to improve reactor performance and productivity. A similar approach could possibly work for red light as well.

The cell size of Chlamydomonas is influenced by light color. Continuous blue light is known to delay cell division which signifies that cells continue to grow in size as biomass is accumulating [51] and [52]. Otherwise stated, a larger cell size is required for cell division to occur. A blue light receptor is likely to be involved [51]. The opposite was determined for red light. Under red light, cells undergo a division cycle when they have achieved the minimal cell size required for division. In practice, the consequence is that, compared to white light, the average cell size is larger under blue light and smaller under red light [17], [35] and [53]. Cell size and the accompanied geometrical arrangement of the chloroplast, as well as the cellular chlorophyll content, are all factors that may influence light penetration and light scattering. This phenomenon, therefore, complicates modeling reactor productivity. Also our productivity measurements may have been influenced by this unintended effect of blue and red light.

CONCLUSIONS

In this study, we presented areal biomass productivities of high density microalgae cultures exposed to high light intensities of different colors. Tubidostat control ensured that the total amount of absorbed light was equal for each color. Our results demonstrate that, under mass culture conditions, biomass productivity and the biomass specific light absorption rate are inversely correlated as oversaturation of the photosystems leads to a waste of light energy and, therefore, a lower biomass yield on light. Highest biomass productivity, measured under continuous illumination, was obtained employing yellow light (54 g m^{-2} d^{-1}), closely followed by cultures grown under warm white light (50 g m^{-2} d^{-1}). Cultivation under blue, orange-red, and deep red light resulted in biomass productivities of approximately 29 g m^{-2} d^{-1} which is nearly half of the productivity measured for yellow light. The microalgae absorption cross section remained the same under all tested conditions. Our approach with different colors of light to investigate photosystem saturation was interfered by intrinsic biological effects. Cultivation under pure yellow light was impeded. Minimal supplementation of blue light to the cultures in yellow light was determined to stimulate normal growth and increase productivity. Additional research is required to reveal the underlying mechanism that is responsible for the beneficial effects of blue light supplementation. Taking into account possible wavelength deficiencies, white light with a high green or yellow content in addition to a small blue fraction would result in the highest productivity of microalgae mass cultures. This study provides a solid base for further research on decreasing the biomass specific light absorption in order to maximize productivity. Presently, the creation of antenna size mutants that permanently absorb less light per cell is the most promising solution.

ACKNOWLEDGEMENTS

This work is part of the research programme of the Foundation for Fundamental Research on Matter (FOM) which is part of the Netherlands Organization for Scientific Research (NWO). This project was conducted within the research programme of BioSolar Cells, co-financed by the Dutch Ministry of Economic Affairs.

Appendix A. Microalgae growth model for reactor productivity estimation

A kinetic model has been adopted to be able to predict the biomass productivity in a photobioreactor as a function of the local light intensity and light spectrum. The model comprises two compartments: the chloroplast and the cell content outside the chloroplast. In the chloroplast, 3-carbon sugars (triose) are produced. Here, we express photosynthetic sugar production in units of the 1-carbon sugar equivalent CH_2O that was formed. Sugar production in the chloroplast (qcCH2O, mol_{CH2O} mol_x^{-1} s^{-1}) is driven by light energy and can be described by the hyperbolic tangent model of Jassby and Platt[54]:

$$q^c_{CH2O} = q^c_{CH2O,m} \cdot \int_0^d \tanh\left(\frac{\sum_{\lambda=400}^{\lambda=700} Y_{CH2O/ph,m,\lambda} \cdot q_{ph,\lambda}(z) \cdot \Delta\lambda}{q^c_{CH2O,m}}\right) \cdot dz. \tag{A.1}$$

In this equation, qcCH2O m is the maximal sugar production rate ($mol_{CH2O}\ mol_x^{-1}\ s^{-1}$) in the chloroplast; YCH2O/ph m,λ is the maximum yield of sugar on light energy ($mol_{CH2O}\ mol_{ph}^{-1}$) in the event that there would be no oversaturation and qph λ(z) is the biomass specific light absorption rate ($mol_{ph}\ mol_x^{-1}\ s^{-1}$) at reactor depth z, which is defined as follows in Eq. (A.2):

$$q_{ph,\lambda}(z) = a_{x,\lambda} \cdot I_{ph,\lambda}(z) = \frac{I_{ph,\lambda}(z - dz) - I_{ph,\lambda}(z)}{C_x \cdot dz} \tag{A.2}$$

where Iph λ(z) is local light intensity at reactor depth z; Cx is the biomass concentration in the reactor. Iph λ(z) is calculated by the following equation according to the Lambert–Beer law:

$$I_{ph,\lambda}(z) = I_{ph,\lambda}(0) \cdot e^{-a_{x,\lambda} \cdot C_x \cdot z \cdot f_l}. \tag{A.3}$$

In this equation, Iph λ(0) is the incident light intensity at the illuminated surface of the reactor, ax λ ($m^2\ mol_x^{-1}$) is the biomass absorption cross section and fl the light path correction factor. This factor accounts for the influence of light scattering within the microalgae suspension and the fact that the incident light is not perfectly collimated. The effect is that the path travelled by the light rays within the algae suspension is longer than would have been the case for light entering perpendicular to the reactor surface and with the absence of light scattering within the microalgae suspension. A longer light path increases the probability of light absorption and therefore the fl factor increases the accuracy of the Lambert–Beer equation shown above. The fl parameter (Table A.1) was fitted to the actual light intensity measurements for each light color separately using the data shown in Table S1 of the supplementary material. The fl value of sunlight was assumed to be equal to that of warm white light.

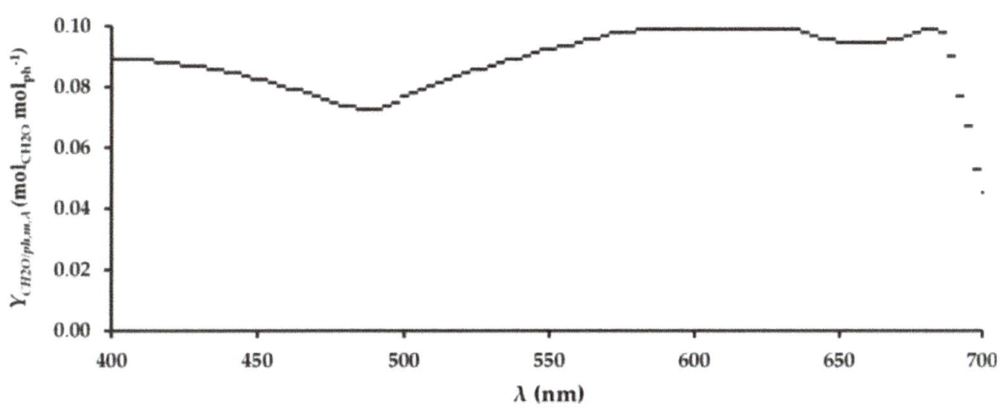

Figure A.1. Maximum yield of sugar on light energy ($mol_{CH2O}\ mol_{ph}^{-1}$) as measured for every wavelength by Emerson and Lewis [2]. The original values have been multiplied by a factor of 1.1 to bring them in line with results obtained by other researchers, as discussed in the text. Numerical values for YCH2O/ph m λ can be found in Table S13 of the supplementary material.

The cell content minus the chloroplast was defined as the second compartment of this model. Here, part of the 3-carbon sugar is used for the production of new biomass at specific growth rate μ. Another part of the sugar is respired in the mitochondria to provide energy in the form of ATP to support the growth reactions and to fulfill the maintenance requirements. The following equation was derived from a substrate balance over the cell minus chloroplast compartment, where the maintenance requirement is fulfilled by substrate consumption according to Pirt [55].

$$\mu = \left(q^c_{CH2O} - m_{CH2O}\right) \cdot Y_{x/CH2O} \tag{A.4}$$

and:

$$\mu_m = \left(q^c_{CH2O,m} - m_{CH2O}\right) \cdot Y_{x/CH2O} \tag{A.5}$$

Here, mCH2O is the biomass specific maintenance rate ($mol_{CH2O}\ mol_x^{-1}\ s^{-1}$); Yx/CH2O is the biomass yield on sugar ($mol_x\ mol_{CH2O}^{-1}$); and μm is the maximum biomass specific growth rate (s^{-1}). Using Eqs. (A.1) and (A.4), μ can be computed for every light intensity Iph at every location inside a flat panel photobioreactor culture illuminated from one side. Values for $a_{x,\lambda}$, Iph in, YCH2O/ph m λqcCH2O m, and mCH2O should be provided (see Table A.1). The integral in Eq. (A.1) is easily solved by numerical integration over a large number of sublayers (N) with depth Δz. This procedure was implemented in a MS Excel based calculation routine and the use of 100 sublayers was demonstrated to result in sufficient accuracy of the model prediction.

The biomass yield on light energy at each point in the reactor (Yx/ph(z), $mol_x\ mol_{ph}^{-1}$) can now also be calculated in the same manner by using the following equation:

$$Y_{x/ph}(z) = \frac{\mu(z)}{q_{ph}(z)} \tag{A.6}$$

The areal biomass productivity (rx, $g\ m^{-2}\ d^{-1}$) in a photobioreactor is the product of the biomass concentration (Cx, $mol_x\ L^{-1}$), reactor depth d , and the average biomass specific growth rate $\bar{\mu}$:

$$r_x = C_x \cdot d \cdot \mu. \quad (A.7)$$

Using numerical integration, for a selected value of Cx, the corresponding rx can be obtained. To maximize productivity, a biomass concentration has to be found such, that at the back of the reactor the maintenance rate (mCH2O) equals the local sugar production rate (qcCH2O(d)). In this situation, the light intensity at the backside is equal to the compensation point and the net photosynthesis rate is equal to zero. This means that at all locations in the photobioreactor, the biomass is positively contributing to biomass productivity and there is no dark zone. Using an iteration method, rx can be maximized by incrementally changing the Cx value for every repetition until it has converged to the optimal value. In this study the MS Excel 'Solver' (GRG nonlinear algorithm) was employed.

In the model it is assumed that under low light conditions, light is used at maximum efficiency. This is expressed using the parameter (YCH2O/ph m λ, which is the maximum yield of sugar on light energy ($mol_{CH2O}\ mol_{ph}^{-1}$). As the maximum yield on light is wavelength dependent, we incorporated the YCH2O/ph m λ values as measured by Emerson and Lewis (1943) into the model. All of their YCH2O/ph m λ values have been multiplied with a factor 1.1 to bring them more in line with the results obtained by a number of other researchers [1], [3], [4], [25] and [46] who also investigated the maximum quantum yield and, overall, ascertained higher values compared to Emerson and Lewis. See Fig. A.1 for the value of YCH2O/ph m λ as a function of wavelength. The fact that the small differences in YCH2O/ph m λ had a minimal effect on the estimated productivity signifies that the effect of YCH2O/ph m λ is of minor importance for the productivity of a reactor exposed to high light conditions. Instead, oversaturation of photosynthesis is believed to be the dominant phenomena.

Table A.1. Overview of the model parameters

Parameter	Value	Unit	Description	References
μm	0.132	h^{-1}	Maximal biomass specific growth rate	Janssen et al. [57]
qcCH2O m	$6.17 \cdot 10^{-5}$	$mol_{CH2O}\ mol_x^{-1}\ s^{-1}$	Maximal 3-carbon sugar production rate; calculated from μm using Eq.(A.5)	
YCH2O/ph m,λ	See Fig. A.1	$mol_{CH2O}\ mol_{ph}^{-1}$	Maximum yield of sugar on light energy	Emerson and Lewis[2]
mCH2O	$3.5 \cdot 10^{-6}$	$mol_{CH2O}\ mol_x^{-1}\ s^{-1}$	Biomass specific maintenance rate	Kliphuis [30]
Yx/CH2O	0.63	$mol_x\ mol_{CH2O}^{-1}$	Biomass yield on 3-carbon sugar using ammonia as N-source	Kliphuis [30]
ax_λ	See Table S2	$m^2\ g^{-1}$	Absorption cross section	This paper, suppl. material
Mx	24	$g\ mol_x^{-1}$	Biomass dry weight to C-mol conversion factor	Duboc [56]
fl	B: 1.17 OR: 1.45 DR: 1.60 Y: 1.57 W: 1.77	–	Light path correction factor	This paper

Appendix B. Model estimation values

Table B1. Overview of all model estimation values of the areal biomass productivity (rx), biomass concentration (Cx), and biomass specific growth rate (μ) at $I_{ph,in}$ = 1500 μmol photons $m^{-2}\ s^{-1}$ for different light colors

Light color	rx ($g\ m^{-2}\ d^{-1}$)	Cx ($g\ L^{-1}$)	D (d^{-1})
Warm white	50.7	2.3	1.56
Sunlight	44.2	2.1	1.48
Yellow	63.2	2.8	1.64
Deep red	33.5	1.3	1.83
Orange-red	47.5	1.9	1.78
Blue	27.4	1.0	1.89

REFERENCES

1. O. Björkman, B. Demmig, Photon yield of O2 evolution and chlorophyll fluorescence characteristics at 77 K among vascular plants of diverse origins, Planta 170 (1987) 489–504.
2. R. Emerson, C.M. Lewis, The dependence of the quantum yield of Chlorella photosynthesis on wave lenghth of light, Am. J. Bot. (1943) 165–178.
3. J. Evans, The dependence of quantum yield on wavelength and growth irradiance, Funct. Plant Biol. 14 (1987) 69–79.
4. S. Malkin, D.C. Fork, Bill Arnold and calorimetric measurements of the quantum requirement of photosynthesis—once again ahead of his time, Photosynth. Res. 48 (1996) 41–46.
5. B. Demmig-Adams, J.J. Stewart, T.A. Burch, W.W. Adams, Insights from placing photosynthetic light harvesting into context, J. Phys. Chem. Lett. (2014).
6. C. Vejrazka, M. Janssen, G. Benvenuti, M. Streefland, R.H. Wijffels, Photosynthetic efficiency and oxygen evolution of Chlamydomonas reinhardtii under continuous and flashing light, Appl. Microbiol. Biotechnol. 97 (2013) 1523–1532.
7. M. Cuaresma, M. Janssen, C. Vílchez, R.H. Wijffels, Horizontal or vertical photobioreactors? How to improve

microalgae photosynthetic efficiency, Bioresour. Technol. 102 (2011) 5129–5137.

8. J.M. Gordon, Tailoring optical systems to optimized photobioreactors, Int. J. Hydrog. Energy 27 (2002) 1175–1184.
9. J.-W.F. Zijffers, M. Janssen, J. Tramper, R.H. Wijffels, Design process of an areaefficient photobioreactor, Mar. Biotechnol. 10 (2008) 404–415.
10. A. Melis, Solar energy conversion efficiencies in photosynthesis: minimizing the chlorophyll antennae to maximize efficiency, Plant Sci. 177 (2009) 272–280.
11. T. de Mooij, M. Janssen, O. Cerezo-Chinarro, J.H. Mussgnug, O. Kruse, M. Ballottari, R. Bassi, S. Bujaldon, F.-A. Wollman, R.H. Wijffels, Antenna size reduction as a strategy to increase biomass productivity: a great potential not yet realized, J. Appl. Phycol. (2014) 1–15.
12. G. Bonente, C. Formighieri, M. Mantelli, C. Catalanotti, G. Giuliano, T. Morosinotto, R. Bassi, Mutagenesis and phenotypic selection as a strategy toward domestication of Chlamydomonas reinhardtii strains for improved performance in photobioreactors, Photosynth. Res. 108 (2011) 107–120.
13. C. De Vitry, F.-A. Wollman, Changes in phosphorylation of thylakoid membrane proteins in light-harvesting complex mutants from Chlamydomonas reinhardtii, BBA Bioenerg. 933 (1988) 444–449.
14. H. Kirst, J.G. García-Cerdán, A. Zurbriggen, A. Melis, Assembly of the light-harvesting chlorophyll antenna in the green alga Chlamydomonas reinhardtii requires expression of the TLA2-CpFTSY gene, Plant Physiol. 158 (2012) 930–945.
15. J.H. Mussgnug, S. Thomas-Hall, J. Rupprecht, A. Foo, V. Klassen, A. McDowall, P.M. Schenk, O. Kruse, B. Hankamer, Engineering photosynthetic light capture: impacts on improved solar energy to biomass conversion, Plant Biotechnol. J. 5 (2007) 802–814.
16. J. Olive, F.-A. Wollman, P. Bennoun, M. Recouvreur, Ultrastructure of thylakoid membranes in C. reinhardtii: evidence for variations in the partition coefficient of the light-harvesting complex-containing particles upon membrane fracture, Arch. Biochem. Biophys. 208 (1981) 456–467.
17. C. Wilhelm, P. Krámer, A. Wild, Effect of different light qualities on the ultrastructure, thylakoid membrane composition and assimilation metabolism of Chlorella fusca, Physiol. Plant. 64 (1985) 359–364.
18. C.W. Kim, M.-G. Sung, K. Nam, M. Moon, J.-H. Kwon, J.-W. Yang, Effect of monochromatic illumination on lipid accumulation of Nannochloropsis gaditana under continuous cultivation, Bioresour. Technol. 159 (2014) 30–35.
19. T.-H. Kim, Y. Lee, S.-H. Han, S.-J. Hwang, The effects of wavelength and wavelength mixing ratios on microalgae growth and nitrogen, phosphorus removal using Scenedesmus sp. for wastewater treatment, Bioresour. Technol. 130 (2013) 75–80.
20. F. Abiusi, G. Sampietro, G. Marturano, N. Biondi, L. Rodolfi, M. D'Ottavio, M.R. Tredici, Growth, photosynthetic efficiency, and biochemical composition of Tetraselmis suecica F&M-M33 grown with LEDs of different colors, Biotechnol. Bioeng. 111 (2014) 956–964.
21. P.S. Schulze, L.A. Barreira, H.G. Pereira, J.A. Perales, J.C. Varela, Light emitting diodes (LEDs) applied to microalgal production, Trends Biotechnol. 32 (2014) 422–430.
22. S.K. Wang, A.R. Stiles, C. Guo, C.Z. Liu, Microalgae cultivation in photobioreactors: an overview of light characteristics, Eng. Life Sci. 14 (2014) 550–559.
23. M. Baba, F. Kikuta, I. Suzuki, M.M. Watanabe, Y. Shiraiwa, Wavelength specificity of growth, photosynthesis, and hydrocarbon production in the oil-producing green alga Botryococcus braunii, Bioresour. Technol. 109 (2012) 266–270.
24. M. Cuaresma, M. Janssen, E.J. Van den End, C. Vílchez, R.H. Wijffels, Luminostat operation: a tool to maximize microalgae photosynthetic efficiency in photobioreactors during the daily light cycle? Bioresour. Technol. 102 (2011) 7871–7878.
25. T. Tanada, The photosynthetic efficiency of carotenoid pigments in Navicula minima, Am. J. Bot. (1951) 276–283.
26. W. Blanken, P. Postma, L. de Winter, R.H. Wijffels, M. Janssen, Predicting microalgae growth, Algal Res. 14 (2016) 28–38.
27. W. Blanken, M. Cuaresma, R.H. Wijffels, M. Janssen, Cultivation of microalgae on artificial light comes at a cost, Algal Res. 2 (2013) 333–340.
28. S. Hutner, L. Provasoli, A. Schatz, C. Haskins, Some approaches to the study of the role of metals in the metabolism of microorganisms, Proc. Am. Philos. Soc. (1950) 152–170.
29. Reference Solar Spectral Irradiance: ASTM G-173, from http://rredc.nrel.gov/solar/ spectra/am1.5/astmg173/astmg173.html.
30. A.M. Kliphuis, A.J. Klok, D.E. Martens, P.P. Lamers, M. Janssen, R.H. Wijffels, Metabolic modeling of Chlamydomonas reinhardtii: energy requirements for photoautotrophic growth and maintenance, J. Appl. Phycol. 24 (2012) 253–266.
31. H. Takache, G. Christophe, J.F. Cornet, J. Pruvost, Experimental and theoretical assessment of maximum productivities for the microalgae Chlamydomonas reinhardtii in two different geometries of photobioreactors, Biotechnol. Prog. 26 (2010) 431–440.
32. B. Demmig, K. Winter, A. Krüger, F.-C. Czygan, Photoinhibition and zeaxanthin formation in intact leaves a possible role of the xanthophyll cycle in the dissipation of excess light energy, Plant Physiol. 84 (1987) 218–224.
33. B.A. Logan, W.W. Adams, B. Demmig-Adams, Viewpoint: avoiding common pitfalls of chlorophyll fluorescence analysis under field conditions, Funct. Plant Biol. 34 (2007) 853–859.
34. E.B. Young, J. Beardall, Photosynthetic function inDunaliella tertiolecta (Chlorophyta) during a nitrogen

starvation and recovery cycle, J. Phycol. 39 (2003) 897–905.

35. Š. Kubín, E. Borns, J. Doucha, U. Seiss, Light absorption and production rate ofbiN Chlorella vulgaris b/iN in light of different spectral composition, Biochem. Physiol. Pflanz. 178 (1983) 193–205.

36. E.R. Mattos, M. Singh, M.L. Cabrera, K.C. Das, Enhancement of biomass production in Scenedesmus bijuga high-density culture using weakly absorbed green light, Biomass Bioenergy 81 (2015) 473–478.

37. H. Takache, J. Pruvost, J.F. Cornet, Kinetic modeling of the photosynthetic growth of Chlamydomonas reinhardtii in a photobioreactor, Biotechnol. Prog. 28 (2012) 681–692.

38. E. Sforza, C. Calvaruso, A. Meneghesso, T. Morosinotto, A. Bertucco, Effect of specific light supply rate on photosynthetic efficiency of Nannochloropsis salina in a continuous flat plate photobioreactor, Appl. Microbiol. Biotechnol. (2015) 1–10.

39. P.G. Falkowski, J. LaRoche, Acclimation to spectral irradiance in algae, J. Phycol. 27 (1991) 8–14.

40. W.S. Chow, A. Melis, J.M. Anderson, Adjustments of photosystem stoichiometry in chloroplasts improve the quantum efficiency of photosynthesis, Proc. Natl. Acad. Sci. 87 (1990) 7502–7506.

41. I. Terashima, T. Fujita, T. Inoue, W.S. Chow, R. Oguchi, Green light drives leaf photosynthesis more efficiently than red light in strong white light: revisiting the enigmatic question of why leaves are green, Plant Cell Physiol. 50 (2009) 684–697.

42. R. Paradiso, E. Meinen, J.F. Snel, P. De Visser, W. Van Ieperen, S.W. Hogewoning, L.F. Marcelis, Spectral dependence of photosynthesis and light absorptance in single leaves and canopy in rose, Sci. Hortic. 127 (2011) 548–554.

43. M. Johkan, K. Shoji, F. Goto, S. Hahida, T. Yoshihara, Effect of green light wavelength and intensity on photomorphogenesis and photosynthesis in Lactuca sativa, Environ. Exp. Bot. 75 (2012) 128–133.

44. H.-H. Kim, G.D. Goins, R.M. Wheeler, J.C. Sager, Green-light supplementation for enhanced lettuce growth under red-and blue-light-emitting diodes, HortSci. 39 (2004) 1617–1622.

45. M. Olle, A. Viršile, The effects of light-emitting diode lighting on greenhouse plant growth and quality, Agric. Food Sci. 22 (2013) 223–234.

46. E. Sforza, E. Barbera, A. Bertucco, Improving the photoconversion efficiency: an integrated photovoltaic-photobioreactor system for microalgal cultivation, Algal Res. 10 (2015) 202–209.

47. W.I. Gruszecki, R. Luchowski, M. Zubik, W. Grudzinski, E. Janik, M. Gospodarek, J. Goc, Z. Gryczynski, I. Gryczynski, Blue-light-controlled photoprotection in plants at the level of the photosynthetic antenna complex LHCII, J. Plant Physiol. 167 (2010) 69–73.

48. B.S. Costa, A. Jungandreas, T. Jakob, W. Weisheit, M. Mittag, C. Wilhelm, Blue light is essential for high light acclimation and photoprotection in the diatom Phaeodactylum tricornutum, J. Exp. Bot. 64 (2013) 483–493.

49. S.W. Hogewoning, G. Trouwborst, H. Maljaars, H. Poorter, W. van Ieperen, J. Harbinson, Blue light dose–responses of leaf photosynthesis, morphology, and chemical composition of Cucumis sativus grown under different combinations of red and blue light, J. Exp. Bot. 61 (2010) 3107–3117.

50. R. Matsuda, K. Ohashi-Kaneko, K. Fujiwara, K. Kurata, Effects of blue light deficiency on acclimation of light energy partitioning in PSII and CO2 assimilation capacity to high irradiance in spinach leaves, Plant Cell Physiol. 49 (2008) 664–670.

51. P. Münzner, J. Voigt, Blue light regulation of cell division in Chlamydomonas reinhardtii, Plant Physiol. 99 (1992) 1370–1375.

52. H. Oldenhof, V. Zachleder, H. Van Den Ende, Blue- and red-light regulation of the cell cycle in Chlamydomonas reinhardtii (Chlorophyta), Eur. J. Phycol. 41 (2006) 313–320.

53. D.G. Kim, C. Lee, S.-M. Park, Y.-E. Choi, Manipulation of light wavelength at appropriate growth stage to enhance biomass productivity and fatty acid methyl ester yield using Chlorella vulgaris, Bioresour. Technol. 159 (2014) 240–248.

54. A.D. Jassby, T. Platt, Mathematical formulation of the relationship between photosynthesis and light for phytoplankton, Limnol. Oceanogr. 21 (1976) 540–547.

55. S. Pirt, The maintenance energy of bacteria in growing cultures, Proc. R. Soc. Lond. Ser. B Biol. Sci. 163 (1965) 224–231.

56. P. Duboc, I. Marison, U. Von Stockar, Quantitative calorimetry and biochemical engineering, Handb. Therm. Anal. Calorim. 4 (1999) 267–365.

57. M. Janssen, M. deWinter, J. Tramper, L.R. Mur, J. Snel, R.H. Wijffels, Efficiency of light utilization of Chlamydomonasreinhardtii under medium-duration light/dark cycles. J Biotechnol. 78 (2000) 123–137.

CHAPTER 8

EVIDENCES OF MANURE DRIVEN AND C:N REGULATED ENHANCED CARBON STATUS AND MICROALGAL PRODUCTIVITY IN MANAGED AQUATIC SYSTEM UNDER SIMULATED GREEN HOUSE CONDITIONS

Jana BB, Kundu D, Datta D, Lahiri S, Bag S, Sarkar D and Bhakta JN

Keywords: Carbon status, Microalgal productivity, Aquatic system, Polyhouse, Soil-carbon, Simulated green house

INTRODUCTION

The climate change is now a scientific reality. The increasing concentrations of greenhouse gases such as carbon dioxide, carbon monoxide, methane and nitrous oxide are the main drivers of the global warming and climate change. The global warming potential of N_2O is 310 times more than CO_2 over a hundred year life span [1]. In India alone, agriculture sector contributes around 17.6% to the total annual anthropogenic emissions of greenhouse gases [2]. As a result, there has been a widespread international movement in order to reduce the emission of greenhouse gases by adopting several climate resilient mitigation and adaptation strategies.

The processes of biological carbon sequestration in the aquatic system framework involves trapping of atmospheric CO_2 through airwater CO_2 exchange, sequestration in water column via microalgae and carbon sink in bottom sediments. Biological CO_2 fixation has been extensively investigated as part of efforts to solve the global warming problem [3]. As microalgae consume high quantities of CO_2 to produce different types of biomass, algae based carbon capture storage technologies are being extensively researched using different strains of microalgae isolated from different habitats [4-6]. The carbon capture potentials of microalgae has been estimated to be about 1.83 kg CO_2per kilogram microalgae [7] and microalgae or pytoplankton support about half of the global primary production

Because of the vast oceanic area, marine phytoplankton has the noble capacity to trap ~50% of the CO_2 that is fixed annually worldwide [9] by sequestering more than 100 million tons of carbon dioxide every day [10]. Microalgae are also cultivated commercially for production of biofuels, biodiesel, etc [7]. The advantages of culturing microalgae in inland and marine sources are: (i) no competition of space for agriculture (ii) harvested algae can be converted into different byproducts such as biodiesel, healthy food, animal feedstock, biogas and fertilizer, etc [6,11].

Wetlands are known to have mitigation effects through their ability to sink carbon, and adaptation effects through their ability to store and regulate water [12]. Wetlands, covering 6% of the world's land surface and 12% of the global carbon pool, play a crucial role in the global cycling of carbon. The wetland soils containing nearly 14.5% of world's soil carbon

Citation: Jana BB, Kundu D, Datta D, Lahiri S, Bag S, et al. (2016) Evidences of Manure Driven and C:N Regulated Enhanced Carbon Status and Microalgal Productivity in Managed Aquatic System under Simulated Green House Conditions. J Earth Sci Clim Change 7:336. doi:10.4172/2157-7617.1000336.

are an important source for carbon sink [13,14] and contribute to the regional and global climate [15,16] by storing as high as approximately 44.6 Tg C yr-1 globally [17]. Intensification of functional activity in managed wetlands particularly in aquaculture units has made them more dynamic having regular input of allochthonous matters in the form of fertilizers, manure and feed. A portion of feed lot, manure, and dead plankton settles in the bottom surface in the form of organic matter and eventually become a part of the soil in the long run. The process of humic formation in bottom surface of water bodies is considered a form of accumulation and storage of carbon containing compounds.

The status of carbon sink or sources affects the biodiversity of freshwater ecosystem across most of the Arctic, and the magnitude, extent, and duration of the impacts have been projected to systemand location dependent [18]. It is reported that the global emission of carbon dioxide from inland waters are similar in magnitude to the carbon dioxide uptake by the ocean and that global burial of organic carbon in inland water sediments exceeds organic carbon sequestration on the ocean floor [19]. Thus, inland waters play even a much important role than ocean in terms of carbon sink via active microbial processes of carbon pump that are responsible for transformation of labile dissolved organic carbon into recalcitrant dissolved organic carbon in the bottom sediment [20]. The interplay between dynamic environmental conditions and metabolic responses determine resource allocation and biogeochemical cycling in aquatic system [9].

Aquaculture ponds sequester as much as 0.21% of the annual global C emissions or about 10 Pg/year (1 Pg = Pentagram = billion ton). It is estimated that about 16.6 million tons of carbon is annually buried in aquaculture ponds worldwide [21] of which maximum sequestration occurs in Asia and particularly in China [22]. At the same time, one of the very strong greenhouse gas N2O-N has been projected to increase by 5.72% in 2030 (3.83 × 1011 g) from its estimated emission of 9.30 × 1010 g in 2009 if the aquaculture industry continues to rise at the current annual growth rate of 7.10% and thereby posing a debate on the issue [1,23]. Nevertheless, the role of aquaculture pond bottom has been greatly emphasized in the global cycling of carbon [24] due to their burial efficiency of organic carbon (buried OC: deposited OC) which was up to 48% (mean) and that was related with the input of allochthonous organic matter in numerous arctic lakes [19,25].

In aquaculture practice, qualitatively different organic manure are used as source of fertilizer for enhancement of fish production via the grazing and detritus food chain that induces organic carbon production through autochthonous and allochthonous sources. Organic matters from dead plankton, organic fertilizers, uneaten feed and excrement of cultured fish sink to the pond bottom and gradually mix with soil particles over time. The small impoundments in farming areas may bury more carbon than the world's oceans [26,27]. This was due to intensification of farming activity per unit area compared to oceans. However, the role of manure driven carbon status in the framework of aquatic system and its state-of-the-art under greenhouse condition has hardly been researched.

It is hypothesized that greenhouse mediated raised temperature would influence the decomposition of chemically and qualitatively different manure that have immense potentials of carbon sink and bring out changes in the structure and functions of aquatic system through explicitly the involved processes of carbon sequestration such as airwater CO_2 exchange, C-sequestration by microalgae and carbon sink in the sediment operating through integrative process of environmental, trophic and microbial activities for carbon pump. The purpose of the present study was to examine the manure driven primary productivity and carbon status in a simulated system under closed and open system within the framework of aquatic environment. The study is pertinent as it envisaged ascertaining the potentials of carbon burial rate of qualitatively different manure mediated through primary production and sink of organic matter as affected by the greenhouse effect.

MATERIALS AND METHODS

Experimental design

The experiment was performed during winter (December, 2013 through March 2014) using thirty six experimental tanks (300 l) that are placed in equal numbers inside and outside the enclosed **polyhouse** simulating the green house condition. All the experimental tanks were provided with 18 cm soil at the bottom, filled with aerated ground water (pH 7.2-7.4) and allowed for a week for establishment. The constructed polyhouse with conventional bamboo structure and standard transparent polythene mimicked greenhouse functions.

The experimental tanks, placed inside and outside the enclosed polyhouse, were allotted for six treatments in triplicate. The treatment conditions were: cattle manure (CM) + saw dust (SD) [T1], poultry droppings (PD) + saw dust (SD) [T2], vermi-compost (VC) + saw dust (SD) [T3], mixed combination of cattle manure, poultry droppings, vermi-compost and saw dust [T4], iso-carbonic states maintained with cattle manure using basal manure of vermin compost [T5] and poultry droppings [T6]. Saw dust was a common ingredient of all the six treatments. The rationale of setting up such iso-carbonic state with cattle manure was to examine the role of saw dust as main carbon source while maintaining the low carbon and low nitrogen for vermicompost and low carbon and high nitrogen for poultry dropping, respectively.

Selection of manure

Organic manure such as cattle manure and poultry droppings were selected because of their ease of availability, low costs and common use in agricultural production in India. Cattle manure (C/N = 19; N/P = 3.4) is common manure used for agriculture as well as for aquaculture. It is estimated that the total carbon dioxide emission from cattle manure is 4.7 t CO_2/year [28]. Poultry litter (C/N = 9; N/P = 2) is a recognized manure for its high nitrogen content [28]. Nitrous oxide and methane are formed during decomposition as a by-product of nitrification/ denitrification and methanogenesis, respectively. Methane emissions from poultry are 0.57 pounds of methane/animal/year. Vermi-compost (C/N = 17.64; N/P = 1.6) is well known for its eco-friendly nature and balanced nutrients [29]. Carbon dioxide, methane and nitrous oxide emission from vermi-compost unit are 1675, 4.76 and 1.17 mg/m^2/hr, respectively. Saw dust (C/N = 511) is wood waste containing a very rich carbonaceous component that are used as a soil amendment especially used to pile up in eco-san toilet to facilitate composting process of human feces [30]. Combustion of saw dust produces 21.3% of CO_2 and 60.3% N_2 [30]. Saw dust waste can also be degraded by indigenous microbial population of bacteria and fungi [31].

Manuring schedule

Required amount of manure was applied in each treatment in the beginning followed by subsequent application in three installments according to the experimental design (**Table 1**).

Table 1: Manuring conditions followed in the experiment

Conditions	Treatments of manure					
	T1	T2	T3	T4	T5	T6
Manures	900 g cow dung + 100 g saw dust	900 g poultry dropping + 100 g saw dust	900 g vermicompost + 100 g saw dust	300 g cow dung + 300 g poultry dropping +300 g vermicompost +100 g saw dust	780 g vermicompost + 100 g saw dust	805 poultry dropping +100 g saw dust
Total amount of manure added (g)	1.0	1.0	1.0	1.0	0.9	0.39
Total input of carbon (g)	331.61	364.01	373.73	356.45	331.39	331.52
Total input of nitrogen (g)	15.41	34.31	18.11	22.61	16.11	11.13
C:N ratio	21.52	10.61	20.64	15.77	21.01	13.96

Collection of data

Samples of surface water of the experimental units were collected from each tank at regular intervals and analyzed for different physicochemical parameters such as temperature, pH, free CO_2, carbonate, bicarbonate, hardness, dissolved oxygen following the standard methods (APHA 2005). Samples of surface sediment were collected thrice from each tank prior to the application of manure, seven days after first installment of manure application and at termination after 100 days of experiment. The samples were analyzed for pH, bulk density, particle density, porosity, water holding capacity, organic carbon and carbon burial rate following the standard protocols [32].

Sedimentation rate

Sedimentation rate of organic matter that accumulated in the bottom surface was determined by calculating the weight differences of empty Petri dish before placing them at the bottom of each tank in the beginning and after the end of experiment when the Petri dish was filled with suspended particles. Precaution was taken while removing the Petri dish from the bottom of the tank so that the amount of accumulated particles in the Petri dish was not washed out or removed. The weight differences of the Petri dish gave the sedimentation rate per unit area.

Soil carbon burial rate

The soil carbon burial rate estimated from sediment accumulation rate, dry bulk density, and percentage organic C in sediment following the calculation as follows:

Climatic regime

Ambient air temperature, light intensity and atmospheric CO2 were recorded from within and outside the polyhouse using digital thermometer, LUX meter and CO_2 meter (Lutron GCH-2018), respectively.

Statistical analysis

All the data were statistically evaluated; analysis of variance in the form of split plot model was applied to find the effects of experimental conditions (open and closed), treatments, time and interactions, if any. In this model, individual tanks were considered as whole plot treatment and time as subplot. This was followed by LSD test at 5% level.

RESULTS

Light intensity and temperature

The mean intensity of light inside the polyhouse was attenuated by 200% with the values ranging from 145 to 340 LUX against 198 to 580 LUX in the open air system. The air temperature has been increased inside the polyhouse (34-44°C) compared to open air system (25- 35°C) as a result of emission of greenhouse gases. The overall mean air temperature was 39°C and 29.5°C, in the closed polyhouse and open system, respectively.

The magnitude of temperature difference was less in water (5.23°C) than in air (9.5°C). Clearly, differences in thermal regime of water were responsible for microbial induced manure decomposition and subsequently productivity of microalgae and their potentials to sequester carbon in the system.

Air carbon dioxide

There has been significant rise in the concentrations (70%) of atmospheric carbon dioxide inside the polyhouse (490-905 ppm) compared to open air system (370 -440 ppm). The mean concentration of air carbon dioxide was 698 ppm in the enclosed polyhouse and 405 ppm in open air, respectively. The elevated mean atmospheric carbon dioxide inside the polyhouse affected the carbon dioxide level of water through absorption and brought out changes in buffering capacity of water as well as carbon sequestration potentials of photosynthesis of microalgae.

Primary productivity of microalgae

The primary productivity of phytoplankton experienced around 10-fold variations ranging from 32.33 $mg/cm^2/hr$ to 307.88 $mg/cm^2/hr$ in different treatments employed. Application of poultry droppings mixed with saw dust (T-2) resulted in highest gross primary productivity (GPP) (**Figure 1**), net primary productivity (NPP) (**Figure 1**) and community respiration (CR) of phytoplankton (**Figure 1**), and lowest in the treatment (T-4) with mixed manure ($F_{5, 35} \geq 24.46$; $P < 0.001$). LSD test revealed that three treatment groups were clearly recognized: highest in T-2, second in T- 3 and T-4 and the lowest in the group comprising T-1, T- 4 and T-6 (**Figure 1**).

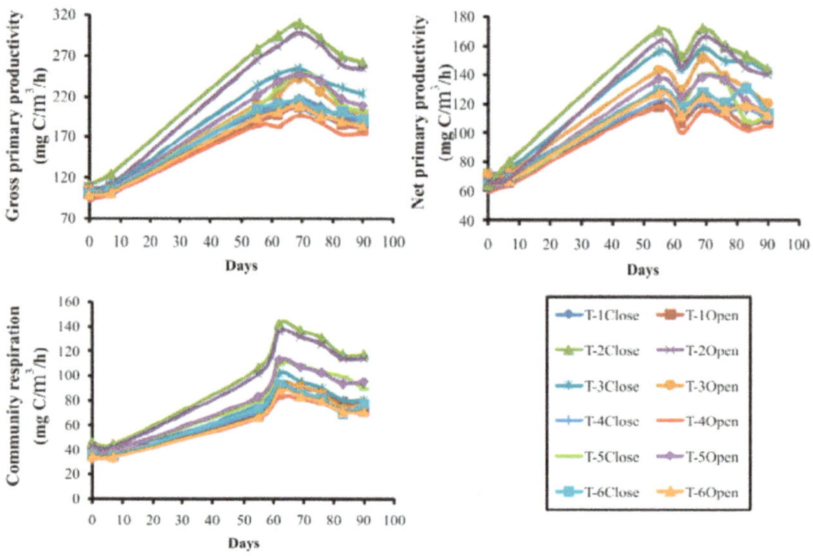

Figure 1: Time scale responses of (a) gross primary productivity of phytoplankton, (b) net primary productivity of phytoplankton and (c) community respiration to different manure treatments under enclosed polyhouse and open condition.

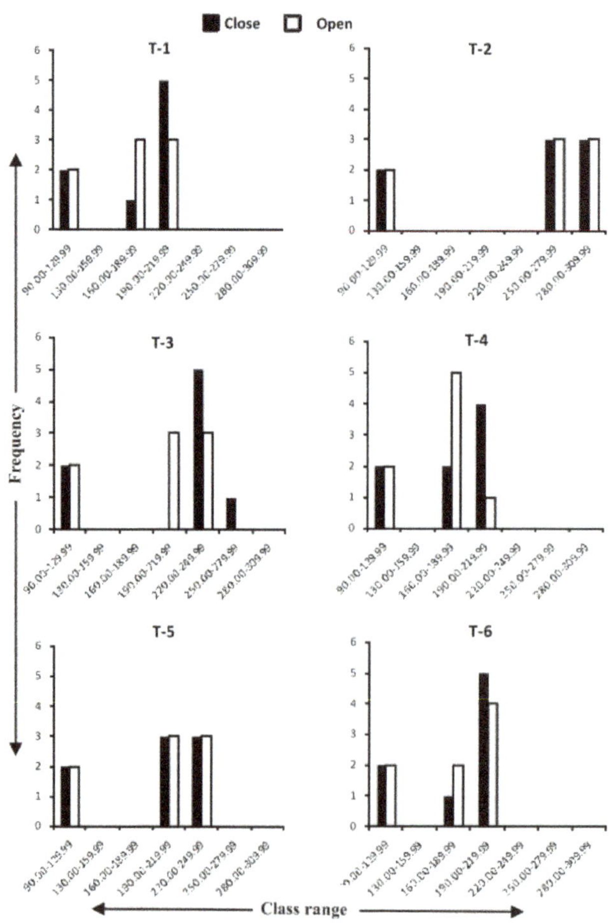

Figure 2: Frequency distribution of gross primary productivity (GPP) of phytoplankton against different class range of values using all treatments together.

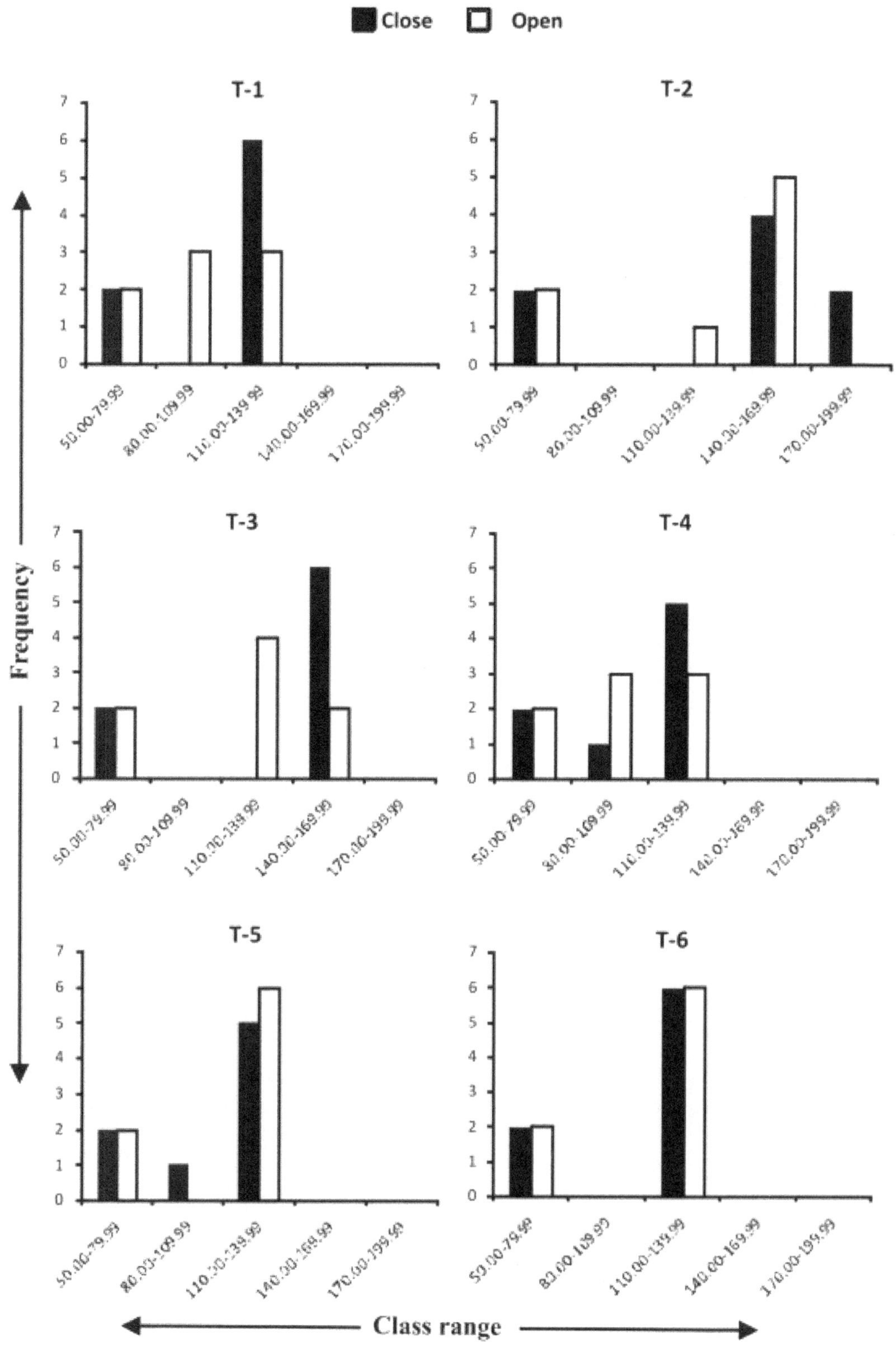

Figure 3: Frequency distribution of net primary productivity (NPP) of phytoplankton against different class range of values using all treatments together.

Frequency distribution of gross primary productivity of phytoplankton (Figure 2), NPP (Figure 3) and CR (Figure 4) showed the occurrence of highest values in T-2 (poultry droppings and saw dust) as evident from more occurrence of higher class (250-279.99 mg/ cm^2/hr). Whereas such higher class range was absent in the remaining treatments (Figure 5). As a result, this group was significantly different ($P < 0.05$) from the next lower group (T-3 and T-4). The third group comprising T-1, T-4 and T-6 (Figure 5) that remained significantly different ($P<0.05$) from the rest two groups was characterized by the absence of productivity class range of 130.00-159.99 mg/cm^2/hr. Responses of net primary productivity of microalgae and community respiration followed the trend similar to that of gross primary productivity.

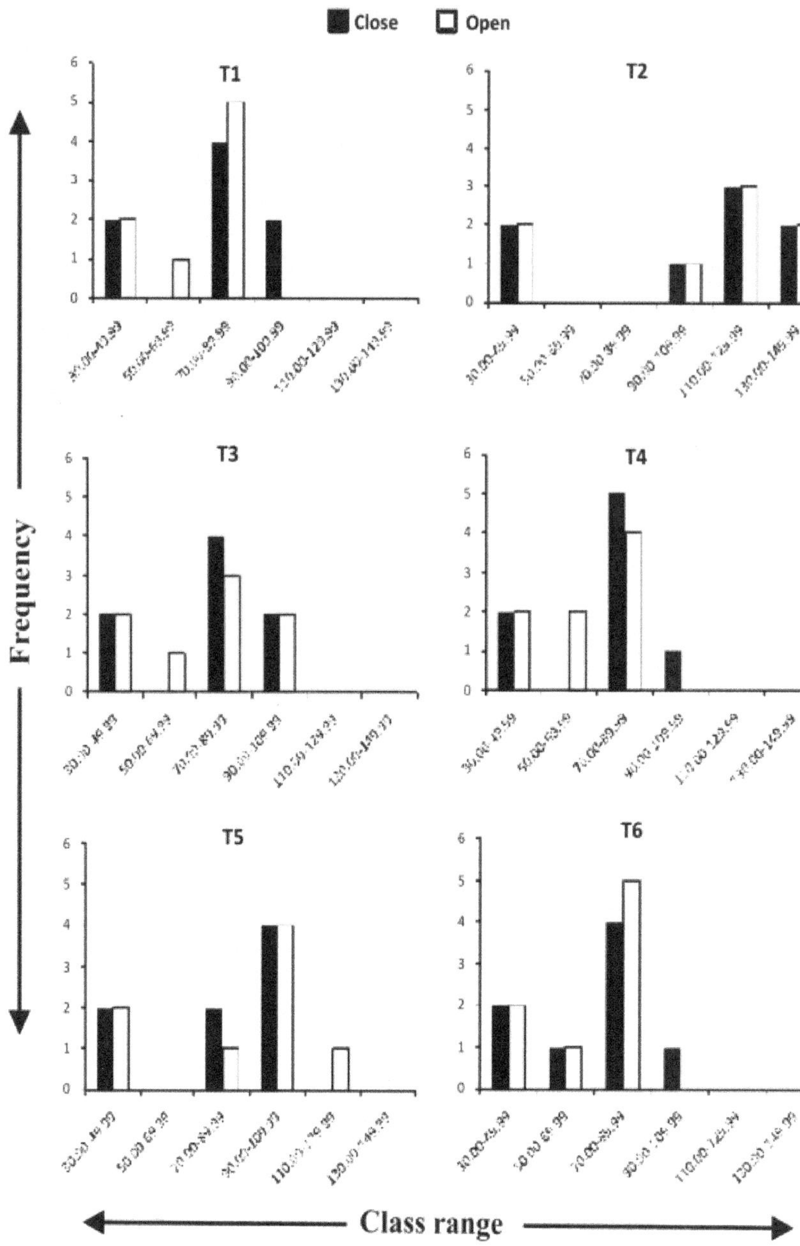

Figure 4: Frequency distribution of community respiration (CR) against different class range of values using all treatments together.

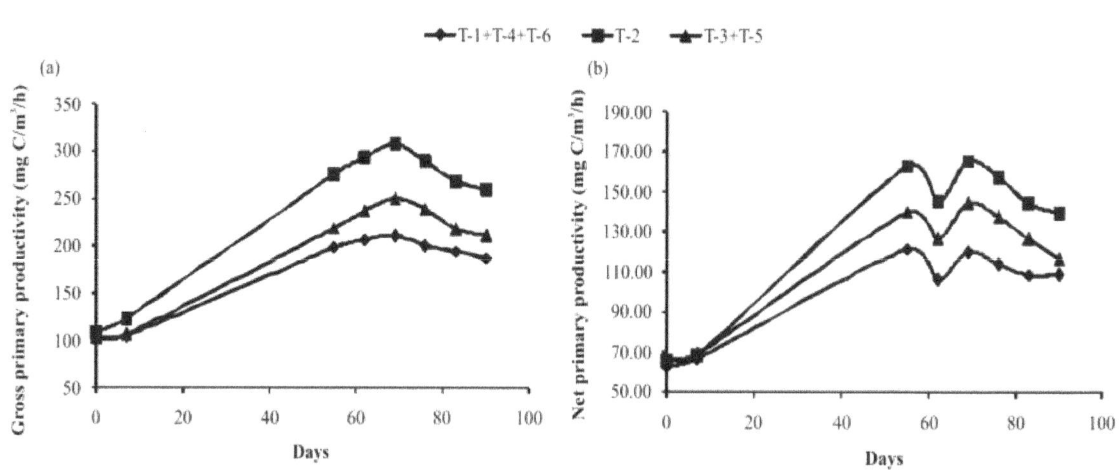

Figure 5: Trend curves of average data for three distinctly different groups of treatments: T-2, (T-3 + T-5) and (T-1 + T- 4 +T-6). The groups were significantly different from each other, whereas the treatments within the group were not.

Temporal courses of variations of GPP experienced almost a sigmoid curve in three groups of treatments. There was no difference in the values of GPP among the treatments during the initial lag phase, but the changes were well marked during the next logarithmic phase (**Figure 1**). Though there was no strong impact of green house (polyhouse) mediated raised temperature on specific treatment, the mean value for all treatments under simulated greenhouse condition remained significantly higher ($P<0.05$) than outside treatments. This shows that GHG emitted raised temperature was responsible for induction of manure driven primary production via microbial decomposition of manure and phytoplankton derived carbon sink in T-2.

Carbon burial rate

Carbon burial rate was less variable (3.57 fold) under simulated greenhouse condition ranging from 33.46 g/m^2 to 118.03 g/m^2 per 100 days compared to open system (23.81 g/m^2 to 102.94 g/m^2 per 100 days) in all treatments. Hence, the overall mean for different treatments attributable to greenhouse effects was 27% higher than under open conditions (**Figure 6**).

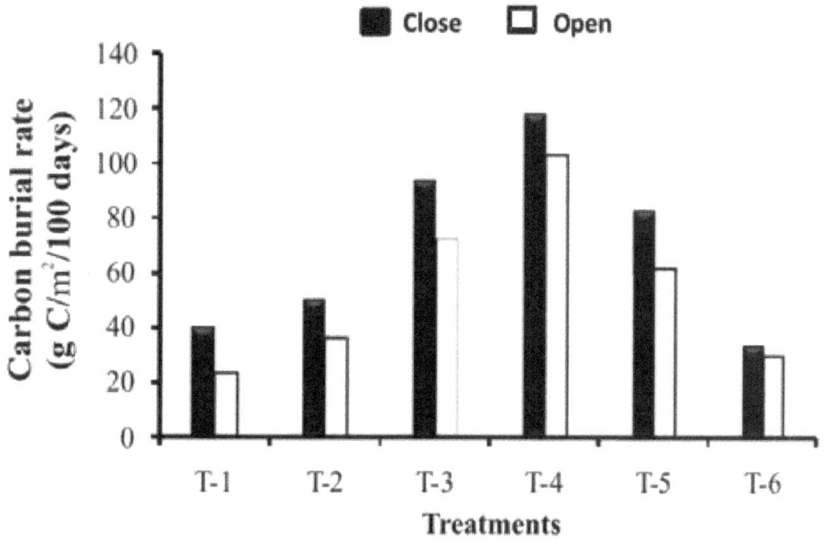

Figure 6: Mean values of carbon burial rates for six different treatments employed under enclosed polyhouse and open condition.

Treatment-wise, application of mixed manure (T-4) resulted in highest carbon burial rate among all the treatments under both the conditions examined. Reduced rates of carbon burial in cattle manure (T-1), poultry droppings (T-2) and iso-carbonic state with poultry dropping (T-6) implied that these treatments were less effective as a depository of carbon due to fast turnover of carbon in the cattle dung manure through biogeochemical cycling of decomposition and mineralization. However, the rate of increase was significantly higher in all treatments under closed system than in open ones (**Figure 6**) implying that increase of carbon sink was explicitly GHG dependent.

Water and soil environment

The pH-free CO_2-CO_3^--HCO_3^-system: Responses of pH to treatments were quite different between the enclosed and outside polyhouse. Fluctuations of water pH were more wide (7.87-8.09) inside the polyhouse compared to outside (8.47-8.80). Prevalence of more acidic condition of water inside the polyhouse than outside ($P<0.05$) was due to rapid rate of microbial decomposition of manure in the former caused by polyhouse mediated raised temperature (**Table 2**). The carbon dioxide- carbonate-bicarbonate systems also differed (ANOVA; $P<0.001$) between the two conditions employed. The concentrations of bicarbonate, total alkalinity and hardness of water were significantly higher ($P<0.05$) in the treatment with poultry dropping and saw dust (T-2) under closed system, whereas lower values were registered in the treatment with cattle manure and saw dust (T-1) under any condition either. This suggests that saw dust and poultry dropping combination caused alkaline state with more calcium carbonate and concomitant increase in the concentration of total alkalinity of water.

Table 2: Range and mean (S.E.) of water quality parameters in different treatments under enclosed polyhouse and open conditions

Parameters	T-1		T-2		T-3		T-4		T-5		T-6	
	Range	Mean ± SE	Range	Mean ± SE	Range	Mean ± SE	Range	Mean ± SE	Range	Mean ± SE	Range	Mean ± SE
Close												
pH	7.4-8.2		7.7-8.2		7.8-8.2		7.7-8.2		7.9-8.2		7.8-8.3	
Free CO_2 (mg/l)	0-16	5.375 ± 0.93	0-20	6.25 ± 1.09	0-16	4.75 ± 0.83	0-18	6.38 ± 1.10	0-24	7.5 ± 1.30	0-30	7.75 ± 1.39
Carbonate (mg/l)	0-24	5.63 ± 0.91	0-28	8.31 ± 1.14	0-20	6 ± 0.80	0-18	4.88 ± 0.71	0-20	5.69 ± 0.82	0-18	6.06 ± 0.77
Bicarbonate (mg/l)	24-294	132.31 ± 14.98	26-370	169.19 ± 19.47	16-312	149 ± 17.35	16-260	132.63 ± 15.02	18-290	150.75 ± 16.91	18-320	148.75 ± 17.13
Alkalinity (mg/l)	28-318	137.94 ± 15.16	42-398	177.5 ± 19.48	23-330	155 ± 17.28	20-276	137.5 ± 14.99	23-301	156.44 ± 16.84	26-338	154.81 ± 17.04
Hardness (mg/l)	224-328	271.25 ± 4.35	260-428	346.625 ± 7.73	268-404	325.81 ± 4.90	244-372	303.25 ± 5.34	224-328	284.5 ± 4.46	248-356	307.5 ± 4.16
TDS (mg/l)	298-565	488.125	420-912	691.1875	325-605	531.0625	320-700	678.5	278-588	486.75	322-645	540.875
Conductivity (µS/cm)	590-755	673.38	570-1260	954.875	853-589	750.9375	610-990	801.9375	575-774	679.75	600-896	758.375
DO (mg/l)	6-8.9	8.16 ± 0.12	0.8-6	2.28 ± 0.28	6.8-8.2	8.15 ± 0.03	6.9-7.9	7.52 ± 0.04	6.8-8.1	7.77 ± 0.04	6.8-9	7.95 ± 0.08
Open												
pH	8.3-8.8		8.3-9.2		8.3-8.8		8.5-9		8.3-8.8		8.3-8.8	
Free CO_2 (mg/l)	0	0	0	0	0	0	0	0	0	0	0	0
Carbonate (mg/l)	3- 22	9.3125 ± 0.74	6- 24	11.56 ± 0.65	4- 18	9.8125 ± 0.59	4- 16	9.8125 ± 0.54	5-20	11.06 ± 0.74	6- 20	11.50 ± 0.62
Bicarbonate (mg/l)	24-289	150.62 ± 16.59	18-322	168.31 ± 18.2	28-305	164.56 ± 17.8	28-308	162.69 ± 17.45	22-295	147.5 ± 16.28	22-310	161.375 ± 18.2
Alkalinity (mg/l)	27-311	159.94 ± 17.23	24-346	179.88 ± 18.7	34-323	174.38 ± 18.3	32-322	172.5 ± 17.95	27-311	158.56 ± 16.91	28-324	172.88 ± 18.84
Hardness (mg/l)	240-364	296 ± 4.82	220-392	311.5 ± 6.58	216-344	275.75 ± 3.99	240-324	289.5 ± 3.46	260-332	305.125 ± 3.02	246-340	291.94 ± 3.96
TDS (mg/l)	260-570	479.3125	259-633	552.6875	240-510	430.9375	246-512	447.9375	280-590	498.75	257-558	493.8125
Conductivity (µS/cm)	575-760	665.8125	570-890	753.0625	535-686	617.375	560-723	647.0625	577-840	713.0625	584-786	695.125
DO (mg/l)	8-10.1	9.32 ± 0.08	1.1-8.1	3.71 ± 0.32	7.4-10.4	9.38 ± 0.12	7.7-10.5	9.39 ± 0.11	7.9-10.7	9.53 ± 0.12	7.6-11.1	9.64 ± 0.14

Total dissolved solids (TDS) and specific conductance: Application of poultry dropping and saw dust (T-2) resulted in significantly higher ($P<0.05$) amount of total dissolved solids (691.19 mg/l) and specific conductivity of water (954.88 µS/cm) compared to other treatments (TDS: 541.81- 486.75 mg/l; Specific conductance: 801.94-673.38 µS/cm) showing almost similar responses among themselves.

GHG mediated raised temperature in T-2 caused 25 or 27% rise in the values of TDS (691.19 mg/l) and Specific conductance (954.88 µS/cm) compared to open system (TDS -552.69 mg/l; SC-753.06 µS/ cm) As expected, specific conductivity measuring the ionic content of a solution was highly related with the TDS reflecting the biological productivity of the system in question.

Dissolved oxygen (DO): The concentration of dissolved oxygen of water ranged from the minimum of 0.8 mg/l to the maximum of 10.9 mg/l in all the treatments during the course of investigation. Lowest values of dissolved oxygen in the poultry dropping and saw dust combination (T-2) reflected the maximum decomposition potential of this combination under both closed and open system.

Soil pH: Application of manure resulted in relatively reduced pH of soil with values ranging from 7.1-7.6 in open conditions against 6.9-7.6 in closed systems. Green house mediated enhanced microbial decomposition was responsible for the more acidic conditions which occurred in more than 84% of the total number of observations (**Table 3**).

Table 3: Range and mean (± SE) of soil quality parameters analyzed in different treatments under enclosed polyhouse and open conditions

Parameters	Close						Open					
	T-1	T-2	T-3	T-4	T-5	T-6	T-1	T-2	T-3	T-4	T-5	T-6
Soil pH	6.9-7.1	7.3-7.4	7.3-7.6	6.9-7.1	7.1-7.3	7.1-7.2	7.1-7.3	7.2-7.5	7.2-7.4	7.2-7.4	7.3-7.6	7.4-7.6
Water holding capacity (%)	90.699	71.543	74.651	71.893	70.546	80.494	86.978	75.436	73.737	72.906	74.387	81.441
Soil sedimentation rate (cm/100days)	0.500	0.500	0.800	0.800	0.700	0.400	0.400	0.400	0.700	0.800	0.600	0.400
Bulk density (gm/cm^3)	0.639	0.747	0.771	0.847	0.826	0.742	0.640	0.732	0.757	0.833	0.797	0.730
Total porosity (%)	57.921	53.416	57.522	60.871	58.240	59.715	55.649	55.210	55.848	60.711	59.276	59.475
Particle density (gm/cm^3)	1.518	1.603	1.814	2.164	1.977	1.841	1.802	1.443	1.634	1.715	2.120	1.957
Organic carbon (mgC/gm)	25.987	29.287	31.073	38.653	33.700	27.640	23.847	27.673	29.787	31.360	28.100	26.333

Soil sedimentation rate: There was no difference in the rate of sedimentation ranging from 0.4 cm to 0.8 cm/100 days in different treatments employed under closed and open systems. The highest rate was observed in mixed treatment (T 4) both in closed and open systems. This was followed by vermi compost and saw dust (T-3) treatment showing the next higher values.

Organic carbon (OC): The organic carbon content of soil ranged from 11.52 mg C/g to 55.1 mg C/g and 16.08 mg C/g to 48.2 mg C/g in the treatments under closed and open system, respectively. The values were significantly higher ($P<0.01$) in all treatments under enclosed polyhouse than outside. Treatment-wise, mixed manure (T-4) and iso-carbonic state with poultry dropping (T-6) showed the highest and lowest values, respectively (**Table 3**). On a temporal scale, increase in the concentration of organic carbon was registered over the passage of time regardless of treatments either in closed or open conditions. However, the rate of increase was found to be much higher under green house condition than outside. The variations of organic carbon content of accumulated sediment were almost similar to that of treatment related organic carbon of soil, being maximal in T-4 followed by T-3.

DISCUSSION

Since the primary productivity of phytoplankton and soil carbon status have been significantly increased in all the tanks regardless of treatment under simulated greenhouse conditions than outside, it is apparent that increased carbon status related with carbon burial rate in the former was the result of manure driven algal productivity and rapid microbial decomposition and subsequent mineralization through biogeochemical cycling bacteria.

Algae inhabit a dynamic photic zone that changes both in quantity and quality of light and posses adaptive mechanisms with high degree of diversity in the photosynthetic apparatus to maximize light harvesting and to limit damage by excessive amount of photosynthetically active radiation and UV radiation in the environment. The green house mediated raised temperature (5.23°C) consequent upon the elevated atmospheric levels of CO_2 (490-905 ppm) under the enclosed polyhouse was responsible for the enhancement of primary productivity of phytoplankton mediated through dissolution of atmospheric CO_2 into water as well as availability of adequate nutrients released from autochthonous and allochthonous sources. Alternatively, this was explained as a result of prevalence of conducive range of both high temperature (34-44°C) and low light intensity (145 to 340 LUX) for the growth of microalgae or reduced photosynthetic performance of microalgae outside due to the effects of high light intensity (198 to 580 LUX) and low temperature (25-35°C). The effect of elevated levels of CO_2 on the photosynthetic performance of microalgae is well understood [33]. The role of light on the photosynthetic efficiency of phytoplankton has been tested by developing an exponential equation model to describe photosynthetic rate of phytoplankton as a function of light intensity for a variety of unicellular microalgae and higher plants where photosynthesis is proportional to $(1 = e^{-\beta I})$. The parameter β $(= I_k^{-1})$ is derived by simultaneous curve fitting method where I is incident quantum-flux density [34]. Chen [35] demonstrated that the optimal, minimal and maximal growth temperatures of phytoplankton decrease with latitude for marine taxa, whereas the minimal and maximal growth temperatures are relatively invariant with latitude for freshwater phytoplankton. Further the thermal breadth is larger for freshwater than marine phytoplankton. Using a mechanistic approach, Huertas et al. [8] provided evidence for different degrees of tolerance to temperature rises as well as an interspecific capacity for genetic adaptation. It is, therefore, reasonable to assume that the enhancement of carbon status in the tank sediment of all treatments under simulated greenhouse conditions was due to greenhouse effect. Using polyhouse experiment in fish ponds, Jana et al. [36] concluded that the benefit of closed polyhouse may be profitably used by the farmers for inducing breeding in tilapia and growth enhancement in other Indian fish during winter months and other region of the country in general.

The ideal C/N ratio for composting is generally considered to be around 30:1 under terrestrial condition [37], whereas it was 11.8:1 for microbial decomposition of fertilizer in aquatic system [38]. About two fold variations of the C/N ratio of different input manure combinations (10.61 to 21.52) used in study reflected their potentials to sequester or source of carbon through microalgal dominated autotrophic or bacterial dominated heterotrophic metabolism in the framework of aquatic system. Accumulated organic matter reflecting organic carbon in the sediment is mineralized to CO_2 or CH_4 by heterotrophic microbes or it is gradually mixed with soil particles or buried in the sediment over time scale. Microbial carbon pump was found to exert a profound role in the transformation of labile organic carbon into recalcitrant organic carbon in the soil [20]. In aquaculture ponds, a layer of sediment with higher concentration of organic carbon of pond soil developed a characteristic profile of well defined strata or horizons [39].

The organic carbon content of soil ranging from 29.58-118.02 ppm was highly correlated ($r = 0.82$; $P < 0.001$) with the carbon burial rates which varied by a factor of 5 in different treatments employed. Further, the green house effects were clearly manifested in the study because of distinctly higher values of soil organic carbon in all treatment tanks inside than outside.

The primary productivity of microalgae was found to be substantially higher within the range of 25 and 40 mg/g of organic carbon of soil in the overall range of 11.52 and 55.1 mg in all treatments. This suggests that this range of organic carbon was conducive to microalgae production regardless of the condition employed. Further, the highest algal productivity accompanied by attenuated carbon burial rate (118.026 g / m^2/100 day) in poultry dropping and saw dust (T-2) treatment combination and the reverse response in the mixed treatment (cattle manure, poultry dropping, vermi-compost and saw dust combination), implied that high manural value of the former which was due to the blend of extremely high C/N ratio of saw dust (100-500) and low C/N ratio of poultry dropping (<10). This shows that the saw dust- almost a discarded material, can be effectively used as manure in aquaculture system in combination with poultry dropping even though the saw dust is rich in carbonaceous substances having very high C/N ratio (fresh-saw dust-200, decomposed saw dust-500). However, it is reported that there was no increase in soil organic matter and soil total carbon due to application of saw dust in agricultural corn field [40].

Substantial decline in carbon burial rate and organic carbon of soil in poultry dropping and saw dust combination was attributed to enhanced manure decomposition due to reduced carbon relative to nitrogen (C/N = 10.61), mineralization and subsequent utilization by microalgal biomass through biogeochemical cycling rather than its sink as residual organic carbon in sediment which was encountered in the mixed treatment (T-4) showing the reverse response with T-2 where nitrogen is more relative to carbon (C/N = 15.77). So, the overall soil carbon status in the former was primarily regulated by CN ratio driven autochthothonous source derived from dead massive algal bloom and less by the allochthonous source of residual manure and vice versa in the later where heterotrophic metabolism and carbon sink dominated by the integrative environmental, trophic and microbial carbon pump. The lowest carbon burial rate in iso-carbonic state (T-6) using reduced load of poultry dropping and saw dust indicated that the carbon sequestrating capacity of the bottom soil was to some extent input load dependent even with low CN ratio (13.96).

Autochthonous DOC accounted for >25% of the total water column DOC pool, and autochthonous OC production attenuated the concentration of dissolved inorganic carbon in the water column and thereby contributing to atmospheric carbon dioxide uptake [41]. It is known that sediment metabolism is dependent upon the allochthonous carbon source [42], and much of the organic matter of autochthonous origin derived from dead plankton, fish excrement, etc. is readily decomposed, mineralized and utilized for ecosystem functioning. Lakes are grouped into lakes with net consumption of CO_2 in the sediment with <5.6 mg/l dissolved organic carbon and lakes with very low dissolved organic carbon (<2.6 mg/l) which were net autotrophic due to dominance of allochthonous organic matter in the water as an energy source for aquatic organism [42]. Diverse soil organic compounds exhibit a wide range of kinetic properties that determine the intrinsic temperature sensitivity of decomposition of specific organic compounds [43] and, hence, have potentials of positive or negative feedback effects on climate.

Experimental evidences suggest that the green house mediated raised temperature primarily coupled with elevated CO_2 absorbed through diffusion in the aquatic phase and brought out changes in the time scale variations of buffering mechanisms showing opposite responses under the closed and open conditions. High temperature induced rapid microbial decomposition under closed condition resulted in gradual development of alkaline condition from its initial acidic mode with high concentrations of free carbon dioxide that, in turn, brought associated changes of water and sediment quality as well as primary productivity of microalgae over the course of time. Exhibiting a strong direct relationship with each other, the pattern of changes of TDS and specific conductance of water remained distinctly higher under simulated greenhouse conditions than open system.

Carbon sequestration capacity of aquaculture ponds could be used as carbon reduction credits against carbon emissions from aquaculture production. Net carbon sequestration (carbon emissionscarbon sequestration) by aquaculture operations could be treated as carbon credits. It is further revealed that the green house effects in the framework of managed aquatic environment were profound because of elevated carbon burial rate under the closed condition than outside. The carbon burial rate in different lakes, aquaculture ponds and reservoirs was highly variable ranging from 0.05 -21.2 t/ha/yr [21]. In a subtropical eutrophic Donghu lake, the carbon sink was estimated to be great with C input as 14905 t and C output as 4950 t and net carbon budget as 9955 t [44]. Further, it is suggested that increasing soil organic carbon (SOC) pool in soils could offset emissions of CO_2 from fossil fuel combustion [21].

As a mechanistic model, it is proposed that substantial rise in carbon sink due to greenhouse effect in the framework of managed aquatic system is attributable to microalgae derived autochthonous carbon sink mediated through integrative functions of raised temperature, carbon dioxide level and manure-microbial driven nutrient release conducive to sequester autochthonous carbon via photosynthesis of microalgae, and the allochthonous carbon sink in heterotrophic dominated framework was mediated through microbial carbon pump that transform labile carbon into immobile or depository carbon in the soil. It is projected that under future global warming scenarios partitioning of primary production into DOC could be enhanced and thus MCP could play an even more important role in carbon sequestration [20].

CONCLUSION

It may be concluded that substantial rise in carbon status in the sediment due to green house effect was attributable to enhanced microalgae derived autochthonous carbon sink mediated through integrative functions of temperature induced microalgal productivity and microbial driven manure decomposition. The dead microalgae and residual manure that settled down in the bottom surface enhanced the carbon status of the system, in question.

ACKNOWLEDGEMENT

This research was component of the ICAR NICRA Project sponsored by the Indian Council of Agricultural Research (to BBJ), New Delhi. We thank Mr. Prodyut Ghosh and Miss Debarati Ghosh and Miss Ankita Bhattacharyya for their help during the preparation of the manuscript.

REFERENCES

1. Hu Z, Lee JW, Chandran K, Kim S, Khandal SK (2012) Nitrous oxide (N2O) emission from aquaculture: A review. Environ Sci Technol 46: 6470-6480.
2. Indian Network for Climate Change Assessment (2010) India: Greenhouse gas emissions 2007. Ministry of Environment and Forests. Government of India, New Delhi.
3. Chisti Y (2007) Biodisel from microalgae. Biotechnolgy Advances 25: 294-306.
4. Sydney EB, Sturm W, Larroche C (2010) Potential carbon dioxide fixation by industrially important microalgae. Bioresource Technology 101: 5892-5896.
5. Ho SH, Chen CY, Lee DJ, Chang JS (2011) Perspectives on microalgal CO2 emission mitigation systems: a review. Biotechnology Advances 29:189-198.
6. Lancu P, Plesu V, Velea S (2012) Flue gas CO2 capture by microalgae in photobioreactor: a sustainable technology. Chemical Engineering Transactions 29: 799-804.
7. Brennan L, Owende P (2009) Biofuels from microalgae: A review of technologies for production, processing, and extraction of biofuels and co-products. Renewable and Sustainable Energy Reviews.
8. Huertas E, Rouco M, Lopez-Rodas V, Costas E (2011) Warming will affect phytoplankton differently: evidence

through a mechanistic approach. Proceedings B, The Royal Society 282: 3534-3543.

9. Toseland A, Daines SJ, Clark JR, Kirkham A, Strauss J, et al. (2013) The impact of temperature on marine phytoplankton resource allocation and metabolism. Nature Climate Change 3: 979-984.

10. Behrenfeld MO, Malley R, Siegel D, McClain C, Samiento J, et al. (2006) Climate driven trends in contemporary ocean productivity. Nature 444: 752-755.

11. De la Rubia MA, Walker M, Heaven S, Banks CJ, Borja R (2010) Preliminary trials of in situ ammonia stripping from source segregated domestic food waste digestate using biogas: effect of temperature and flow rate. Bioresource Technology 101: 9486-9492.

12. US EPA (2009) Synthesis of adaptation options for coastal areas, Washington DC. US. Environmental Protection Agency, Climate Ready Estuaries Programme. EPA 430-F-08-024.

13. Ferretti DF, Miller JB, White JWC, EtheriddgeDM, Lassey KR, et al. (2005)Unexpected Changes to the Global Methane Budget over the Past 2000 Years. Science 309: 1714-1717.

14. IPCC (2014) Climate change -2014: impacts, adaptation, and vulnerability: Part-B: Regional Aspects. Cambridge University Press.

15. Cole JJ, Prairie YT, CaracoNF, McDowell WH, TranvikLJ, et al. (2007) Plumbing the global carbon cycle: integrating inland waters into the terrestrial carbon budget. Ecosystems 10:171-184.

16. Sobek S, Tranvik LJ, Prairil YT, Kortelainen P, Cole JJ (2007) Patterns and regulation of dissolved organic carbon: An analysis of 7500 widely distributed lakes. Limnology &Oceanogr52: 1208-1219.

17. Chmura GL, Anisfeld SC, Cahoon DR, Lynch JC (2003) Global carbon sequestration in tidal, saline wetland soils. Global Biogeochemical Cycles 17: 1111.

18. Wrona FJ, Prowse TD, Reist JD, Hobbie JE, Levesque LMJ, et al.(2006) Climate change effects on biota, ecosystem structure and function. Ambio35: 359-369.

19. Tranvik LJ, Downing JA, Cotner JB,Loiselle SA, Striegl RG, et al. (2009)Lakes and reservoirs as regulators of carbon cycling and climate. Limnolgy& Oceanography54: 2298-2314.

20. Jiao N, Zheng Q (2011)The microbial carbon pump: from genes to ecosystems. Applied Environmental Microbiol77: 7439-7444.

21. Boyd CE, Wood WW, Chaney PL, Queiroz JF (2010) Role of aquaculture pond sediments in sequestration of annual global carbon emissions. Environmental Pollution 158: 2537-2540

22. Verdegem MC, Bosma RH (2009) Water withdrawal for brackish and inland aquaculture, and opinions to produce more fish in ponds with present water use.Water Policy11: 52-68.

23. Zheng J (2010) Temperature sensitivity of soil organic matter decomposition and the influence of soil carbon pools and attributes. Iowa State University, USA.

24. Dean WF, Gorham E (1998) Magnitude and significance of carbon burial in lakes, reservoirs and peatlands. Geology 26: 535-538.

25. Sobek S, Anderson NJ, Bernasconi SM, Del Sontro T (2014) Low carbon burial efficiency in arctic lake sediments. Journal of Geophysical Research: Biogeosciences, 119: 1231-1243.

26. Downing JA, Prairie YT, Cole JJ, Durante CM,TranvikLJ,et al. (2006) The global distribution of lakes, ponds, and impoundments. Limnology and Oceanography 51: 2388-2397.

27. Downing JA, Cole JJ, Middelburg JJ, Striegl R, Durante CM, et al. (2008) Sediment organic carbon burial in agriculturally eutrophic impoundments over the last century. Global Biogeochemical Cycles 22: 1-10.

28. Food and Agriculture Organization of the United Nations (FAO) (2007) Fisheries Technical Paper 366. Aquaculture Feed and Fertilizer Resource Atlas of the Philippines.

29. MistryJ, Mukhopadhyay AP, Baur GN (2015) Status of NPK in vermicompost prepared by two common weed and two medicinal plants. Journal of Advanced Studies in Agricultural Biological and Environmental Sciences 2: 25-38.

30. Jenkins J (2005) The Carbon/Nitrogen Ratio. The Humanure Handbook. (3rd Ed) Chelsea Green Publishing: Grove City, USA

31. Eze VC, Uzoaru N, Agwung-Fobella D (2011) Isolation and characterization of microorganisms involved in degradation of saw dust waste in rivers state, Nigeria. Asian Journal of Science and Technology 1: 44-48.

32. Jackson ML (1967) Soil Chemical Analysis. Prentice Hall of India, New Delhi.

33. Wetzel RG (2001) Limnology. (3rd Ed) Academic Press, Cambridge.

34. Peterson DH, Perry MJ, Bencala KE, Talbot MC (1987) Phytoplankton productivity in relation to light intensity: A simple equation. Estuarine, Coastal and Shelf Science 24: 813-832.

35. Chen B (2015) Patterns of thermal limits of phytoplankton. J Plankton Res.

36. Jana BB, Lahiri S, Nandy S, Ghosh P, Bag SK, et al. (2015) Polyhouse driven raised temperature induces cichlid tilapia to breed during winter. Indian Journal of Animal Science 85: 108-115.

37. Dickson N, Richard T, Kozlowski R (1991) Composting to reduce the waste stream. Cornell Composting-Science and Engineering, Cornell University, Ithaca, USA.

38. JanaBB, Chakraborty P, Biswas JK, Ganguly S(2001) Biogeochemical cycling bacteria as indices of pond fertilization: importance of CNP ratios of input fertilizers. J Applied Microbiology 90: 733-740.

39. Munsiri P, Boyd CE, Hajek BF (1995) Physical and chemical characteristics of bottom soil profiles in ponds at Auburn, Alabama, USA, and a proposed method for describing pond soil horizons. Journal of the World Aquaculture Society 26: 346-377.

40. Tran HM (2005) Quantifying the effects of saw dust application and soil chemical and physical properties and corn yield. University of Tennessee.

41. Watanabe K, Tomohiro K (2015)How organic carbon derived from multiple sources contributes to carbon sequestration processes in a shallow coastal system? Global Changes Biology21: 2612-2623.

42. Algesten G, Sobek S, Bergstrom AK, Jonsson A, Tranvik, LJ, et al. (2005)Contribution of Sediment Respiration to Summer CO2 Emission from Low Productive Boreal and Subarctic Lakes Microbial Ecology 50: 529-535.

43. Davidson EA, Jenssen IA (2006) Temperature sensitivity of soil carbon decomposition and feedbacks to climate change. Nature 440: 165-173.

44. Yang H, Xing Y, Xie PNL, Rong K (2008) Carbon source/sink function of a subtropical, eutrophic lake fetermined from an overall mass balance and gas exchange and carbon burial balance. Environmental Pollution 15: 559-568.

CHAPTER 9

AN APPROACH TO MICROALGAL PRODUCTION SYSTEMS FOR COMMODITIES

Renil Anthony and Troy Runge

Microalgae have been actively researched in recent years as a feedstock for commodities such as fuels, biosolvents [1], animal feed [2], and bio-based plastics and polymers [3]. Microalgae are a promising biomass feedstock due to a wide range of product possibilities, high growth rates, and low resource demands. Although there are small to mid-size commercial microalgae production systems for high value human nutritional supplements worldwide, large-scale commercialization of microalgal systems for commodities are limited by the low economic returns on the products generated. Continuing research and testing of different production strategies and scenarios represents the best path to achieve technical and economic feasibility of microalgae to bioproducts.

Currently, commercial production of microalgae totals nearly 15,000 t/y with the majority of the material produced for human consumption, and almost all cultivated in outdoor open ponds [2]. Microalgae Spirulina, Chlorella, Dunaliella, and Haematoccocus are few of the commercially produced strains with major production facilities in China, Japan, USA, India, and Australia [2]. High value nutritional and pharmaceutical products such as ω-3 fatty acids, antioxidants, vitamins, and proteins derived from these strains justify the economics of commercial production and processing. Unfortunately, the economic viability of microalgae for lower margin commodities needs to be improved considerably. Since commodity prices are unlikely to rise significantly, the best strategy is to lower algal feedstock production costs by improving cultivation and harvesting practices.

Citation: Anthony R, Runge T (2014) An Approach to Microalgal Production Systems for Commodities. Oceanography 2: 116. doi:10.4172/2332- 2632.1000116.

Commercial cultivation of microalgae in outdoor open ponds or raceways is considered to have lower costs and thus more economically attractive than closed system photobioreactors [2]. Such open ponds systems can be built upstream of wastewater treatment plants or beside agricultural facilities. Nutrient growth media requirements in these ponds can be fulfilled using municipal wastewater or agricultural residue [4]. Additionally, waste streams of carbon dioxide from flue gas or fermenters can be utilized as a source of inorganic carbon for enhanced growth [5]. However, a considerable challenge faced by open pond systems is to constantly maintain the desired strain cultures. Such contamination problems can be overcome by carefully selecting microalgae strains that thrive in adverse conditions such as marine algae, thermophilic algae, and algae that can grow in extreme pH conditions and then simulating those conditions to inhibit wild type algae growth.

Facile harvesting of microalgae represents a significant technical challenge and is one of the key processing steps that impact production economics. Simple harvesting methods need to be developed to concentrate microalgae to greater than 3% solids prior to using conventional dewatering methods such as centrifugation or filtration to allow for sufficient equipment throughput to minimize drying costs. Microalgae can be agglomerated by bioflocculation or by pH adjustment of growth media through control of the CO_2 supply [6,7]. The use of coagulant or flocculant additives should be avoided as much as possible to support downstream processes and keep processing costs low.

Based on the desired bioproduct from microalgae, decisions should be made regarding drying, pretreatment, and biomass conversion. Drying requirements for products such as animal feed or fertilizer should be met with waste heat sources or solar energy whenever possible, taking advantage of local resources and geographical location of the facility. For other products that require further processing and conversion of the biomass, the goal should be to utilize a biorefinery platform to produce several products by fractionating the algal biomass without drying [8]. To this effect, recent studies employing subcritical co-solvent extraction [9], in situ transesterification [10], acid-base hydrolysis [11], and polymeric membranes [12] have been successful at laboratory scale. To be acceptable as a sustainable feedstock, the economics of low value products from microalgae should be credited with the offset costs of wastewater remediation and CO_2 capture, with these credits and processing costs allocated to all the products generated by fractionating microalgae.

Finally, genetic engineering of microalgae is another path towards lower processing costs by achieving superior biochemical composition or in some instances by the direct secretion of products desired [13]. Similarly, heterotrophic and mixotrophic microalgae are also being tested to utilize carbohydrate feedstocks to produce lipid content as high as 60% dry weight [14,15]. Although these systems incorporate carefully controlled closed systems, such as photobioreactors, the experience gained is invaluable in developing commercial-scale microalgal systems.

In summary, microalgae are organisms with high photosynthetic efficiency that can utilize and remediate waste streams and have the potential to generate an array of important commodities. However, this potential can only be unlocked by developing effective low cost production systems. The attainment of sustainable bioenergy from microalgae requires significant R&D to develop simple and efficient processes. Although the establishment of commercial microalgal systems is a complex task, a concerted effort will ensure feedstock sustainability stimulating local and regional development.

REFERENCES

1. Ellis JT, Hengge NN, Sims RC, Miller CD (2012) Acetone, Butanol, and Ethanol Production from Wastewater Algae. Bioresour Technol 111: 491-495.
2. Benemann J (2013) Microalgae for Biofuels and Animal Feeds. Energies 6: 5869-5886.
3. Rahman A, Anthony RJ, Sathish A, Sims RC, Miller CD (2014) Effects of Wastewater Microalgae Harvesting Methods on Polyhydroxybutyrate Production. Bioresour Technol.
4. Hoffmann JP (1998) Wastewater Treatment with Suspended and Nonsuspended Algae. J Phycol 34: 757-763.
5. Lam MK, Lee KT, Mohamed AR (2012) Current Status and Challenges on Microalgae-Based Carbon Capture. International Journal of Greenhouse Gas Control 10: 456-469.
6. Liu J, Zhu Y, Tao Y, Zhang Y, Li A, et al. (2013) Freshwater Microalgae Harvested via Flocculation Induced by pH Decrease. Biotechnol Biofuels 6: 98.
7. Wu Z, Zhu Y, Huang W, Zhang C, Li T, et al. (2012) Evaluation of Flocculation Induced by pH Increase for Harvesting Microalgae and Reuse of Flocculated Medium. Bioresour Technol 110: 496-502.
8. Anthony RJ, Ellis JT, Sathish A, Rahman A, Miller CD, et al. (2013) Effect of Coagulant/flocculants on Bioproducts from Microalgae. Bioresour Technol 149: 65-70.
9. Chen M, Liu T, Chen X, Chen L, Zhang W, et al. (2012) Subcritical Co-Solvents Extraction of Lipid from Wet Microalgae Pastes of Nannochloropsis sp. Eur J Lipid Sci Technol 114: 205-212.
10. Cao H, Zhang Z, Wu X, Miao X (2013) Direct Biodiesel Production from Wet Microalgae Biomass of Chlorella pyrenoidosa through In Situ Transesterification. BioMed Research International.
11. Sathish A, Sims RC (2012) Biodiesel from Mixed Culture Algae via a Wet Lipid Extraction Procedure. Bioresour Technol 118: 643-647.
12. Yoo G, Yoo Y, Kwon J-H, Darpito C, Mishra SK, et al. (2013) An Effective, CostEfficient Extraction Method of Biomass from Wet Microalgae with a Functional Polymeric Membrane. Green Chemistry 16: 312-319.
13. Radakovits R, Jinkerson RE, Darzins A, Posewitz MC (2010) Genetic Engineering of Algae for Enhanced Biofuel Production. Eukaryotic Cell 9: 486- 501.
14. Hu B, Min M, Zhou W, Du Z, Mohr M, et al. (2012) Enhanced Mixotrophic Growth of Microalga Chlorella sp. on Pretreated Swine Manure for Simultaneous Biofuel Feedstock Production and Nutrient Removal. Bioresour Technol 126: 71-79.
15. Perez-Garcia O, Escalante FM, de-Bashan LE, Bashan Y (2011) Heterotrophic Cultures of Microalgae: Metabolism and Potential Products. Water Res 45: 11-36.

CHAPTER 10

MICROALGAE AS NANOFACTORY FOR PRODUCTION OF ANTIMICROBIAL MOLECULES

Luis G Ramirez-Merida, Leila Queiroz Zepka, Cristiano Ragagnin de Menezes and Eduardo Jacob-Lopes

Keywords: Nanofactory, Prophylactic therapy, Ubiquitous, Antimicrobial effect

MICROALGAL NANOFACTORIES

Microbial contamination has caused countless deaths and large public health problems in the world population. The presence of microbial pathogens in foods [1], intensive care unit [2], prosthesis infection [3], among others, has led to the development of methods for the control and prevention of microbial infections. The extensive use of antibiotics has facilitated the continued emergence and spread of resistant organisms, causing global alarm due to the public health threat they represent. Added to this, biofilms formation on biotic and abiotic structures that are difficult to eliminate which causes greater resistance to the removal, generating chronic problems of reinfection [4].

Looking for solutions, some possible strategies have been based on the rotation control of certain antibiotics like employing compounds or complements derived from plants extract, the primary or secondary cell metabolites, that submit multiple mechanisms of action in medical and industrial applications [5,6]. However, the chemical interaction of these biocontrol agents with organic or inorganic elements produces decreased of antimicrobial action.

In this regard, nanotechnology offers a wide range of possibilities as to development of new materials with antimicrobial properties that may affect medical and industrial areas. Despite this, physical and chemical methods for making these nanomaterials cause problems of toxicity limiting their applications in clinical fields. Therefore, for integration of nanotechnology at health systems, it is necessary to use biological routes from microorganisms and plants for the synthesis of nanomaterials [7].

Microalgae offer certain advantages to become a powerful nanofactory. In recent years, the use of microalgae has focused on biofuel production and carbon dioxide fixation [8,9]. However, the diversity of microalgae metabolic compounds has caught the attention due to nutraceutical and pharmacological contribution. Although the antimicrobial activity of microalgae and/or bioproducts have been poorly investigated, there are a great diversity of species and metabolic compounds that recently received important consideration as a new source of novel bactericidal and fungicidal activity (Figure 1).

Citation: Ramirez-Merida LG, Zepka LQ, de Menezes CR, Jacob-Lopes E (2015) Microalgae as Nanofactory for Production of Antimicrobial Molecules. J Nanomed Nanotechnol S6-004. doi:10.4172/2157-7439.S6-004.

Chemical compound	Chemical structure	Microbial strains targets	Ref.
Malyngolide	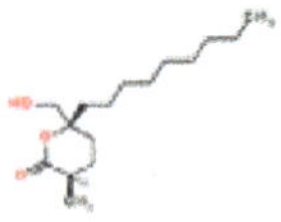	*M. smegmatis*, *S. pyogenes*, *S. aureus*, *Pseudomonas* sp.	[10]
Halimedatrial	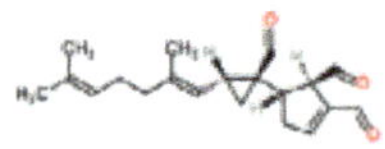	*S. marinorubra*, *V. splendida*, *V. leiognathi*, *V. harveyi*.	[11]
Phlorotannins	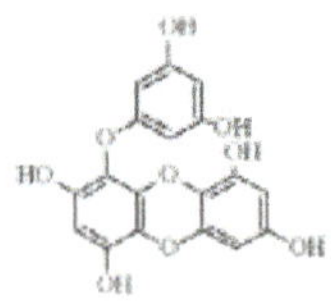	*T. rubrum*, *S. aureus*, MRSA*, *C. jejuni*, *E. coli*, *S. enteritidis*, *S. typhimurium*, *V. parahaemolyticus*, *B. subtilis*, *Acinetobacter* sp., *K. pneumonia*	[12]
(4E,7S)-7 methoxytetradec-4-enoic acid	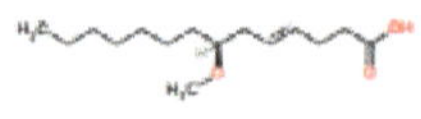	*S. aureus*, *B. subtilus*	[13]
Pahayokolide A	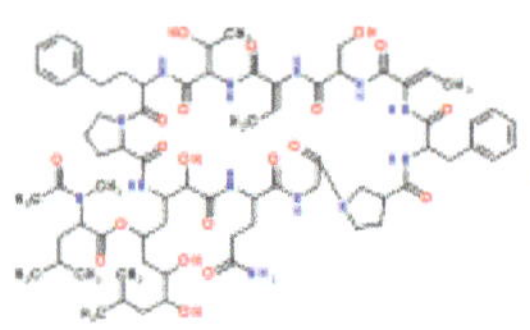	*B. megaterium*, *B. cereus*, *S. cerevisiae*	[14]
Eicosapentaenoic acid		*Photobacterium* sp., *B. cereus*, *B. weihenstephanensis*, *S. aureus*, *S. epidermidis*, MRSA*	[15]
Hexadecatrienoic acid		*L. anguillarum*, *P. citreus*, *S. aureus*, *S. epidermidis*	[16]
Palmitoleic acid	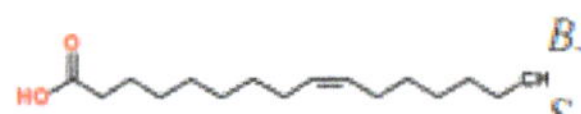	*B. cereus*, *B. weihenstephanensis*, *S. aureus*, *S. epidermidis*, MRSA*	[16]

Figure 1: Chemical structure of microalgae metabolic compounds with antimicrobial activity. *MRSA: multidrug-resistant S. aureus.

Microalgae cells can be used for the synthesis of various inorganic nanoparticles, ranging in size from 1-10 nm [10-17]. The microalgae have certain advantages as nanofactories due to fast growth, formation of metabolic compounds and proteins that reduce silver, gold ions and other metals that helps stabilize metallic nanoparticles [18]. Many species of microalgae have the ability to adsorb and entrap heavy metal ions by binding to carboxyl groups of cell biomass [18], meanwhile, polyphosphates and polysaccharides groups, intra- and extracellular, participate in chelation and oxidation of metals [19].

Microalgae can perform extracellular synthesis of metal nanoparticles, besides of eliminate some steps in the process of biorefinery, which it brings greater importance for biomedical and biotechnological applications.

The antimicrobial activity of metal nanoparticles has been demonstrated [20]. The silver nanoparticle acts as a broad-spectrum antibiotic. They produce rapid killing kinetics, even on multidrugresistant pathogens [21]. The mechanism of action is much debated, but is believed that DNA loses its ability to replicate and membrane proteins are inactivated by interaction of Ag+ ions [22]. The union of Ag+ ions causes denaturation of cellular proteins besides the respiratory chain is disengaged causing cellular collapse [23]. Moreover, the metals in the presence of oxygen can induce the generation of reactive oxygen species generating oxidative stress [24].

Many of nanoparticles and bioproducts synthesized by microalgae can be coupled into nanofibers and nanotubes. Thus, this new nanomaterial would enhance the antimicrobial action. These nanofibers can be marketed as prophylactic treatment by incorporation into wound healing bandages, surgical equipment, prosthesis, food packaging, among others.

Therefore, the integration of biotechnological processes that solve medical problems in society would make a breakthrough for biomedical market.

The productive development of microalgae cultivation with suitable photobioreactors [25], could generate high yields in production of metal nanoparticles and/or biomass bioproducts that can be integrated into nanofibers or nanotubes with antimicrobial activity, helping to prevent reinfection by microorganisms, it would be an ideal process.

Microalgae can achieve a promising future as antimicrobials with exploring potential areas for bioprospecting. This coupled with the physiological advantages; it allows incorporation into nanotechnology helping to have antimicrobial molecules capable of solving problems of contamination and infection not solved by common antibiotics.

REFERENCES

1. Ramírez MLG, Moron SA, Alfieri GAY, Gamboa O (2009) Prevalence of Listeria monocytogenes in fresh tomatoes (Lycopersicumesculentum) and coriander (Coriandrumsativum) in three markets of Valencia, Venezuela. ArchivosLatinoamericanos de Nutrición 59: 318-324.
2. Edwards JD, Herzig CT, Liu H, Pogorzelska-Maziarz M, Zachariah P, et al. (2015) Central line-associated blood stream infections in pediatric intensive care units: Longitudinal trends and compliance with bundle strategies. Am J Infect Control 43: 489-493.
3. Dutkiewicz A, Malinowski P (2012) Aortobifemoralprothesis infection. Polish Annals of Medicine 19: 129-133.
4. Burmølle M, Ren D, Bjarnsholt T, Sørensen SJ (2014) Interactions in multispecies biofilms: do they actually matter? Trends Microbiol 22: 84-91.
5. Ramírez Mérida LG, Morón de Salim A, Catinella R, Castillo L (2012) Bacteriostatic and/or bactericidal extract of Aloe vera gel on cultures of Listeria monocytogenes. Arch LatinoamNutr 62: 73-78.
6. Barra Caracciolo A, Topp E, Grenni P (2015) Pharmaceuticals in the environment: biodegradation and effects on natural microbial communities. A review. J Pharm Biomed Anal 106: 25-36.
7. Narayanan KB, Sakthivel N (2011) Green synthesis of biogenic metal nanoparticles by terrestrial and aquatic phototrophic and heterotrophic eukaryotes and biocompatible agents. Adv Colloid Interface Sci 169: 59-79.
8. Francisco EC, Balthazar D, Jacob-Lopes E, Franco TT (2010) Microalgae as feedstock for biodiesel production: carbon dioxide sequestration, lipid production and biofuel quality. Journal of Chemical Technolgy and Biotechnology 85: 395-403.
9. Jacob-Lopes E, Franco TT (2013) Fromoil refinery to microalgalbioreinery. Journal of CO2 Utilization 2: 1-7.
10. Cardllina JH, Moore RE, Arnold E, Clardy J (1979) Structure and absolute configuration of malyngolide, an antibiotic from the marine blue-green alga LyngbyamajusculaGomont. The Journal of Organic Chemistry 44: 4039-4042.
11. Paul VJ, Fenical W (1983) Isolation of halimedatrial: chemical defense adaptation in the calcareous reef-building alga halimeda. Science 221: 747-749.
12. Eom SH, Kim YM, Kim SK (2012) Antimicrobial effect of phlorotannins from marine brown algae. Food ChemToxicol 50: 3251-3255.
13. Faulkner DJ1 (1987) Marine natural products. Nat Prod Rep 4: 539-576.
14. Berry JP, Gantar M, Gawley RE, Wang M, Rein KS (2004) Pharmacology and toxicology of pahayokolide A, a bioactive metabolite from a freshwater species of Lyngbya isolated from the Florida Everglades. Comparative Biochemistry and Physiology Part C: Pharmacology, Toxicology and Endocrinology 139: 231-238.
15. Desbois AP, Mearns-Spragg A, Smith VJ (2009) A fatty acid from the diatom Phaeodactylumtricornutum is antibacterial against diverse bacteria including multi-resistant Staphylococcus aureus (MRSA). Mar Biotechnol (NY) 11: 45-52.
16. Desbois AP, Lebl T, Yan L, Smith VJ (2008) Isolation and structural characterisation of two antibacterial free fatty acids from the marine diatom, Phaeodactylumtricornutum. ApplMicrobiolBiotechnol 81: 755-764.
17. Govindaraju K, Basha SK, Kumar VG, Singaravelu G (2008) Silver, gold and bimetallic nanoparticles production using single-cell protein (Spirulinaplatensis) Geitler. Journal of Materials Science 43: 5115-5122.
18. Hameed A, Hasnain S (2005) Cultural characteristics of chromium resistant unicellular cyanobacteria isolated from local environment in Pakistan. Chinese Journal of Oceanology and Limnology 23: 433-441.
19. De Philippis R, Sili C, Paperi R, Vincenzini M (2001) Exopolysaccharide-producing cyanobacteria and their possible exploitation: A review. Journal of Applied Phycology 13: 293-299.
20. Merin DD, Prakash S, Bhimba BV (2010) Antibacterial screening of silver nanoparticles synthesized by marine micro algae. Asian Pacific Journal of Tropical Medicine 3: 797-799.
21. Rai MK, Deshmukh SD, Ingle AP, Gade AK (2012) Silver nanoparticles: the powerful nanoweapon against multidrug-resistant bacteria. J ApplMicrobiol 112: 841-852.
22. Feng QL, Wu J, Chen GQ, Cui FZ, Kim TN, et al. (2000) A mechanistic study of the antibacterial effect of silver ions on Escherichia coli and Staphylococcus aureus. J Biomed Mater Res 52: 662-668.
23. Lok CN, Ho CM, Chen R, He QY, Yu WY, et al. (2006) Proteomic analysis of the mode of antibacterial action of silver nanoparticles. J Proteome Res 5: 916-924.
24. Park HJ, Kim JY, Kim J, Lee JH, Hahn JS, et al. (2009) Silver-ion-mediated reactive oxygen species generation affecting bactericidal activity. Water Res 43: 1027-1032.
25. Ramírez-Merida LG, Zepka LQ, Jacob-Lopes E (2015) Why does the Photobioreactors Fail? Journal of Bioprocessing &Biotechniques 5:7.

CHAPTER 11

CELL DISRUPTION AND PRESSURIZED LIQUID EXTRACTION OF CAROTENOIDS FROM MICROALGAE

Josef Taucher, Sascha Baer, Philipp Schwerna, Dominik Hofmann, Magdalena Hümmer, Rainer Buchholz and Anna Becker

Keywords: *H. pluvialis*, *C. sorokiniana*, *C. zofingiensis*, Carotenoids, Cell disruption, Pressurized liquid extraction, HPLC

INTRODUCTION

Microalgae, including cyanobacteria, play an increasing role in science and industry, due to their wide range of commercial and potential novel products [1-3]. They can be cultivated using sunlight (energy source) and waste streams, such as recycled media from other processes (e.g., aquaculture, biogas digestate - nutrient source) and industrial flue gas (anorganic carbon source), that positively contribute to cost reduction and ecological balance [4-9]. One of the most important, and on the market best established, high value products from microalgae are carotenoids. Carotenoids are a group of structurally highly diverse terpenoid pigments (more than 750 have been isolated), which can be divided into carotenes and oxygenated derivatives of carotenes, so called xanthophylls [10-12]. They act as light-harvesting pigments, absorb light in a range of λ=400-550 nm and transfer the light to chlorophyll [12,13]. Furthermore, they stabilize the structure and the functionality of the photosynthetic apparatus [11,12], e.g., by quenching chlorophyll triplet states, dissipating excess energy, or scavenging reactive oxygen species [11,14]. In general, applications of carotenoids can be divided into three main groups: i) as natural dyes in food and feed industry, ii) as feed additives in aquaculture and poultry farming, iii) as well as in the pharmaceutical sector and in cosmetics, due to their antioxidative properties [10,11,15]. The most commonly used carotenoids are β-carotene, astaxanthin and lutein that can be produced by various microalgae such as Dunaliella salina, Haematococcus pluvialis, Chromochloris zofingiensis, Chlorella sorokiniana, etc. [1,11,15]. The overall market for carotenoids was estimated to US$ 1.2 billion in 2010 and was predicted to increase up to US$ 1.4 billion in 2018 [15]. Market prices vary from 300 - 3,000 US$·kg^{-1} and 2,500-10,000 US$·kg-1 in case of β-carotene and astaxanthin, respectively [16,17]. Although there are many described algae species, until now only few have found application in biotechnological processes (e.g., H. pluvialis, D. salina, Arthrospira platensis, Chlorella species) [18]. This is due to still relatively high production costs making them incapable of competing with the chemical industry while only delivering one class of value products [19]. To approach this issue, integrated processes allowing almost complete utilization of the produced biomass, waste streams and nutrients are needed. Derived from the oil-based refinery, in which the raw material is more or less utilized completely, such integrated biotechnical applications are called biorefinery processes. Particularly, an optimal design of the downstream processing is essential for feasibility of such systems as it may contribute up to 90% of the total costs [20].

In order to establish an efficient isolation procedure for potential products for the future industrial application, at first an optimization of the extraction in labscale has to be performed for various strains of microalgae. This is not only necessary during the screening phase to assure comparability of the results, but also in later stages of the process design as the collected data and experience can be used for example for upscaling. In case of intracellular products not only extraction parameters such as: selected solvent, temperature, duration, cycle number are crucial but also an efficient cell disruption. At the same time, when considering further upscaling of the process, other aspects such as costs and feasibility for industrial application have to be considered. Thus, the main aim of the work presented here was the optimization of the cell disruption

Citation: Taucher J, Baer S, Schwerna P, Hofmann D, Hümmer M, et al. (2016) Cell Disruption and Pressurized Liquid Extraction of Carotenoids from Microalgae. J Thermodyn Catal 7: 158. doi:10.4172/2158-7544.1000158.

and extraction procedure for carotenoids for future application as an important isolation step of one of the relevant product classes in a microalgae based biorefinery.

MATERIALS AND METHODS

Strains and Media

In this work, three microalgae strains, provided by the Culture Collection of Algae at Goettingen University in Germany (SAG), were used: Haematocouccus pluvialis SAG 34-1b, Chromochloris zofingiensis SAG 211-14 and Chlorella sorokiniana SAG 211-8k. Basal medium (with 1 g·L^{-1} peptone and without soil extract) suggested by SAG was used for photoautotrophical cultivation of H. pluvialis. In case of C. zofingiensis and C. sorokiniana, a modified medium of Arnon et al. [21] described by Cordero et al. [10], was used for photoautotrophical cultivation.

Cultivation

Photobioreactor screening modules (PSM) were used for cultivations under monoseptic conditions, as previously described [22-24]. The aeration was kept constant at 0.45 L·min^{-1}, enriched with 3% CO_2, and the temperature was 25°C. The light intensity (photon flux density PFD) was measured with a submersible spherical light sensor (US-SQS/L, ULM-500, Walz, Germany) in the center of the PSM filled with water. H. pluvialis and C. sorokiniana were cultivated with a light intensity of 700 µmol·m-2·s-1. In case of C. zofingiensis the PFD was increased after six days from 470 up to 700 µmol·m-2·s-1 to boost the carotenoid production. All three microalgae were kept in batch mode for two weeks.

Determination of Algae Carotenoid Yield

After two weeks of cultivation, the harvested and freeze-dried biomass was disrupted/homogenized using various methods described below and extracted via pressurized liquid extraction (PLE) at the given conditions. Extracts were evaporated to dryness under the nitrogen and redissolved in 2.5 mL of a mixture of acetone and MeOH (9:1, v:v) before saponification. The latter was started by addition of 0.5 mL of 0.05 M NaOH in MeOH and incubated for 3 h (room temperature, darkness) to eliminate carotenoid esters [25-27]. Next, 3 mL of petroleum ether were added, the mixture was washed with 3 mL 10% NaCl, centrifuged (3,500 rpm, 1 min, 5810R, Eppendorf; Hamburg, Germany) and the aqueous phase was discarded. The washing step was repeated twice. The resulting organic phase was evaporated to dryness under N2 and the resulting substance was redissolved in 1 mL of MTBE. Extracts were analyzed via HPLC-UV/Vis (Prominence HPLC System, Shimadzu, Japan) according to conditions reported previously by Sander with the following modifications [28]. The chromatographic separation was performed on a reversed phase column (YMC carotenoid C30, 150 × 4.6 mm, 3 µm) with a mobile phase consisting of two solvent mixtures of methanol, MTBE and water (A: 81:15:4, B: 8:89:3, v:v:v) under following gradient conditions. The concentration of 2% B was kept constant for 11 minutes, then increased to 40% B in 7 minutes and held for 6.5 minutes, followed by an increase to 100% B (in 2.5 min, held for 3 min) and successive reequilibration to 2% B (in 3 min, held for 7 min). The flow rate was 1 mL·min^{-1}, injection volume was 5 µL and column temperature was kept at 22°C. The detection of the analytes was done at 445, 452 and 475 nm for lutein, beta-carotene and astaxanthin, respectively. All these steps were performed in triplicates.

For calibration, a stock solution of astaxanthin (Adipogen International Inc., San Diego, USA) was prepared by dissolving 10 mg in 33.3 mL of DMSO, resulting in a concentration of 300 µg·mL^{-1}. Stock solutions of lutein (CaroteNature GmbH, Ostermundigen, Switzerland) and β-carotene (Sigma Aldrich, St. Louis, USA) had the same concentration by solving 1 and 10 mg in 3.3 and 33.3 mL of MTBE, respectively. All analytes were calibrated using mixed standards prepared by appropriate dilution of stock solutions in a range of 0.05 to 100 µg·mL^{-1}.

Cell Disruption Methods

In the scope of this work various cell disruption methods were evaluated based on the optical determination of the disruption yield and total carotenoid yield after pressurized liquid extraction (Dionex ASE 350, 1 cycle, 30 min, 100 bar, 60°C, solvent:acetone:MeOH, 9:1 v:v for H. pluvialis and DCM for C. zofingiensis and C. sorokiniana). The disruption yield C was calculated as ratio of intact and disrupted cells by counting in a Neubauer-improved hemocytometer (equation 1).

$$C[\%] = \frac{cells_{disrupted}}{cells_{total}} \times 100\% = \frac{cells_{disrupted}}{cells_{disrupted} + cells_{intact}} \times 100\%$$

The following disruption methods, performed according to the parameters summed up in Table 1, were compared: labscale high pressure homogenizer (French Pressure Cell Press, SLM instruments, Inc., Urbana, USA), labscale ball mill (TissueLyser LT, Qiagen, Venlo, Netherlands), Ultra Turrax (T 18 Basic, IKA-Werke, Staufen, Germany), repeated freeze and thaw cycles, freeze-drying (Alpha 1-4, Christ, Osterode, Germany) and ultra-sonication (UW 2200, Bandelin electronic, Berlin, Germany). All samples were freeze-dried again before extraction in order to remove water residues and assure comparability.

Table 1: Parameters for various cell disruption methods tested. Each method was performed with dried biomass of H. pluvialis (30 mg), C. zofingiensis, C. sorokiniana (40 mg)

Cell disruption	Parameters
High pressure homogenizer	Biomass resuspended in 10 mL of H_2O_u, p=1000 bar, cycles n=3
Ball mill	Biomass, frozen at -80°C for 1 h, 1 cycle, 50 Hz, 2 min
Freeze and thaw	Biomass resuspended in 10 mL of H_2O_u, freezing -20°C, thawing at RT for 30 min, cycles n=3
Freeze-drying	Biomass resuspended in 10 mL of H_2O_u frozen at -80°C overnight, P=0.4 mbar, overnight
Ultra-sonication	Biomass resuspended in 10 mL of H_2O_u, maximum intensity, 10 min, one cycle
Ultra Turrax	Biomass resuspended in 10 mL of H_2O_u, 24,000 rpm, 10 min, on ice

The methods were tested on freeze-dried biomass of three algae with different morphological characteristics: (I) H. pluvialis (30 mg, encysted state, 30-50 μm, rigid cell walls, [29,30], (II) C. zofingiensis (30 mg, 2-15 μm, [31]) and (III) C. sorokiniana (30 mg, 2-4. μm, [32]). The optical evaluation of the disruption yield was exemplarily applied on H. pluvialis.

Optimization of the Pressurized Liquid Extraction Method

Selection of the solvent for PLE of carotenoids from algae biomass (30 mg dried biomass, 1 cycle, 30 min, 100 bar, 60°C) was performed based on the experiments with biomass of H. pluvialis after Ultra Turrax-disruption (conditions see Table 1) using 6 different solvents: acetone, ethanol, ethyl acetate, n-hexane, dichloromethane and a mixture of acetone:MeOH (9:1, v:v).

Next, the influence of temperature on the efficiency of the carotenoid extraction at first between 20 and 60°C (1 cycle, 30 min, 100 bar, solvent: acetone:MeOH, 9:1, v:v) and later between 60 and 120°C (1 cycle, 30 min, 100 bar, solvent: DCM) was tested with the biomass of C. zofingiensis (30 mg, freeze-dried, no disruption) and H. pluvialis (40 mg, after Ultra Turrax-disruption).

After appointing the optimum extraction temperature, various extraction durations - 5, 10, 20 and 30 min - were investigated, on the same biomass of H. pluvialis that was used before, in order to test the possibility of shortening the extraction time for increased sample throughput. Additionally, the effect of using multiple extraction cycles (4 × 5 min) was evaluated.

RESULTS AND DISCUSSION

Cell Disruption

A downstream process for recovery of intracellular products in general consists of the following unit operations: harvesting, dehydrating, product isolation (cell disruption and extraction) and product purification [18]. To gain high product yields, both efficient cell disruption and extraction is necessary. For instance in the work of Ceron et al. the recovery of lutein from freeze-dried biomass of S. almeriensis with various cell disruption methods was up to 60% higher compared to a method without cell disruption [33]. Cell disruption can be performed by a multitude of established methods, generally divided into non-mechanical methods, such as alkaline lysis, organic solvents, freezing, osmotic shock, etc. or mechanical methods, e.g., homogenizers, bead mills, ultra-sonication [18,34].

In the work presented here, four mechanical (ball mill, Ultra Turrax, high pressure homogenizer and ultra-sonication) as well as two nonmechanical methods (repeated freeze and thaw cycles, lyophilization) were used for the three microalgae H. pluvialis, C. zofingiensis, C. sorokiniana. To evaluate the efficiency of the mentioned methods an optical determination of the disruption yield C, based on counting disrupted and intact cells after cell disruption, was used (see equation 1). The disruption yield C of freeze-dried H. pluvialis biomass is shown in (Figure 1). The highest disruption yield - around 80% - of the thick encysted cell walls was reached by using mechanical methods such as ball mill, Ultra Turrax and high pressure homogenizer, while other mechanical (ultra-sonication) and non-mechanical methods, such as repeated freeze and thaw cycles resulted in disruption yields of $\leq$ 10%. In the next step the disrupted biomass was used for determination of the total carotenoid extraction yield via HPLC-UV/vis analysis. The obtained total carotenoid yield values are normalized to the highest reached value for each alga (in case of H. pluvialis - high pressure homogenizer) (Figure 2). In general, the normalized total carotenoid values are in good agreement with the calculated disruption yield. The highest carotenoid extraction yield was measured after disruption with the high pressure homogenizer (4.21 μg total carotenoid·mg^{-1} dry weight (d.w.), taken as 1 for normalization purpose), ball mill (3.56 μg total carotenoid·mg^{-1} d.w.) and Ultra Turrax (3.42 μg total carotenoid·mg^{-1} d.w.). Repeated freeze-and-thaw-cycles resulted in 240-fold lower total carotenoid concentration (0.02 μg total carotenoid·mg^{-1} d.w.) in comparison to high pressure homogenizer. Similar observations for H. pluvialis are described in the work of Mendes-Pinto et al. In this work mechanical disruption via a cell homogenizer resulted in nearly factor four higher total carotenoid amounts compared to nonmechanical methods like alkaline lysis, spray drying or treatment with base or enzymes. Also autoclaving of the biomass showed nearly the same effectivity as mechanical disruption [35]. However, even though the method used here for the determination of the disruption yield gives a very fast and a relatively good estimation of the disruption effectivity, it can only be performed with the statistically selected parts of the samples and only for some algal species. In case of H. pluvialis, the method can be easily applied and gives reliable data due to big cell diameters during encysted stadium [36]. For the other two investigated microalgae C. zofingiensis and C. sorokiniana with small cell diameters, it is inapplicable. Furthermore, the optical impression of the cell integrity should not lead to the conclusion that carotenoids cannot be extracted, as minor damage might not be visible but might still allow extraction of carotenoids. All in all, this method used here for H. pluvialis is a quick and easy one and results in a rough estimation, however due to the above listed limitations, further evaluation based on the total carotenoid yield measured by HPLC is necessary.

In the next step, the best three disruption methods selected in the experiments with H. pluvialis were tested on the biomass of C. zofingiensis (2-15 μm), showing the best results again for the high pressure homogenizer (2.87 μg total carotenoid·mg^{-1} d.w.) and ball mill (2.81 μg total carotenoid·mg^{-1} d.w.). In contrast to the other two species, for C. sorokiniana - the smallest tested cells (2-4.5 μm) - the effectivity of repeated freeze and thaw cycles of the biomass showed comparable results (1.95 μg total carotenoid·mg^{-1} d.w.) to the results reached with the use of high pressure homogenizer (1.97 μg total carotenoid·mg^{-1} d.w.) and ball mill (2.15 μg total carotenoid·mg^{-1} d.w.). Furthermore, also other methods, such as lyophilization and ultra-sonication, showed better results for this species (0.57 and 0.31 normalized carotenoid yield) in comparison to H. pluvialis (0.005 and 0.08 normalized carotenoid yield) (Figure 2). This is in good correspondence to the data of Prakabaran et al. that showed comparably good results achieved by osmotic shock and bead mill for lipid extraction for Chlorella sp. [37]. As a consequence, it can be assumed that the effectivity of nonmechanical disruption methods increases inversely proportionally to the cell size and the rigidity of cell walls of the examined species.

In order to establish an efficient isolation procedure for industrial application, the suitability of the various investigated methods for scale-up has also to be taken into account. Because of high power consumption per unit of biomass volume ultrasonication and lyophilization are considered to be suitable only for small-scale applications [18,38]. Capital cost for repeated freeze and thaw cycles are relative low compared to the other used methods. However, the fact that this is a very time-consuming as well as the difficulty to maintain the product quality under such conditions has to be addressed. In contrast to this, mechanical methods i.e., high pressure homogenization and bead mills are commercially available and applied [18]. Furthermore, these methods showed the best disruption yields resulting in the highest carotenoid recoveries for all three tested algae. Finally, the right choice of the best cell disruption method strongly depends on the used microalgae, the product of interest as well as the scale of production (quantity and quality) [18,34].

Pressurized Liquid Extraction of Carotenoids

One of the most important factors affecting the extraction yield of carotenoids via PLE is the selection of the suitable extraction solvent (Figure 3). The application of n-hexane for H. pluvialis resulted in the isolation of the lowest measured carotenoid amount (1.59 $\mu g \cdot mg^{-1}$ d.w.), followed by ethyl acetate (1.98 μg total carotenoid$\cdot mg^{-1}$ d.w.), ethanol (2.15 μg total carotenoid$\cdot mg^{-1}$ d.w.), acetone (2.43 μg total carotenoid$\cdot mg^{-1}$ d.w.) and acetone:methanol (9:1, v:v) (3.43 μg total carotenoid$\cdot mg^{-1}$ d.w.). The use of DCM showed the best results with 4.72 μg total carotenoid$\cdot mg^{-1}$ d.w. - a 3-folds higher yield than reached with hexane. However, this general finding should not be generalized for individual carotenoids. In fact, β-carotene gave comparable results for most applied solvents whereas the yield for lutein extraction could be improved by factor 2.9 and for astaxanthin by factor 3.2 by using DCM compared to hexane. This can be explained by the increased solubility of more polar compounds such as astaxanthin or lutein in dichloromethane (i.e., astaxanthin 30 $g \cdot L^{-1}$ in DCM, 0.2 $g \cdot L^{-1}$ in acetone [27]). Increased solubility of the very unpolar β-carotene in hexane (30 $mg\ L^{-1}$, described by Craft and Soares) compared to the other solvents did not show significant influence [39]. Due to the low yield in the examined species (0.2 $mg\ L^{-1}$ in the extract) all tested solvents were able to extract it completely. Jamie et al. also reported better extraction yields for PLE extraction from H. pluvialis using ethanol in comparison to n-hexane, which is in consent with the results presented above [40]. Mixing dichloromethane with other solvents showed no additional advantages (data not shown here), while Denery showed the best results for a mixture of DCM and methanol (1:3, v:v) for extraction of astaxanthin from H. pluvialis compared to acetone, ethanol and mixtures of acetone and methanol (7:3, v:v) [41]. In contrast to this, various mixtures of acetone and methanol are commonly used in literature as extraction solvents [10,25,42]. Furthermore, other extraction media such as oil [43,44] or CO_2 for supercritical fluid extraction [45,46] have also been described in literature but not tested in this work. Further tests are necessary to evaluate these methods.

The first experiments with temperature variations for PLE showed that the yield of the carotenoid extraction can be improved by increasing the temperature from room temperature to 60°C. Therefore, further tests with even higher temperatures were conducted here in order to check if additional improvement is possible, as the solubility itself is a thermodynamic equilibrium, influenced by the temperature. It turned out, that the carotenoid recovery yield for the selected solvent decreased again with temperatures above 80°C, probably due to the thermal destruction of the analytes (Figure 4). At 120°C, the total yield of carotenoids was 35% lower than after extraction at 60°C (3.08 compared to 4.72 μg total carotenoid$\cdot mg^{-1}$ d.w.). The optimum could be found around 60°C. That corresponds well with findings of Mustafa et al. who tested PLE with ethanol for α- and β-carotene from carrots showing 42% reduction of β-carotene recovery by using 180°C in comparison to 60°C [47]. Denery et al. did not find a reduction of astaxanthin recovery by PLE from H. pluvialis using acetone between 20°C and 100°C but a reduction of lutein and total pigment amounts at higher temperatures [41]. In contrast to this, Jaime et al. found increasing extraction yields for temperatures up to 200°C for carotenoid extraction with ethanol from H. pluvialis [40]. From the somewhat contradicting results reported by the above mentioned groups, it can be assumed that optimum temperature depends on the solvent that is used for the extraction and has to be carefully selected for specific applications. In the case of the work presented here, future tests with smaller temperature intervals might reveal a more exact optimum.

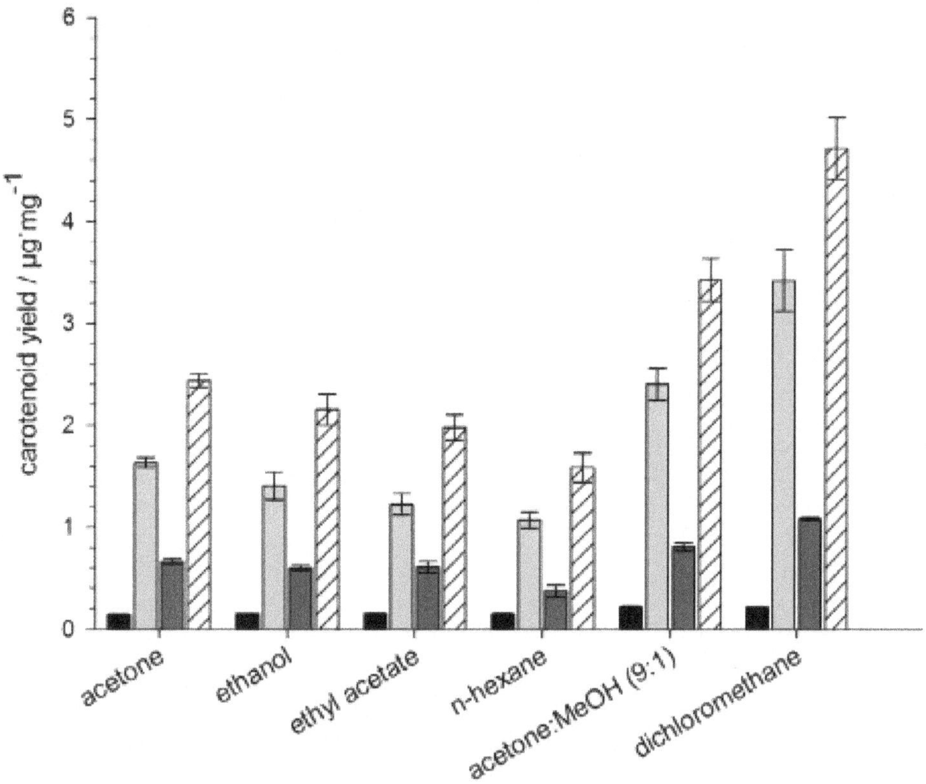

Figure 3: Yield of individual and total carotenoids (black bar: β-carotene; light gray bar: astaxanthin; dark gray bar: lutein; dashed bar: total carotenoids) of H. pluvialis measured by HPLC UV/Vis resulting from pressurized liquid extraction with different solvents: 30 mg of freeze-dried H. pluvialis biomass were disrupted using Ultra Turrax, extracted via PLE (60°C, 30 min, 100 bar) by six selected solvents, saponified and analysed by HPLC -UV/Vis on a C30 column.

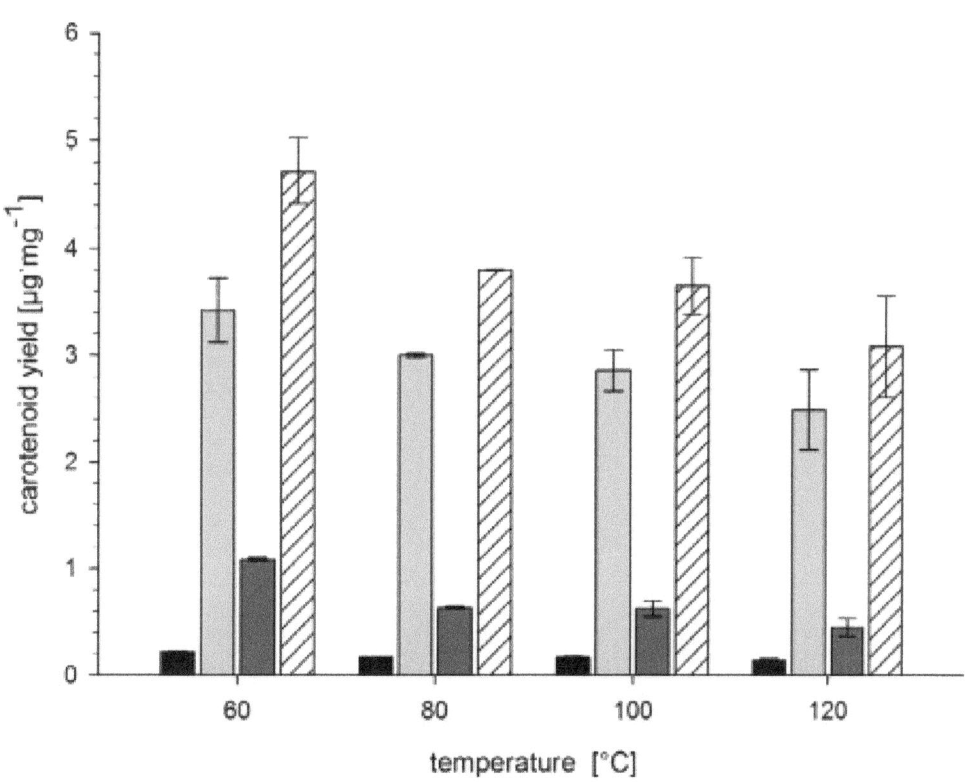

Figure 4: Yield of individual and total carotenoids (black bar: β-carotene; light gray bar: astaxanthin; dark gray bar: lutein; dashed bar: total carotenoids) of H. pluvialis measured by HPLC UV/Vis resulting from pressurized liquid extraction with different temperatures: 30 mg of freeze-dried H. pluvialis biomass were disrupted using Ultra Turrax, extracted via PLE (30 min, 100 bar, DCM) by 4 different temperatures, saponified and analysed by HPLC -UV/Vis on a C30 column.

The optimum duration of the extraction process depends on the time needed to build up the equilibrium between the analyte concentration in the sample matrix and the solvent. Five minutes showed to be insufficient for carotenoid extraction in the tested setup, resulting in only 79% of the maximal total carotenoid yield (3.74. instead of 4.72 µg total carotenoid·mg^{-1} d.w.) and 58% of the lutein recovery (0.63 instead of 1.08 µg lutein·mg^{-1} d.w.), whereas a nearly complete carotenoid level could be reached applying 10 minutes and more (Figure 5). Implementing additional extraction cycles did not improve the results, as 99% of the analytes were extracted within the first cycle (Figure 6). The carotenoid yield in the extracts of the third and fourth cycle were below the detection limit (0.0125 µg·mg-1 for astaxanthin and lutein, 0.025 µg·mg-1 for ß-carotene). This suggests that the equilibrium between solvent and biomass is established after 10 minutes and according to the solubility data was also far away from saturation [27,39]. Performing multiple steps is often preferred in literature, e.g., Mustafa et al. showed that the optimum PLE of β-carotene with ethanol at 60°C to be five 2 min cycles after 5 min preheating [47]. On the other hand, this highly depends on the solvent and the used matrix.

With the parameters selected (high pressure homogenizer, PLE at 60°C for 10 min, in 1 cycle, solvent: DCM), a total carotenoid yield of 4.78 µg total carotenoid·mg^{-1} d.w. and a total astaxanthin yield of 3.69 µg astaxanthin·mg^{-1} d.w. could be determined for H. pluvialis, which is 0.4% of the total cell weight. The values that can be found in literature range from 0.5 to 4% astaxanthin depending on the strain, culture conditions and cultivation mode [31,34,41,42,48]. Values of up to 98 µg·mg^{-1} d.w. have been reported for heterotrophic cultivations [49]. The lower carotenoid yield presented in this study can be easily explained by the fact that no optimization regarding the carotenoid production was performed at this stage of the work. As numerous possible modifications of the process to induce astaxanthin production is described in literature (nitrogen starvation, light stress, salinity, iron addition, strain selection, genetic modifications, heterotrophic growth on acetate), further optimizations of the cultivation is possible and planed for future work [50-56]. High pressure homogenizer and ball mill showed the best disruption for C. zofingiensis, the parameters for the PLE were based on the findings regarding the extraction of carotenoids from H. pluvialis. Applying an optimized isolation process for carotenoids to the biomass of C. zofingensis 1.48 µg lutein·mg^{-1} d.w. and 1.29 µg astaxanthin·mg^{-1} d.w. could be measured. Values reported for this species by Del Campo and Liu range between 4 and 1.3 to 6.5 µg·mg^{-1} d.w. for lutein und astaxanthin, respectively [25,31]. The lower yield for lutein measured in this work can be explained by the harvesting of the biomass at the end of the cultivation while the highest yield was reported for an early stage of cultivation by Del Campo et al. [25]. Thus further optimization of the whole production process including cultivation is necessary. Possible approaches were published in literature [50,57,58]. For C. sorokiniana the highest yield of lutein (2.08 µg lutein·mg^{-1} d.w.) was measured after the ball mill disruption. In comparison, Cordero et al. reported 4.2 µg lutein·mg^{-1} d.w. in wildtype algae and up to 7 µg·mg^{-1} d.w. in genetically modified strains, also suggesting that further optimization of the cultivation process is necessary [10].

CONCLUSION

Due to the fact that downstream processing is one of the decisive elements, it has to be very carefully optimized in order to reach the maximum recovery yield of the products at lowest possible costs. For this reason, various cell disruption methods and parameters for pressurized liquid extraction of the carotenoids from various algae species were examined in labscale in this work. The influence of the cell size and cell wall properties on the success of cell disruption methods was shown: while mechanical methods, i.e., high pressure homogenizer or ball mill were most efficient for all three tested microalgae, the effectivity of soft disruption increased with decreasing cell size and rigidity. Selecting DCM as the extraction solvent at 60°C improved the extraction yield of total carotenoids. The extraction time could be set to 10 minutes, with no additional extraction cycles, improving the sample throughput.

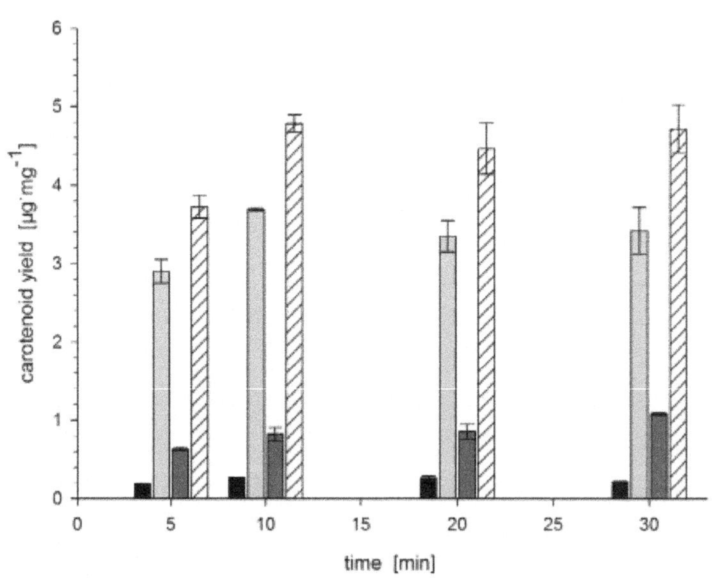

Figure 5: Yield of individual and total carotenoids (black bar: β-carotene; light gray bar: astaxanthin; dark gray bar: lutein; dashed bar: total carotenoids) of H. pluvialis measured by HPLC UV/Vis resulting from pressurized liquid extraction with different extraction durations: 30 mg of freeze-dried H. pluvialis biomass were disrupted using Ultra Turrax, extracted via PLE (60°C, 100 bar, DCM) for 4 different durations, saponified and analysed by HPLC -UV/Vis on a C30 column.

The best extraction yield for carotenoids from above described species could be achieved by PLE using DCM at 60°C in a 10 min cycle (n=1) with lyophilized and high pressure disrupted biomass. The carotenoid yield was lower than the highest values reported in literature since when this work was carried out the cultivation conditions were not optimized for the highest product yield. Obviously, culture optimization will be as crucial as optimizing the downstream process. Nevertheless, the proposed sample preparation procedure allows reliable carotenoid determination in algae biomass that is necessary for further optimization of the whole process for economic use of microalgae for this application.

ACKNOWLEDGEMENTS

The authors are grateful for financial support of the Federal Ministry of Education and Research (BMBF) and the Bavarian State Ministry of the Environment and Consumer Protection, due to which this work could be performed.

REFERENCES

1. Borowitzka MA (2013) High-value products from microalgae - their development and commercialisation. J Appl Psychol 25: 743-756.
2. Yaakob Z, Ali E, Zainal A, Mohamad M, Takriff MS, et al. (2014) An overview: biomolecules from microalgae for animal feed and aquaculture. J Biol Res (Thessalon) 21: 6.
3. Pulz O, Gross W (2004) Valuable products from biotechnology of microalgae. Appl Microbiol Biotechnol 65: 635-648.
4. Borowitzka MA, Moheimani N (2013) Sustainable biofuels from algae. Mitigation and Adaptation Strategies for Global Change 18: 13-25.
5. Thukral N, Sharma M (2015) Microalgae as Future Fuel: Real Opportunities and Challenges. J ThermodynCataly 6.
6. Rodolfi L, Chini Zittelli G, Bassi N, Padovani G, Biondi N, et al. (2009) Microalgae for oil: strain selection, induction of lipid synthesis and outdoor mass cultivation in a low-cost photobioreactor. Biotechnol Bioeng 102: 100-112.
7. Brennan L, Owende P (2010) Biofuels from microalgae - A review of technologies for production, processing, and extractions of biofuels and co-products. Renewable and Sustainable Energy Reviews 14: 557-577.
8. Raja R, Hemaiswarya S, Kumar NA, Sridhar S, Rengasamy R, et al. (2008) A perspective on the biotechnological potential of microalgae. Crit Rev Microbiol 34: 77-88.
9. Bhatnagar A, Chinnasamy S, Singh M, Das KC (2011) Renewable biomass production by mixotrophic algae in the presence of various carbon sources and wastewaters. Applied Energy 88: 3425-3431.
10. Cordero BF, Obraztsova I, Couso I, Leon R, Vargas MA, et al. (2011) Enhancement of lutein production in Chlorella sorokiniana (Chorophyta) by improvement of culture conditions and random mutagenesis. Mar Drugs 9: 1607-1624.
11. Guedes AC, Amaro HM, Malcata FX (2011) Microalgae as sources of carotenoids. Mar Drugs 9: 625-644.
12. Varela JC, Pereira H, Vila M, León R (2015) Production of carotenoids by microalgae: achievements and challenges. Photosynth Res 125: 423-436.
13. Carvalho AP, Silva SO, Baptista JM, Malcata FX (2011) Light requirements in microalgal photobioreactors: an overview of biophotonic aspects. Appl Microbiol Biotechnol 89: 1275-1288.
14. Demmig-Adams B, Adams WW 3rd (2002) Antioxidants in photosynthesis and human nutrition. Science 298: 2149-2153.
15. Sun Z, Liu J, Bi YH, Zhou ZG (2014) Microalgae as the Production Platform for Carotenoids. OMICS Group International.
16. Spolaore P, Joannis-Cassan C, Duran E, Isambert A (2006) Commercial applications of microalgae. J Biosci Bioeng 101: 87-96.
17. Rosenberg JN, Oyler GA, Wilkinson L, Betenbaugh MJ (2008) A green light for engineered algae: redirecting metabolism to fuel a biotechnology revolution. Curr Opin Biotechnol 19: 430-436.
18. Richmond A, Hu Q (2013) Handbook of Microalgal Culture.2nd edn. John Wiley & Sons, Ltd., Oxford, UK.
19. Olaizola M (2003) Commercial development of microalgal biotechnology: from the test tube to the marketplace. Biomol Eng 20: 459-466.
20. Straathof A (2011) The Proportion of Downstream Costs in Fermentative Production Processes. Moo-Young, Murray. Comprehensive Biotechnology2: 811-814.
21. Arnon D, McSwain BD, Tsujimoto HY, Wada K (1974) Photochemical activity and components of membrane preparations from blue-green algae. I. Coexistence of two photosystems in relation to chlorophyll a and removal of phycocyanin. Biochim Biophys Acta 357: 231-245.
22. Heining M, Sutor A, Stute SC, Lindenberger CP, Buchholz R, et al. (2015) Internal illumination of photobioreactors via wireless light emitters: a proof of concept. J Appl Psychol 27: 59-66.
23. Klein BC, Walter C, Lange HA, Buchholz R (2012) Microalgae as natural sources for antioxidative compounds. J Appl Psychol 24: 1133-1139.
24. Geier SC, Huyer S, Praebst K, Husmann M, Walter C, et al. (2012) Outdoor cultivation of Chlamydomonas reinhardtii for photobiological hydrogen production. J Appl Psychol 24: 319-327.
25. Del Campo JA, Rodríguez H, Moreno J, Vargas MA, Rivas J, et al. (2004) Accumulation of astaxanthin and lutein in Chlorella zofingiensis (Chlorophyta). Appl Microbiol Biotechnol 64: 848-854.
26. Yuan JP, Chen F (1998) Chromatographic Separation and Purification of trans-Astaxanthinfrom the Extracts of Haematococcus pluvialis. J Agric Food 46: 3371-3375.
27. Johnson EA, An GH (1991) Astaxanthin from Microbial Sources. Crit Rev Biotechnol11: 297-326.
28. Sander LC, Sharpless KE, Craft NE, Wise SA

(1994) Development of engineered stationary phases for the separation of carotenoid isomers. Anal Chem 66: 1667-1674.

29. Kobayashi M, Katsuragi T, Tani Y (2001) Enlarged and astaxanthin-accumulating cyst cells of the green alga Haematococcus pluvialis. J Biosci Bioeng 92: 565-568.

30. Boussiba S (2000) Carotenogenesis in the green alga Haematococcus pluvialis: Cellular physiology and stress response. Physiol Plant 108: 111-117.

31. Liu J, Sun Z, Gerken H, Liu Z, Jiang Y, et al. (2014) Chlorella zofingiensis as an alternative microalgal producer of astaxanthin: biology and industrial potential. Mar Drugs 12: 3487-3515.

32. Lizzul AM, Hellier P, Purton S, Baganz F, Ladommatos N, et al. (2014) Combined remediation and lipid production using Chlorella sorokiniana grown on wastewater and exhaust gases. Bioresour Technol 151: 12-18.

33. Cerón MC, Campos I, Sánchez JF, Acién FG, Molina E, et al. (2008) Recovery of lutein from microalgae biomass: development of a process for Scenedesmus almeriensis biomass. J Agric Food Chem 56: 11761-11766.

34. Del Campo JA, García-González M, Guerrero MG (2007) Outdoor cultivation of microalgae for carotenoid production: current state and perspectives. Appl Microbiol Biotechnol 74: 1163-1174.

35. Mendes-Pinto MM, Raposo M, Bowen J, Young AJ, Morais R, et al. (2001) Evaluation of different cell disruption processes on encysted cells of Haematococcus pluvialis: effects on astaxanthin recovery and implications for bio-availability. J Appl Psychol 13: 19-24.

36. Grewe C (2009) Untersuchungen zurAstaxanthin-Biosynthese in den Grünalgen Scenedesmus sp. und Haematococcus pluvialis.

37. Prabakaran P, Ravindran AD (2011) A comparative study on effective cell disruption methods for lipid extraction from microalgae. Lett Appl Microbiol 53: 150-154.

38. Ranjith Kumar R, Hanumantha Rao P, Arumugam M (2015) Lipid Extraction Methods from Microalgae: A Comprehensive Review. Frontiers in Energy Research 2: 457.

39. Craft NE, Soares JH (1992) Relative Solubility, Stability and Absorptivity of Lutein and @-Carotenein Organic Solvents. J Agric Food Chem 40: 431-434.

40. Jaime L, Rodríguez-Meizoso I, Cifuentes A, Santoyo S, Suarez S, et al. (2010) Pressurized liquids as an alternative process to antioxidant carotenoids' extraction from *Haematococcus pluvialis* microalgae. LWT - Food Science and Technology 43: 105-112.

41. Denery JR, Dragull K, Tang C, Li QX (2004) Pressurized fluid extraction of carotenoids from Haematococcus pluvialis and Dunaliella salina and kavalactones from Piper methysticum. Analytica Chimica Acta 501: 175-181.

42. Sarada R, Vidhyavathi R, Usha D, Ravishankar GA (2006) An efficient method for extraction of astaxanthin from green alga Haematococcus pluvialis. J Agric Food Chem 54: 7585-7588.

43. Nonomura AM (1987) Process for producing a naturally-derived carotene/oil composition by direct extraction from algae.

44. Fernández-Sevilla JM, Acién Fernández FG, Molina Grima E (2010) Biotechnological production of lutein and its applications. Appl Microbiol Biotechnol 86: 27-40.

45. Herrero M, Cifuentes A, Ibañez E (2006) Sub- and supercritical fluid extraction of functional ingredients from different natural sources: Plants, food-by-products, algae and microalgae: A review. Food Chem 98: 136-148.

46. Careri M, Furlattini L, Mangia A, Musci M, Anklam E, et al. (2001) Supercritical fluid extraction for liquid chromatographic determination of carotenoids in Spirulina Pacifica algae: a chemometric approach. J Chromatogr A 912: 61-71.

47. Mustafa A, Trevino LM, Turner C (2012) Pressurized hot ethanol extraction of carotenoids from carrot by-products. Molecules 17: 1809-1818.

48. Cuellar-Bermudez SP, Aguilar-Hernandez I, Cardenas-Chavez DL, Ornelas-Soto N, Romero-Ogawa MA, et al. (2015) Extraction and purification of high-value metabolites from microalgae: essential lipids, astaxanthin and phycobiliproteins. Microb Biotechnol 8: 190-209.

49. Domínguez-Bocanegra AR, Guerrero Legarreta I, Martinez Jeronimo F, Tomasini Campocosio A (2004) Influence of environmental and nutritional factors in the production of astaxanthin from Haematococcus pluvialis. Bioresour Technol 92: 209-214.

50. Orosa M, Valero JF, Herrero C, Abalde J (2001) Comparison of the accumulation of astaxanthin in Haematococcus pluvialis and other green microalgae under N-starvation and high light conditions. Biotechnol Lett 23: 1079-1085.

51. Aflalo C, Meshulam Y, Zarka A, Boussiba S (2007) On the relative efficiency of two- vs. one-stage production of astaxanthin by the green alga Haematococcus pluvialis. Biotechnol Bioeng 98: 300-305.

52. Borowitzka MA, Huisman JM, Osborn A (1991) Culture of the astaxanthin-producing green alga Haematococcus pluvialis 1. Effects of nutrients on growth and cell type. J Appl Psycholy 295-304.

53. Boussiba S, Avigad V (1991) Astaxanthin Accumulation in the Green Alga Haematococcus pluvialis. Plant Cell Physiol 32: 1077-1082.

54. Boussiba S, Bing W, Yuan JP, Zarka A, Chen F, et al. (1999) Changes in pigments profile in the green alga Haeamtococcus pluvialis exposed to environmental stresses. Biotechnology Letters 21: 601-604.

55. Fábregas J, Otero A, Maseda A, Domínguez A (2001) Two-stage cultures for the production of astaxanthin from Haematococcus pluvialis. J Biotechnol 89: 65-71.

56. Cai M, Li Z, Qi A (2009) Effects of iron electrovalence and species on growth and astaxanthin production of Haematococcus pluvialis. Chinese Journal of Oceanology and Limnology 27: 370-375.

57. Sun N, Wang Y, Li YT, Huang JC, Chen F, et al. (2008) Sugar-based growth, astaxanthin accumulation and carotenogenic transcription of heterotrophic Chlorella zofingiensis (Chlorophyta). Process Biochemistry 43: 1288-1292.

58. Wang Y, Peng J (2008) Growth-associated biosynthesis of astaxanthin in heterotrophic Chlorella zofingiensis (Chlorophyta). World J Microbiol Biotechnol 24: 1915-1922.

CHAPTER 12

ATTENUATION OF NEGATIVE IMPACTS BY MICRO ALGAE AND ENRICHED ARTEMIA SALINA ON PENAEUS MONODON AND LITOPENAEUS VANNAMEI LARVAL CULTURE

Karthik R, Ramalingam K, Yuvaraj D, Vanitha MC and Muthezhilan R

Keywords: Shrimp seeds, Micro algae, *Artemia salina*, Digestive enzyme, Water quality

INTRODUCTION

Shrimp farming is one of the most important aquaculture practices worldwide especially in Asia due to their high economic value. It is estimated that approximately more than 5 million metric tons of shrimp are annually produced but the current global demand for both the wild and farmed shrimps is approximately more than 6.5 million metric tons per annum [1]. Artificial culture of shrimps in grow out ponds has been intensified to keep shrimp production on par to the demand. As a sequel semi intensive and extensive methods of culture brought various ecological, economical and social issues [2]. In general, during the shrimp growth after metamorphosis it is considerably affected by the gradual change from planktonic to benthic existence coinciding with changes in the alimentation. During early post larval development, high mortality was noticed due to the changes in the gut associated digestive enzyme production levels [3,4]. The first feeding during the growth of any cultivable organism is the most 'critical phase' of their life cycle for their survival. Hence developing a new technology and new live feed may offer great hope for the future with a promise for blue revolution in the century to match the green revolution. Protein is the major component in the natural food of penaeids shrimps. Thus, feeding the penaids with protein rich live diets such as, phytoplankton such as microalgae (2–20 μm) and zooplankton such as rotifers (50– 200 μm) and brine shrimp, Artemia salina (200–300 μm) can increase the gut associated digestive enzymes [5]. Micro algae and Artemia salina like zooplankton help to stabilize and improve the water quality improve the oxygen production, promote pH stabilization (the action of some excreted biochemical compounds along with the induction of behavioral processes like initial prey catching) and regulate disease causing bacterial population and above all the probiotics and stimulate immunity in the host animal [6]. The microalgae most frequently used in aquaculture include Chlorella, Tetraselmis, Isochrysis, Pavlova, Phaeodactylum, Chaetoceros, Nannochloropsis, Skeletonema and Thalassiosira [7]. Moreover, Artemia salina are biologically uncontaminated readily available and acceptable larval feed and established as a standard live feed for over 85% of marine species. The present study was aimed to examine the digestive enzymes (amylase and protease) activity, survival rate (%), average growth (mm), water quality parameters and bacterial (Vibrio) load in Penaeus monodon and Litopenaeus vannamei shrimp culture from zoea to postlarvae 20 stages by feeding with five different microalgae such as, Isochrysis galbana, Cheatoceros calcitrans, Tetraselmis sp, Chlorella sp and Nannochloropsis sp and algae enriched Artemia salina nauplii.

MATERIALS AND METHODS

Microalgae

The five different microalgae such as Isochrysis galbana, Cheatoceros calcitrans, Tetraselmis sp, Chlorella sp and Nannochloropsis sp were obtained from AMET Microbial Culture Collection Centre, Department of Marine Biotechnology, AMET Unidversity.

Experimental animal

The shrimp (nauplii 24 h) of Penaeus monodon and Litopenaeus vannamei were obtained from a commercial shrimp hatchery

Citation: Karthik R, Ramalingam K, Yuvaraj D, Vanitha MC, Muthezhilan R (2015) Attenuation of Negative Impacts by Micro Algae and Enriched Artemia Salina on Penaeus Monodon and Litopenaeus Vannamei Larval Culture. J Aquac Res Development 6: 365. doi:10.4172/2155-9546.1000365.

located in Marakanam, Kanchipuram District, Tamil Nadu, India. They were kept in seawater with aeration for a period of 6 h in order to avoid any stress to the animals and then used for the experiments.

Experimental design

The experiments were conducted at the Department of Marine Biotechnology, AMET University, Chennai. A total of twelve 100 L glass tanks (Six tanks for Penaeus monodon and another six tanks for Litopenaeus vannamei) were added with 70 L of filtered seawater at 32% salinity and kept at ambient temperature (28 ± 1°C) and aerated continuously. The seeds were transferred in the tank with a stocking density of 75 nauplii per litre.

Feeding Schedule from Zoea to Post Larvae

The Zoea of Penaeus monodon and Litopenaeus vannamei kept in the experimental tanks were fed with five different microalgae such as Isochrysis galbana, Cheatoceros calcitrans, Tetraselmis sp, Chlorella sp and Nannochloropsis sp. On 1st day of the Zoea (Z) of I-III stages were fed thrice with 30×10^4 cells/ml of algal cells. On the 2nd day of the Mysis (M) (I-III/ Postlarve 1) of Penaeus monodon and Litopenaeus vannamei were fed thrice with 40×10^4 cells/ml of algal cells (Tables 1 and 2). From 3rd day up to 20th day they were fed thrice with 3–8 No/ml of Artemia salina nauplii enriched with different microalgae. The 24 h old Artemia salina nauplii enriched with Isochrysis galbana, Cheatoceros calcitrans, Tetraselmis sp, Chlorella sp and Nannochloropsis sp at 20, 30, 40, 50 and 60×10^4 cells/ml at 16 h and 9 h respectively for a period of 24 h as fed to Penaeus monodon and Litopenaeus vannamei. Filtered seawater was exchanged daily and the debris settled at the bottom was siphoned out without disturbing the animals. This experiment was conducted for a period of 20 days (PL20). At the end of the experiments the animals from all the experimental tanks were randomly selected and the survival rate and the average length were recorded.

Digestive Enzyme Activities of Shrimps Fed With Different Microalgae

The shrimps were collected on 1st day (Zoea III stages) after naupliiar stage, 2nd day (Mysis III), 5th day (PL5), 10th day (PL10), 15th day (PL15) and 20th day (PL20) from all the experimental tanks and they are subjected for different enzyme analysis.

Microbiological Analysis

Water samples were taken from the Penaeus monodon and Litopenaeus vannamei larval rearing tanks (all the experimental tanks) during various larval stages of shrimps (i.e. nauplii, zoea, mysis and post larvae) and the Vibrio sp load was enumerated by growth on TCBS agar.

RESULTS AND DISCUSSION

In general, traditional and non-scientific shrimp farms depend upon the shrimp seeds caught from the wild or those entered with the tides for stocking. Such seeds collected naturally from tides may have fortuitous infection with microbes and bound to affect shrimp culture. There is a demand for healthy and quality seeds throughout the year. Thus the successful shrimp culture is dependent upon stocking disease free, healthy seeds in the hatcheries [8]. The applications of microalgae for aquaculture are associated with nutrition and other biological activities [9]. The study of digestive enzymes constitutes a denominator role to consider the nutritional requirements of specific stages of development in shrimp. In the present study, the Zoea to Postlarvae 20 of Penaeus monodon and Litopenaeus vannamei were fed with five different microalgae such as, Isochrysis galbana, Cheatoceros calcitrans, Tetraselmis sp, Chlorella sp and Nannochloropsis sp and algae enriched Artemia salina. The protease activity at all the different stages of P. monodon and L. vannamei, the animals fed with Cheatoceros calcitrans and enriched Artemia salina exhibited a maximum protease activity of (P. monodon 17.02 ± 0.01 U/ml and L. vannamei 16.55 ± 0.08 U/ ml) at pH 8 and a minimum total activity of (P. monodon 16.19 ± 0.01 U/ml and L. vannamei 16.13 ± 0.13 U/ml) at pH 6 followed by animals fed with Chlorella sp and enriched Artemia salina respectively. A decrease of protease activity was found from PL6 to PL20 stages of animals fed with other microalgae in all the groups. Same results were obtained when the animals fed with Artemia salina enriched with five different microalgae (Tables 3-7). Madhumathi and Rengasamy [3,4] also observed similar results, while studying the ontogenetic changes in the digestive enzyme of Penaeus monodon at Z3 to PL20 stages, when fed with C. calcitrans and other four different microalgae and enriched Artemia salina.

Regarding, the amylase activity all the different stages of P. monodon and Litopenaeus vannamei at different time intervals the Artemia nauplii fed with Cheatoceros calcitrans and enriched exhibited a maximum protease total activity at 15 minutes time interval (P. monodon 1.10 ± 0.04 U/ml and L. vannamei 1.02 ± 0.15 U/ml) and it slightly decreased from 30 and 45 minutes (Tables 8-12). However, there was a gradual increase in the enzyme activity up to PL5, and then it decreased in PL10. Again it increased during the stages of PL15, the enzyme activity decreased during the stages of PL20.

Understanding the connection between the water quality and aquatic productivity is absolutely essential for optimum growth and production. The quality of water during the culture period will go down mainly due to the accumulation of organic wastes produced by the animals and unutilized feed. Thus, in this study, during the entire culture period some important water quality parameters such as, salinity, pH, temperature, dissolved oxygen, and ammonia were monitored. Salinity is the most important factor influencing many functional responses of the organisms as metabolism, growth, migration, osmotic behavior, reproduction etc. The marine organisms maintain their internal salt concentration (salt concentration of blood and body fluids) by osmoregulation. In shrimp hatchery the recommended salinity range is 28–35 ppt [10]. In the present study the salinity was maintained at 30 ppt in all the experimental tanks. Similarly Krishnaprakash [11] also maintained the salinity for the larval rearing of P. monodon at 31 ppt. During the culture period the pH was found to be 8.1 to 8.2 in the entire experimental tanks and the temperature was found to be in the range of 31 ± 1.1°C in both the P. monodon and L. vannamei reared experimental tanks. Similarly Kannupandi et al. [10] also reported that the required range of pH for shrimp larval culture is 8.2-8.5.

Dissolved oxygen (DO) in the larvae/adult animals rearing water is an important factor for the respiration of aquatic organisms and also to maintain favorable chemical and hygienic environment of the water body. Moreover, it also controls many oxidation reactions and maintains aerobic conditions in water. Because, when oxygen level is very low and anaerobic conditions exist, nitrate is reduced into ammonia, which will be toxic to the larvae or adult in the culture water and it also increases the pH. Furthermore, low-level of oxygen tension hampers metabolic performances in shrimp larvae and it will reduce the growth and finally cause huge mortality [12]. So, in this study the oxygen level in the culture medium was maintained in the desired range by aeration continuously during the whole culture period. In P. monodon culture while checking the dissolved oxygen it was found to be higher in experimental tank 2 (6.19 ± 0.17 ml/litre) followed by tank 4 (6.10 ± 0.17 ml/litre). Moreover the ammonia was found to be low in experimental tank 2 (0.14 ± 0.11 mg/litre) followed by tank 4 (0.15 ± 0.13 ml/litre). In L. vannamei culture whereas checking the dissolved oxygen it was found to be higher in experimental tank 2 (6.10 ± 0.10 ml/litre) followed by tank 4 (6.08 ± 0.18 ml/litre). Moreover the ammonia was found to be low in experimental tank 2 (0.14 ± 0.34 mg/ litre) followed by tank 4 (0.16 ± 0.44 ml/litre).

In the present study, whereas monitoring the survival rate of the nauplii of P. monodon and L. vannamei, it was found to be higher in all the experimental tanks (Figures 1 and 2). At the end of the culture period (PL20) the average length of all the post larvae was found to be maximum higher in the experimental tank 2 where the animals fed with Cheatoceros calcitrans and enriched Artemia salina (Figures 3 and 4). Madhumathi and Rengasamy [3,4] also observed the similar results, like high survival rate, length and weight of Zoea – PL I of P. monodon fed with C. calcitrans and PL I-PL20 stages fed with C. calcitrans enriched Artemia naupli diets. They also stated that, the Zoea to PLI of shrimps had high protein content of 51% followed by carbohydrate 6% and lipid 50% when it was fed with C. calcitrans.

In normal, diseases in aquaculture practices are mostly caused by luminous bacteria Vibrio harveyi, and it has been referred to as the largest economic loss in the shrimp aquaculture due to mass mortalities. Furthermore, in general pathogenic microbes always remain in the water as opportunistic pathogens and they attack the larvae only when the larvae are weak due to environmental stress or nutritional deficiency [13]. During larval rearing stages different microbes may enter the hatchery system through the eggs and live feeds such as algae and Atremia sp. In the present study, while checking the vibrio load at different stages of P. monodon and L. vannamei larvae and cultured water it was found to be comparatively lower in tank 2 where the animals fed with Cheatoceros calcitrans and enriched Artemia salina than other groups (Figures 5-8).

CONCLUSION

Since the demand for shrimp products in world markets continues to increase, the continuous supply of healthy, inexpensive and robust shrimp seed stocks to the farmers is important to maintain production of adult shrimps. Microalgae stay an important part of the aquaculture production chain particularly for hatchery shrimp feed, in spite of expensive culture installation and high production cost. From the results, it is concluded that, the use of Cheatoceros calcitrans and enriched Artemia salina will promote the successful production of shrimp larvae in hatcheries and reduce the potential negative impact of shrimp farming on the environment like, organic matter accumulation, ammonification, eutrophication and water toxicity and increase the productivity of the farms to the benefit of local economies.

REFERENCES

1. Karthik R, JaffarHussain A, Muthezhilan R (2014) Effectiveness of Lactobacillus sp (AMET1506) as Probiotic against Vibriosis in Penaeusmonodon and Litopenaeusvannamei Shrimp Aquaculture. Biosciences Biotechnology Research Asia11:297-305
2. FAO (2012a) The State of World Fisheries and Aquaculture 2012. Food and Agricultural Organization of the United Nations.
3. Madhumathi M, Rengasamy R (2011a) Effect of Algal Enriched Artemiasalina . On the Growth and Digestive Enzyme Activity of Penaeusmonodon from Zoea to Postlarval 20. American-Eurasian J Agronomy 4: 28-33
4. Madhumathi M, Rengasamy R (2011b) Evaluation of nutritional Parameters and carotenoid pigment from Penaeusmonodon of Zoea – PL 20 stages fed with live algal diet and Artemia enriched algal diet. Recent Research in Science and Technology 3: 76-82
5. Annon(2000)The state of world fisheries and aquaculture. Food and agriculture organisation of the United Nations, Rome, Italy.
6. Raja R, Hemaiswarya S, Rengasamy R (2007) Exploitation of Dunaliella for β-carotene production. Applied Microbiology and Biotechnology 74:517-523
7. Spolaore P, Joannis-Cassan C, Duran E, Isambert A (2006) Commercial application of microalgae. J Bioscience and Bioengineering 101: 87-96
8. Soundarapandian P, Babu R (2010) Effect of Probiotics on the Hatchery Seed Production of Black Tiger Shrimp, Penaeusmonodon (Fabricius). IntAnimVeterinAdvan 2: 9-15
9. Hemaiswarya S, Raja R, Kumar RR, Ganesan V, Anbazhagan C (2011) Microalgae: a sustainable feed source for aquaculture. World J Microbiology and Biotechnology 27:1737-1746.
10. Kannupandi T, Soundarapandian P, Anantharaman P (2002) Hatchery operation manual for Penaeusmonodon Fabricius.
11. Krishnaprakash R (2007) Studies on the production of high quality shrimp (Penaeusmonodon Fabricius) seed in a commercial hatchery.
12. Gilles Le M (2001) Environmental factor affect immune response and resistance in crustaceans.
13. Sivakumar N, Sundararaman M, Selvakumar G, Varalakshmi P,Balasubramaniem (2014)Lactobacillus SP . a potent probiotic for disease free shrimp aquaculture. Int J Recent Scientific Research5: 1031-1045

CHAPTER 13

PERFORMANCE OF ZERO WATER DISCHARGE (ZWD) SYSTEM WITH NITRIFYING BACTERIA AND MICROALGAE CHAETOCEROS CALCITRANS COMPONENTS IN SUPER INTENSIVE WHITE SHRIMP (LITOPENAEUS VANNAMEI) CULTURE

Suantika G, Lumbantoruan G, Muhammad H, Azizah FFN and Aditiawati P

Keywords: Zero Water Discharge system, nitrifying bacteria, Chaetoceros calcitrans, Litopenaeus vannamei, Super intensive

INTRODUCTION

Crustaceans are sources of high quality protein and its production contributes significantly to the gross national product of Indonesian fishery sector. National production of crustaceans in Indonesia reached up to 639,589 tonnes in 2013 with L. vannamei contributed about 60% of the total production (386,314 tonnes) [1]. From January to June 2014, Indonesia shrimp export reached up to 13,000 tons per month (≈ almost 156,000 for a year) [2], with penaeid shrimp as the dominating crustacean commodity, standing at approximately 80% of the total shrimp export [3]. Because of its significant contribution to the Indonesia's economic growth, the increment of shrimp production has become a national policy and recently the production is targeted to reach around 700,000 tonnes. Although the production tends to increase annually, shrimp farm productivity in Indonesia is still a big question mark, since this increment could probably caused by the extensification of shrimp land farming. In Indonesia and other countries (Vietnam, India, Ecuador, China, Malaysia, Philippines) most shrimp farming are still cultured traditionally, using external outdoor earthen pond, with less attention paid on diseases and water quality control [4]. It is not surprising that the system cannot maintain an adequate water quality and might lead to disease outbreak. Later, this condition might contribute to unpredictable shrimp yield production [5]. One major water quality parameter that contributes to these conditions is the accumulation of dissolved N-inorganic compound NH_4^+ (ammonium) [6]. To overcome this unfavorable condition, farmers generally conduct a frequent water renewal. Unfortunately, due to biosecurity and environmental reasons, water renewal is not considered practically. Furthermore, water renewal method is also known to cause unstable water quality [7]. Many efforts had been conducted to develop and improve a new approach in shrimp rearing strategy (e.g.: probiotic, biofloc, RAS). To this point, the results are still unpredictable in terms of production stability. Recently, the use of Zero Water Discharge by using nitrifying bacteria and Chlorella has been applied on cultivation of giant freshwater prawn (Macrobrachium rosenbergii) [8]. The application of zero water discharge was reported to help increase survival rate by 10- 20%. Based on this promising result, the development of zero water discharge system as an alternative zoo technique in super intensive white shrimp culture needs to be evaluated further. Zero water discharge provides an improvement for batch system and a modification of green water technology, especially in optimizing the nutrient cycle (inorganic and particulate nitrogen). The ammonium, nitrite, and nitrate

Citation: Suantika G, Lumbantoruan G, Muhammad H, Azizah FFN, Aditiawati P (2015) Performance of Zero Water Discharge (ZWD) System with Nitrifying Bacteria and Microalgae Chaetoceros calcitrans Components in Super Intensive White Shrimp (Litopenaeus vannamei) Culture. J Aquac Res Development 6:359. doi:10.4172/2155-9546.1000359.

level can be maintained at acceptable level by adding nitrifying bacteria and microalgae Chaetoceros calcitrans. These microorganisms have a positive effect on increasing water quality. Nitrifying bacteria can convert ammonium to nitrite (Nitrosomonas) and nitrite to nitrate (Nitrobacter) [9], while biomass of microalgae has an important role in nitrate reduction and as food supplement for shrimps [10].

MATERIALS AND METHOD

Activation and cultivation of nitrifying bacteria C. calcitrans

Culture of nitrifying bacteria was obtained from the Laboratory of Microbiology SITH ITB. Nitrifying bacteria was activated in liquid medium of Winogradsky and scaled up to a volume of 100 L as expressed in Figure 1 [8]. Culture of C. calcitrans that were used in this study was obtained from the Laboratory of Aquatic Ecosystem Analysis SITH ITB. C. calcitrans was activated in Guillard medium and scaled up to a volume of 40 L as illustrated in Figure 2 [11].

Acclimatization of White Shrimp Post Larvae

White shrimps PL-10 used in this study were obtained from PT. Suri Tani Pemuka Indramayu, West Java and acclimatized at room temperature (25 ± 1°C) at the Laboratory of Aquatic Ecosystem Analysis, School of Life Sciences and Technology, Institut Teknologi Bandung (SITH ITB), Indonesia to reach the PL-15 stadia.

Conditioning of Zero Water Discharge System

The study was conducted in cylindrical fiber tank filled with 300 L sterilized and filtered seawater at 30 ppt salinity. As illustrated in Figure 3, ZWD system consists of four compartments: (1) microbial component: nitrifying bacteria and microalgae C. calcitrans which play a role in nitrogen cycle of the culture, (2) aeration line to provide and maintain dissolved oxygen level and allow culture homogenization,

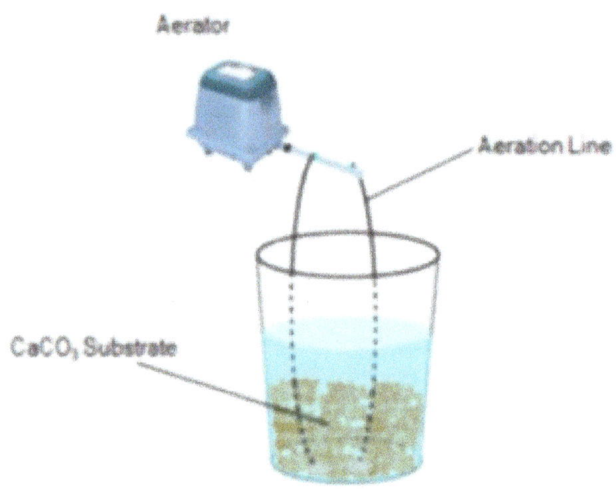

Figure 1: The bioreactor used for upscalling the nitrifying bacteria.

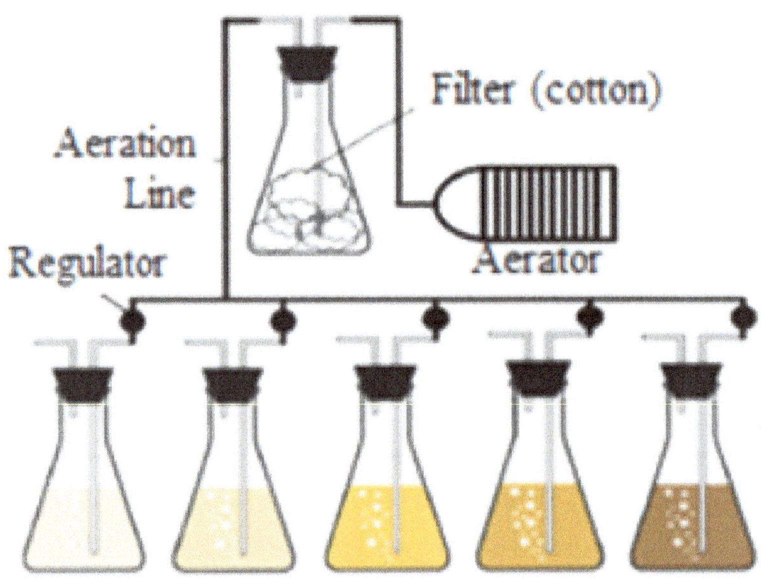

Figure 2: Scheme of batch system of C. calcitrans.

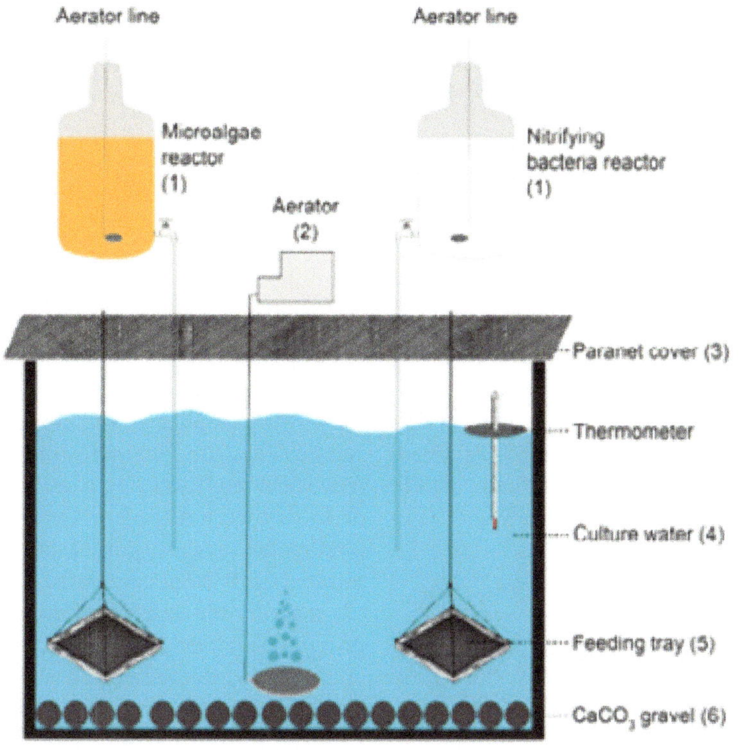

Figure 3: Schematic overview of ZWD system for white shrimp culture (L. vannamei).

(3) paranet cover to reduce light illumination and enable the system to promote better growth of diatom C. calcitrans and nitrifying bacteria, (4) filtered and sterilized culture water as shrimp growth media, (5) feeding tray to control amount of feed delivered daily, and (6) $CaCO_3$ gravel that is used as substrate for nitrifying bacteria colonization and as pH buffering agent.

Before starting the culture, the ZWD system was conditioned by inoculating 10 L of nitrifying bacteria (at 10^6 CFU/ml) and 12 L of C. calcitrans (at 10^6 cells/ml) into the ZWD systems and total volume is 300 L. The ammonium breakdown capacity of nitrifying bacteria was conditioned by adding 3 g of NH_4Cl (≈ 10 mg/L of ammonium) to the system. The conditioning was performed until the NH_4^+ and NO_2^- level decreased to approximately 0 mg/L.

Treatment Implementation

There were two treatments done in triplicate for this research: conventional culture (without any addition of nitrifying bacteria and C. calcitrans and 80% of the water volume was renewed once in four weeks) as the control, and ZWD system (with the addition of nitrifying bacteria and C. calcitrans). In order to compensate water losses due to siphoning and evaporation, every two weeks six liters (2% of culture volume) new seawater was added into the culture system. After the conditioning, PL-15 shrimp was stocked to the conventional and ZWD culture with stocking density of 120 ind/0.3m^3 which refers to the super intensive stocking density (≈400 ind/m^3) [12]. Every two weeks, 5 L of nitrifying bacteria and 12 L of microalgae C. calcitrans were added at a density of ± 10^6 CFU/ml and ± 10^6 cells/ml, respectively, into the ZWD system. The experiment was conducted for 90 days culture period.

Feeding Management

The amount of feed was adjusted weekly based on measurement of mean body weights, estimation of survival rate, and feeding rate

(1). The formula for feeding calculation is:

$$\Sigma \text{ feed (gr)} = SD \times MBW \times FR \times SR$$

where: SD is stocking density (individual/tank), MBW is mean body weight (gr), SR is Survival rate (%), FR is feeding rate (%)

Feed was placed on feeding tray and monitored frequently to provide information of daily delivered feed accuracy (Table 1).

Table 1: Feeding scheme for super intensive of white shrimp culture at 25 ± 1°C

Mean Body Weight (gr)	Feeding Rate (%)	Survival Rate (%)	Feeding Tray Monitoring Intervals (hours)
<1	10.0	100	3.5
1-3	8.0	98	3.5
3-5	6.0	96	2.5
5–7	5.0	94	2
7–9	4.0	92	2

The size and condition of shrimp can be checked and their consumption rate estimated based on the left-over feed in the tray. The feeding scheme was modified from Tacon [13]. Feed were delivered five times a day at 08:00 am, 11:00 am, 14:00 pm, 17:00 pm and 20:00 pm.

Physical and Chemical Water Parameters Measurement

Physical parameters: DO and temperature were measured everyday using a digital meter Hach® 40 qd, and pH level using a pH meter Eutech Instruments. Chemical parameters: ammonium, nitrite, nitrate and orthophosphate were measured twice a week by using Nessler, Diazotation, Nitrate HCl, and Stannous Chloride methods, respectively [14].

Biological and Microbiological Parameter Measurement

Biological parameters: Total weight, mean body weight (MBW), specific growth rate (SGR), survival rate (SR) and feed conversion ratio (FCR) were measured during culture period. The survival rate of the shrimp was calculated by using the following equation [8]:

$$SR=N_t/N_o \times 100\%$$

where: SR is survival rate , N_o is initial shrimp number, N_t is final shrimp number, t is culture period (day).

Shrimp specific growth rate was calculated by using the following equation:

$$SGR\ (\%/day)=[\ln (W_2/W_1)/(T_2-T_1)\times 100]$$

where: SGR is specific growth rate, W_1 is initial body weight (g) at time T_1 (day), and W_2 is final body weight (g) at time T_2 (day)

Microbiological parameters were measured by total bacterial count or total plate count method [15]. Total bacteria was counted every week by plating water sample on nutrient agar (NA) medium, while total Vibrio sp. count was conducted every four weeks using thiosulphate citrate bile salt agar (TCBS) medium [16].

RESULTS AND DISCUSSION

Conditioning of Zero Water Discharge System

During conditioning period, it was observed that nitrifying bacteria was able to convert 10 mg/L of ammonium to 0 mg/L in 5 days (average of breakdown capacity was 2 mg/L per day) as shown in Figure 4. The conversion of ammonium to nitrite can be seen from the increment of nitrite concentration which reached 3 mg/L on day 5 and remained steady until day 15 of culture period (Figure 4). By day 17, the nitrite level had plummeted to nearly 0 mg/L. As a final product of nitrification, nitrate had been accumulated, in which the level was around 60 mg/L by the end of culture period (day 17). Conditioning process becomes a crucial step prior to shrimp stocking and culture. This step enables ZWD system to convert toxic NH_4^+ and NO_2^- into less harmful substances NO_3^- . From the result, it can be seen that ammonium oxidizing bacteria was activated on day 5 of conditioning period. However, longer conditioning period of 17 days was taken to activate nitrite oxidizing bacteria. For nitrite oxidizing bacteria, longer activation period is due to slower growth or doubling time of nitrite oxidizing bacteria compared to ammonium oxidizing bacteria [9]. As the second step of nitrification process, NO_2^- oxidation takes place after NH_4^+ oxidation reaction. This consecutive reaction also contributes to slower activation process for nitrite oxidizing bacteria since the reaction depends heavily on NO_2^- availability of the first reaction [9].

Even though the nitrification process was able to be achieved during conditioning period, a long conditioning period (17 days) should be considered as a constraint in ZWD system because it will require a longer culture period to run one shrimp production cycle. One alternative solution to overcome this disadvantage is a better activation process of nitrite oxidizing bacteria in stock culture during maintenance, prior to being inoculated into ZWD system.

Physical and Chemical Parameters Measurement

The range of ammonium, nitrite and nitrate concentrations documented in conventional and ZWD rearing technique are presented in Table 2. Ammonium, nitrite and nitrate concentrations were still in tolerance levels for white shrimp in both rearing strategies [17-19]. The trend for ammonium, nitrite and nitrate concentration during 90 days of culture period is presented in Figure 5. It is noticeable that the ammonium, nitrite and nitrate level tended to slightly increase. In ZWD system, ammonium, nitrite and nitrate levels ranged between 0.07–0.69 mg/L, 0–3.15 mg/L, 1.04–42.9 mg/L, respectively, while in conventional (batch) system, it ranged between 0.20–0.59 mg/L, 0–3.2 mg/L, 1.38–14.17 mg/L, respectively. Based on these levels, all dissolved N-inorganic levels measured during 90 days culture period were not significantly different in both rearing systems ($p>0.05$). In spite of this condition, the breakdown capacity of ZWD system on NH_4^+ and NO_2^- levels were higher compared to conventional (batch) system. This is because a higher amount of feed (≈ 1178.28 g) was delivered to ZWD system; around 44% higher than that of conventional system (≈ 656.15 g). This contributed to a higher organic matter accumulation and therefore had an effect on higher ammonium (NH_4^+) accumulation in the system [20]. Based on [21] estimation, it was assumed that the amount of feed given to shrimp in ZWD for 90 days could produce a total NH_4^+ of 55.20 mg/L in shrimp culture which is equivalent to 0.61 mg/L per day. In contrast, total ammonia produced by shrimp for 90 days in conventional system was 30.73 mg/L, which is equivalent to 0.34 mg/L per day (total of 656.15 g feed delivery). From this estimation, it can be stated that the ammonium breakdown capacity with average ammonium conversion rate of 2 mg/L/day was higher in ZWD system than that of conventional culture system.

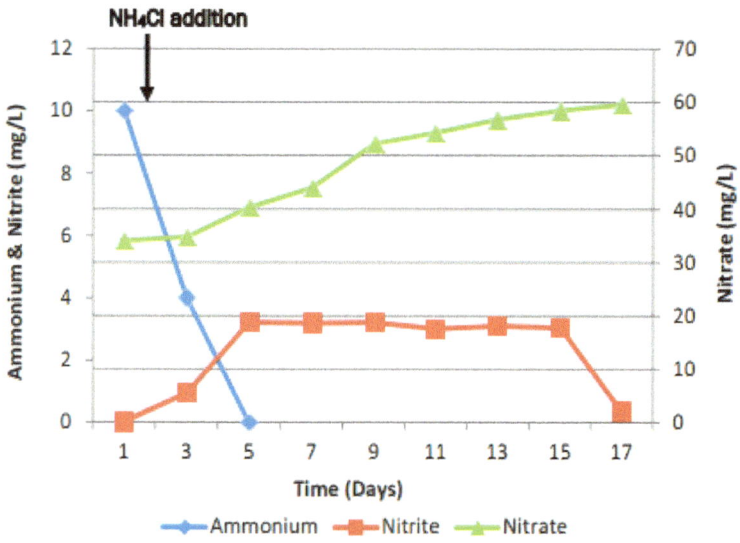

Figure 4: Performance of nitrifying bacteria based on ammonium, nitrite and nitrate concentration in ZWD conditioning.

Table 2: Range of ammonium, nitrite and nitrate concentration in conventional and ZWD for 90 days culture period

Parameter	Conventional	ZWD	Limit
Ammonium (mg/L)	0.20-0.59	0.07-0.69	≤ 3.95
Nitrite (mg/L)	0-3.2	0-3.15	≤ 25.7
Nitrate (mg/L)	1.38-14.17	1.04-42.9	≤ 232

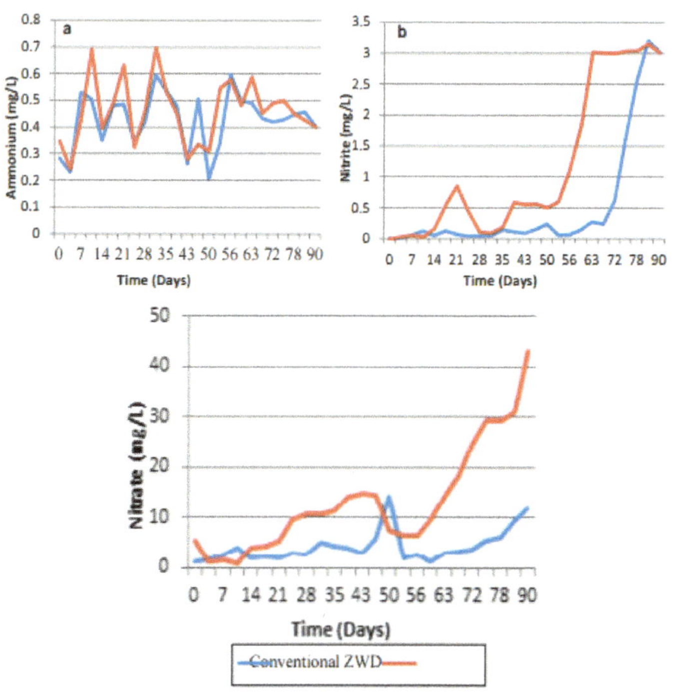

Figure 5: (a) Ammonium, (b) nitrite and (c) nitrate levels on conventional and ZWD treatment during 90 days of culture period.

Better ammonium and nitrite conversion rate were determined by measuring higher nitrate accumulation after 90 days culture period (Figure 5c). At the end of culture period, nitrate level of 42.9 mg/L at ZWD system was significantly higher ($p<0.05$) compared to conventional system (14.17 mg/L). This level of final nitrification product indicates that ammonia oxidizing bacteria (AOB) and nitrite oxidizing bacteria (NOB) performs well in the system. They contributed in maintaining a stable and low toxic NH_4^+ and NO_2- level in culture water, despite the feed delivery was approximately 180% higher in ZWD system.

Besides shrimp productivity, the accumulation of NH_4^+ and NO_2^- levels in ZWD should be taken as a serious consideration, since the presence of these toxic substances increases the amount of feed needed for better growth and to avoid shrimp's system failure. Based on this fact, an adequate system conditioning process (Figure 3) becomes a critical step prior to system operation. This conditioning will provide us with certainty on ZWD performance, NH_4^+ and NO_2^- breakdown capacity in particular.

Relatively low NH_4^+ and NO_2^- levels also probably had an effect on marine diatom C. calcitrans inclusion of ZWD system. C. calcitrans can uptake both NO_3^- and NH_4^+ as N source during assimilation process, thus, the level of NH_4^+ and NO_2^- can be maintained at an acceptable level. Keeping the ammonium, nitrite and nitrate in a low and stable level is important for shrimp cultivation. An excessive amount of ammonium might reduce growth, increase oxygen consumption and ammonia-N excretion, alter concentration of hemolymph protein and free amino acid levels, and even cause mortality [22], whereas a high level of nitrite could induce methemoglobin formation; which might result in hypoxia and cyanosis [23] and later contribute to retardation of shrimp growth and death [24].

Predicted microbial loop during 90 days culture period can be described as in Figure 6. In general, nutrient cycle processes that are involved in ZWD systems include nitrogen and carbon cycle. It can be seen that excessive feed input will not only contribute to shrimp growth, but also to the accumulation of organic matter in water, together with shrimp excreted metabolism [25]. The organic matter will then be utilized by heterotrophic bacteria and converted into NH_4^+ through ammonification process. Furthermore, NH_4^+ ions will be oxidized to form toxic substance NO_2^- by chemolithotroph ammonium oxidizing bacteria and NO_2^- will be converted into less toxic substance NO_3^- by chemolithotroph nitrite oxidizing bacteria, respectively [26]. Finally, NO_3^- as one of macronutrients will be utilized through assimilation process by photoautotrophic organism, microalgae C. calcitrans, which serves as shrimp live feed and plays a role in increasing O_2 level in water [27-28]. Microalgae biomass also contributes to organic matter increment in water, together with chemolithotrophic and heterotrophic bacteria.

Among three microbial components that have a role in ZWD microbial loop, only two (photoautotrophic and chemolithotrophic bacteria) were applied in this ZWD system, while the heterotrophic component still has not been used yet. However, for further development of the ZWD system, application of functional heterotrophic component will be taken as crucial step. Hopefully, the use of the three components altogether could help stabilizing and lowering the microbial diversity, as well as enhancing the ammonification, nitrification, and assimilation processes in culture water.

In both systems, other physicochemical parameters (i.e. pH, temperature and salinity) range between 7.63-8.8, 25.96-30.63°C and 27.6-38.3 ppt, respectively. These figures are still in tolerance range for white shrimp culture [29-31].

Appropriate culture ZWD condition will create a better environment and growth for the shrimp, even though it might contribute to the decrement of dissolved oxygen budget due to different shrimp survival (90.82% compared to 27.22%) and higher organic input. However, in this system, the DO level decrement was less than 1 mg/L (7.42 ± 0.52 mg/L in conventional and 6.81 ± 0.5 mg/L in ZWD) after 90 days culture period. Despite there was no significant effect of DO level decrement, this result should be taken as a serious consideration in further application and development of ZWD system, especially if the culture process is going to be conducted for a duration of more than 3 months (4 months of normal culture period), in which the shrimp density and the accumulation of organic matter might reduce the oxygen budget of the system. From the result, it seemed that the dissolved oxygen budget could be covered by oxygen production of microalgae and constant aeration in ZWD system [32].

Microbiological Parameter Measurement

Total bacteria counted in conventional and ZWD for 12 weeks culture period is given in Figure 7.

Based on the results, total bacteria count during the trial tended to increase in both rearing system. In ZWD, maximum number of Total Bacterial Count and total Vibrio sp. were 1×10^{10} CFU/ml and 4.8×10^2 CFU/ml, respectively, while in conventional system the figures stood at 6.7×10^{10} CFU/ml and 1.05×10^2 CFU/ml, respectively Table 3. There were no significant differences on Total Bacterial Count and total Vibrio sp. obtained during culture period between ZWD and conventional culture system ($p>0.05$).

It is noticeable that excessive organic matter accumulation in ZWD system did not significantly increase the number of bacteria, despite it has been widely known that the availability of higher organic content could promote the growth of heterotrophic bacteria [33]. The presence of nitrifying bacteria and C. calcitrans was well documented in controlling the growth of opportunistic bacteria and indigent Vibrio sp. in shrimp culture. Nitrifying bacteria can compete with other bacteria in terms of space and oxygen [34] and marine diatom secrete fatty acids and esters that can act as antibacterial compounds [35]. It is not surprising that the total Vibrio sp. in ZWD system was below pathogenicity level of 106 CFU/ml [36], even when shrimp culture in ZWD contained higher shrimp density and higher input of feed.

A low total number of bacteria in conventional system are suspected due to regular dilution by water renewal (80 % every four weeks) and also due to less organic matter load of feed and biological wastes (less shrimp density) accumulated during the culture period.

Biological Parameter Measurement

Shrimp culture productivity was observed based on mean body weight (MBW), total weight, survival rate (SR), specific growth rate (SGR) and feed conversion ratio (FCR) measurement Table 4.

In ZWD system, better shrimp culture performance was obtained after 90 days culture period. Measurements of biological parameters, including mean body weight (MBW) of 8.24 ± 0.84 g, total weight of 923.38 ± 42.15 g, Survival Rate (SR) of 90.82 ± 2.5%, and specific growth rate (SGR) of 7.7 ± 0.11%, were significantly higher in ZWD than those of conventional culture system (MBW: 5.45 ± 0.28 g, Total Weight: 160.48 ± 6.62 g, SR: 27.22 ± 2.09%, and SGR 7.24 ± 0.05). The FCR in ZWD (1.27) were significantly lower than that of conventional culture system (4.10) ($p<0.05$). It is of note that the value of two critical culture parameters, SR and FCR, were still in tolerance range of shrimp culture, in which the SR and FCR are expected to be between 51 to 91% and 1.5 to 2.6, respectively [13]. Better culture performance obtained in ZWD culture system was due to better and stable water quality maintained during culture period, even though more feed amount was delivered to support shrimp growth at higher shrimp density.

Beside the advantages of nitrifying bacteria application on water quality (NH_4^+ and NO_2^-), the use of microbial component, marine diatom C. calcitrans, has also contributed in creating a favorable condition by providing shading effect and alternative feed for shrimp [37]. Both of these factors have a significant role in reducing cannibalism due to less water visibility (minimize predation during moulting) and as a supplement diet for shrimp

Besides the information on physico-chemical, microbiological and biological parameters, a cost estimation of shrimp production using both systems also become an important step that needs to be conducted before further application of this new culture approach in aquaculture industry. Estimation was calculated in assumption to produce 100 kg shrimp during 90 days culture period, based on shrimp culture performance (total weight, FCR, MBW, SR) obtained from this research, seed cost, feed cost and water consumption as shown in Table 5.

The calculation was made based on biological parameters obtained during 90 days culture period. Better shrimp growth and survival implied in better and lower FCR in ZWD which resulted in profit by 2,375,240 IDR. Based on this analysis, the use of zero water discharge system for super intensive shrimp rearing technology can improve the profit of the culture compared to the conventional batch system with the same initial shrimp density. This economic advantage was due to better culture performance on SR, MBW, and FCR measured in ZWD system. During culture period, the use of ZWD system also consumed less water volume (52.1 m^3) compared to conventional batch system (78 m^3) to produce 100 kg of shrimps as shown in Table 4. This condition shows that the use of ZWD system will not only contribute in sustaining a more cost effective system, but also in minimizing water quality disturbance and reducing wastewater release into the environment.

CONCLUSION

Based on this research, Zero Water Discharge system can be used as an alternative for shrimp culture since ZWD was able to enhance white shrimp culture performance (water quality, SR, growth and FCR).

ACKNOWLEDGEMENT

We would like to thank Mr. Tirsa and PT. Suri Tani Pemuka unit Indramayu- Japfa Group for the shrimp larvae used in this research.

REFERENCES

1. http: //www.djpb.kkp.go.id/statistik.php?id=2
2. http: //www.globefish.org/shrimp-november-2014.html
3. Wickins JF, Daniel OL(2002) Crustacean Farming: Ranching and Culture.
4. Rosenberry R (1999) World Shrimp Farming 1999, No.12. Shrimp News International, San Diego, USA. Page 320.
5. Tacon AGJ (1996)Nutritional studies in crustaceans and the problems of applying research findings to practical farming systems. Aquaculture Nutrition 2: 165-174
6. http: //www.nrdc.org/living/shoppingwise/meals_mass/destruction_shrimp.asp
7. Pearl GG (2000) Rendering role in biosecurity and emerging diseases. Render 29(2): 46-54
8. Suantika G, Astuti DI, Arief RR, Rusni M, Turendro OR (2012) Use of zero water discharge technology through the application of nitrifying bacteria and textile vertical substrate in grow-out phase of Macrobrachiumrosenbergii De Man. J Aquacult Res Dev 3: 139
9. Kuhn DD, Drahos DD, Marsh L, Flick GJ (2010) Evaluation of nitrifying bacteria product to improve nitrfication efficacy in rercirculating aquaculture system. Aquaculture Engineering 43: 78-82
10. Avnimelech Y(2005) Tilapia harvest microbial flocs in active suspension research pond. Glob Aquacult Advocate 8: 57-58
11. Suantika G, Astuti DI, Aditiawati P, Sofyan Y (2009) Pengaruh kepadatan awal inokulum terhadap kualitas kultur Chaetocerosgracilis (Schutt) pada sistem batch. Jurnal Matematika dan Sains 14(1)
12. Tacon AGJ, Jory DE, Nunes AJP (2013) Shrimp feed management: issues and perspectives,On-farm feeding and feed management in aquaculture.
13. http: //www.fao.org/fishery/culturedspecies/Litopenaeusvannamei/en
14. APHA (1999) Standard Methods for the Examination of Water and Wastewater.
15. Cappucino J, Sherman N (2011) Microbiology: A Labolatory ManualBenjamin Cummings.
16. http: //himedialabs.com/TD/M189.pdf
17. Lin YC, Chen JC (2001) Acute toxicity of ammonia on Litopenaeusvannamei Boone juveniles at different salinity levels. Journalof Experimental Marine Biology and Ecology 259: 109-119

    1. LIN YC, CHEN JC (2003) ACUTE TOXICITY OF NITRITE ON LITOPENAEUSVANNAMEI BOONE JUVENILES AT DIFFERENT SALINITY LEVELS. AQUACULTURE 224: 193-201

18. Tsai SJ, Chen JC (2002) Acute toxicity of nitrate on Litopenaeusvannamei Boone juveniles at different salinity levels. Aquaculture 213:163-170.
19. Hargreaves JA, Craig ST (2004) Managing Ammonia in Fish Ponds
20. FAO (2014) Small-scale aquaponic food production: Integrated fish and plant farming.
21. Chen JC, Chen SF(1992)Effect of nitrite on growth and molting of Penaeusmonodon juveniles. Comp BiochemPhysiol 101(3): 453-458
22. Jensen FB (2003) Nitrite disrupts multiple physiological functions in aquatic animals. Comp Biochem Physiol 135: 9-24
23. Chen JC, Lin CY (1992) Oxygen consumption and ammonia-N excretion of Penaeuschinensis juveniles exposed to ambient ammonia at different salinity levels. Comp Biochem Physiol 102C: 287-291
24. Burford MA, Longmore AR (2001) High ammonium production from sediments in hypereutropic shrimp ponds. Mar Ecol Prog Ser 224: 187-195
25. Castillo-Soriano FA, Ibarra-Junquera V, Escalante-Minakata P, Mendoza-Cano O, Ornelas-Paz JJ, et al. (2013) Nitrogen dynamics model in zero water exchange, low salinity intensive ponds of white shrimp, Litopenaeusvannamei, at Colima, Mexico. Lat Am J Aquat Res41(1): 68-79
26. Hu H, Zhang X (2008)Nitrite utilization by Chaetocerosmuelleri under elevated CO2 concentration. World J Microbiol Biotechnol 24: 891–894
27. Collos Y (1998) Nitrate uptake, nitrite release and uptake, and new production estimates. Mar Ecol Prog Ser 171: 293-301
28. Briggs M, Funge-Smith S, Subasinghe R, Philips M (2004) Introduction and movement of Penaeusvannamei and Penaeusstylirostris in Asia and the Pacific
29. Powers LW, Bliss DE (1983) The Biology of Crustacea 8, Enviromental Adaptations
30. Suwoyo HS, Mangampa M (2010) Aplikasi probiotik dengan konsentrasi berbeda pada pemeliharaan udang vanname (Litopenaeusvannamei).
31. Iba WA, Rice MA, Wikfors GH (2014) Microalgae in Eastern Pacific White Shrimp, Litopenaeusvannamei (Boone 1931) hatcheries: A Review on Roles and Culture Environments. Asian Fisheries Science 27: 212-233
32. Ebeling JM, Timmons MB, BisogniJJ(2006) Engineering analysis of the stoichiometry of photoautotrophic, autotrophic, and heterotrophic removal of ammonia-nitrogen in aquaculture systems. Aquaculture 257: 346-358
33. Michaud L, Blancheton JP, Bruni V, Piedrahita R (2006) Effect of particulate organic carbon on heterotrophic bacterial populations and nitrification efficiency in biological filters. Aquaculture Engineering 34(3): 224-233
34. Lebeau TL, Robert JMR (2003) Diatom cultivation and biotechnologically relevant products. Appl Microbiol Biotechnol 60: 624-632
35. Pena LD, Celia RL, Mlagros GP (2001)Luminescent Vibriosis Associated with Mortality in Pond-Cultured Shrimp Penaeusmonodon in the Philippines: Species Composition. Fish Pathology 36 (3): 133-138
36. Saravanan S, Biju SKJ (2008) Moulting and Behaviour Changes in Freshwater Prawn

CHAPTER 14

BIOSORPTION OF ARSENIC BY LIVING AND DRIED BIOMASS OF FRESH WATER MICROALGAE

Sibi G

Keywords: Biosorption, Arsenic removal, Microalgae, *Scenedesmus*, *Pandorina*

INTRODUCTION

Aquatic ecosystems are mainly affected by heavy metals and represent a potential risk to the health of humans and ecosystems [1]. Heavy metal ions in the environment are biomagnified in the food chain and are accumulated in tissues. Although low concentrations of some heavy metals are metabolically important to many living organisms, at higher levels they can potentially be toxic [2]. Arsenic (As) is a component of many industrial raw materials, products and wastes. Elevated levels of arsenic in drinking water have been implicated in human diseases and mortality [3]. Chronic exposure to arsenic causes neurological and haematological toxicity [4]. Arsenic impacts the major organs and is a potential carcinogen [5-7].

The most common arsenic species observed in the environment are the trivalent form arsenite As (III) and pentavalent form arsenate As (V) in which As (III) is more toxic than As (V) [8]. Because arsenic readily changes valence state and reacts to form species with varying toxicity and mobility, effective treatment of arsenic can be challenging. Arsenic treatment technologies require peroxidation step to form As (V) from As (III) [9] but the cost and secondary product formation during other conventional methods reduce the practices [10]. Use of biological processes provides a means for cost-effective removal of metals for the treatment of metal contaminated waters.

Microbes are ideal candidates to decrease the heavy metal ion concentration from ppm to ppb levels [11]. Microalgae are known to sequestrate heavy metals due to their cell wall constituents which act as binding sites for metals [12-17]. Bioaccumulation of metals by algae may create a feasible method for remediating water contaminated with metals [18,19]. It is well established that several marine and fresh water algae are able to take up various heavy metals selectively from aqueous media and to accumulate these metals within their cells [20- 22].

Microalgae have been shown to accumulate arsenic and could potentially remediate through adsorption and biotransformation of inorganic arsenic [23-25]. In this study, the emphasis has been laid to know the efficiency of fresh water algal species in removing the arsenic from aqueous solutions. The research was to simultaneously determine the differences in arsenate and arsenite sorption capacities of living and dried algal biomass.

MATERIALS AND METHODS

Sample Collection and Identification

Algal samples were collected from Bangalore fresh water habitats (13°04'N, 77°58'E) and washed several times with tap water and then with deionized water before analysed by using microscope. The family and genus identified with reference to the biology of algae [26]. The identified algae were Chlorophyceae (Botryococcus, Chlamydomonas, Chlorella, Gonium, Pandorina, Scenedesmus, Spirogyra, Volvox), Cyanophyceae (Oscillatoria, Spirulina) and Euglenophyceae (Phacus).

Citation: Sibi G (2014) Biosorption of Arsenic by Living and Dried Biomass of Fresh Water Microalgae. J Bioremed Biodeg 5:249. doi: 10.4172/2155-6199.1000249

Chemicals

All chemicals used in this study were of analytical reagent grade. Stock solutions of 100 mg/L concentration each of As (III) and As (V) were prepared by dissolving $NaAsO_2$ and $Na_2 HAsO_4 .7H_2 O$ in deionized water respectively. Solutions for adsorption and metal analyses were prepared by appropriate dilution of the freshly prepared stock solution.

Cultivation

The identified algae were cultivated in appropriate media BG 11 media (Oscillatoria), Bold 1NV media (Spirogyra), modified Bold 3N media (Pandorina, Volvox, Phacus), Proteose medium (Scenedesmus, Gonium, Chlorella), Soil extract medium (Chlamydomonas) and soil water media (Spirulina) [27,28].

Growth Inhibition Studies

Growth inhibition bioassays for the influence of As (III) and As (V) at a concentration of 10 mg/L on algal isolates (10 mg/l) were studied. The final cell density and biomass after arsenic exposure for a period of 15 days were compared to controls in dilution water (media). Samples from each flask was taken and fixed with Lugol's iodine solution to measure the cell density using haemocytometer. The mean number of cells produced at each concentration after exposing period was expressed as percentage growth reduction with respect to control. The algal biomass was determined by the spectrophotometric transmission of algal suspension (Schimadzu UV-2600) at 550 nm. From the results, arsenic tolerant microalgae were chosen and used for biosorption studies.

Arsenic Biosorption Studies

Living biomass: Cells in the exponential growth phase were obtained by centrifuging at 2500 rpm for 7 minutes and inoculated into flasks with fresh media at a concentration of 1 x 10^4 cells/ml. The flasks were shaken by hand and randomly placed in a growth cabinet (27 ± 1°C, 12:12 h light/dark cycle, Philips TL 40W cool white fluorescent lighting, 140 µmol photons/m^2 /s). The quantity of biomass used for the biosorption studies varied from 1- 2x10^5 cells/ml.

Non-living biomass: The algal biomass was sun-dried and then dried in oven at 50°C for 24 h. The dried algal biomass was shredded, ground in a mortar and an average size of 500-600 µm was used for biosorption experiments at a concentration of 0.5 g/l.

Arsenic concentration: Both living and non-living algal biomass were inoculated separately into 250 ml conical flasks containing 100 ml of media supplemented with As (III) and As (V) with initial concentrations ranging from 10 to 50 mg/l. The flasks were kept under illumination at 2500 lux for 12 hr light - dark cycle at 24 ± 2°C. Metalfree and algae-free blanks were used as control groups. Test flasks were rotated and shaken twice daily to ensure sufficient gas exchange. Separation of biomass from metal bearing solution after appropriate incubation time was achieved through centrifugation at 10,000 rpm for 10 minutes.

pH study: Measurements for the effects of varied pH on Ar (III) and (V) biosorption performance were chosen by adding 0.1M NaOH or 0.1M HCl to achieve pH values of 2.0 through 10.0.

Temperature study: Measurements for the effects of increased temperature on arsenic adsorption performance by the algal isolates were studied at temperature between 23.0 to 35.0 ± 2°C.

Biomass study: Various biomass concentrations of living and nonliving algal cells in the range of 0.2, 0.4, 0.6, 0.8 and 1 g/l were used to study the adsorption of arsenic.

Apparatus

All measurements were carried out on Agilent 280FS AA spectrometer (Agilent Technologies, USA). The organic phase was aspirated directly without elution into the flame and the signals related to As (III) and As (V) was recorded at 193.7 and 197.2 nm (lamp current 10mA). The data were analyzed by using Spectra AA software.

Biosorption Kinetics Studies

The metal biosorption (q) by the algae and bioremoval efficiency (R) were calculated by the following formulae [29,30].

$$q = \frac{(C_i - C_f)V}{M}$$

$$R(\%) = \frac{(C_i - C_f)}{C_i} x100$$

Where q = metal adsorption (mg/g); M = dry mass of a lgae (g); V = volume of i nitial metal ion solution (L); R = bioremoval efficiency (%); Ci = initial concentration of metal in aqueous solution (mg/l); Cf = final concentration of metal in aqueous solution (mg/l).

Biosorption equilibrium models: The biosorption equilibrium isotherm was obtained by the Freundlich model (Equation 1) and the Langmuir model (Equation 2) respectively [31].

$$q = K_f C_e^{1/n} \quad (1)$$

Where, K_f and n are the distribution coefficient and a correction factor, respectively and C_e is equilibrium concentration of heavy metal (mg/l). By plotting the linear form of Eq. (1), lnq = 1/n lnC_e + lnK_f , the slope is the value of 1/n and the intercept is equal to lnK_f

$$q_{eq} = \frac{q_{max} bC_e}{1 + bC_e} \quad (2)$$

The linear form of Langmuir is

$$C_{eq} / q_{eq} = 1 / q_{max} b + C_{eq} / q_{max} \quad (3)$$

Where q_{max} is the Langmuir constant (mg/g) reflecting the maximum adsorption capacity of the metal ion per unit weight of biomass to form a complete monolayer on the surface bound at high C_{eq}. Langmuir constant b(l/mg) represents a ratio of adsorption rate constant to desorption rate constant, which also gives an indication of the affinity of the metal for binding sites on the biosorbent. qmax and b can be determined from the linear form of Langmuir equation (3) by plotting C_{eq}/q_{eq} vs. C_{eq}. The specific metal biosorption q was calculated using the following equation

$$q_e\,(mg/g) = \left[\frac{C_i - C_e}{M}\right] X V \quad (4)$$

Where q_e is the specific metal biosorption (mg metal/g biomass), V is the volume of metal solution (l), C_i and C_e are the initial and equilibrium concentration of metal (mg /l), respectively, and m is the dry weight of the biomass (g).

Arsenic Biosorption Potential

Based on the arsenic biosorption equilibriums (mg/g), optimum conditions for pH, temperature and biomass were determined and further studied to determine arsenic biosorption potential of microalgae based on contact time. The living and non-living algal biomass (0.8 g/l) was suspended in 100 ml of Ar (III) and (V) solution (30 mg/l) in 250 ml flasks containing a pH of 4.0. The cell/metal suspensions were shaken (150 rpm) at 32°C. Samples were taken from the solutions at different time intervals from 0-60 hours and analysed for metal biosorption.

RESULTS AND DISCUSSION

Growth pattern of isolated microalgae in the presence of arsenic has revealed that the degree of growth inhibition by arsenic varied widely between the isolates. The growth rate of Spirogyra, Volvox, Phacus, Gonium and Chlamydomonas was inhibited by As (III and V) in 5 days at 10 mg/l concentration. Oscillatoria and Spirogyra reached stationary phase in 8-10 days followed by decline phase. The growth rate of Pandorina was high up to 11 days where as Chlorella and Scenedesmus had exponential phase up to 13 days. These five genera have been considered as arsenic tolerant microalgae for further studies. The arsenic tolerant microalgae were grown in the presence As (III) and As (V) with a initial concentrations of 10-50 mg/l during which increased concentration of arsenic in the growth medium has led to deceased adsorption rate. The environmental setups had varying pH, temperature and biomass which produced variable results irrespective of the species studied.

Table 1: Biosorption of AS (III) (mg/g) under varying pH

AS(III)	pH	Living					Dried				
		Chlorella	*Oscillatoria*	*Scenedesmus*	*Spirogyra*	*Pandorina*	*Chlorella*	*Oscillatoria*	*Scenedesmus*	*Spirogyra*	*Pandorina*
10 mg/l	2.0	4	3	5	4	4	4	4	5	5	4
	4.0	7	6	7	6	7	7	6	7	6	6
	6.0	7	6	7	6	7	7	5	7	6	6
	8.0	5	4	5	4	5	5	4	5	5	5
	10.0	4	3	4	3	4	4	3	4	3	5
20 mg/l	2.0	8	6	9	8	7	9	8	9	9	9
	4.0	11	9	13	12	12	12	11	14	13	15
	6.0	10	9	12	12	14	11	10	14	13	14
	8.0	8	5	8	6	6	7	4	8	7	7
	10.0	5	3	5	3	3	4	2	5	3	3
30 mg/l	2.0	10	9	11	8	8	9	9	10	9	8
	4.0	20	18	23	20	21	21	18	22	20	21
	6.0	19	17	22	20	20	21	18	21	18	19
	8.0	9	7	8	9	8	9	8	8	9	9
	10.0	5	3	5	4	4	5	4	5	4	4
40 mg/l	2.0	10	9	12	11	12	12	9	13	11	11
	4.0	17	16	20	18	19	17	15	19	18	17
	6.0	13	11	14	12	12	14	8	13	13	12
	8.0	11	7	9	8	9	9	7	8	9	8
	10.0	9	6	8	7	7	9	6	7	6	6
50 mg/l	2.0	12	7	11	9	9	12	8	10	9	9
	4.0	14	10	15	13	12	14	11	14	12	12
	6.0	11	8	12	10	9	10	7	11	9	9
	8.0	10	7	9	9	8	9	7	9	8	8
	10.0	8	6	7	8	7	8	6	8	8	8

Table 2: Biosorption of AS (V) (mg/g) under varying pH

AS(V)	pH	Living					Dried				
		Chlorella	*Oscillatoria*	*Scenedesmus*	*Spirogyra*	*Pandorina*	*Chlorella*	*Oscillatoria*	*Scenedesmus*	*Spirogyra*	*Pandorina*
10 mg/l	2.0	5	4	5	4	4	5	3	5	4	5
	4.0	6	4	6	6	6	7	5	7	6	7
	6.0	6	4	5	6	5	6	4	6	5	6
	8.0	4	3	4	4	4	3	2	4	4	4
	10.0	3	1	2	2	2	3	2	3	3	3
20 mg/l	2.0	9	7	9	8	8	10	8	11	9	10
	4.0	12	10	13	11	11	13	12	14	12	13
	6.0	11	9	12	10	10	11	12	12	10	10
	8.0	9	7	9	8	8	9	8	9	8	8
	10.0	6	4	6	5	6	7	4	6	6	5
30 mg/l	2.0	13	11	14	12	12	14	12	17	15	16
	4.0	19	16	20	17	18	20	18	21	19	20
	6.0	17	14	17	13	15	17	12	16	15	14
	8.0	11	9	10	9	9	11	9	11	10	9
	10.0	7	5	7	6	6	7	5	6	6	6
40 mg/l	2.0	11	9	12	11	12	12	9	13	11	11
	4.0	14	12	16	14	15	15	13	18	15	15
	6.0	12	10	13	11	12	12	9	15	13	12
	8.0	9	7	9	7	7	10	8	10	9	9
	10.0	8	6	7	7	6	7	7	8	7	7
50 mg/l	2.0	15	11	15	13	14	16	10	16	14	14
	4.0	19	15	21	18	18	20	14	21	17	17
	6.0	16	13	17	15	16	15	12	17	13	14
	8.0	15	11	13	12	11	13	10	13	10	10
	10.0	9	8	10	9	10	10	9	11	9	9

The pH of the solution has played a key role in the biosorption of arsenite and arsenate by the microalgae. Aqueous solutions containing As (III) and As (V) were prepared with varying pH ranging from 2.0 to 10.0. Arsenic sorption decreased with increasing pH and the maximum arsenic removal occurred at pH 4.0 for both living and dried biomass. The highest As (III) uptake (Q) of 23 mg/g was recorded with living biomass of Scenedesmus at a initial concentration of 30 mg/l followed by Pandorina (21 mg/g). Maximum uptake of 22 mg/g As (III) was found with the dried biomass of Scenedesmus which was followed by Chlorella and Pandorina (Table 1). The metal sorption of As (V) was highest in living biomass of Scenedesmus (20 mg/g) and the trend was reflected in dried biomass of Chlorella and Pandorina at 30 mg/g (Table 2)

Various temperatures in the range of 23°C - 35°C were used to study the metal uptake by the algal isolates. The results indicated that 32°C was found optimum in which maximum adsorption has taken place and there were no significant changes in the metal uptake at temperature below 29°C and above 32°C. The living biomass of Scenedesmus and Chlorella has adsorbed maximum As (III) from the aqueous solution followed by Spirogyra and Pandorina at 30 mg/l initial concentrations. An uptake of 25 mg/g was recorded with dried biomass of Scenedesmus followed by Chlorella and Pandorina (24 mg/g) under similar conditions (Table 3). The uptake levels of As (V) were comparatively lower than As (III) with a maximum sorption of 16 mg/g by living biomass of Scenedesmus and Oscillatoria. The dried biomass of Scenedesmus and Chlorella has up taken 19 and 17 mg/g of As (V) at 30 mg/l levels (Table 4).

Table 3: Biosorption of AS (III) (mg/g) under varying temperature

AS(III)	Temp	Living					Dried				
		Chlorella	*Oscillatoria*	*Scenedesmus*	*Spirogyra*	*Pandorina*	*Chlorella*	*Oscillatoria*	*Scenedesmus*	*Spirogyra*	*Pandorina*
10 mg/l	**23°C**	**4**	**2**	**3**	**2**	**3**	**3**	**3**	**4**	**2**	**3**
	26°C	4	3	5	4	4	4	4	4	3	3
	29°C	7	5	7	5	6	6	5	7	5	5
	32°C	7	6	6	6	7	6	6	6	5	6
	35°C	6	5	6	5	5	5	4	7	4	5
20 mg/l	23°C	7	5	7	6	7	8	7	10	7	8
	26°C	9	8	10	9	10	11	9	12	9	10
	29°C	11	11	14	12	13	15	12	16	13	14
	32°C	12	12	15	13	14	15	12	17	14	14
	35°C	9	9	12	10	11	13	11	14	11	11
30 mg/l	23°C	10	9	11	10	11	14	10	12	11	12
	26°C	12	11	15	13	13	17	13	18	14	15
	29°C	20	17	21	18	20	24	21	25	22	23
	32°C	21	17	21	20	20	24	23	25	23	24
	35°C	16	13	18	16	17	21	18	22	19	19
40 mg/l	23°C	11	9	12	10	10	11	10	12	11	11
	26°C	14	19	13	11	13	13	11	15	12	12
	29°C	16	12	15	14	14	17	13	18	12	14
	32°C	18	13	17	15	15	19	14	18	14	15
	35°C	16	12	17	14	15	18	12	17	12	14
50 mg/l	23°C	13	9	11	11	10	14	10	12	12	12
	26°C	15	10	14	13	12	14	10	16	17	16
	29°C	17	14	28	16	17	21	19	22	20	20
	32°C	20	17	21	18	20	24	21	25	22	23
	35°C	16	25	20	17	18	21	19	23	20	20

Table 4: Biosorption of AS (V) (mg/g) under varying temperatures

AS(V)	Temp	Living					Dried				
		Chlorella	*Oscillatoria*	*Scenedesmus*	*Spirogyra*	*Pandorina*	*Chlorella*	*Oscillatoria*	*Scenedesmus*	*Spirogyra*	*Pandorina*
10 mg/l	23°C	6	5	6	5	6	5	4	6	5	5
	26°C	6	6	7	7	6	6	5	5	7	6
	29°C	7	7	7	6	6	6	5	6	6	7
	32°C	8	6	8	6	6	7	5	7	6	5
	35°C	7	7	8	6	5	7	5	6	6	5
20 mg/l	23°C	8	6	8	7	7	8	7	8	8	7
	26°C	8	7	9	7	7	9	7	9	8	8
	29°C	10	11	12	9	10	10	9	9	10	10
	32°C	11	13	13	11	12	11	11	11	11	11
	35°C	11	11	12	10	10	11	10	10	10	10
30 mg/l	23°C	12	11	11	11	12	12	10	12	11	11
	26°C	13	11	13	13	14	14	12	16	13	13
	29°C	14	12	15	14	13	14	10	17	14	12
	32°C	15	16	16	14	15	17	15	19	16	16
	35°C	14	12	17	16	15	14	13	18	15	16
40 mg/l	23°C	8	6	8	6	7	8	6	8	7	7
	26°C	10	8	11	8	9	9	8	11	9	9
	29°C	13	11	13	10	11	12	10	14	10	12
	32°C	16	11	15	13	14	17	11	17	12	14
	35°C	14	10	13	11	11	14	9	14	10	11
50 mg/l	23°C	9	6	9	8	8	8	6	8	7	9
	26°C	12	9	13	10	11	11	8	11	10	12
	29°C	16	13	17	14	16	17	11	17	15	17
	32°C	18	13	19	15	17	19	14	20	17	20
	35°C	15	11	16	13	14	16	13	15	13	15

Table 5: Biosorption of AS (III) (mg/g) under varying biomass concentrations

AS(III)	Biomass g/l	Living					Dried				
		Chlorella	*Oscillatoria*	*Scenedesmus*	*Spirogyra*	*Pandorina*	*Chlorella*	*Oscillatoria*	*Scenedesmus*	*Spirogyra*	*Pandorina*
10 mg/l	0.2 g	4	3	5	4	4	5	3	4	5	4
	0.4 g	9	7	10	9	9	11	8	11	10	11
	0.6 g	15	12	16	15	15	16	13	14	13	15
	0.8 g	22	18	23	21	21	22	19	21	17	19
	1.0 g	19	16	21	18	17	17	16	19	19	20
20 mg/l	0.2 g	5	4	6	6	5	6	4	6	5	6
	0.4 g	12	9	12	11	11	13	10	15	13	15
	0.6 g	17	15	18	17	18	18	15	17	16	17
	0.8 g	26	23	29	26	27	28	25	27	23	25
	1.0 g	20	19	24	21	21	22	20	23	22	23
30 mg/l	0.2 g	5	5	5	5	5	5	5	4	5	4
	0.4 g	15	11	16	12	13	13	9	14	11	12
	0.6 g	18	15	18	16	17	18	14	19	15	15
	0.8 g	23	24	26	23	27	25	24	28	25	26
	1.0 g	20	20	21	19	22	22	20	23	20	21
40 mg/l	0.2 g	6	5	7	7	6	6	4	6	6	6
	0.4 g	16	11	17	16	18	16	13	17	15	15
	0.6 g	20	19	21	21	22	22	20	23	21	21
	0.8 g	28	26	32	29	30	30	28	31	26	29
	1.0 g	23	22	27	24	26	27	25	29	25	27
50 mg/l	0.2 g	7	6	7	7	7	7	5	6	6	7
	0.4 g	16	13	18	17	17	17	13	17	16	16
	0.6 g	21	20	23	21	23	23	17	23	21	22
	0.8 g	28	27	33	29	30	31	28	31	27	30
	1.0 g	24	23	29	25	26	28	25	29	26	28

The biomass dosage was varied from to 0.2 to 1.0 g/l. Increased biomass has increased the arsenic sorption and the maximum uptake of arsenic was obtained at a biomass concentration of 0.8 g/l. At 30 mg/l initial concentration of As (III), maximum uptake was seen with living biomass of Pandorina (27 mg/g) followed by Scenedesmus (26 mg/g) whereas the dried biomass of Scenedesmus (28 mg/g) has recorded the highest AS (III) uptake over Pandorina (26 mg/g) (Table 5). The trend from temperature variations was repeated in biomass dosage in As (V) sorption with living cells of Oscillatoria and Pandorina has exhibited maximum uptake of 18 mg/g which relatively lower than As (III) uptake under similar conditions. The dried biomass of Pandorina has maximum As (V) sorption (17 mg/g) followed by Oscillatoria (16 mg/g) (Table 6).

Table 6: Biosorption of AS (V) (mg/g) under varying biomass concentrations

AS(V)	Biomass g/l	Living					Dried				
		Chlorella	*Oscillatoria*	*Scenedesmus*	*Spirogyra*	*Pandorina*	*Chlorella*	*Oscillatoria*	*Scenedesmus*	*Spirogyra*	*Pandorina*
10 mg/l	0.2 g	2	2	2	1	2	2	1	3	1	3
	0.4 g	4	2	4	3	4	4	2	3	3	4
	0.6 g	5	3	6	3	5	5	3	6	4	5
	0.8 g	5	4	7	5	7	6	4	7	5	6
	1.0 g	6	5	6	4	7	6	4	6	5	6
20 mg/l	0.2 g	3	2	3	2	3	3	2	3	2	2
	0.4 g	5	3	5	3	4	4	3	5	3	5
	0.6 g	6	6	7	6	6	6	6	7	6	6
	0.8 g	9	9	10	8	10	10	9	11	9	11
	1.0 g	9	9	10	9	11	9	9	10	9	11
30 mg/l	0.2 g	7	5	8	6	7	8	6	8	5	7
	0.4 g	10	8	12	8	11	10	9	11	8	11
	0.6 g	12	10	15	11	15	11	11	13	11	14
	0.8 g	14	18	16	16	18	13	16	15	15	17
	1.0 g	13	17	14	15	18	13	16	13	13	16
40 mg/l	0.2 g	11	11	12	11	11	12	11	12	11	10
	0.4 g	16	15	18	16	16	14	15	19	14	15
	0.6 g	21	16	24	21	25	21	16	22	22	23
	0.8 g	25	18	27	21	28	22	18	25	23	26
	1.0 g	24	19	26	20	26	20	18	24	22	25
50 mg/l	0.2 g	15	14	17	15	15	16	13	15	14	15
	0.4 g	26	23	26	22	23	25	23	24	22	22
	0.6 g	34	30	34	31	33	31	31	34	30	30
	0.8 g	36	32	36	34	36	33	33	35	33	33
	1.0 g	34	32	33	33	33	32	31	30	31	32

The optimum pH, temperature and contact time obtained from equilibrium isotherm experiments were used to determine the arsenic biosorption of both living and non-living algal biomass based on time dependent manner. The time dependent biosorption of arsenic by living and dried biomass, samples equivalent to 0.8g/l were placed in solutions of As (III) and As (V) adjusted to pH 4.0 and 32°C temperature. The results of arsenic sorption as a function of time under the controlled conditions have been depicted in Figure 1-4. The kinetic studies revealed that arsenic biosorption increases with time and recorded highest at 36 hrs contact time. However the soption rate decreased progressively and became almost constant during the 60 hrs contact with arsenic. Scenedesmus has showed fast kinetic of As (III) binding and was adsorbed appreciable quantities of As (III) (20 mg/g) during the first 24 -36 hrs from the solution. The second highest uptake of As (III) was observed with Pandorina (21 mg/g) followed by Chlorella and there was no significant changes in the metal uptake after 36 hrs of contact time by the isolates. Similarly highest As (V) was recorded with Scenedesmus with a metal uptake of 20 mg/g followed by Chlorella and Pandorina after 36 hrs of contact time. Oscillatoria and Spirulina have recorded moderate arsenic uptake throughout the study.

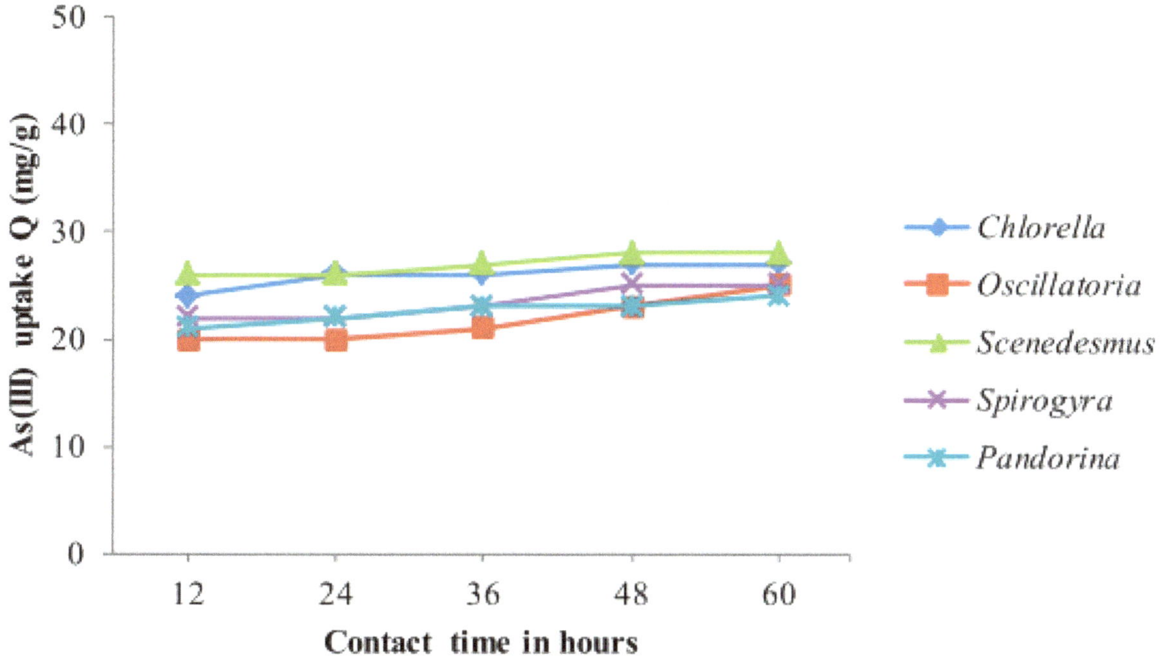

Figure 1: Effect of contact time on biosorption of As (III) at pH 4.0; temperature 32°C; As concentration 30 mg/l; living biomass 0.8 g/l.

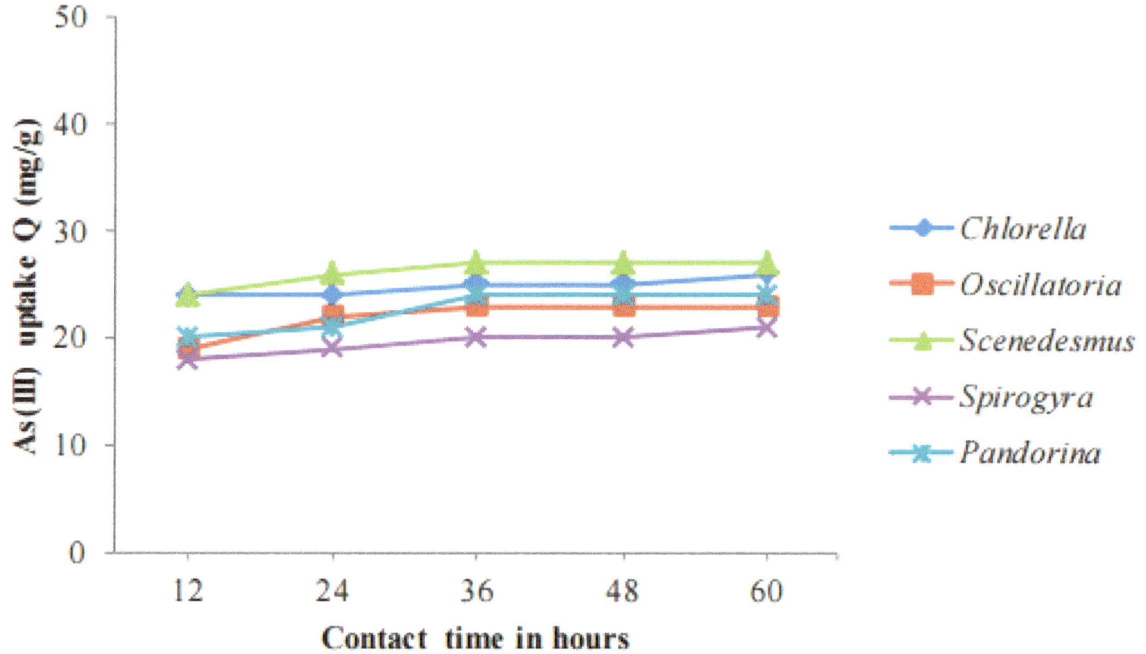

Figure 2: Effect of contact time on biosorption of As (III) at pH 4.0; temperature 32°C; As concentration 30 mg/l; dried biomass 0.8 g/l.

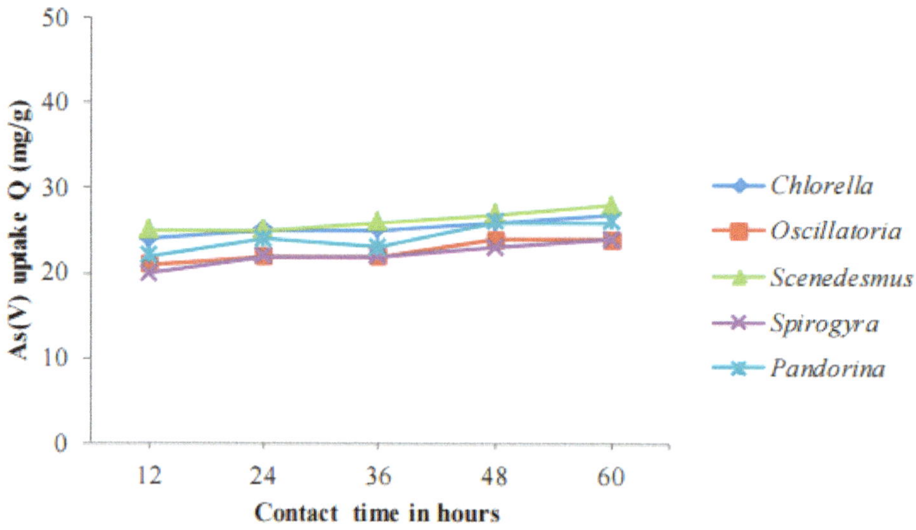

Figure 3: Effect of contact time on biosorption of As (V) at pH 4.0; temperature 32°C; As concentration 30 mg/l; living biomass 0.8 g/l.

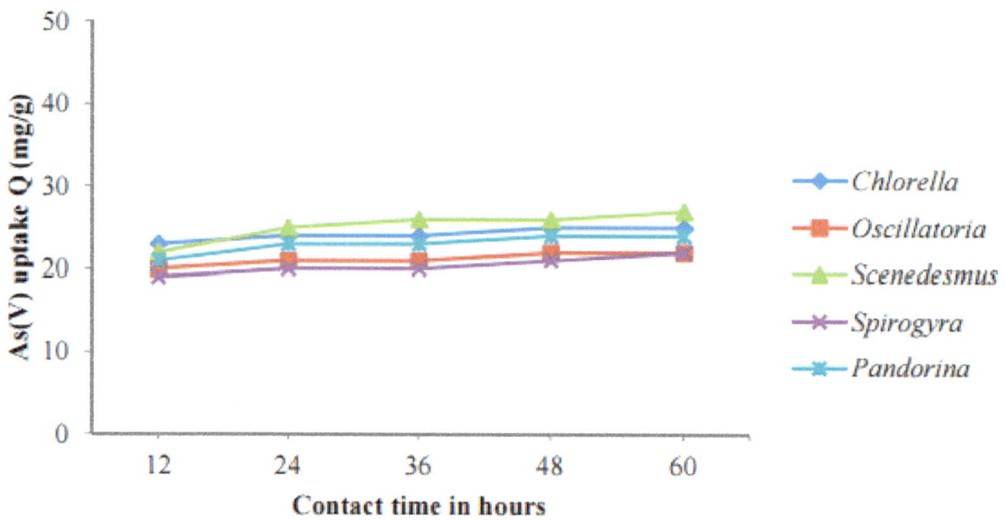

Figure 4: Effect of contact time on biosorption of As (V) at pH 4.0; temperature 32°C; As concentration 30 mg/l; dried biomass 0.8 g/l.

Algae are known for their capability to accumulate heavy metals as they are required as essential nutrients [32] and have been explored for metal removal [33]. The level of metal removal by a microalga depends on biomass concentrations, pH and contact time [34-36]. In this experiment, arsenic sorption values of dried biomass were high in comparison with living biomass. This could be due to the larger surface area due to the destruction of cell membranes during the dried biomass preparation. Non-living microbial cells have lower sensitivity to toxic metal ions concentrations over living cells which offer to use them at adverse operating conditions [37]. The metal sorption capacity was drastically reduced at pH values above 6.0, which could be due to the formation of insoluble metal hydroxides. Further, it was noted that the arsenic uptake was lowered at pH below 4.0. The temperature of the solution could influence the metal biosorption of living cells [37] and culture temperatures have profound effects on the chemical composition of the algal cells. The biosorption of As increased with temperatures from 23°C-32°C and the results indicated that elevated temperatures tend to increase the biosorptive properties of isolated microalgae with an optimum temperature of 32°C. However the variations in uptake of As (III) and As (V) based on temperature need to be investigated. It was observed that higher biomass levels have reduced the adsorption amount which was influenced by the formation of aggregates at higher concentrations that has ultimately resulted in reduced biosorption area [38].

The accumulation of heavy metals in algae involves and rapid uptake initially followed by slower uptake [37,39-41]. The same trend was observed in the experiments as the metal sorption rate was higher in first 36 hours followed by significantly slower uptake in the next hours. Further, As (III) and As (V) sorption was more at initial time (0- 36 hrs) followed by almost constant after 36 hrs of contact time.

Based on the pH studies, As (III) was effectively removed by both living and dried biomass of Scenedesmus and Pandorina whereas As (V) was removed by Scenedesmus and Chlorella. Significant levels of As (III) was removed by the living biomass of Spirogyra at varying pH. Under varying temperatures, Scenedesmus, Chlorella and Pandorina were efficient in removing As (III) and As (V) from the solutions.

In general, Scenedesmus, Pandorina and Chlorella were found as potential arsenic removers from the aqueous solutions under varying environmental conditions. Heavy metal removal capacities of Scenedesmus was reported in earlier studies [42-45]. Arsenite and arsneate tolerance levels of Chlorella were studied earlier [23,24,46,47]. Spirogyra was efficient in As (III) removal than As (V) and the previous studies reported the arsenic removal by Spirogyra [48]. Samal et al. [49] has revealed the arsenic removing capacity of Oscillatoria in their studies and in this study dried biomass of Oscillatoria was significant at 30 mg/l of As (V).

In this study, removal rate of As (III) was higher than the As (V) by the microalgae under the experimental conditions. The trivalent arsenite is more toxic than the pentavalent arsenate and this study reveals the potential use of microalgae in removing the toxic As (III). The dried biomass was found effective for arsenic sorption than the living biomass. This study emphasizes that microalgae are ideal candidates to be further exploited for the removal of other heavy metal ions.

REFERENCES

1. Rai PK (2010) Phytoremediation of heavy metals in a tropical impoundment of industrial region.Environ Monit Assess 165: 529-537.
2. Phillips DJH (1995)The chemistries and environmental fates of trace metals and organochlorines in aquatic ecosystems. Mar Pollut Bull. 31: 193-200.
3. Wang S, Zhao X (2009) On the potential of biological treatment for arsenic contaminated soils and groundwater.J Environ Manage 90: 2367-2376.
4. Byron WR, Bierbower GW, Brouwer JB, Hansen WH (1967) Pathologic changes in rats and dogs from two-year feeding of sodium arsenite or sodium arsenate. ToxicolApplPharmacol 10: 132-147.
5. Naqvi SM, Vaishnavi C,Singh H (1994) Toxicity and metabolism of arsenic in vertebrates. Arsenic in the environment, Part II: Human Health and Ecosystem effects, Edited by Jeome O. Nriagu. Pp: 55-91.
6. Morton W, Starr G, Pohl D, Stoner J, Wagner S, et al. (1976) Skin cancer and water arsenic in Lane County, Oregon.Cancer 37: 2523-2532.
7. Duker AA, Carranza EJ, Hale M (2005) Arsenic geochemistry and health.Environ Int 31: 631-641.
8. Cullen WR, Reimer KJ(1989) Arsenic speciation in the environment. Chem. Rev. 89: 713-764.
9. Kim MJ, Nriagu J (2000) Oxidation of arsenite in groundwater using ozone and oxygen.Sci Total Environ 247: 71-79.
10. Wang S, Mulligan CN (2006) Natural attenuation processes for remediation of arsenic contaminated soils and groundwater.J Hazard Mater 138: 459-470.
11. Wang J, Chen C (2006) Biosorption of heavy metals by Saccharomyces cerevisiae: a review.BiotechnolAdv 24: 427-451.
12. Gong R, Ding Y, Liu H, Chen Q, Liu Z (2005) Lead biosorption and desorption by intact and pretreatedspirulina maxima biomass.Chemosphere 58: 125-130.
13. Davis TA, Volesky B, Mucci A (2003) A review of the biochemistry of heavy metal biosorption by brown algae. Water Res 37: 4311-4330.
14. Sheng PX, Ting YP, Chen JP, Hong L (2004) Sorption of lead, copper, cadmium, zinc, and nickel by marine algal biomass: characterization of biosorptive capacity and investigation of mechanisms.J Colloid Interface Sci 275: 131-141.
15. Schiewer S, Volesky B (2000) Biosorption Processes for Heavy Metal Removal in: D.R. Lovely (Ed.), Environmental Microbe Metal Interactions, ASM Press, Washington, DC, 2000, Pp. 329-362.
16. Yu Q, Matheickal JT, Kaewsarn P(1999) Heavy metal uptake capacities of common marine macro-algal biomass, Water Res. 33:1534-1537.
17. Holan ZR, Volesky B (1994) Biosorption of lead and nickel by biomass of marine algae.BiotechnolBioeng 43: 1001-1009.
18. Darnal DW, Greene B, Henzl MT, Hosea JM, McPherson RA, et al. (1986) Selective recovery of gold and other metal ions from an algal biomass.Environ SciTechnol 20: 206-208.
19. Nakajima A, Horikoshi T, Sakagushi T (1981) Recovery of varnlum by immobilized microorganisms. Eur. J. Appl. Microbiol. Biotechnol. 16: 88-91.
20. Afkar A, Ababna H, Fathi AA (2010) Toxicological response of the green alga Chlorella vulgaris to some heavy metals. Am. J. Environ. Sci. 6: 230-237.
21. Kumar D, Gaur JP (2011) Metal biosorption by two cyanobacterial mats in relation to pH, biomass concentration, pretreatment and reuse.BioresourTechnol 102: 2529-2535.
22. Chen CY, Chang HW, Kao PC, Pan JL, Chang JS (2012) Biosorption of cadmium by CO(2)-fixing microalga Scenedesmusobliquus CNW-N.BioresourTechnol 105: 74-80.
23. Levy JL, Stauber JL, Adams M, Maher W, Kirby JK, et al. (2005) Toxicity, biotransformation, and mode of action of arsenic in two freshwater microalgae (Chlorella sp. and Monoraphidiumarcuatum). Environmental Toxicology and Chemistry. 24: 2630-2639.
24. Maeda S, Nakashima S, Takeshita T, Higashi S (1985) Bioaccumulation of arsenic by freshwater algae and the application to the removal of inorganic arsenic from an aqueous phase. Part II. By Chlorella vulgaris isolated from arsenic-polluted environment. SeparSciTechnol 20: 153-161.
25. Giddings JM, Eddlemon GK (1977)The effects of microcosm size and substrate type on aquatic microcosm behaviour and arsenic transport. Arch. Environ. Contam. Toxicol. 6: 491-505.
26. Round FE (1973) The Biology of the Algae (2nd ed). Edward Arnold (Publishers) Ltd, UK, Pp. 1-23.
27. Anderson RA (2005) Algal culture techniques. 1st edition. Elsevier Academic press. California, USA.
28. Stanier RY, Kunisawa R, Mandel M, Cohen-Bazire G (1971) Purification and properties of unicellular blue-

green algae (order Chroococcales).Bacteriol Rev 35: 171-205.

29. Volesky B (1992) Removal of heavy metals by biosorption. In: Ladisch MR, Bose A (eds). Harnessing biotechnology for the 21stCentury.Amer. Chem. Soc., Washington DC, Pp. 462-466.

30. Zhang L, Zhao L, Yu Y, Chen C (1998) Removal of Pb2+ from aqueous solution by non-living Rhizopusnigricans. Water Res. 32: 1437-1444.

31. Volesky B (1990) Removal and recovery of heavy metals by biosorption. In: Volesky, B. (Ed.), Biosorption of Heavy Metals. CRC Press, Boca Raton, Fla. Pp. 7-43.

32. Gadd GM (1990) Heavy metal accumulation by bacteria and other microorganisms. Cell Mol Life Sci 46:834-840.

33. Wilde EW, Benemann JR (1993) Bioremoval of heavy metals by the use of microalgae.BiotechnolAdv 11: 781-812.

34. Aksu Z, Donmez G(2006) Binary biosorption of cadmium(II) and nickel(II) onto dried Chlorella vulgaris: Co-ion effect on mono-component isotherm parameters. Process Biochem. 41: 860-868.

35. Solisio C, Lodi A, Soletto D, Converti A (2008) Cadmium biosorption on Spirulinaplatensis biomass. BioresourTechnol 99: 5933-5937.

36. Tang YZ, Gin KYH, Aziz MA (2002) Equilibrium model for cadmium adsorption by green algae in a batch reactor. J Environ Eng 128:304-312.

37. Arica MY, Tuzun I, Yalc E¸Ince O,Bayramoglu G(2005) Utilisation of native, heat and acid-treated microalgae Chlamydomonasreinhardtii preparations for biosorption of CrI) ions, Process Biochemistry, 40:2351-2358.

38. Ahuja P, Gupta R,Saxena RK(1999) Zn2+ biosorption by Oscillatoriaanguistissima, Process Biochemistry, 34: 77-85.

39. Hamdy AA (2000) Biosorption of heavy metals by marine algae.CurrMicrobiol 41: 232-238.

40. Vinoj Kumar V, Kaladharan P(2006)Biosorption of metals from contaminated water using seaweed, Current Science, 90: 1263-1267.

41. Bates SS, Tessier A, Campbell PGC, Buffle J (1982) Zinc adsorption and transport by Chlamydomonasvariabilis and Scenedesmussubspicatus (Chlorophyceae) grown in semi-continuous culture. J. Phycol. 18: 521-529.

42. Jahan K, Mosto P, Mattson C, Frey M, Derchak L (2006) Microbial removal of arsenic. Aater, air and Soli ollution: Focus 6: 71-82.

43. Monteiro CM, Castro PML, Malcata FX(2009) Use of the microalga Scenedesmusobliquus to remove cadmium cations from aqueous solutions. World J MicrobiolBiotechnol. 25: 1573-1578.

44. Ajayan KV, Selvaraju M, Thirugnanamoorthy K (2011) Growth and heavy metals accumulation potential of microalgae grown in sewage wastewater and petrochemical effluents.Pak J BiolSci 14: 805-811.

45. Kizilkaya B, Turkera G, Akgulb R, Doganc F(2012) Comparative Study of Biosorption of Heavy Metals Using Living Green Algae Scenedesmusquadricauda and Neochlorispseudoalveolaris: Equilibrium and Kinetics. Journal of Dispersion Science and Technology. 33: 410-419.

46. Jiang Y, Purchase D, Jones H, Garelick H (2011) Effects of arsenate (AS5+) on growth and production of glutathione (GSH) and phytochelatins (PCS) in Chlorella vulgaris.Int J Phytoremediation 13: 834-844.

47. Rao PH, Kumar RR, Raghavan BG, Subramanian VV, Sivasubramanian V(2011) Is phycovolatilization of heavy metals a probable (or possible) physiological phenomenon? An in situ pilot-scale study at a leather-processing chemical industry. Water Environ Res. 83: 291-297.

48. Kumar JI, Oommen C (2012) Removal of heavy metals by biosorption using freshwater alga Spirogyra hyalina.J Environ Biol 33: 27-31.

49. Samal AC, Bhar G, Santra SC (2004) Biological process of arsenic removal using selected microalgae.Indian J ExpBiol 42: 522-528.

CHAPTER 15

ALGOMICS FOR THE DEVELOPMENT OF A SUSTAINABLE MICROALGAE BIOREFINERY

Mustapha Benmoussa

Keywords: Microalgae, Biomass, Biorefinery, Biodiesel, Algomics, Metabolomics

INTRODUCTION

Many advantages are associated with replacing fossil fuel by biodiesel. It is produced from renewable biomass. It can be continuously produced comparatively to an expected depletion of fossil fuel. Biodiesel is a biodegradable and non-toxic product, and safe to be shipped. The use of biodiesel is associated by a significant reduction of greenhouse gases, carbon dioxide, and sulfur dioxide en air emission. It reduces also asthma-causing soot, and can be produced domestically [1]. Consumers in the US used over 2 billion gallons of biodiesel in 2015. This record consumption reduced 18.2 million metric tons of carbon dioxide production in the US (www.EPA.GOV). Many land crops biomass such as soybean, corn, rapeseeds and sugarcane can be used to produce biodiesel. Moreover, the demand on biofuels resulted in an increase in using land crops for energy production. US fuels companies reported producing 2.09 billion gallons of biodiesel in 2015 [2]. The use of land crops for renewable energy and the global population growth are major challenging concerns of the global food and energy security. These data showed a continuous increase of biofuels production. It started to partially divert land crops from their original uses as human food and animal feed to nonconventional fuels production. A noncompetitive to food and sustainable bioenergy source could be more feasible and globally accepted.

The production of biodiesel from waste cooking oil and animal fat is an environmentally friendly process, and ovoids city waste treatment and handling. However, this alternative process cannot significantly affect the total biofuels production and will not be able to respond to the global biofuel demand. The only crop that could use salt, brackish wastewater, and be produced on non-arable land, and could be source of food and renewable energy is microalgae. This fabulous micro-crop has the potential to yield biomass 50 times more than any other land crop biomass. Moreover, it converts CO_2 and H_2O with sunlight into chemical energy, and yields high oil content sometimes more than 50% of the dry weight. Microalgae biomass is produced yearly. Hu et al. reported an extensive review on microalgae for biofuel [3].

The challenge is how to make microalgae biorefinery profitable and sustainable. High-value micro-crop cell components could make microalgae biorefinery commercially feasible. A better understanding of high-value microalgae cell components and lipids metabolism could be integrated into yield and quality improvement programs [4]. Algomics is the application of omics technologies to study algae including genomics, transcriptomics, proteomics and metabolomics are tools that could be used to further cell metabolism knowledge within a short time than few decades before [5-7].

GENOMICS

Genomics became a powerful tool after the development of advanced automatic sequencing technology during the last two decades. It is complemented by bioinformatics to assemble DNA sequences and analyze gene structure and expression. The human genome research projects demonstrated the power

Citation: Benmoussa M (2016) Algomics for the Development of a Sustainable Microalgae Biorefinery. Single Cell Biol 5:132. doi:10.4172/2168-9431.1000132.

of omics in general and genomics in particular. Both scientists and clinicians can study the role and even interaction between different genetic factors to develop therapeutic strategies. In the case of microalgae, a full genome sequencing of strains with potential commercial applications will enable scientists to use other omics technologies for a better understanding of high-value cell components and lipids metabolism and the overall cell behavior under stress and non-stress condition. Many strains genomes were recently sequenced including Botryococcus braunii Showa, Botryococcus braunii UTEX 572, Chlorella sp. NC64A, Chlamydomonas reinhardtii, Coccomyxa sp. C-169, Dunaliella salina, Ostreococcus tauri, Micromonas pusilla and Nannochloropsis oceanica (http://genome.jgi.doe.gov). Comparative genomic analysis contributed in the understanding of some mechanism of microalgae adaptation to extreme environmental conditions [8]. Homologous recombination is another tool to study functional genomics [9]. They were able to generate knockouts of nitrate (NR) and nitrite (NiR) reductase genes in Nannochloropsis sp. in one single transformation step. This result provided evidence that the strain is haploid. Nannochloropsis sp is a microalgae strain known by its high growth rate. The use of this genetic transformation technique on Nannochloropsis sp could enable scientists to explore further microalgae functional genomics and develop a realistic knowledge for future gene manipulation and establish different procedures in the biorefinery to improve the yield and quality of high-value byproducts and oil. Chemical genomics is another way to characterize bioactive compounds specific trafficking pathways via endomembrane trafficking system [10].

Post-Genomics

To achieve targeted goals, post-genomics tools could complement the genomics data. Many post-genomics studies were performed on strains with sequenced genome. However, the de novo approach developed by Guamieri et al. could be performed on strains with unsequenced genome [11]. It is a tool to bypass the genome sequencing step and is discussed below.

Transcriptomics

Transcriptomics is involved after the transcription of genes to RNA. It is the study of a single or group of genes expression dynamic. In case of uncharacterized genes, Aoki et al. developed a new gene coexpression database for microalgae (http://alcodb.jp) [12]. The gene coexpression analysis is powerful when there are some uncharacterized genes to be profiled and analyzed. It is based on the similarity of gene expression profiles and in the case of expressed genes "functionally and coordinately related". The database was developed using green Chlamydomonas reinhardtii and red Cyanidioschyzon merolae. C. reinhardtii is generally used as the model green algae in post-genomics analysis.

The de novo transcriptome approach reported by Guamieri et al. [11] started by transcriptome sequencing of Chlorella vulgaris UTEX 395, followed by a comparative global transcriptomic and proteomic study. A combination of available software and new bioinformatics tools they developed were performed for transcriptome sequencing data assembly. It offers the opportunity to perform a comprehensive system biology investigation of microalgae species with unsequenced genome.

Proteomics and Metabolomics

Proteomics and metabolomics are involved after the translation of mRNA to protein. Both omics could be used to identify peptides and their profiling. The post-transcriptional regulation of microalgae cells reaction to environmental changes could be investigated using proteomics and metabolomics approaches. It is well known that photosynthesis efficiency is one of the bottlenecks in the physiology and metabolism of any photoautotrophic cell. It could affect the growth rate, byproducts and lipids yield. In this regard, Battchikova et al. reported a proteomics review on cyanobacterial photosynthesis process [13]. A study reported by Nguyen et al. targeting purified oil bodies of C. reinhardtii under nitrogen deprivation by proteomic profiling showed a number of proteins involved in vesicle trafficking and transport [14].

Comparative proteomic analysis is another efficient way to understand the cell metabolism and identify key proteins involved biosynthesis pathways and trafficking systems [15]. Garnier et al. performed a comparative proteomic analysis on wild and mutant Tisochrysis lutea strains under two metabolic conditions. Proteomes analysis showed a differential expression of proteins between both strains under nitrogen starvation conditions [16]. Identified proteins demonstrated the influence of nitrogen starvation on lipids metabolism, carbohydrate catabolism and carbon homeostasis. Proteomics results could be complemented with an overall transcriptomic study to understand genes expression regulation. The procedure helps targeting genes of interest to manipulate yield and quality of lipids accumulated in oil bodies of microalgae cells.

Metabolomics is another approach to complement genomics and other post-genomics technologies for gaining insights into functional biology systems. The metabolomics investigation could be performed on proteins and non-proteins with specific role in metabolites biosynthesis pathways. Generally, there are many metabolomics techniques including protein and metabolites profiling, specific target analysis and metabolic fingerprinting. Since there is diversity in cell metabolites and their physicochemical properties, the metabolic analysis could require more than one approach. There is a range of highly sophisticated and sensitive technologies available to identify peptides and metabolites and their dynamic in the overall cell metabolism.

Although lipids metabolism network is complex, it should be one of the major targets using proteomics and metabolomics. Lipidomics is an approach to understand the process of fatty acid synthesis and lipids accumulation oil bodies in the microalgae cells, and then develop strategies to improve the yield and quality. For example, a link between the dynamic metabolic profiles obtained by the metabolomic study and the improvement of lipids biosynthesis during microlagal cultivation [17]. The study was performed on Clamydomonas sp. JSC4 using different growth condition including nitrogen depletion and light intesity, CE/MS metabolite analysis and 13C-labeling for the profiling and turnover.

CONCLUSION

Omics and bioinformatics are highly advanced techniques in the study of algae cells could be performed in short time. It is a unique opportunity to carryout large-scale studies and to develop a realistic understanding of cell physiology and metabolism. Algomics data would provide an understanding of oil and high value metabolites biosynthesis pathways under stress and non-stress conditions. This further knowledge could be used to manipulate the cell factory and improve oil and high-value products yield and quality [18,19]. There is a review on recent advances made in algal genetic manipulation [20], and there are many research centers and University facilities that offer services in this regard. The outcome would be the development of commercially feasible microalgae biorefineries and boost the microalgae industry.

REFERENCES

1. Biodiesel: Fat to Fuel. United States Environmental Protection Agency. Accessed on: 10 Mar 2016.
2. Renewable Energy Group reports Q4, full-year 2015 financials. Biodiesel. Accessed on:10 Mar 2016.
3. Hu Q, Sommerfeld M, Jarvis E, Ghirardi M, Posewitz M, et al. (2008) Microalgal triacylglycerols as feedstocks for biofuel production: perspectives and advances. Plant J 54: 621-639.
4. Hannon M, Gimpel J, Tran M, Rasala B, Mayfield S (2010) Biofuels from algae: challenges and potential. Biofuels 1: 763-784.
5. Jamers A, Blust R, DeCoen W (2009) Omics in algae: paving the way for a systems biological understanding of algal stress phenomena? Aquat Toxicol 92: 114-121.
6. Ma NL, Lam SS, Zaidah R (2015) The Application of Algae for Cosmeceuticals in the Omics Age. Genomics, Proteomics and Metabolomics in Nutraceuticals and Functional Foods (2ndedn). 476-488.
7. Misra N, Panda PK, Parida BK (2013) Agrigenomics for microalgal biofuel production: An overview of various bioinformatics resources and recent studies to link OMICS to bioenergy and bioeconomy. Omics 17: 537-549.
8. Blanc G, Agarkova I, Grimwood J, Kuo A, Brueggeman A, et al. (2012) The genome of the polar eukaryotic microalga Coccomyxasubellipsoidea reveals traits of cold adaptation. Genome Biol 13: R39.
9. Kilian O, Benemann CS, Niyogi KK, Vick B (2011) High-efficiency homologous recombination in the oil-producing alga Nannochloropsis sp. Proc Natl Acad Sci USA 108: 21265-21269.
10. Rodriguez-Furlán C, Hicks GR, Norambuena L (2014) Chemical Genomics: Characterizing Target Pathways for Bioactive Compounds Using the Endomembrane Trafficking Network. Methods Mol Biol 1174: 317-328.
11. Guarnieri MT, Nag A, Smolinski SL, Darzins A, Seibert M, et al. (2011) Examination of triacylglycerol biosynthetic pathways via de novo transcriptomic and proteomic analyses in an unsequenced microalga. PLoS One 6: e25851.
12. Aoki Y, Okamura Y, Ohta H, Kinoshita K, Obayashi T (2016) ALCOdb: Gene Coexpression Database for Microalgae. Plant Cell Physiol 57: e3.
13. Battchikova N1, Angeleri M2, Aro EM2 (2015) Proteomic approaches in research of cyanobacterial photosynthesis. Photosynth Res 126: 47-70.
14. Nguyen HM, Baudet M, Cuine S, Adriano JM, Barthe D, et al. (2011) Proteomic profiling of oil bodies isolated from the unicellular green microalga Chlamydomonasreinhardtii: with focus on proteins involved in lipid metabolism. Proteomics 11: 4266-4273.
15. Benmoussa M, Chandrashekar A, Ejeta G, Hamaker BR (2015) Cellular Response to the high protein digestibility/high-Lysine (hdhl) sorghum mutation. Plant Sci 241: 70-77.
16. Garnier M, Carrier G, Rogniaux H, Nicolau E, Bougaran G, et al. (2013) Use comparative proteomics on a selected microalgae provides candidates for biofuel production. EuPA, Saint-Malo, France.
17. Ho SH, Nakanishi A, Ye X, Chang JS, Chen CY, et al. (2015) Dynamic metabolic profiling of the marine microalga Chlamydomonas sp. JSC4 and enhancing its oil production by optimizing light intensity. Biotechnol Biofuels 8: 48.
18. Dunahay TG, Jarvis EE, Dais SS, Roessler PG (1996) Manipulation of microalgal lipid production using genetic engineering. In Seventeenth Symposium on Biotechnology for Fuels and Chemicals. Humana Press 57: 223-231.
19. Guarnieri MT, Pienkos PT (2015) Algal omics: unlocking bioproduct diversity in algae cell factories. Photosynth Res 123: 255-263.
20. Rasala BA, Gimpel JA, Tran M, Hannon MJ, Miyake-Stoner SJ, et al. (2013) Genetic engineering to improve algal biofuels production. In Algae for Biofuels and Energy. Springer 5: 99-113.

CHAPTER 16

MASS CULTIVATION FROM A KOREAN RACEWAY POND SYSTEM OF INDIGENOUS MICROALGAE AS POTENTIAL BIOFUEL FEEDSTOCK

Hong JW, Kim OH, Kim H, Jo SW, Cho HW and Yoon HS

Keywords: Microalgae, Biomass, Biorefinery, Biodiesel, Algomics, Metabolomics

INTRODUCTION

The world's demand for renewable and sustainable energy resources is increasing exponentially because of changes in climate and energy shortage problems. This situation has given rise to the development of numerous new technologies such as biomass, geothermal, solar, tidal, and wind energies. Among these resources, microalgae are now considered one of the most attractive candidates for biofuel production due to their higher photosynthetic efficiency and oil yield compared to terrestrial sources [1,2]. A number of microalgae strains with desired characteristics for biofuel production have been found and/ or developed [3,4]. However, the attempts to grow these microalgae in outdoor open pond systems have not always been successful due to rapid contamination with bacteria, predatory zooplanktons, and other algal species [5,6]. To overcome these challenges, indigenous microalgae strains have been cultivated for large scale production since these endemic wild types are well adapted to their local conditions and, therefore, they are able to outcompete other indigenous algal strains [7-9]. In this study, biomass productivity and characterization data obtained from commercial-scale microalgal cultivation from outdoor raceways over 1.25 years are presented.

Citation: Hong JW, Kim OH, Kim H, Jo SW, Cho HW, et al. (2016) Mass Cultivation from a Korean Raceway Pond System of Indigenous Microalgae as Potential Biofuel Feedstock. Oil Gas Res 2:108. doi:10.4172/ogr.1000108.

MATERIALS AND METHODS

Raceway (RW) Pond system

A freshwater microalgal cultivation system was constructed in August 2013 at the Chilgok-gun agricultural Technology Center (36°02'N,128°22'E); Dongan-ri, Yakmok-myeon, Chilgok-gun, Gyeongsangbukdo, South Korea). Two identical 675.0 m2 open raceway ponds (RW #1 and #3) were constructed (Figures 1 and 2). A semitransparent film cover was added to RW #1 to compare the RWs in terms of biomass yield and consumption of resources under unfavorable Korean weather conditions (monsoon and winter seasons). A 526.5 m2 RW (uncovered, RW #2) also was built as an auxiliary pond for seed culture cultivation and harvest purposes. A coagulation-flotation system was added to the facility for large-scale microalgae harvesting.

Cultivation of Microalgae

Initially, indigenous Coelastrella grown in small-scale raceways was inoculated into the 675.0 m^2 RWs and species succession was monitored both microscopically and molecularly. Commercial water soluble fertilizers, Eco-Sol (N-P-K: 25-9-18, Dongbu Farm Hannong, Ulsan, South Korea) and monopotassium phosphate (N-P-K: 0-52-34, Sang Rok Chemical, Daegu, South Korea), were added to each RW for a final concentration of 15.0-30.0 mg total nitrogen (TN)/L and 3.0-6.0 mg total phosphorus (TP)/L. A continuous mode of cultivation was carried out at a velocity of 25.0-30.0 cm/sec approximately two thirds of the algal culture was harvested from the pond and replaced with the same volume of underground water. The remaining culture was used as inoculum and appropriate amounts of nutrients were added. From October 2014 on, approximately 10.0 L/min carbon dioxide (CO2) was aerated into the RWs through Venturi tubes to enhance biomass productivity.

Microalgal Culture Monitoring

The microalga cultures were sampled every three days and inspected at 1000X magnification on a Nikon Eclipse E100 Biological Microscope (Tokyo, Japan). In addition, optical density at 680 nm on an X-ma 1200 V spectrophotometer (Human, Seoul, South Korea) and dry mass were measured. Representative species of each season were axenically isolated and added to our laboratory culture collection. Molecular identification was carried out using NS1/NS8 and ITS1/ITS4 primer sets [10]. Consumption of TN and TP (Tables 1-4) was analyzed by using HS-TN(CA)-L and HS-TP-L water test kits (Humas, Daejeon, South Korea). In addition, dissolved oxygen, pH, and temperature in the RWs (Tables 2-4) were monitored every three hours with an automatic water quality meter (WQC-24, DKK-TOA, Tokyo, Japan).

Microalgal Biomass Harvest

When the microalgal culture reached late exponential or early stationary phase, biomass was harvested by chemical coagulation with 17% polyaluminum chloride (Kumsung E and C, Ansan, South Korea) followed by dissolved air flotation (Dongshin enTech, Yangsan, South Korea). The harvested biomass was stored at -20°C until utilization.

Characterization of microalgal biomass

The biomass samples were freeze-dried, pulverized with a mortar and pestle, and sieved through ASTM No. 230 mesh (opening=63 μm). Total lipid content was determined by the sulfo-phosphovanillin colorimetric method [11]. Ultimate analysis was conducted to determine the carbon (C), hydrogen (H), nitrogen (N), and sulfur (S) contents using a Flash 2000 elemental analyzer (Thermo Fisher Scientific, Milan, Italy). Proximate analysis was carried out to measure ash content using a DTG-60A thermal analyzer (Shimadzu, Kyoto, Japan). The oxygen (O) content was calculated by subtracting the ash and CHNS contents from the total and gross calorific value (GCV, hereafter CV) was estimated by the following equation developed by Given et al. [12]: CV=0.3278C+1.419H+0.09257S-0.1379O+0.637 (MJ/kg).

RESULTS AND DISCUSSION

Dominant Microalgal Species

In this study, various species were allowed to dominate naturally and reach their stable states to avoid competition and increase productivity [13,14]. Since South Korea experiences four distinct seasons, the changes in the dominant species could be attributed to the daylight time and temperature variations of each season. As presented in Figure 3, Coelastrella dominated in the summer of 2014, Chlorella and Acutodesmus were most common in the winter and spring of 2014, and then Pseudopediastrum became dominant in the summer of 2015. It seemed that the changes in CO_2 allowed for the Pseudopediastrum to outcompete the Coelastrella.

Identification of Dominant Microalgae

The dominant microalgae were isolated and named as strains KNUA036, 037, 038, and 039. All the isolates were further identified by small subunit 18S ribosomal RNA (rRNA) and internal transcribed spacer (ITS) sequence data analyses (Table 1). For strain KNUA036, molecular characterization by 18S rRNA and ITS sequences indicated that the isolate belonged to the genus Micractinium, but the key compensatory base changes (CBS) in the ITS2 secondary structure confirmed that the isolate was a member of the genus Chlorella [15]. Due to its extremely simple morphology, no conclusion could be drawn by the morphological features of strain KNUA036. Strain KNUA037 was identified as a member of the genus Coelastrella by 18S rRNA and morphological characterization results. Even though its closet match inferred from the ITS sequence was Chlamydomonas moewusii, strain KNUA037 did not have any common features of the genus Chlamydomonas such as an eyespot and flagella. This may be due to the lack of sequence data in GenBank for Coelastrella ITS genes, so no identification could be made with this. It was found that strains KNUA038 and 039 belonged to the genera Acutodesmus and Pseudopediastrum, respectively. Their molecular and morphological identification results were in agreement. DNA sequences obtained in this study were deposited in the database of the National Center for Biotechnology Information (NCBI) under accession numbers KT883906-KT883913 (Table 1). Numerous studies have shown that Chlorella and Scenedesmus (some of them have recently been reclassified as Acutodesmus [16] are able to synthesize C14:0, C16:0, C18:1, C18:2, and C18:3 fatty acids, which can be used as biodiesel [17]. Therefore, strains KNUA036 and 038 could have potential to serve as a biodiesel feedstock. Not much is known about the possibility of deriving biofuels from Coelastrella and Pseudopediastrum. Further researches using the pure cultures of these dominant microalgae should be followed.

Microalgal Productivity and Biomass Characterization

From June 2014 to August 2015, the microalgal biomass had an overall average productivity of approximately 7.0 g dry weight (DW)/ m2/day, a lipid content of 12.2%, and a CV of 17.5 MJ/kg (Table 4). Yearround cultivation, including during the winter season, was possible only in RW #1. The average biomass productivity of RW #3 was higher than that of RW #1. Nonetheless, the average lipid content and CV of RW #1 were slightly higher than those of RW #3 (Figure 4 and Table 4). This is probably due to the growth phases at the time of harvest [18]. In many cases, more actively growing cells were harvested in RW #3 compared to RW #1 to avoid biomass reduction and contamination since aging algal cultures tend to enter the death phase quickly. The highest monthly productivity was achieved in RW #3 in August 2015, but the highest monthly lipid content and CV were found in RW #1 in December 2014 and in RW #1 in February 2015, respectively. The long rainy weather in August 2014 led to substantial reductions in biomass productivity in both RWs. In addition, rotifers caused two sudden algal population crashes in RW #1 leading to the lowest monthly productivity measured of 1.1 g DW m^2/day in September 2014. However, under favorable weather conditions in the spring and summer of 2015, the productivities of RW#1 and RW#3 were approximately 8.2-9.0 g DW/ m^2/day and 14.2-15.1 g DW/m^2/day, respectively. The average CV from June 2014 to September 2014 in both RWs was only 10.9 MJ/kg, which is lower than that of terrestrial energy crops (17.0-20.0 MJ/kg) [19-21]. Nevertheless, after the addition of extra CO_2, the value (19.7 MJ/kg) became close to the CVs of the terrestrial biomass resources. Therefore, it can be concluded that when a higher concentration of CO_2 is available, higher productivity, lipid content, and CV are attainable. Since industrial flue gas could serve as a CO_2 source at no or a very little cost [22], the additional CO_2 supply was not considered as an expense in this study. In addition, after extraction of C14 to C18 fatty acids, the resulting microalgae biomass could be used as energy pellets or organic fertilizer.

Nutrient Consumption and CO_2 Fixation

The nutrient consumption was largely dependent on the biological (dominant species) and operational (culturing period) parameters. The overall average consumption of TN and TP for each run was approximately 13.1 mg/kg and 2.5 mg/kg, respectively (Table 5), suggesting that N/P-rich wastewater could be used as a growth medium [23,24]. It has been reported that 1.0 kg of dry algal biomass utilizes 1.83 kg CO2 [25]. In this study, a total of 2612.7 kg of dry microalgal biomass, which is equivalent to 4781.2 kg CO_2, was produced. South Korea introduced carbon emission trading on January 12, 2015 and carbon is currently traded at the price of US$ 9.7 per ton of CO_2 equivalent on the market. This opens up new opportunities for the microalgae-based industry to profit from carbon trading.

CONCLUSIONS

This study demonstrated the potential of commercial-scale microalgal biomass production for biofuels under Korean geoclimatic conditions. Naturally occurring microalgae were allowed to dominate in the RW ponds to establish more reliable cultures. The species composition along with nutrient availability strongly affected the biomass productivity and characteristics. In this study, the CV was calculated to understand the potential of microalgal biomass as a biofuel feedstock and the overall microalgal biomass CV (17.5 MJ/kg) approached the CV found for terrestrial energy crops in other studies. Although more innovative work is still needed to enhance the biomass and lipid productivity, the present work showed that microalgae hold great promise as a potential biofuel source, more so than crop plants.

ACKNOWLEDGEMENTS

This work was supported by the Advanced Biomass R&D Center (ABC) of the Global Frontier Project funded by the Ministry of Science, ICT and Future Planning (2015M3A6A2065698), South Korea. This project was also supported by the Freshwater Microalgae-based Bioenergy Research and Development Project from Chilgok-gun, Gyeongsangbuk-do, South Korea. Hong JW and Kim OH contributed equally to this work.

REFERENCES

1. Huntley ME, Redalje DG (2007) CO2 mitigation and renewable oil from photosynthetic microbes: a new appraisal. Mitigation and Adaptation Strategies for Global Change 12: 573-608.
2. Li Y, Horsman M, Wu N, Lan CQ, Dubois-Calero N (2008) Biofuels from microalgae. Biotechnology Progress 24: 815-820.
3. Brennan L, Owende P (2010) Biofuels from microalgaea review of technologies for production, processing, and extractions of biofuels and co-products. Renewable and Sustainable Energy Reviews 14: 557-577.
4. Hannon M, Gimpel J, Tran M, Rasala B, Mayfield S (2010) Biofuels from algae: challenges and potential. Biofuels 1: 763-784.
5. Wang H, Zhang W, Chen L, Wang J, Liu T (2013) The contamination and control of biological pollutants in mass cultivation of microalgae. Bioresource Technology 128: 745-750.
6. Carney LT, Lane TW (2014) Parasites in algae mass culture. Frontiers in Microbiology 5.
7. Mutanda T, Ramesh D, Karthikeyan S, Kumari S, Anandraj A, et al. (2011) Bioprospecting for hyper-lipid producing microalgal strains for sustainable biofuel production. Bioresource Technology 102: 57-70.
8. Odlare M, Nehrenheim E, Ribé V, Thorin E, Gavare M, et al. (2011) Cultivation of algae with indigenous species potentials for regional biofuel production. Applied Energy 88: 3280-3285.
9. Rawat I, Kumar RR, Mutanda T, Bux F (2013) Biodiesel from microalgae: a critical evaluation from laboratory to large scale production. Applied Energy 103: 444-467.
10. Innis MA, Gelfand DH, Sninsky JJ, White TJ (1990) PCR Protocols: A Guide to Methods and Applications. Academic Press, San Diego.
11. Mishra SK, Suh WI, Farooq W, Moon M, Shrivastav A, et al. (2014) Rapid quantification of microalgal lipids in aqueous medium by a simple colorimetric method. Bioresource Technology 155: 330-333.
12. Given PH, Weldon D, Zoeller JH (1986) Calculation of calorific values of coals from ultimate analyses: theoretical basis and geochemical implications. Fuel 65: 849-854.
13. Smith VH, Sturm BS, Billings SA (2010) The ecology of algal biodiesel production. Trends in Ecology and Evolution 25: 301-309.
14. Kazamia E, Aldridge DC, Smith AG (2012) Synthetic ecology a way forward for sustainable algal biofuel production?. Journal of Biotechnology 162: 163-169.
15. Hoshina R, Iwataki M, Imamura N (2010) Chlorella variabilis and Micractiniumreisseri sp. nov. (Chlorellaceae, Trebouxiophyceae): Redescription of the endosymbiotic green algae of Paramecium bursariaPhycologicalResearch 58: 188-201.
16. Mata TM, Martins AA, Caetano NS (2010) Microalgae for biodiesel production and other applications: a review. Renewable and Sustainable Energy Reviews. 14: 217-232.

17. Krienitz L, Bock C (2012) Present state of the systematics of planktonic coccoid green algae of inland waters. Hydrobiologia 98: 295-326.

18. Ryckebosch E, Bruneel C, Muylaert K, Foubert I (2012) Microalgae as an alternative source of omega-3 long chain polyunsaturated fatty acids. Lipid Technology 24: 128-130.

19. Demirbaş (1997) A Calculation of higher heating values of biomass Fuels. 76: 431-434.

20. Ross AB, Jones JM, Kubacki ML, Bridgeman T (2008) Classification of macroalgae as fuel and its thermochemical behavior. Bioresource Technology 99: 6494-6504.

21. Naik S, Goud VV, Rout PK, Jacobson K, Dalai AK (2010) Characterization of Canadian biomass for alternative renewable biofuel. Renewable Energy 35: 1624-1631.

22. Chisti Y (2008) Biodiesel from microalgae beats bioethanol. Trends in Biotechnology 26: 126-131.

23. Park JBK, Craggs RJ (2011) Nutrient removal in wastewater treatment high rate algal ponds with carbon dioxide addition. Water Science and Technology 63: 1758-1764.

24. Boelee NC, Temmink H, Janssen M, Buisman CJ, Wijffels RH (2012) Scenario analysis of nutrient removal from municipal wastewater by microalgal biofilms. Water 4: 460-473.

25. Slade R, Bauen A (2013) Micro-algae cultivation for biofuels cost, energy balance, environmental impacts and future prospects. Biomass and Bioenergy 53: 29-38.

CHAPTER 17

PEPTIDE-MEDIATED MICROALGAE HARVESTING METHOD FOR EFFICIENT BIOFUEL PRODUCTION

Yoshiaki Maeda, Takuma Tateishi, Yuta Niwa, Masaki Muto, Tomoko Yoshino, David Kisailus and Tsuyoshi Tanaka

Keywords: Biofuel production, Harvesting, Sedimentation, Frustulin, Cell-surface display, Silica-affinity peptide, Diatom, Fistulifera solaris JPCC DA0580

BACKGROUND

With an increased demand for a sustainable energy supply, biofuel production has attracted much attention [1–4]. Microalgal biodiesel production has been investigated to meet such demands due to its advantageous features (e.g., global carbon dioxide fixation, no competition with food sources, significantly greater biomass yield than higher plants, and oil accumulation at a high level inside the cells) [5, 6]. Several eukaryotic microalgae can accumulate triacylglycerol (TAG) in high levels as a form of oil droplets, and some promising oil producers have been intensively studied to understand the TAG biosynthesis [7, 8]. In such pursuits, we recently identified an oleaginous diatom, Fistuliferasolaris JPCC DA0580 from our marine microalgal culture collection [9, 10]. A beneficial feature of this diatom for biodiesel production is its high total lipid content (~60 %, w/w) [11], leading to high oil productivity [12]. Bench-scale (~200 L) outdoor mass cultivation has been studied using raceway and column bioreactors [13]. Steady growth of F. solaris with areal oil productivity of 3.23 $g/m^2/day$ was confirmed and little bacterial contamination was observed.

However, even though promising microalgal strains are used, the efficiency of energy production using microalgal biomass is not currently recognized to be sufficient. Energy production efficiency is usually expressed by the ratio of energy production to energy consumption (i.e., termed energy profit ratio). Although the current ratio derived from the microalgal biomass is desired to be more than 1 (i.e, energy production is higher than the consumption), in practice, it is difficult to achieve [14]. Towards this goal, tremendous efforts have been made to enhance TAG accumulation in target microalgal cells by means of cultivation techniques [15] as well as metabolic engineering [16]. However, researchers determined that such approaches directing enhancement of energy production are not sufficient due to the significant energy consumption in their processes. Therefore, decreasing this energy consumption during the microalgal biomass processing is critical to the practicality of microalgal biofuels.

Recently, careful investigation of process design parameters revealed that the harvesting process (i.e., collection of microalgal biomass from the liquid medium) is one of the steps responsible for a large portion of energy consumption [17, 18]. Conventional centrifugation or filtering systems cannot attain the strictly designed process goals aimed at minimizing energy consumption [18, 19]. Several alternative processes for microalgal harvesting have been proposed [19, 20]. Among them, flocculation is a possible harvesting strategy without consuming large amounts of energy. Specifically, flocculant agents (e.g., multivalent cations and cationic polymers) can readily induce aggregation by interacting with the negatively charged microalgal cell surface [19]. However, the cost of flocculants is currently an issue. Another proposed strategy is pH-responsible cell flocculation [21, 22]. However, large quantities of harsh acids or alkalis would be necessary to control the pH of culture media when mass cultivation is performed for practical production. The next challenge is to add an auto- and controllable-flocculation function to the microalgal cell surface so that a smarter harvesting process

Citation: Maeda Y, Tateishi T, Niwa Y, et al. Peptide-mediated microalgae harvesting method for efficient biofuel production. Biotechnology for Biofuels. 2016;9:10. doi:10.1186/s13068-015-0406-9.

can be established. In this strategy, interactive protein factors, which result in self-aggregation and lead to bridging the cells for flocculation are attached on the surface of the microalgal cells. Using an inducible expression system [23, 24], timing the control of harvesting is also possible. To achieve this goal, it is essential to establish a technique for protein display on the surfaces of microalgae, in particular, the oleaginous diatom F. solaris.

Protein displays on the diatom cells have been rarely demonstrated. To date, only a non-oleaginous diatom,Thalassiosira pseudonana, had displayed proteins of interest in the form of fusion proteins with the native cell wall (termed frustule) proteins [25, 26]. Thus, identification of frustule-associated proteins as anchor molecules is imperative for the success of diatom-cell-surface display technologies. However, there are only a few diatoms whose frustule-associated proteins are well studied. Frustule-associated proteins are species-specific, and thus the challenge is that they need to be identified from individual diatoms prior to establishment of the display system.

In this study, we identified frustule-associated proteins, termed frustulins, from the genome of the oleaginous diatom, F. solaris [27]. A selected frustulin was evaluated as an anchor for protein display by fusing with green fluorescent protein (GFP), and the protein display efficiency was compared to that using the different frustule-associated protein identified in our previous study [28]. Furthermore, cell harvesting experiments were performed to demonstrate whether displayed protein-mediated harvesting is possible, in which a silica-affinity peptide (SAP) was displayed on the diatom cells, and silica particles were added to induce flocculation. The engineered peptide-mediated smart harvesting strategy established in this study should be useful in efficient biofuel production.

RESULTS AND DISCUSSION

Primary Structure and Gene Expression Level of Frustulins of *F. solaris*

In the present study, putative frustulin-like proteins were screened from F. solaris JPCC DA0580 genome using α1-frustulin of Cylindrothecafusiformis [29] and ε-frustulin of Navicula pelliculosa [30] as the query sequences, whereas our previous study had utilized only α1-frustulin for the homology search, and resulted in revealing seven putative proteins [28]. As a result of the present homology search, 15 proteins exhibited sequence similarities to these queries. However, four proteins lack the signal peptide II [30], which is a typical pre-sequence of frustule-associated proteins and responsible for directing the polypeptide into silica deposition vesicle and were thus excluded. The remaining 11 proteins met the sequence criteria of frustulins (see "Methods" section), and thus, were referred to as members of the frustulin family (Fig. 1 and Additional file 1: Fig. S1). The clustering analysis revealed that all of the selected 11 frustulins (assigned from frustulin1–frustulin11) were closely related to ε-frustulin of N. pelliculosa rather than α1-frustulin of C. fusiformis (Additional file 1: Fig. S2). This is reasonable because F. solaris is phylogenetically close to N. pelliculosa [9]. The ε-frustulin consists of a pre-sequence, acidic cysteine-rich (ACR) domains containing two hexapeptide motifs (i.e., CQGDCD and VPGCSG), polyglycine domains, and a tryptophan-rich domain [30]. The frustulins found in F. solaris JPCC DA0580 share some typical features of the pre-sequence, ACR domains, and a tryptophan-rich domain (indeed the tryptophan content is as low as 4 %, Additional file 1: Fig. S3), but some F. solaris-specific features were also observed as described below. The cysteine residues in the hexapeptide motifs, which are considered to have calcium-binding properties of frustulin [29, 30], are well conserved between ε-frustulin and F. solaris-frustulins, while other amino acids in the motif are substitutable. In addition, some of F. solaris-frustulins do not have polyglycine domains between the ACR domains. Instead, the ACR domains are connected by proline-rich domains, which are found in α-frustulins rather than ε-frustulin. Note that frustulin3, 4, and 9 have neither polyglycine domains nor proline-rich domains (Additional file 1: Fig. S1). It could suggest that the two linker domains might not be essential for a frustulin function. When the functional motifs were explored from the frustulin sequences using an InterProScan 5 program, a ricin-type β-trefoil lectin domain (i.e., ricin B lectin domain, InterPro entry IPR000772) was found at individual tryptophan-rich domains, while its function still remains obscure.

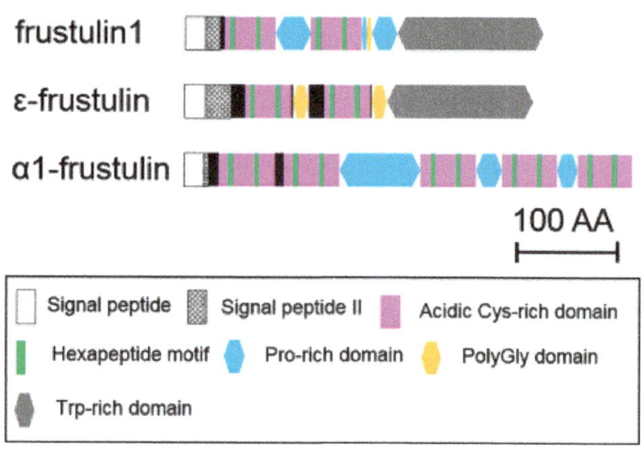

Figure 1: Schematic primary structures of frustulins. Frustulins 1 (formerlyG20405) found in the genome of F. solaris JPCC DA0580, ε-frustulin of N. pelliculosa are presented.

To confirm the gene expression level of the frustulin proteins, reads per killobases per million (RPKM) value, acquired in our transcriptome analysis [31], were compared (Additional file 1: Table S1). During the cell growth, frustulin1, 2, 5, and 8 were substantially expressed (all RPKM values >10). Among them, frustulin1 demonstrated high RPKM values without significant fluctuation. Based on this, we employed frustulin1 as an anchor molecule for diatom cell-surface display applications.

Diatom Cell-Surface Display

To evaluate frustulin1 as a molecular anchor of the cell-surface display system, the recombinant frustulin1-GFP fusion protein was designed and expressed in F. solaris JPCC DA0580. Prior to observation, the fusion protein expression was confirmed with Western blotting using an anti-GFP antibody (Additional file 1: Fig. S4). Subsequently, confocal microscopy of the resulting transformant cells revealed the substantial GFP expression surrounding the cells (Fig. 2). The lateral view of the re-constructed 3D image demonstrated intensive fluorescence around the central region of the girdle face. Meanwhile, a slightly weaker fluorescence was also observed in the valve regions. When fluorescent intensity was profiled across the girdle face, the central region exhibited approximately double the intensity than the valve regions (Additional file 1: Fig. S5a–d). A similar tendency was reproducibly observed in multiple cells. The uneven distribution of fluorescence derived from frustulin1-GFP at the central part of the girdle face may be due to the structural features of the diatom frustule. Two parts of cell wall, the large epitheca and small hypotheca, face each other, and each theca overlaps at the center of the girdle face [29, 30]. This structure was also directly found in the F. solaris cells [9, 28]. If the frustulin1 is evenly distributed on the cell wall, the overlapped region would emit a stronger GFP fluorescence (as observed here). This is consistent with a previous report where an even distribution of frustulin on the whole surface of the cell wall of another diatom (C. fusiformis) was demonstrated by staining the cell using fluorescein isothiocyanate (FITC)-labeled anti-frustulin antibody [32]. This feature could be beneficial for harvesting because frustulin1 can be expected to display interactive proteins on the whole surface of the diatom cell, and consequently mediate better flocculation rather than localizing in a specific small area. The fluorescence signal from the wild-type diatoms was undetectable under the same observed conditions. Interestingly, some transformant cells had the isolated dots which emitted significant fluorescence at the valve face (Additional file 1: Fig. S5e). However, the precise localization and its function needs to be determined.

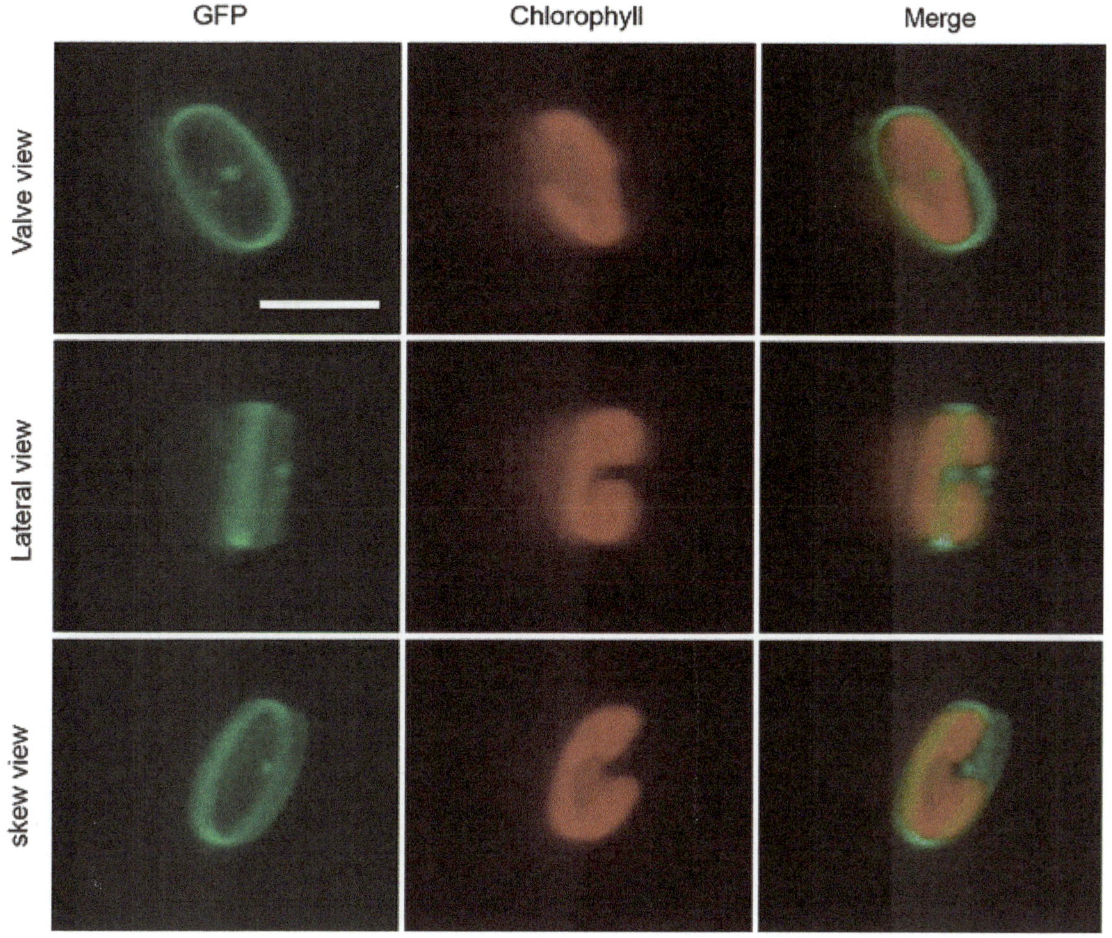

Figure 2: Confocal microscopic images of F. solaris JPCC DA0580 transformant expressing frustulin1-GFP. Scale bar 5 μm.

Further characterization of frustulin1 was performed to evaluate the potential of this protein as an anchor for protein display on the diatom cell. First, the expression efficiency was observed and compared to the other frustule-associated protein (i.e., G7408) identified in a previous study [28]. G7408 had been previously demonstrated to localize at the frustule. However, the transformant cells harboring the G7408-GFP expression gene readily lost the expression capacity only after a few subculturing rounds (Fig. 3a), perhaps due to the native gene silencing. In contrast, all of the transformant cells harboring the frustulin1-GFP expression gene maintained fluorescence even after repeating the subculturing (Fig. 3b). This result clearly indicates that frustulin1 is more suitable for surface protein-mediated harvesting because, in principle, only the cell displaying the engineered protein can be collected.

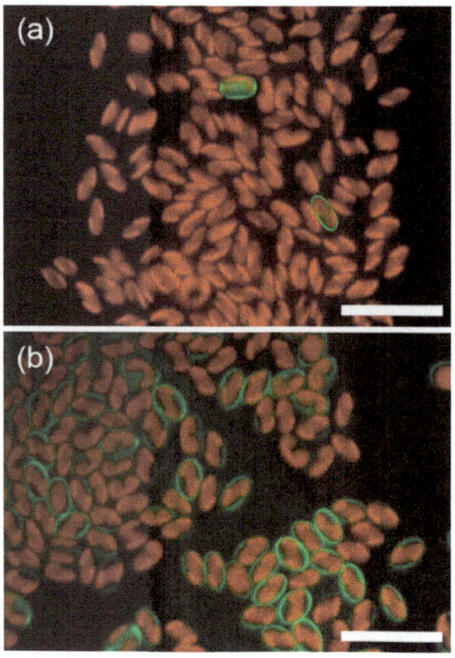

Figure 3: Fluorescent images of transformant cells harboring the expression genes of G7408-GFP **a** or frustulin 1-GFP **b**. Scale bar 20 μm.

To achieve surface protein-mediated harvesting of microalgal cells, it is essential for the interactive proteins to not only localize at the cell wall but also to be exposed to the outermost surface of the cell, otherwise they cannot induce flocculation. Fluorescence localization of frusutlin1-GFP at the cell wall is of not sufficient evidence of its exposure because it is well known that the extracellular polymeric substances exist at the outside of the diatom cell wall [33], beneath which frustulin1-GFP might be embedded. To confirm this, an antibody-binding assay was performed. When the antibody was added to the intact transformant cells, the binding signal was obviously higher than that from the wild-type cells (Fig. 4). This result indicated that the frustulin1-GFP was displayed on the outermost surface of the diatom cell to which the antibody proteins can be directly accessible.

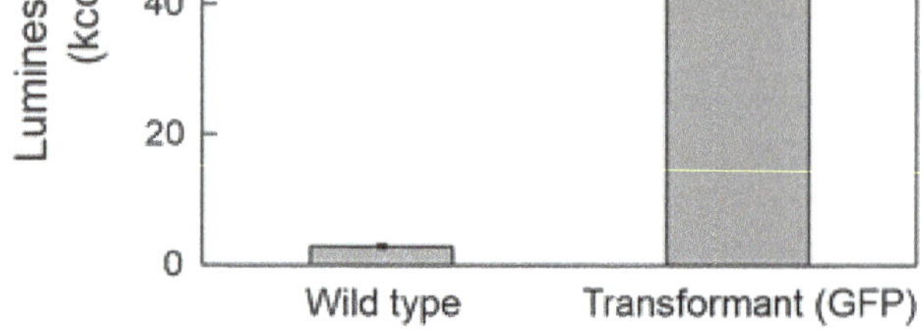

Figure 4: Antibody-binding assay to F. solaris wild-type and transformant expressing frustulin1-GFP. An alkaline phosphatase (ALP)-conjugated anti-GFP antibody was used. The luminescent substrate was added, and the ALP reacted with it to generate luminescence. Error bars standard deviation (n = 3).

The data described above enable potential cell harvesting via a diatom cell-surface display system which uses frustulin1 as an anchor molecule. To the best of our knowledge, this is the first report of a cell-surface display system for the oleaginous diatom toward the efficient bioenergy production process.

Silica-Affinity Peptide-Mediated Cell Harvesting Using Silica Particles

As a proof of concept, we investigated whether it is possible for the diatom cell-surface display system established here to mediate algal biomass harvesting. In this study, a silica-affinity peptide (SAP; MSPHPHPRHHHT) [34] was displayed on the diatom cell surface. SAP was fused at the C-terminus of frustulin1-GFP via a $(G_4S)_3$ peptide linker; i.e., the fusion protein was expressed as a form of frustulin1-GFP-$(G_4S)_3$-SAP, in which GFP allowed us to confirm the successful display by fluorescent observation (Supplementary Fig. S6a). The transformant cells displaying GFP-SAP were co-cultivated with the silica particles. No sedimentation was observed during the co-cultivation due to aeration that contributed to the supply of CO_2 for photosynthesis and agitation of the culture. Once the aeration stopped, silica particles immediately precipitated within 1 min, and simultaneously, the culture medium turned transparent, indicating the precipitation of the transformant cells with the particles (Fig. 5a). In contrast, little precipitation was observed when wild-type cells were co-cultivated with the silica particles (Fig. 5a). Microscopic observation of the precipitated silica particles revealed that a substantial number of the transformant cells were associated with the precipitated particles, while significantly less adhesion to wild-type cells was confirmed (Additional file 1: Fig. S6b). This result indicates that the SAP display is responsible for cell adhesion to the silica particles, and subsequently leads to cell flocculation. When, instead of co-cultivation, silica particles were added after completing the diatom cultivation, little flocculation was observed (Additional file 1: Fig. S6c). This could be because silica particles precipitated too rapidly or had limited interaction with the cells versus the time with which they aggregated. Thus, it was hard to enable sufficient interaction with the particles and all of the diatom cells. This issue might be addressable by changing the mixing condition of the particles and cells. In this co-flocculation system, it is likely that the silica particles, which carry a negative charge in f/2 medium (pH ~7), might electrostatically interact with transformant cells displaying SAP (MSPHPHPRHHHT, theoretical pI = 9.59), which have a significant fraction of basic amino acid residues. Therefore, the silica particles likely form bridges between cells, effectively increasing their mass and enabling facile sedimentation. However, details of the interaction mechanisms have yet to be determined. Flocculation efficiency was increased by increasing the amount of silica particles in cultures, reaching 85 % when 5 or 10 g/L of silica particles was added transformant culture (Fig. 5b). This rate is comparable to other flocculation studies [19, 22, 35]. When silica particles were not supplied, no flocculation occurred with both the transformant and wild-type cells, indicating that silica particles are necessary for cell harvesting. It should be noted that the transformant displaying GFP-SAP exhibited steady cell growth and oil accumulation as much as the wild-type system (Additional file 1: Fig. S7). Lipid extraction from the transformant cells was readily achieved as is the case with wild-type cells. These results indicate that the cell-surface display system established in this study does not have a significantly negative impact on the biomass and lipid productivity, as well as downstream processes after harvesting.

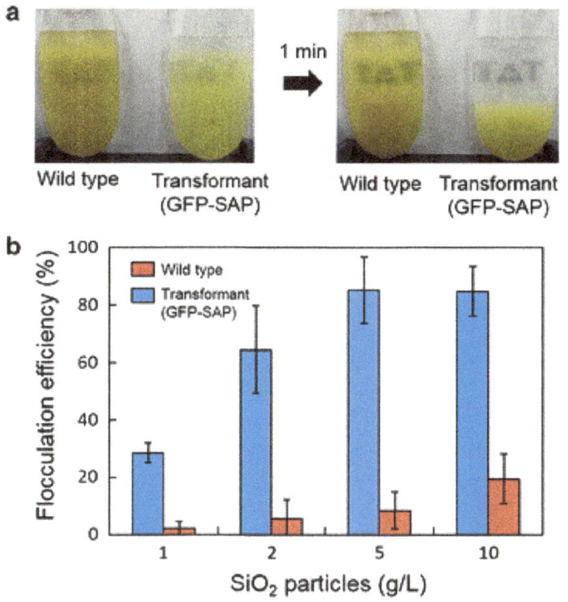

Figure 5: Silica-affinity peptide (SAP)-mediated cell harvesting for the F. solaris transformant expressing frustulin1-GFP-$(G_4S)_3$-SAP. **a** Photographs of culture media of the wild-type (left) and transformant displaying GFP-SAP (right) co-incubated with silica particles before and after flocculation. Co-flocculation of diatom transformant cells and silica particles (2.5 g) were observed within 1 min. **b** Flocculation efficiency of the transformant (blue bars) and wild-type (red bars) with silica particles. Diatom cells were co-cultivated with different concentrations of silica particles in 500 ml cultures. Culture media containing cells and particles were collected at 5 cm below the water surface before and after flocculation, and flocculation efficiency was calculated by Eq. 1. Error bars standard deviation (n = 3).

A striking feature of this silica-inducible and SAP-mediated cell flocculation is not to regulate the component of growth medium. Silica particles do not significantly affect the components of the medium because the f/2 medium contains sodium silicate and it inherently generates silica precipitates. This feature is beneficial, in particular, for repeated batch cultures, a culture technique for enhanced biomass production, and repeatedly utilizes a part of the culture medium after cell harvesting for subsequent batch cultures [13]. After SAP-mediated harvesting, the remaining culture components are available for the repeated batch cultures without modification. In contrast, other previously proposed flocculation methodologies (e.g., pH change [21, 22], supplemental chemical [19] and biological [35, 36] flocculation agents) drastically altered the properties of culture media, and neutralization of those impacts needs additional harsh chemicals or separation steps, both of which add cost. Our harvesting strategy does not require such additional chemicals or processes. Additionally, silica particles are inexpensive as compared with other flocculation-inducing salts and polymers such as ferric chloride, aluminum sulfate, ferric sulfate, chitosan, and polyacrylamide [19]. Comparable doses of these chemical reagents are needed for flocculation of microalgal biomass. After harvesting the diatom cells, lipid extraction is performed by adding an organic solvent (e.g., n-hexane). Even if silica particles were not removed from the harvested biomass, hexane extraction would not be inhibited. Therefore, we propose that this is more efficient, environmentally benign, and an inexpensive method for biomass harvesting than conventional flocculation methodologies.

Besides diatoms, cell-surface display systems were established for cyanobacteria [37–39]. Although systems for green algae have not yet been thoroughly investigated, cell wall proteins, which can serve as anchors in a cell-surface system, were well studied [40]. These studies suggest that the harvesting strategy proposed in this study could be available for a broad range of microalgae producing useful materials.

CONCLUSIONS

Using a newly identified frustulin as an anchor molecule, we successfully demonstrated GFP display on the outermost surface of the oleaginous diatom, F. solaris JPCC DA0580. An affinity peptide, SAP, was displayed on the cells by fusing with the frustulin, and effective flocculation was demonstrated by co-cultivating with silica particles. Our initial trial showed considerable promise for biomass harvesting. The next challenge would be implementation of this system into large-scale bioreactors used in outdoor mass cultivation. The feasibility of this harvesting method should be statistically evaluated in large-scale outdoor cultivation systems prior to practical realization, which we will investigate in the near future. We propose that the peptide-mediated microalgal harvesting strategy established in this study will pave the way towards future highly efficient biofuel production.

METHODS

Strain and Culture Conditions

The marine pennate diatom F. solaris JPCC DA0580 cells were maintained in f/2 medium [41] dissolved in artificial seawater. The transformed cells were cultivated in the same medium with the addition of antibiotics G-418 (500 µg/ml). Cultures (500 ml of f/2 medium in flat-shaped flasks) were incubated at 25 °C under continuous, cool-white fluorescent lights at 140 µmol/m^2/s with aeration (0.8 l of air containing 2 % CO_2/l/min). Genes were cloned in Escherichia coli TOP10 (Life Technologies, Tokyo, Japan) cultured in a lysogeny broth (LB) medium containing 50 µg/ml ampicillin at 37 °C.

Homology Search of Frustulin

Two known frustulin genes, α1-frustulin derived from C. fusiformis [29] and ε-frustulin from N. pelliculosa[30] were used as query sequences for protein basic local alignment search tools (Protein BLAST) against the draft genome sequence of F. solaris JPCC DA0580 [27]. The E-value cut off was set to 10^{-7}. Subsequently, it was confirmed whether the amino acid sequences of the putative frustulins meet the criteria that frustulin possesses at least three of five sequence features: (1) a pre-sequence domain which contains an endoplasmic reticulum (ER) signal peptide and another signal peptide (termed SPII) aimed at targeting the silica deposition vesicle, (2) acidic cysteine-rich (ACR) domains, (3) proline-rich domains, (4) polyglycine domains, and (5) a tryptophan-rich domain previously described [30].

Plasmid Construction and Transformation

An expression vector for frustulin1 (formerly G20405, DDBJ/EMBL/GenBank database accession numberAB854058) fused with GFP was constructed using general gene fusion techniques [42, 43]. We used the plasmid pSP-GFP/GAPDH containing an enhanced green fluorescence protein (GFP) gene, a glyceraldehyde 3-phosphate dehydrogenase (GAPDH) gene promoter and a fucoxanthin chlorophyll a/c-binding protein A fucoxanthin chlorophyll a/c-binding protein A (fcpA) terminator, in which an EcoRI site exists between the promoter and GFP gene [44]. First, the EcoRI site was changed to the EcoRV site using a Strategene Quik Change II XL site-Directed Mutagenesis Kit (Agilent Technologies, Santa Clara, CA, USA) and a primer pair (5′-CAA ACA ATC ATC GAT ATC ATG GTG AGC AAG GG-3′ and 5′-CCT TGC TCA CCA TGA TAT CGA TGA TTG TTT GTT G-3′). The modified pSP-GFP/GAPDH was then digested by EcoRV. The frustulin1 gene was amplified with PCR using a cDNA library prepared from F. solaris. Total RNA was extracted from F. solaris using a Concert Plant RNA Reagent (Life Technologies, Tokyo, Japan) and purified with RNeasy Mini Kit (Qiagen, Hilden, Germany). After quantification of total RNA using an Agilent RNA 6000 Nano Assay Kit (Agilent Technologies, Santa Clara, CA, USA), cDNA was obtained with a SMARTer™ RACE cDNA Amplification Kit (Takara Bio, Otsu, Japan) from 1 μg of total RNA as the template. PCR amplification was performed using the specific primer set for fso:g20405 (5′-ATG ACA GTC CTT CAG CTT TTA CTC AAG GTC-3′and 5′-GTA TTT GTT CCA GTA CGA TGT GTT GCT GC-3′). The PCR product was ligated with the modified pSP-GFP/GAPDH. For the flocculation experiment, a $(G_4S)_3$ linker and silica-affinity peptide (SAP, MSPHPHPRHHHT) [34] were added at the C-terminus of frustulin-GFP. The constructed plasmids were introduced into the F. solaris, respectively, by microparticle bombardment using the Biolistic PDS-1000/He Particle Delivery System (Bio-Rad Laboratories, Inc., Hercules, CA, USA) as described previously [45].

SDS-PAGE and Western Blotting

F. solaris (5×10^7 cells) were collected by centrifugation, suspended in 100 μL of 1 % (w/v) sodium dodecyl sulfate (SDS) in aqueous solution, and boiled for 10 min. After centrifugation, supernatant was collected, and SDS sample buffer was added (final concentration of 62.5 mM Tris–HCl, pH 6.8, 5 % 2-mercaptoethanol, 2 % SDS, 5 % sucrose, and 0.002 % bromophenol blue). Denatured proteins were separated by SDS–polyacrylamide gel electrophoresis sodium dodecyl sulfate–polyacrylamide gel electrophoresis (SDS-PAGE) using a 10 % (w/v) gel. Bio-Safe Coomassie G-250 Stain (Bio-Rad laboratories, Inc., Hercules, CA, USA) was used for gel staining. For Western blotting, the separated proteins were transferred to a polyvinylidene difluoride membrane. GFP was then detected using alkaline phosphatase (ALP)-labeled anti-GFP antibody (Rockland Immunochemicals Inc., Gilbertsville, PA, USA, 1/5000 dilution from stock in PBS containing 0.05 % Tween 20). BCIP/NBT-Blue (Sigma, St. Louis, MO, USA) was used as the ALP substrate for visualization.

Fluorescence Microscopy

Transformant cells were observed using a fluorescent microscope BX51 (Olympus Corporation, Tokyo, Japan) with a U-MGFPPHQ filter set for GFP and a U-MWIG2 filter set for chlorophyll. Confocal microscopy was performed using Fluoview FV1000 (Olympus Corporation, Tokyo, Japan) with an Alexa Fluor 488 filter set for GFP and an Alexa Fluor 568 set for chlorophyll fluorescence. The images were obtained every 0.2 μm along the z-axis to re-construct three-dimensional (3D) images.

Antibody-Binding Assay

F. solaris JPCC DA0580 wild-type and flustulin1-GFP expressing transformant cells (5.0×10^6 cells) were centrifuged at 8500g for 5 min, and re-suspended in 50 μL of ALP-conjugated anti-GFP antibody (1 μg/ml, Rockland, Gilbertsville, USA) solution diluted in Tris-buffered saline containing 0.05 % tween 20 (TBST) and incubated for 30 min. The cells were then washed three times with TBST and mixed with 50 μL of Lumi-Phos® 530 (Wako Pure Chemical Industries, Osaka, Japan). After 5 min, luminescence intensity was measured using a microplate reader SH-9000 (Corona electric Co., Ltd., Ibaraki, Japan).

Silica-Affinity Peptide (SAP)-Mediated Harvesting

SAP-mediated cell harvesting experiments were performed using polydisperse silica particles (approximately 1–20 μm) that are purchased from Wako Pure Chemical Industries (Osaka, Japan). F. solaris wild-type and the transformant displaying GFP-SAP (5.0×10^5 cells/ml as initial cell concentration) were incubated for 6 days in the absence or presence (1, 2, 5 and 10 g/l) of silica particles at 25 °C under continuous, cool-white fluorescent lights at 140 μmol/m^2/s with aeration (0.8 l of air containing 2 % CO_2/l/min). Subsequently, the flat-shaped flasks (stand perpendicularly) containing the cells and silica particles were vigorously shook by hands. The flocculation process was recorded using a video camera. To evaluate the flocculation efficiency, cell- and particle-suspensions were collected before and after flocculation using a pipette where the pipet tips reached were placed 5 cm below the water surface. The flocculation efficiency was calculated by following (Eq. 1).

$$\text{Flocculation efficiency (\%)} = (1 - A/B)\ (\%) \qquad 1$$

where A and B are cell concentrations after and before flocculation, respectively. Cell concentrations were determined by cell counting using a microscope and hemocytometer.

Additional Files

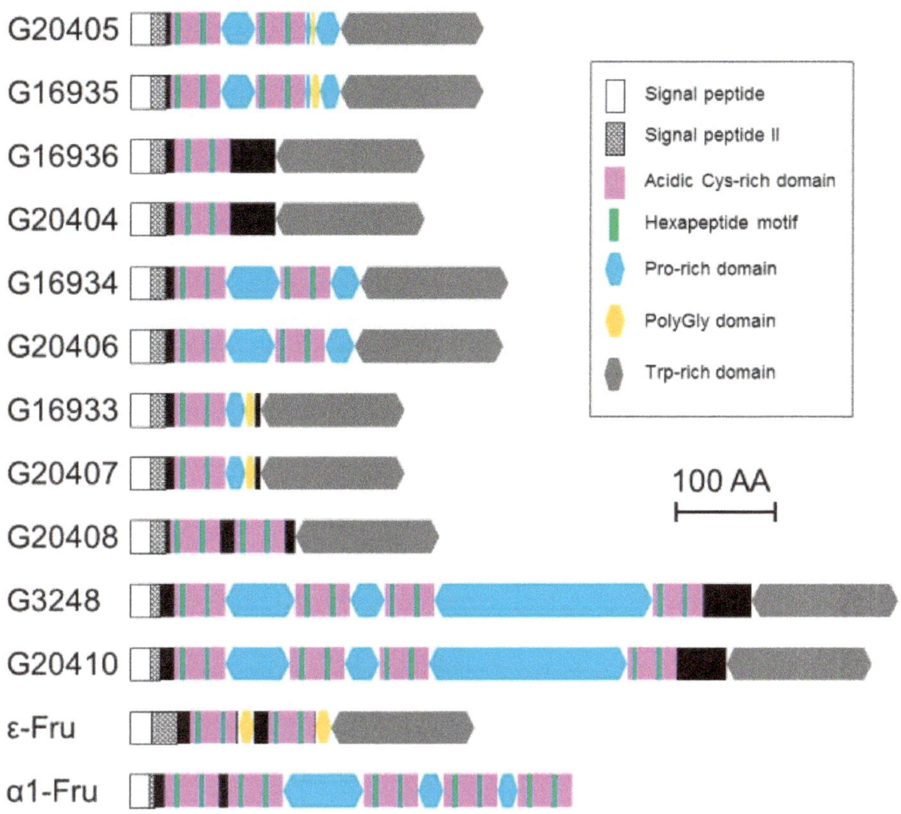

Figure S1: Schematic of primary structures of frustulins found in the genome of Fistulifera solaris JPCC DA0580.

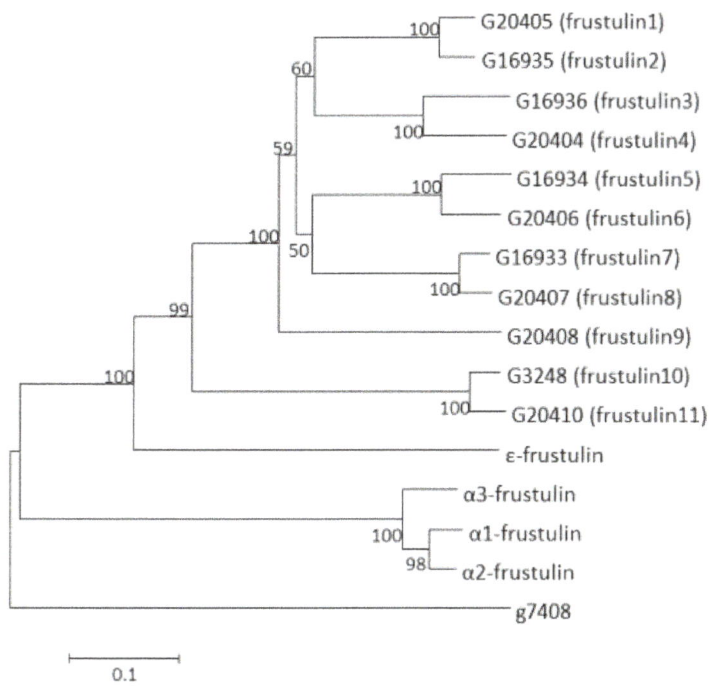

Figure S2: Comparison of the amino acid sequences of frustulins identified from F. solaris, ε-frustulin from Navicula pelliculosa and α1-, α2-, α3-frustulin derived from Cylindrotheca fusiformis. The phylogenetic tree was constructed using the neighbor-joining method. Bootstrap trials were replicated 1000 times. Gaps were treated as missing data.

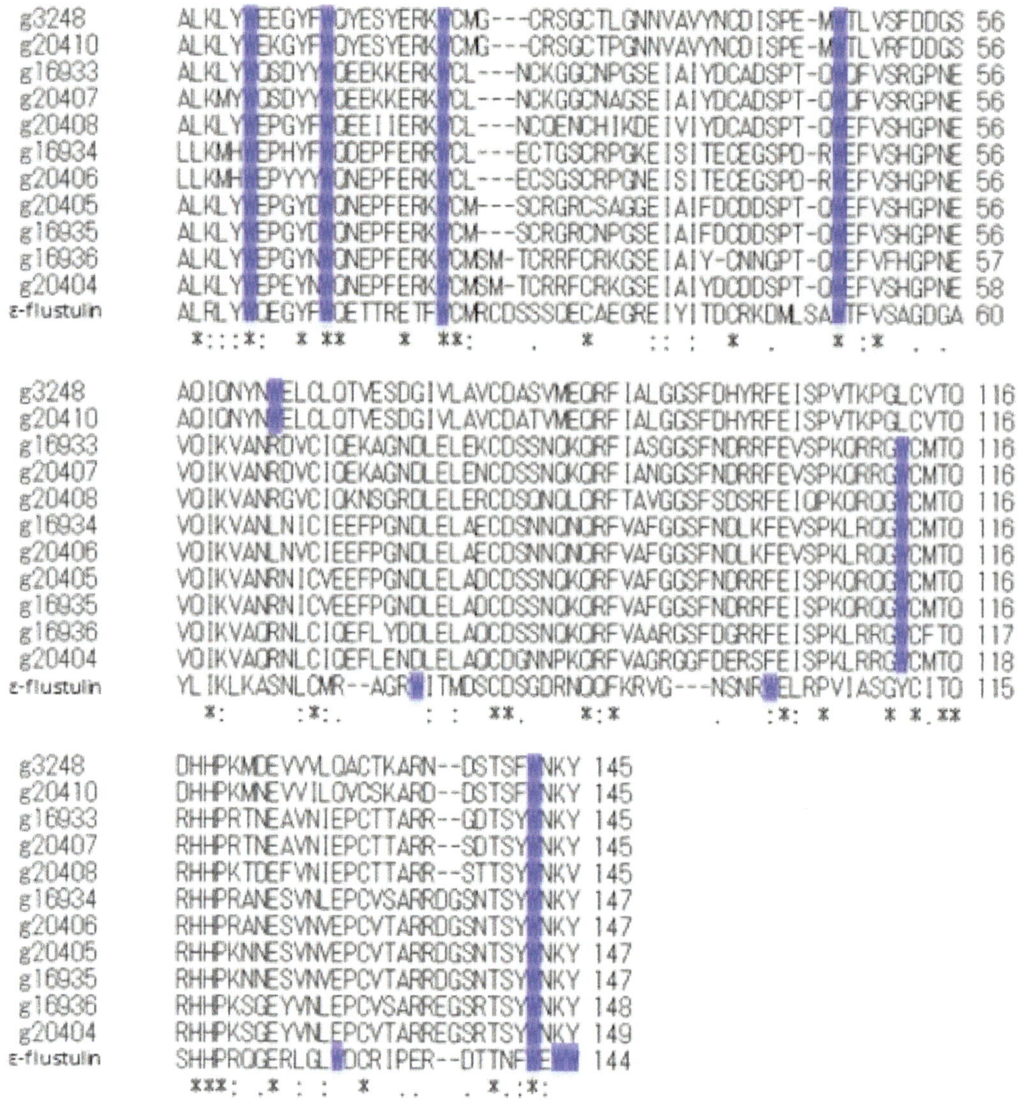

Figure S3: Comparison of the amino acid sequences of the tryptophan-rich domains of frustulins identified from F. solaris and ε-frustulin from Navicula pelliculosa. ClustalW was used for the comparison. Characters highlighted in blue are tryptophan.

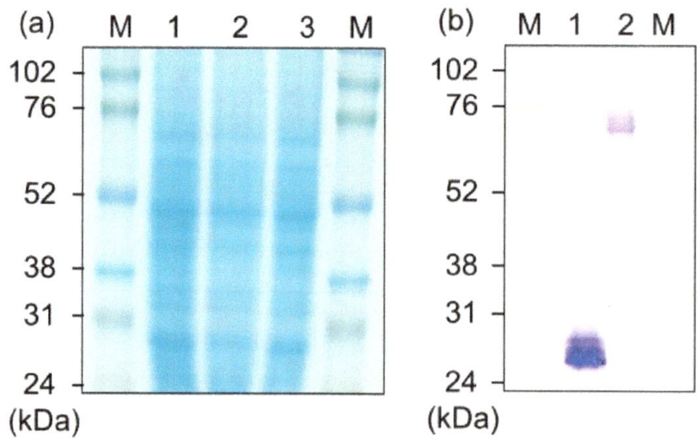

Figure S4: SDS-PAGE and Western blotting analyses of the proteins extracted from the cells of F. solaris JPCC DA0580. (a) The proteins extracted from wild type (Lane 1), the transformant expressing GFP (Lane 2), and the transformant expressing frustulin1-GFP (Lane 3) with the SDS-boiling method were applied to the 10% polyacrylamide gel. It confirms that nearly equal amount of proteins were tested, but specific bands could not be found from transformants, which could be due to the low sensitivity of Coomassie brilliant blue staining. (b) The separated proteins by SDS-PAGE were then transferred to the membrane, and subjected to Western blotting with ALP-labeled anti-GFP antibody for more sensitive detection of GFP. GFP (Lane 1) and frustulin1-GFP were observed. M indicates molecular markers.

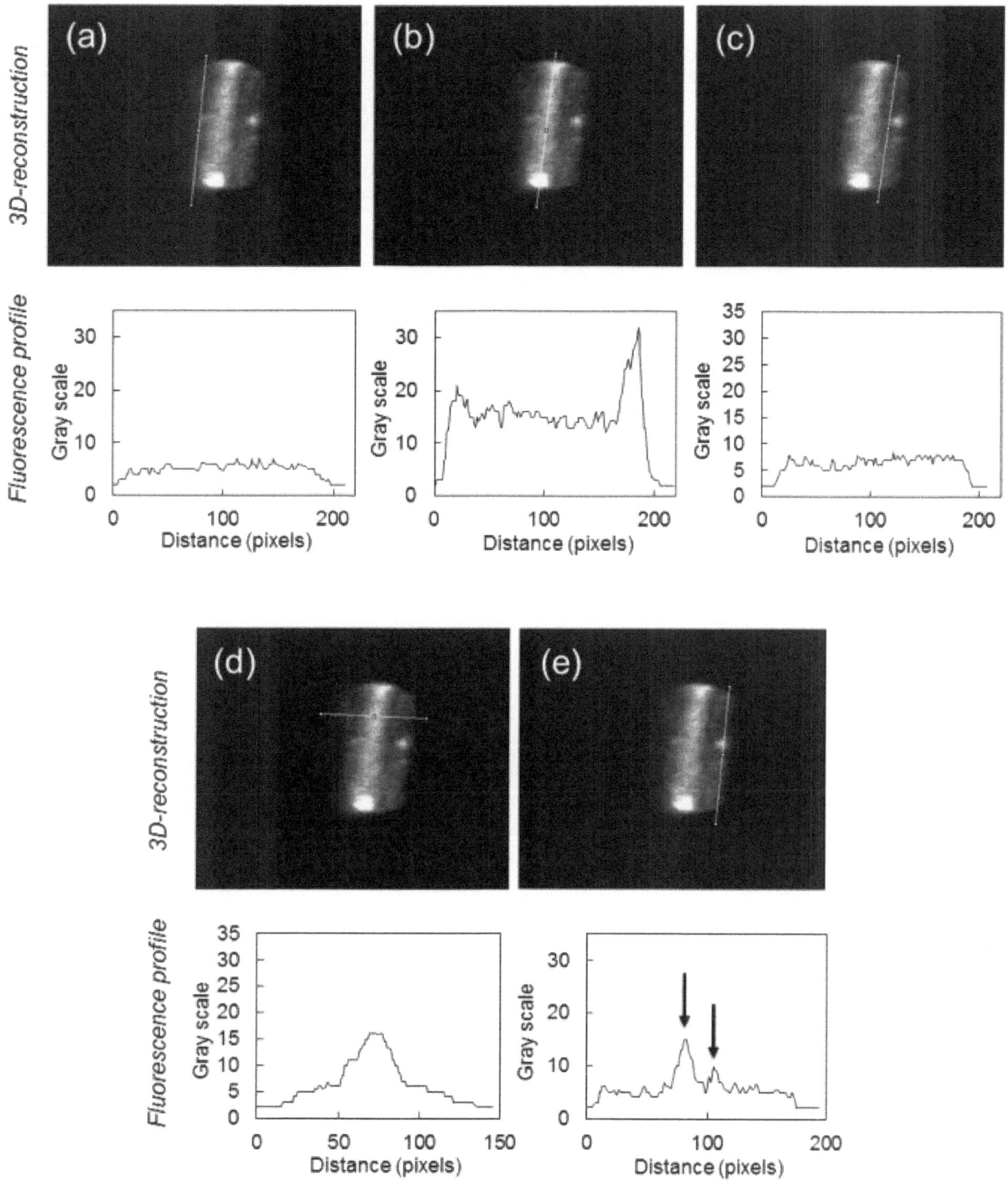

Figure S5: Fluorescent signal profiling on the transformant cell expressing frustulin1-GFP. The profiling was performed along the valve regions (a, c), a central region of the girdle face (b), across the girdle face, and isolated points showing intensive signal (e, the isolated points were shown by arrows). Yellow lines are the profiled regions.

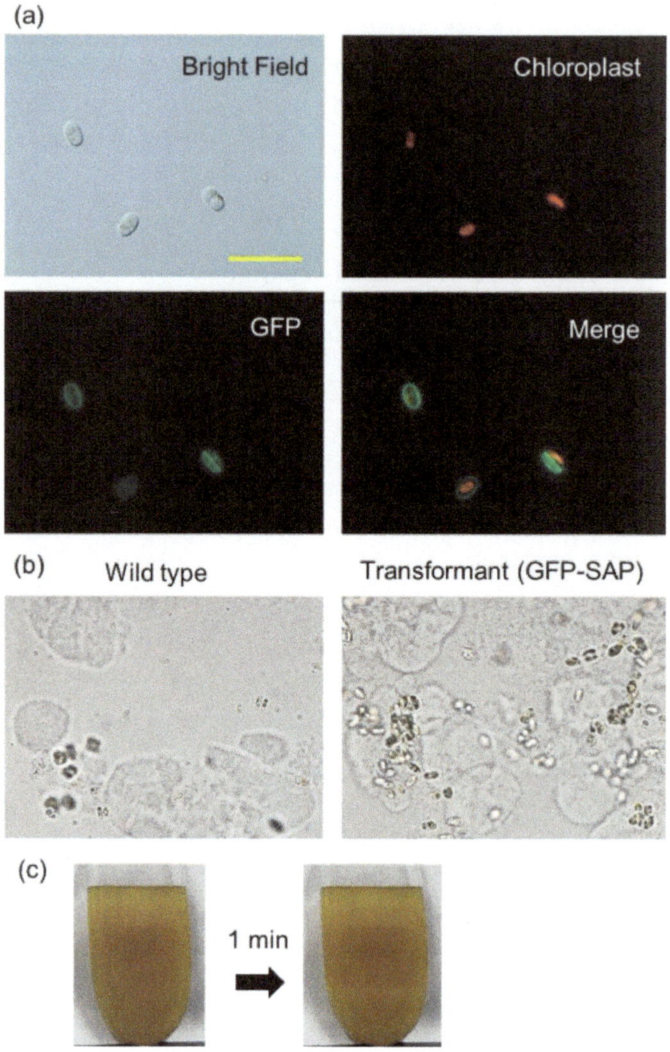

Figure S6: SAP-mediated flocculation by co-cultivation of the diatom transformant and silica particles. (a) Bright field and fluorescent observations of the transformant cells expressing frustulin1-GFP-(G4S)3-SAP. (b) Precipitated silica particles (2.5 g) collected from the bottom of the flat-shaped flask containing the wild type or transformant cells displaying GFP-SAP cultures. (c) Silica particles were added after completing the transformant diatom cultivation, instead of co-cultivation.

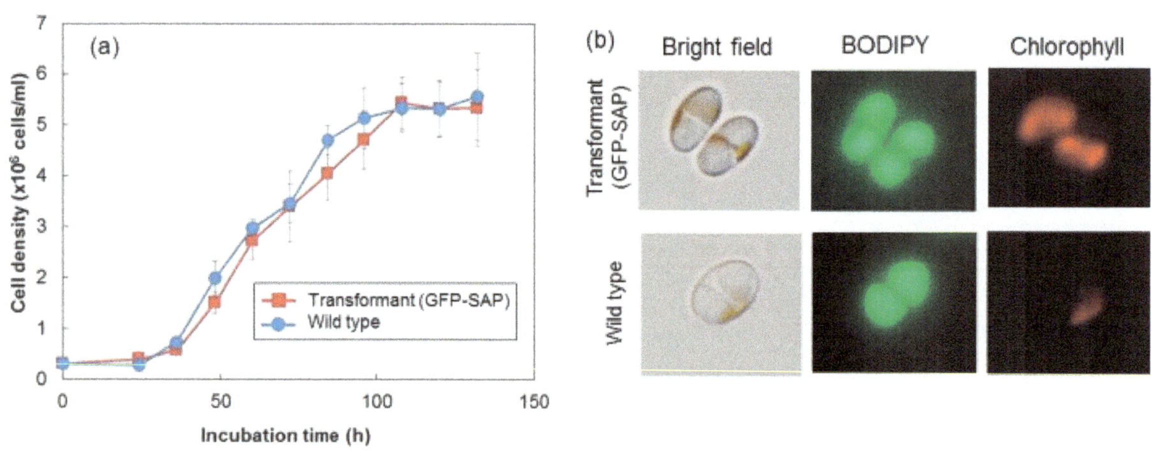

Figure S7: Comparison of cell growth and lipid accumulation between the F. solaris wild type and transformant displaying GFP-SAP. (a) Cell growth comparison based on cell counting. (b) Lipid accumulation comparison based on BODIPY 505/515 (Molecular Probes; 25 mg/mL, 2% DMSO [v/v] in water) staining. BODIPY staining was performed by mixing cell culture and the dye (99:1 volume ratio) and incubated for 10 min at room temperature. The stained cells were observed by fluorescence microscopy (BX51; Olympus). Green labeled-oil bodies occupied almost of the area of both wild type and the transformant cells.

Table S1: Expression level of frustulins of F. solaris JPCC DA0580 incubated in the f/2 medium. Transcriptome analysis was performed using a Genome Analyzer IIx (Illumina) every 48 hours during the cultivation. RPKM values were calculated

Gene ID	Name	48 hours	96 hours	144 hours
fso:g20405	Frustulin 1	67.69	54.61	54.96
fso:g16935	Frustulin 2	94.62	24.55	48.72
fso:g16936	Frustulin 3	0.07	5.69	19.49
fso:g20404	Frustulin 4	0.79	17.33	38.38
fso:g16934	Frustulin 5	25.27	12.83	12.43
fso:g20406	Frustulin 6	8.69	8.32	9.69
fso:g16933	Frustulin 7	14.05	4.88	11.75
fso:g20407	Frustulin 8	36.21	31.62	45.14
fso:g20408	Frustulin 9	3.37	64.94	45.87
fso:g3248	Frustulin 10	2.34	0.33	2.9
fso:g20410	Frustulin 11	2.22	0.27	1.56

ACKNOWLEDGEMENTS

This work was supported by JST, CREST Grant number 09154495, and JSPS KAKENHI Grant Numbers 26289309 (Grant-in-Aid for Scientific Research B), and 15H01594 (Grant-in-Aid for Scientific Research on Innovative Areas). The Global Innovation Research Organization (GIR) at TUAT also supports this international joint research.

REFERENCES

1. Brethauer S, Wyman CE. Review: continuous hydrolysis and fermentation for cellulosic ethanol production. Bioresour Technol. 2010;101:4862–4874. doi: 10.1016/j.biortech.2009.11.009.
2. Chisti Y. Biodiesel from microalgae. Biotechnol Adv. 2007;25:294–306. doi: 10.1016/j.biotechadv.2007.02.001.
3. Show K, Lee D, Tay J, Lin C, Chang J. Biohydrogen production: current perspectives and the way forward. Int J Hydrogen Energy. 2012;37:15616–15631. doi: 10.1016/j.ijhydene.2012.04.109.
4. Wyman CE. Alternative fuels from biomass and their impact on carbon dioxide accumulation. Appl Biochem Biotechnol. 1994;45:897–915. doi: 10.1007/BF02941858.
5. Smith VH, Sturm BS, Denoyelles FJ, Billings SA. The ecology of algal biodiesel production. Trends Ecol Evol. 2010;25:301–309. doi: 10.1016/j.tree.2009.11.007.
6. Ho SH, Ye X, Hasunuma T, Chang JS, Kondo A. Perspectives on engineering strategies for improving biofuel production from microalgae—a critical review. Biotechnol Adv. 2014;32:1448–1459. doi: 10.1016/j.biotechadv.2014.09.002.
7. Matsunaga T, Matsumoto M, Maeda Y, Sugiyama H, Sato R, Tanaka T. Characterization of marine microalga, Scenedesmus sp. strain JPCC GA0024 toward biofuel production. Biotechnol Lett. 2009;31:1367–1372. doi: 10.1007/s10529-009-0029-y.
8. Rodolfi L, Chini Zittelli G, Bassi N, Padovani G, Biondi N, Bonini G, et al. Microalgae for oil: strain selection, induction of lipid synthesis and outdoor mass cultivation in a low-cost photobioreactor. Biotechnol Bioeng. 2009;102:100–112. doi: 10.1002/bit.22033.
9. Matsumoto M, Mayama S, Nemoto M, Fukuda Y, Muto M, Yoshino T, et al. Morphological and molecular phylogenetic analysis of the high triglyceride-producing marine diatom, Fistulifera solaris sp. nov. (Bacillariophyceae) Phycol Res. 2014;62(4):257–268. doi: 10.1111/pre.12066.
10. Matsumoto M, Sugiyama H, Maeda Y, Sato R, Tanaka T, Matsunaga T. Marine diatom, Navicula sp. strain JPCC DA0580 and marine green alga, Chlorella sp. strain NKG400014 as potential sources for biodiesel production. Appl Biochem Biotechnol. 2010;161:483–490. doi: 10.1007/s12010-009-8766-x.
11. Liang Y, Maeda Y, Yoshino T, Matsumoto M, Tanaka T. Profiling of fatty acid methyl esters from the oleaginous diatom Fistulifera sp. strain JPCC DA0580 under nutrition-sufficient and -deficient conditions. J Appl Phycol. 2014;26:2295–2302. doi: 10.1007/s10811-014-0265-y.
12. Satoh A, Ichii K, Matsumoto M, Kubota C, Nemoto M, Tanaka M, et al. A process design and productivity evaluation for oil production by indoor mass cultivation of a marine diatom, Fistulifera sp. JPCC DA0580. Bioresour Technol. 2013;137:132–138. doi: 10.1016/j.biortech.2013.03.087.
13. Sato R, Maeda Y, Yoshino T, Tanaka T, Matsumoto M. Seasonal variation of biomass and oil production of the oleaginous diatom Fistulifera sp. in outdoor vertical bubble column and raceway-type bioreactors. J Biosci Bioeng. 2014;117:720–724. doi: 10.1016/j.jbiosc.2013.11.017.
14. Razon LF, Tan RR. Net energy analysis of the production of biodiesel and biogas from the microalgae: haematococcus pluvialis and Nannochloropsis. Appl Energy. 2011;88:3507–3514. doi: 10.1016/j.apenergy.2010.12.052.
15. Gardner R, Peters P, Peyton B, Cooksey KE. Medium pH and nitrate concentration effects on accumulation of triacylglycerol in two members of the Chlorophyta. J Appl Phycol. 2011;23:1005–1016. doi: 10.1007/s10811-010-9633-4.

16. Trentacoste EM, Shrestha RP, Smith SR, Gle C, Hartmann AC, Hildebrand M, et al. Metabolic engineering of lipid catabolism increases microalgal lipid accumulation without compromising growth. Proc Natl Acad Sci. 2013;110:19748–19753. doi: 10.1073/pnas.1309299110.
17. Sander K, Murthy GS. Life cycle analysis of algae biodiesel. Int J Life Cycle Assess. 2010;15:704–714. doi: 10.1007/s11367-010-0194-1.
18. Uduman N, Qi Y, Danquah MK, Forde GM, Hoadley A. Dewatering of microalgal cultures: a major bottleneck to algae-based fuels. J Renew Sustain Energy. 2010;2:012701. doi: 10.1063/1.3294480.
19. Molina Grima E, Belarbi EH, Acien Fernandez FG, Robles Medina A, Chisti Y. Recovery of microalgal biomass and metabolites: process options and economics. Biotechnol Adv. 2003;20:491–515. doi: 10.1016/S0734-9750(02)00050-2.
20. Pragya N, Pandey KK, Sahoo P. A review on harvesting, oil extraction and biofuels production technologies from microalgae. Renew Sustain Energy Rev. 2013;24:159–171. doi: 10.1016/j.rser.2013.03.034.
21. Horiuchi J, Ohba I, Tada K, Kobayashi M, Kanno T, Kishimoto M. Effective cell harvesting of the halotolerant microalga Dunaliella tertiolecta with pH control. J Biosci Bioeng. 2003;95:412–415. doi: 10.1016/S1389-1723(03)80078-6.
22. Liu J, Zhu Y, Tao Y, Zhang Y, Li A, Li T, et al. Freshwater microalgae harvested via flocculation induced by pH decrease. Biotechnol Biofuels. 2013;6:98. doi: 10.1186/1754-6834-6-98.
23. Poulsen N, Chesley PM, Kroger N. Molecular genetic manipulation of the diatom Thalassiosira pseudonana (Bacillariophyceae) J Phycol. 2006;42:1059–1065. doi: 10.1111/j.1529-8817.2006.00269.x.
24. Poulsen N, Kroger N. A new molecular tool for transgenic diatoms: control of mRNA and protein biosynthesis by an inducible promoter-terminator cassette. FEBS J. 2005;272:3413–3423. doi: 10.1111/j.1742-4658.2005.04760.x.
25. Poulsen N, Berne C, Spain J, Kroeger N. Silica immobilization of an enzyme through genetic engineering of the diatom Thalassiosira pseudonana. Angew Chem Int Ed Engl. 2007;46:1843–1846. doi: 10.1002/anie.200603928.
26. Marshall KE, Robinson EW, Hengel SM, Pasa-Tolic L, Roesijadi G. FRET imaging of diatoms expressing a biosilica-localized ribose sensor. PLoS ONE. 2012;7:e33771. doi: 10.1371/journal.pone.0033771.
27. Tanaka T, Maeda Y, Veluchamy A, Tanaka M, Abida H, Marechal E, et al. Oil accumulation by the oleaginous diatom Fistulifera solaris as revealed by the genome and transcriptome. Plant Cell. 2015;27:162–176. doi: 10.1105/tpc.114.135194.
28. Nemoto M, Maeda Y, Muto M, Tanaka M, Yoshino T, Mayama S, et al. Identification of a frustule-associated protein of the marine pennate diatom Fistulifera sp. strain JPCC DA0580. Mar Genomics. 2014;16:39–44. doi: 10.1016/j.margen.2014.01.006.
29. Kroger N, Bergsdorf C, Sumper M. A new calcium binding glycoprotein family constitutes a major diatom cell wall component. EMBO J. 1994;13:4676–4683.
30. Kroger N, Bergsdorf C, Sumper M. Frustulins: domain conservation in a protein family associated with diatom cell walls. Eur J Biochem. 1996;239:259–264. doi: 10.1111/j.1432-1033.1996.0259u.x.
31. Liang Y, Maeda Y, Sunaga Y, Muto M, Matsumoto M, Yoshino T, et al. Biosynthesis of polyunsaturated fatty acids in the oleaginous marine diatom Fistulifera sp. strain JPCC DA0580. Mar Drugs. 2013;11:5008–5023. doi: 10.3390/md11125008.
32. Kroger N, Lehmann G, Rachel R, Sumper M. Characterization of a 200-kDa diatom protein that is specifically associated with a silica-based substructure of the cell wall. Eur J Biochem. 1997;250:99–105. doi: 10.1111/j.1432-1033.1997.00099.x.
33. Hecky R, Mopper K, Kilham P, Degens E. The amino acid and sugar composition of diatom cell-walls. Mar Biol. 1973;19:323–331. doi: 10.1007/BF00348902.
34. Naik RR, Brott LL, Clarson SJ, Stone MO. Silica-precipitating peptides isolated from a combinatorial phage display peptide library. J Nanosci Nanotechnol. 2002;2:95–100. doi: 10.1166/jnn.2002.074.
35. Muradov N, Taha M, Miranda AF, Wrede D, Kadali K, Gujar A, et al. Fungal-assisted algal flocculation: application in wastewater treatment and biofuel production. Biotechnol Biofuels. 2015;8:24. doi: 10.1186/s13068-015-0210-6.
36. Powell RJ, Hill RT. Rapid aggregation of biofuel-producing algae by the bacterium Bacillus sp. strain RP1137. Appl Environ Microbiol. 2013;79:6093–6101. doi: 10.1128/AEM.01496-13.
37. Chungjatupornchai W, Fa-Aroonsawat S. Biodegradation of organophosphate pesticide using recombinant cyanobacteria with surface- and intracellular-expressed organophosphorus hydrolase. J Microbiol Biotechnol. 2008;18:946–951.
38. Chungjatupornchai W, Kamlangdee A, Fa-Aroonsawat S. Display of organophosphorus hydrolase on the cyanobacterial cell surface using Synechococcus outer membrane protein a as an anchoring motif. Appl Biochem Biotechnol. 2011;164:1048–1057. doi: 10.1007/s12010-011-9193-3.
39. Ferri S, Nakamura M, Ito A, Nakajima M, Abe K, Kojima K, et al. Efficient surface-display of autotransporter proteins in cyanobacteria. Algal Res. 2015;12:337–340. doi: 10.1016/j.algal.2015.09.013.
40. Wang SB, Hu Q, Sommerfeld M, Chen F. Cell wall proteomics of the green alga Haematococcus pluvialis (Chlorophyceae) Proteomics. 2004;4:692–708. doi: 10.1002/pmic.200300634.
41. Guillard RR, Ryther JH. Studies of marine planktonic diatoms. I. Cyclotella nana Hustedt, and Detonula confervacea (cleve) Gran. Can J Microbiol. 1962;8:229–239. doi: 10.1139/m62-029.
42. Maeda Y, Sunaga Y, Yoshino T, Tanaka T. Oleosome-associated protein of the oleaginous diatom Fistulifera

solaris contains an endoplasmic reticulum-targeting signal sequence. Mar Drugs. 2014;12:3892–3903. doi: 10.3390/md12073892.

43. Sunaga Y, Maeda Y, Yabuuchi T, Muto M, Yoshino T, Tanaka T. Chloroplast-targeting protein expression in the oleaginous diatom Fistulifera solaris JPCC DA0580 toward metabolic engineering. J Biosci Bioeng. 2015;119:28–34. doi: 10.1016/j.jbiosc.2014.06.008.

44. Nojima D, Yoshino T, Maeda Y, Tanaka M, Nemoto M, Tanaka T. Proteomics analysis of oil body-associated proteins in the oleaginous diatom. J Proteome Res. 2013;12:5293–5301. doi: 10.1021/pr4004085.

45. Muto M, Fukuda Y, Nemoto M, Yoshino T, Matsunaga T, Tanaka T. Establishment of a genetic transformation system for the marine pennate diatom Fistulifera sp. strain JPCC DA0580-a high triglyceride producer. Mar Biotechnol. 2013;15:48–55. doi: 10.1007/s10126-012-9457-0.

CHAPTER 18

MICROALGAE HARVESTING METHODS FOR INDUSTRIAL PRODUCTION OF BIODIESEL

Mariam Al hattab, Abdel Ghaly and Amal Hammouda

Keywords: Microalgae, Harvesting, Dewatering, Physical treatment, Chemical treatment, Electrophoresis processes

INTRODUCTION

Microalgae are photosynthetic microorganisms that are abundant in nature and capable of growing in various environments [1]. Microalgae biomass can be used to produce numerous value added products such as biofuels (biodiesel, bioethanol, biogas and biohydrogen) [2], fish feed [3], animal feed [4], human food supplements such as vitamins A, B1, B2, B12, C, E, nicotinate, biotin, folic acid and pantothenic acid), Omega 3 fatty acids (Eicosapentaenoic acid (EPA), Docosahexaenoic acid (DHA)) and chlorophyll [5,6] and skin care products such as antiaging creams, anti-irritant creams and skin regenerate creams [3,7,8]. Various microalgae strains contain high amounts of proteins (43-71% of dry matter) compared to meat (43%), soybeans (37%), milk (26%) and rice (8%). They synthesize a wide range of amino acids essential for humans and animals which make them great for use in food supplements [9,10]. Microalgae carbohydrates (10-30% of dry matter) are synthesized in the forms of sugars, starch and polysaccharides which are easy to digest [9]. The oil content in the cells can make up 25-77% of the dried biomass weight [11]. Microalgae are regarded as the best candidate for the production of biodiesel as they do not compete with edible crops [1,12] and can produce between 20,000 to 80,000 L of oil per acre per year which is 7-31 times greater than that produced by the best terrestrial crop (palm tree) [13]. Application of biorefinery concept to produce biodiesel and other value added products will enhance the economics of biodiesel production.

However, processing microalgae into biodiesel and other value added products requires culturing of the microalgae, recovery of the microalgae biomass and the extraction and downstream processing of the oil and other value added products [14]. However, the major obstacle for using microalgae biomass on an industrial-scale for the production of value added products is the dewatering step [15,16]. Microalgae cultures need to be concentrated because they exist as a dilute suspension containing 0.1-2.0 g of dried biomass per litre [15,17]. Dewatering microalgae accounts for 20-30% of the total costs associated with microalgae production and processing [17,18]. The cost of the extraction, purification and extraction processes decrease with increased biomass concentration [15-17].

Therefore, in order to achieve economically viable biodiesel production, microalgae recovery needs to be made less costly. Different methods for solid-liquid separation can be employed to dewater/ concentrate the microalgae culture to 10-450 g/L. Such methods include sedimentation, vacuum filtration, cross flow filtration, pressure filtration, decanter centrifugation, disc stack centrifugation, dissolved air flotation, dispersed air flotation, micro bubble generation organic flocculation, inorganic flocculation, bio-flocculation, auto-flocculation) and electrolytic coagulation, electrolytic flocculation and electrolytic flotation.

The aim of this study was to review the current methods used for harvesting and concentrating microalgae and perform a comparative analysis in order to determine the most efficient economically dewatering methods for large scale processing

Citation: Al hattab M, Ghaly A, Hammoud A (2015) Microalgae Harvesting Methods for Industrial Production of Biodiesel. J. Fundam Renewable Energy Appl 5:154. doi:10.4172/2090-4541.1000154.

of microalgae biomass for production of biodiesel and value added products.

PHYSICAL HARVESTING METHODS

Numerous physical methods for microalgae dewatering processes have been used to retrieve the microalgae cells from their liquid suspension. These can be divided into four categories: sedimentation, filtration, centrifugation and flotation.

Sedimentation

In this technique, the solids and liquids are separated from one another by gravitational forces as shown in Figure 1 [19]. Different materials are separated from one another based on the density of the material and/or particle size. A larger difference in density would result in faster sedimentation rates while a smaller difference in densities and/or smaller particle size would require longer time to settle out by gravitational forces [20].

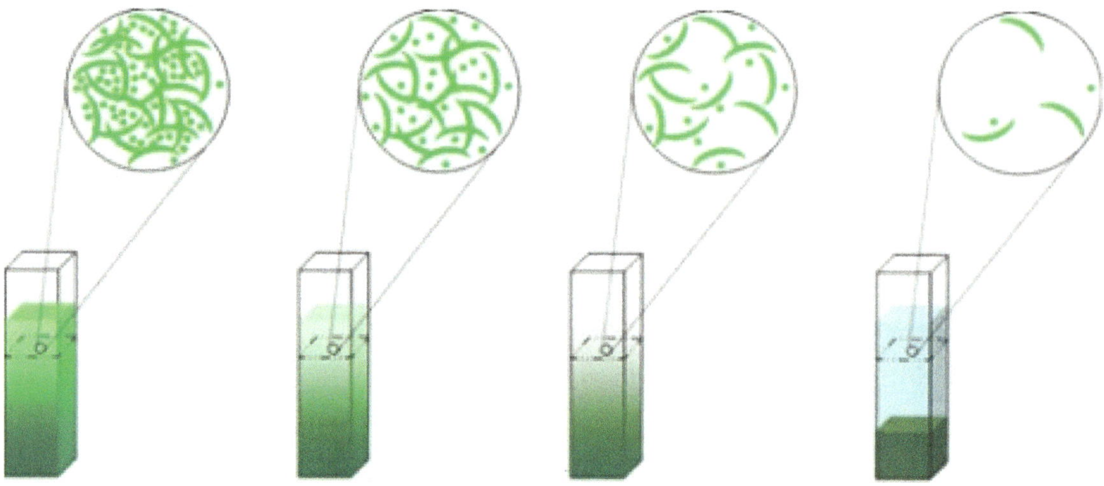

Figure 1: Sedimentation process of microalgae over time [19].

Type of Settling Tanks

Lamella separator (Figure 2a) and sedimentation tanks (Figure 2b) are used to separate solids from liquid [21,22]. Lamella separators offer a greater settling area than conventional thickeners as a result of plate orientation [23]. Lamella tanks work by inserting the microalgae biomass through the inlet. The liquid floats to the surface (effluent) and the biomass is caught onto the slanted plates. With time, the biomass settles down to the bottom of the tank and can be collected through the harvest outflow. Sedimentation tanks are cylindrical with a funnel shaped bottom so that the settled microalgae are concentrated near the outlet. The outlet is placed at the bottom of the tank so that the collection of the settled microalgae can easily be recovered. The tank is equipped with a pump that carries the microalgae biomass from the cultivation tank into the sedimentation tank through the inlet. These tanks work by allowing the denser solids to settle to the bottom of the tank, leaving the clear water at the surface. Once the settling process is complete, the algae can be retrieved from the tank through the outlet.

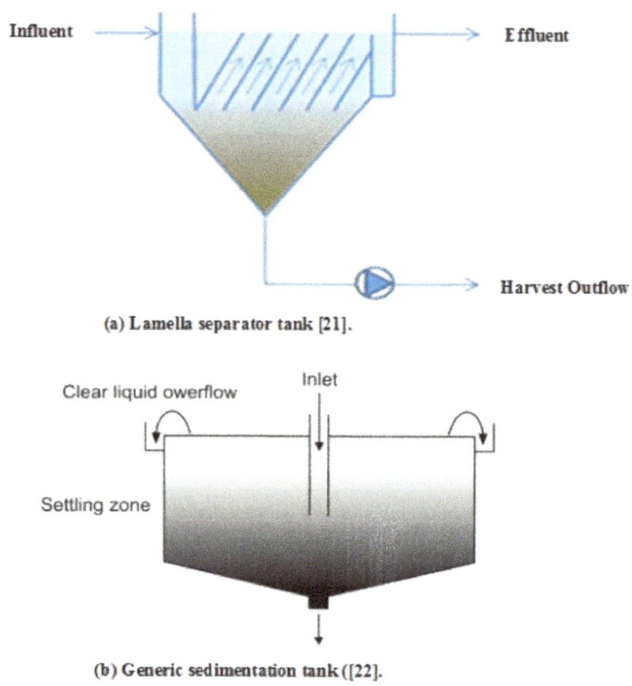

Figure 2: Sedimentation tanks.

Factors Affecting Sedimentation

The factors influencing the settlement rates of microalgae include: density and particle size, temperature, aging of the cells, light intensity and time [24-26].

Density and particle size: The density of marine microalgae varies from 1030 to 1100 kg/m^3 and the density of freshwater microalgae varies from 1040 to 1140 kg/m^3 [27-29]. Granados et al. [30] reported that the densities of fresh water (1000 kg/m^3) and salt water (1025 kg/ m^3) are similar to that of microalgae and as a result the rate of settlement of algae is low. Murphy and Allen [31] stated that it is a challenge to remove microalgae biomass from the liquids because of the identical densities of the cells and media.

Cole and Buchak [32] indicated that the rate of settlement is dependent on the type of microalgae present and found the green microalgae to have an average settling rate of 0.1 m/d. Peperzak et al. [33] noted that the sedimentation rate of 24 different microalgae species (ranging in size from 10-1000 μm) varied from 0.4 to 2.2 m/d and there was no correlation between the size of the cells and the sinking rates. Milledge and Heaven [20] reported a settlement rate of 0.1 m/d for *Chlorella* species in freshwater. Yang et al. [34] reported an algae settling rate in the range of 0.1-0.3 m/d. Choi et al. [35] noted that the sedimentation rate of large and small sized algae were 2.6 cm/h and less than 1.0 cm/h, respectively.

Temperature: Knuckey et al. [25] noted that the temperature of 4°C settled a wide array of microalgae species after 24 h when the pH was adjusted in the range of 8-8.5. Davis et al. [36] noted slower settling rates of microalgae in colder waters as a result of increased viscosity. Harith et al. [24] tested the effect of varying temperature (4-27°C) and the presence and absence of light on sedimentation rates of Chaetoceros calcitrans at a pH of 8. The highest efficiency (9%) was obtained at day 8 at a temperature of 27°C in the dark. Greenwell et al. [37] noted that harvesting microalgae by sedimentation in high temperature areas will deteriorate the cells.

Cell age: Danquah et al. [15] noted that the settling rate for microalgae harvested during the high growth phase (4-10 days) was lower than that harvested during low growth phase (10-12 days). Choi et al. [35] reported that the settling rate of algae significantly increased in the stationary growth phase of microalgae. Manheim and Nelson [38] noted that in the exponential growth phase (day 6) there was little to no settling in Scenedesmus sp. observed over 2 h period, but the greatest removal efficiency was noted in the stationary phase (day 15). They also noted that the settling rate for C. vulgaris species in the exponential phase was 6 times greater than the late stationary phase. Peperzak et al. [33] reported that the settling rate at 15 and 20 weeks for *Phaeocystis globosa* and *Eucampia zodiacus* were 0.5 and 1.0 m/day and 0.7 and 1.0 m/day, respectively.

Light: Danquah et al. [15] noted that the absence of light increased the settling rate during high growth and low growth phases. The supernatant obtained during the high growth phase contained 0.57 g/L of biomass in the presence of light and 0.39 g/L in the absence of light, while the supernatant obtained during the low growth phase contained 0.28 g/L in the presence of light and 0.17 g/L in the absence of light. Schlenk et al. [39] noted that the concentration of microalgae in the light and dark conditions were 1075 cells/mL and 775 cells/mL, respectively. On the other hand, Harith et al. [24] reported that the settling rates observed in the presence and absence of light in *Chaetoceros calcitrans* were similar.

Time: The concentration of microalgae by sedimentation requires long settling times that are greater than 24 h. Park et al. [40] noted long retention times of 1-2 days for algae recovery in large-scale settling tanks. Harith et al. [24] reported that increasing the settling time to 15 days increased the settling efficacy to 94%. Griffiths et al. [41] noted that the percentage of biomass recovery after 24 h of settling for *S. platensis, C. fusiformis, T. suecica, Nannochloropsis* and *Scenedesmus* were 95, 96, 80, 59 and 86%, respectively. Wang et al. [42] noted that the biomass recovery for the species *S. dimorphus* and *C. vulgaris* after 2 h of gravitational settling was 80 and 55%, respectively.

Advantages and Disadvantages

Although sedimentation tanks are effective in concentrating microalgae suspensions to 1.5% total suspended solids (TSS), they are not widely use in the industry. The costs associated with gravitational settling are low, but the reliability of this method without the use of flocculating agents is also low [16,43]. The settling time required is much longer than other processes [44] and energy is required for pumping the slurry [16]. Gonzalez-Fernandez and Ballesteros [45] stated that this method is time consuming and the composition of the cells can change. Mata et al. [46] stated that the cell concentrations obtained by sedimentation are low. Ras et al. [47] indicated that harvesting microalgae biomass by sedimentation alone is not the most efficient method since the cell recovery rates of 60-65% are low.

Filtration

This type of algae harvesting method uses a medium that is permeable so that it can retain the algae biomass while allowing the liquid to pass through. This technique requires a pressure difference across the filter which can be driven by vacuum, pressure or gravity. The membrane filters can be classified based on the size of the pores into macro filtration (greater than 10 μm), micro-filtration (0.1-10 μm), ultrafiltration (0.02-0.20 μm) and reverse osmosis (less than 0.001 μm) [48]. The pressure required to force the fluid across the membrane decreases as the pore size of the membrane is increased. Filtration techniques can concentrate microalgae cells in the suspension upto 5-18% and the operating costs vary from $10 to $20/gal. The harvesting efficiency using filtration methods ranges from 20% to 90% [49].

Vacuum Filtration

Vacuum filtration separates solids from liquid media by capturing the solid particles onto a filter while pulling the liquid through by suction from the filter (Figure 3). Microalgae range in size from 2 to 30 μm indicating that a micro-filtration membrane is suitable for vacuum filtration [17,48]. Milledge and Heaven [20] stated that the macro-filtration membranes can be used for large microalgae cells or if the algae cells are flocculated together. Uduman et al. [16] reported that the vacuum filtration harvesting technique is most suited for large microalgae cells (greater than 10 μm). Stucki et al. [54] separated *Spirulina platensis* species using vacuum filtration equipped with regenerated cellulose membrane with a pore size of 0.45μm.

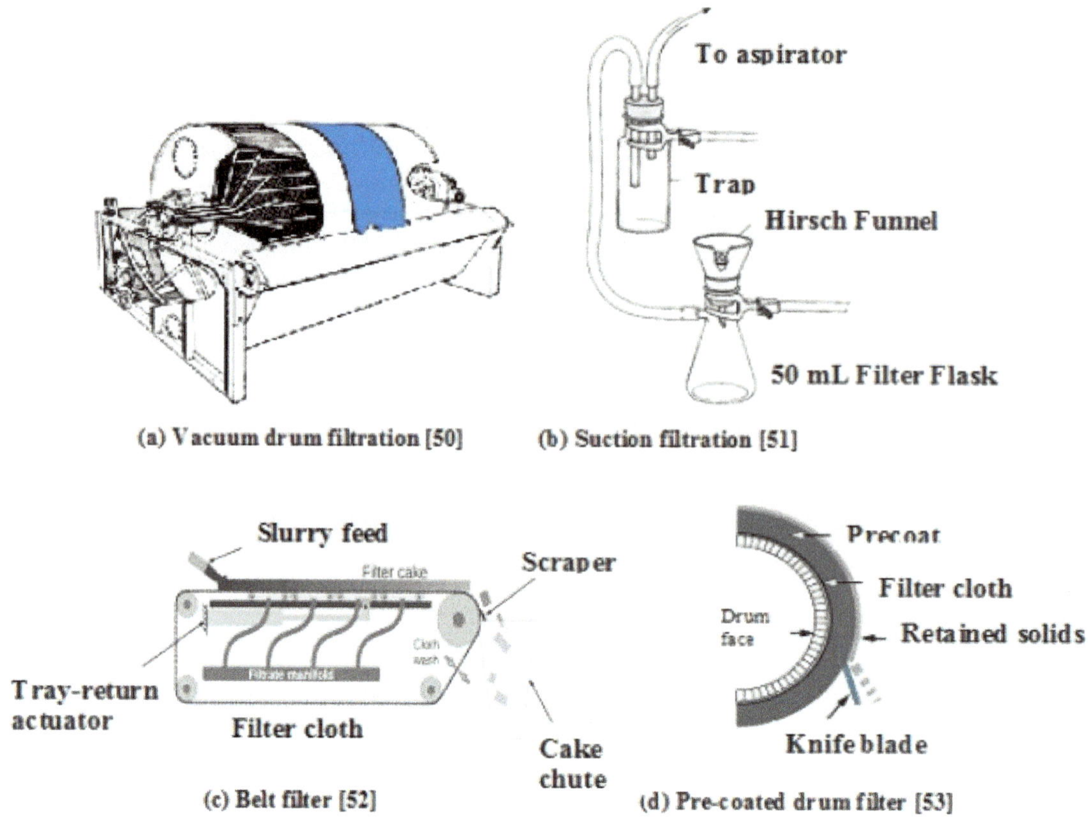

Figure 3: Types of vacuum filtration.

Type of vacuum filters: There are five different filter membranes that can be used in vacuum filtration. They are vacuum drum filter, suction filter, filter thickener, belt filter and starch precoated drum filter. Mohn [23] noted that the suction filter, starch precoated drum filter and belt filter were suitable for concentrating the *Coelastrum* microalgae species to a range of 5-37%. The author found the drum filters were not effective harvesting techniques as a result of clogging. Filter thickeners were not recommended as a result of low solid contents (3-7%) and high energy requirements. Ferrentino et al. [55] noted that vacuum filtration equipped with a Buchner funnel and cellulose fiber filters were effective in the recovery of microalgae from solution.

Successful recovery of microalgae cells has been noted using filtration equipped with diatomaceous earth as a filter aid to avoid the clogging of the filter [17]. Gudin and Chaumont [56] reported that precoated drum filter with filter aid (diatomaceous earth) is effective in harvesting the microalgae *Chlamydomonas reinhardtii.* Molina Grima et al. [17] evaluated filters made of cellulose fibers and sand filters and obtained unsatisfactory results but found that diatomaceous earth filter effectively recovered the micro sized *Dunaliella* species. Uduman et al. [16] noted an exceptional recovery of *Dunaliella* species using diatomaceous earth aided filter. Brennan and Owende [48] stated that the use of diatomaceous earth as a filter aid can effectively remove microalgae cells from medium.

Energy consumption: Shelef et al. [43] reported energy consumption in the range of 0.1-5.9 kWh/m^3 depending on the type of filter used. Mohn [23] noted that vacuum filtration consumed 5.9 kWh/ m^3 of energy in order to concentrate the suspended solids in solution to 18-27%. They also reported that the energy required to dewater the *C. proboscideum* using suction filter (8% SS), belt filter (9.5% SS) and filter thickener (5-7% SS) was 0.1, 0.45 and 1.6 kWh/m^3, respectively. Umesh [57] noted that harvesting the microalgae strain *Spirulina fusiformis*under vacuum filtration with a coarse pores medium was low in cost ($83.3/ m^2) and capable of harvesting 23 kg/ m^2 kwh. Shelef et al. [43] reported that the pressure drop required for vacuum operations is in the range of 70-80 kPa. Milledge and Heaven [20] reported a power consumption of 0.25 kWh/m^3 for microalgae harvest using vacuum belt filter. Mohn [58] reported an energy consumption value of 3 kWh/ m^{3}for microalgae harvesting, using vacuum drum filtration.

Advantages and disadvantages: The advantages of using vacuum filtration technique for harvesting microalgae are the preservation of the cells after the recovery process [59]. The effectiveness of the filtration process is dependent on the membrane size and the microalgae cell size. Harvesting microalgae by filtration is more efficient than the sedimentation technique, but drawbacks of this method include membrane replacement and/or periodical washing of the membrane to avoid clogging the membrane pores [45]. However, drawbacks are associated with large energy requirements and costs associated with periodic replacement of membrane as a result of clogging [17,46,60- 62]. Arar and Collins [63] recovered microalgae for chlorophyll extraction using vacuum filtration at 6 in. Hg (20 kPa) and noted that higher pressures and prolonged filtration (beyond 10 min) may damage the cells. Uduman et al. [16] noted that filtration technique is suitable for larger microalgae cells, but inadequate for recovery of microalgal species. Rossi et al. [64] noted that rapid clogging of the membrane resulted using the ultrafiltration membrane technique. Flocculation assisted filtration processes would lower the energy requirements, but additional costs for the flocculent would be encountered [20]. Molina Grima et al. [17] recovered microalgae biomass using filtration method and concluded that this harvesting method is not economically viable for large scale production.

Pressure Filtration

Pressure filtration is a technique used for separating particles form a liquid suspension into a compacted form. It works by separating the liquid from the particles (that are collected onto the filter) by means of pressure [65]. The flow of fluids through the filter is created by raising the pressure above the atmospheric pressure to create a pressure differential across the filter [66]. This process is operated in batches that are most often fed from and discharged to a continuous process. A surge tank is required upstream to the filter and one is required for the collection of the filtrate [67].

Type of pressure filtration: Pressure filtration harvesting can be achieved by plate-and-frame filter presses or by using a pressure vessel that is equipped with filters as shown in Figure 4 [50]. The plate-andframe filter presses works by forcing the liquid in the microalgae suspension through the filter using high pressure. A series of rectangular plates that are mounted in a vertical position, face to face, make up the press system. A fitted filter cloth is applied to each of the plates and they are held together with one another by force under pressure. The fluid that contains the algae is pumped into the gaps between the plates and the pressure is applied in order to force the liquid through the plate outlets and filter cloths. After separation, the dewatered microalgae cake is recovered [68].

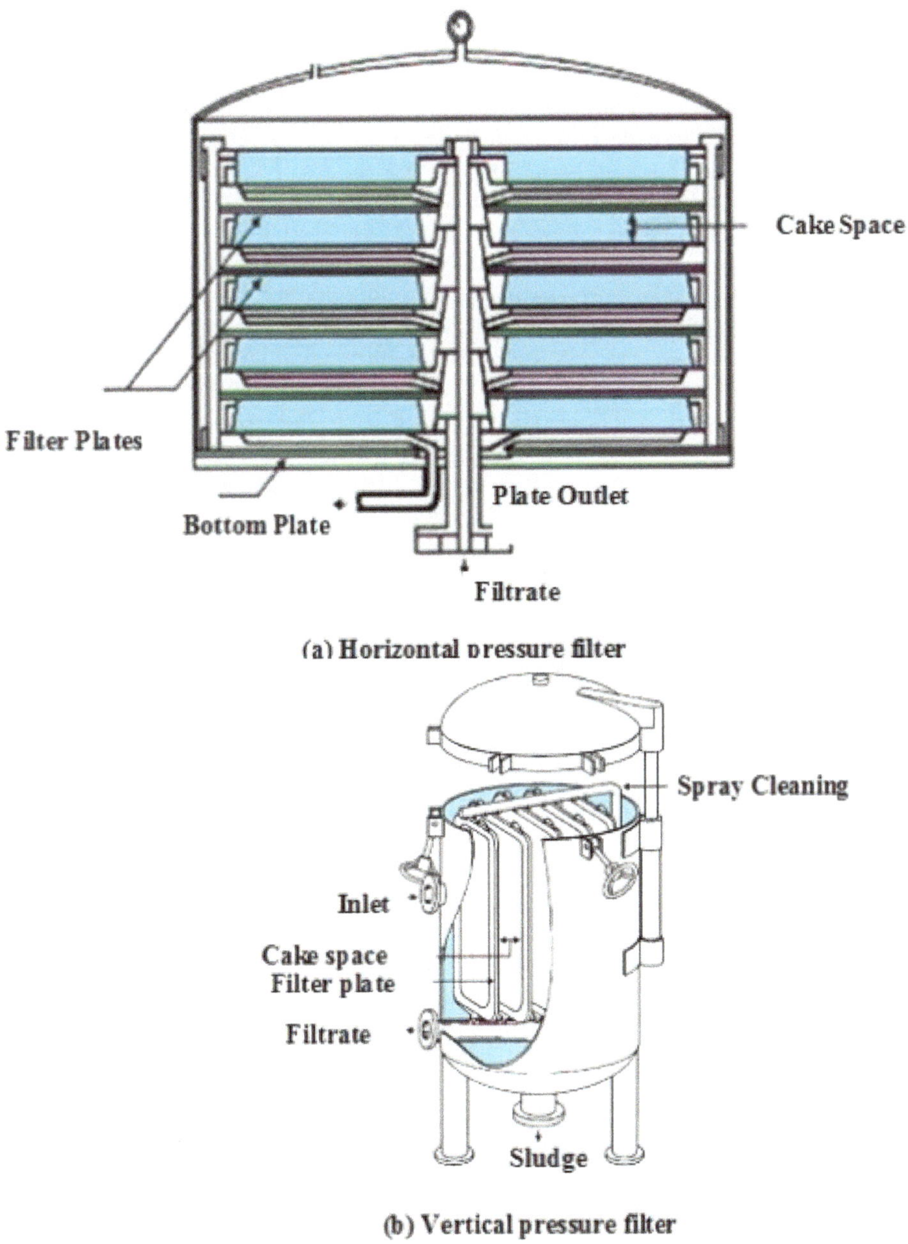

Figure 4: Pressure filter [50].

Energy consumption: This method can be considered energy efficient since a minimal amount of energy is required upon assessment of the output product and the amount of initial feedstock added [15]. However, the effectiveness of the method is dependent on the type of algae species.

Molina Grima et al. [17] noted that the amount of energy required to harvest 22-27% (w/v) of *C. paroboscideum* species using pressure filters is 0.88 kWh/m^3. Nagle and Lemke [69] noted an 8% concentration of microalgae (up to 0.5%) using a filter press that has 20 plates and frame (30 cm in diameter), equipped with filter paper that has a pore size of 5 µm. Harun et al. [8] noted that the microalgae *Dunaliella* and *Chlorella*species were too small to be recovered by pressure filtration. Mohn [23] reported that the energy consumed for harvesting *C. proboscideum*using cylindrical sieve (7.5% suspended solid concentration) and filter basket (5% suspended solid concentration) was 0.3 kWh/m^3 and 0.2 kWh/m^3, respectively. He also found that pressure filtration was not suitable for the species Scenedesmus, *Dunaliella* and *Chlorella*, but was satisfactory for other larger microalgae species such as *Coelastrum proboscideum and Spirulina platensis.*

Advantages and disadvantages: Some of the advantages of using pressure filtration are: the cakes collected (composed of the particles in the liquid suspension) have low moisture content, the soluble recovery from the cake is high, re-circulating the filtrate for 1-2 min will clean the filter, high degree of clarity in solutions can be achieved and alloy and synthetic materials can be used to construct the filters and the internal parts [66,70]. The disadvantages of using this technique include: the difficulty in washing the filter medium which increases when the solid is sticky, the internals are difficult to clean in food-grade applications and the difficulty in viewing the condition of the filter due to vessel encapsulation [70].

Cross Flow Filtration

Harvesting microalgae cells in large volumes can be effectively done using cross flow filtration a shown in Figure 5 [71]. In this technique, the sample flows tangentially across a membrane. The particles larger in size than the membrane pores are retained and referred to as the retentate. Smaller particles pass through the membrane with the liquid solution and are referred to as the permeate.

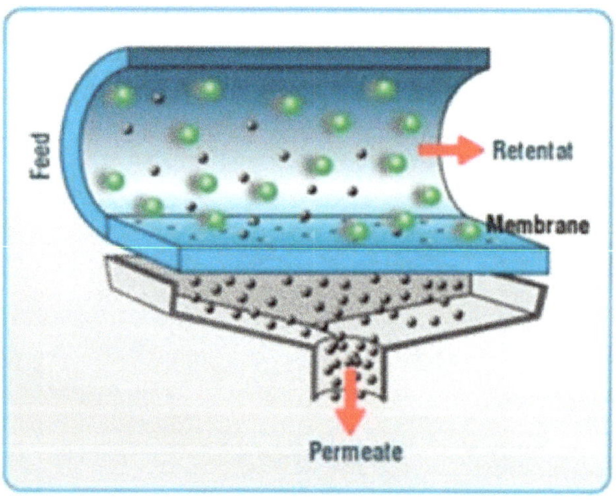

Figure 5: Cross flow filtration diagram [71].

Membrane type: Ultrafiltration or microporous membranes are the type of filter membranes used in this technique. These membranes are available with a wide range of pore sizes and molecular weight retentions. Polymer membranes have a long operating life when used at suitable cross flow velocity conditions and low transmembrane pressures. Petrusevski et al. [59] used a cross filtration with a membrane pore size of 0.45 µm and achieved a biomass recovery efficiency of 70-89%. Rossignol et al. [72] found that polymer membranes were effective in recovering the marine microalgae species Haslea ostraria and *Skeletonema costatum,* but the performance depended on the hydrodynamic conditions,properties of the microalgae and the concentration of the microalgae cells.

Uduman et al. [16] reported that the initial flux for microfiltration Figure 5: Cross flow filtration diagram [71]. membranes were much higher than those of ultrafiltration, but they clogged more easily. Zhang et al. [73] used a cross-flow ultrafiltration membrane with a cross-flow velocity of 0.17 m/s and noted an increase in algae concentration from 0.104% to 92.5% at the membrane surface with a harvesting efficiency value of 46.01 g/m^2/h. Rossi et al. [74] tested 14 various inorganic membranes and noted that the ultrafiltration membrane ATZ-50 kDa illustrated the best performance and concentrated the *Arthrospira platens* species by a factor of 20. Rossi et al. [64] used a cross-flow filtration technique equipped with an organic ultrafiltration membrane (polyacrylonitrile, 40kDa) and concentrated *Arthrospira platens* by a factor of 10. Rossignol et al. [72] concentrated the species *Skeletonema costatum* using cross-flow ultrafiltration with a flux of 30 l/h for 12 h. Rose et al. [75] effectively concentrated the species *Dunaliella salina* by cross-flow ultrafiltration with flux rates of 30-40 l/h. Walsh et al. [76] concentrated the species *Thalassiosira pseudonana* to 2.3 L from 2840 L which was composed of 2.33x10^{12} cell/L using microfiltration membrane system. Ahmed et al. [77] noted that the resistance of the cross-flow microfiltration decreased as the cross-flow velocity increased from 0.13 to 4 m/s while harvesting *Chlorella sp.* species.

Energy consumption: Rossignol et al. [72] reported that the energy consumption using cross-flow filtration techniques can range from 3kWh/m^3 to 10kWh/m^3 depending on the feed characteristics, the system design and the pressure used. Danquah et al. [15] noted that cross-flow filtration can also be used in sensitive suspensions and is a cheap method for concentrating suspended solids in the range of 2.5- 8.9 % with an energy consumption in the range of 0.38-2.06 kWh/m^3. Crittenden et al. [78] reported that the energy consumption for a crossflow filtration technique was 5kWh/m^3. Danquah et al. [15] reported that the amount of energy required for dewatering microalgae to a concentration of 8.88% (w/v) was 2.06kWh/m^3.

Advantages and disadvantages: Cross flow microalgae filtration is advantageous over other conventional harvesting methods such as sedimentation, flocculation and centrifugation because it results a in complete removal of debris and microalgae cells [16]. The equipment are considered to be cheap because the costs are only associated with pumping and replacement of membranes [72]. The structure and properties of the recovered microalgae are preserved using this filtration technique [59]. However, large scale recovery of algae cells using this method can be limited due to fouling and frequent replacement of the membrane [16].

Centrifugation

This type of removal mechanism is widely used in beverage, food and pharmaceutical industries. Centrifugation is a process in which a centrifugal force is used to enhance the separation of solids. Spinning the suspension creates the pressure differential necessary for particle separation from the liquid suspension. Thus, the efficiency of the recovery process is dependent on the centrifugal force [17].

Types of Centrifuges

The two types of centrifugation used for harvesting microalgae are: disc stack and decanter centrifuges.

Disc stack centrifuge: The most common industrial centrifuge used today in commercial plants producing high value products and algal biofuel is the disc stack type centrifuge. It consists of a shallow cylindrical bowl that has numerous stacks of metal cones (discs) which are closely spaced together as shown in Figures 6 [79,80]. Separation of the materials is based on densities. The mixture is placed on the centre of a stack of discs and the lighter phase of the mixture remains on the inside towards the centre while the denser phase is displaced outwards to the underside of the discs. This technique separates materials of different densities by layering them [81]. It is most suited for separating materials with particle sizes in the range of 3-30 μm and for concentrations of suspensions that has solid content ranging from 2 to 25% [20].

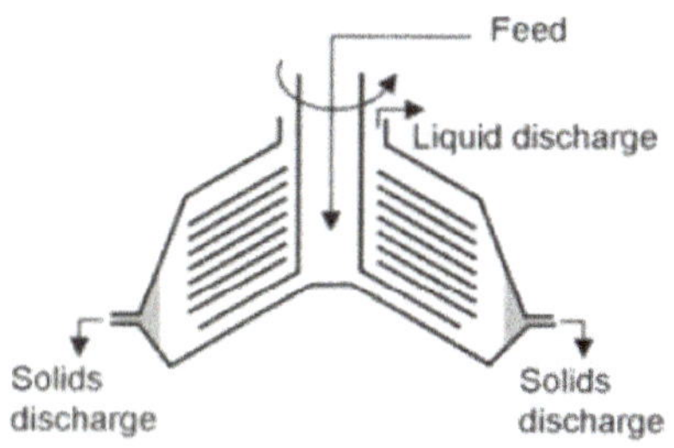

(a) Borowitzka and Moheimani [79]

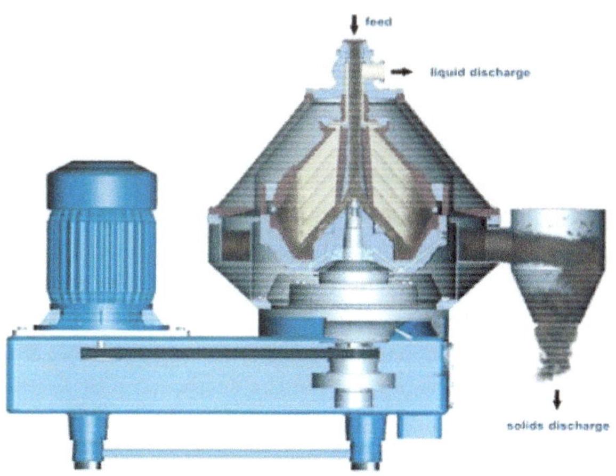

(b) Enviropro [80]

Figure 6: Disc stack centrifuge.

Heasman et al. [82] evaluated the cell recovery efficiency of nine different microalgae species using a disk stack centrifuge and noted a recovery efficiency greater than 95% at a force of 13,000g. They also noted that the recovery efficiency declined with a decrease in the gravitational force to 60% and 40% at gravitational forces of 6,000g and 1,300g, respectively. Sim et al. [83] noted a 90 % microalgae removal efficiency using a disc stake type of centrifuge. Vasudevan et al. [84] achieved an 18% microalgae concentration using a disc centrifuge. Mackay [85] used a disc centrifuge operating at a force of 4,000-15,000g for a biomass suspension containing 0.2-20% v/v algae cells. Chojnacka et al. [86] harvested the *Spirulina sp.* using a disc type centrifuge operating at 6,000 rpm for 5 min.

Decanter centrifuge: Decanting centrifugation is based on the concept of using a special settling tank in which the solids in suspension are forced to fall down due to the gravitational forces [87]. The decanter centrifuge (Figure 7) operates continuously by pumping the cultivated microalgae biomass into the centrifuge bowl whereby the suspended particles in solution are forced to the bottom of the bowl. The liquid left after the particles have been extracted is passed through the overflow pipe [88].

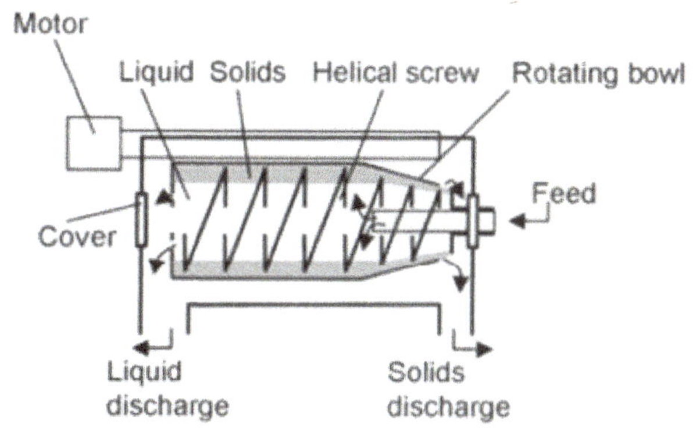

Figure 7: Decanter centrifuge [79].

Molina-Grima et al. [17] noted that concentration of microalgae biomass using a decanter centrifuge is better than other harvesting methods. Dassey and Theegala [89] achieved a harvesting efficiency of 28.5% for microalgae at a flow rate of 18 L/min using continuous flow decanter centrifuge. Smith and Charter [87] reported that the clarity of the liquid produced after separation was not as great as that achieved using disc-stack centrifugation. Vasudevan et al. [84] reported that a 12% microalgae concentration was achieved using a decanter centrifuge. Mackay [85] reported that the decanter centrifuge operates using a force of 4000-10000 g and is effective for slurries with a biomass content of 5-80%. Vasudevan et al. ([84] stated that microalgae biomass needs to undergo an initial thickening step such as dissolved air flotation in order to concentrate microalgae suspensions (0.02-0.05 weight %) to 2-3% before using decanter centrifugation.

Energy Consumption

Disc-stack centrifuge: The energy consumption reported in the literature for the disc stack centrifuge varied from 0.53 kWh/m^3 to 5.5 kWh/m^3. Alfa Laval [90] used a disc type centrifuge for dewatering microalgae and achieved a 16% with a power consumption of 0.53 kWh/m^3. Mohn [23] noted a 12% suspended solids concentrate of the microalgae species *Scendesmus* using a disc-stack centrifuge with an energy consumption of 1 kWh/m^3. Goh [91] harvested microalgae grown in pig waste using disc centrifugation with an energy consumption of 1.4kWh/

m³. Sharma et al. [92] noted that the disc stack centrifuge consumed 5.5kWh/m³ for Chlorella sp. harvesting.

Advantages and Disadvantages

Disc-stack centrifuge: The advantages of using disc-stack centrifuge for harvesting microalgae is their high removal efficiency compared to other industrial centrifuges. The concentration of the feed for these units is typically in the range of 0.5-10% w/w. This type centrifuge handles high flow rates and is capable of separating fine (0.1- 100 μm) particles [93]. This device can be used to separate solid from liquid in continuous, semi-continuous and batch operation. Some of the disadvantages of this type of centrifuge include: low dry substance content in the discharge system, mechanically complex, costly and the small space between the closely stacked discs makes it harder to clean and may require chemicals for cleaning [92,93].

Decanter centrifuge: The dewatered biomass using the decanter centrifuge is much more concentrated than that achieved using the disc centrifuge. However, the decanter centrifuge is more suited for suspensions with higher solid particles and is unsuitable for microalgae suspensions [58]. This type of centrifuge is most suited for separating materials with particle sizes greater than 15 μm and solid suspensions containing higher than 15% [20]. It operates at inertial forces that are less than 6000 g. The disadvantages of using this method for microalgae harvesting are: highly concentrated feeds (typically in the range of 4-40% w/w) is required, the liquid leaving the system may not be clear due to the presence of fines, processing finer particles may result in poor flow properties of the thickened solids and cause mechanical difficulties, much more energy intensive than disc centrifuges and the high costs associated with the equipment required for processing large volume [17,20,93-95]. This type of centrifuge has been estimated to consume 3000 kWh/ton of dry alga biomass.

Flotation

The flotation technique for microalgae harvesting takes advantage of the low density of microalgae [28]. This technique is classified as a physiochemical gravity separation process in which gas bubbles pass through a liquid-solid suspension causing the microalgae to float to the surface by adhering to the gaseous bubbles [43,96]. The aeration also assists in removing the volatile organic compounds that are in the solution which provide cleaner residual water [43]. The efficiency of this method depends on the suspended particles instability, higher air-particle contact corresponds to a lower instability [97]. In flotation technique, the size of the particle is of importance, the smaller the particle size the more likely it will be lifted to the top of the medium by the bubbles. Solutions with particle sizes of less than 500 μm can be used in flotation [98].

Types of Air Flotation

The flotation processes are grouped by the method that is used for the bubble formation into: dispersed air flotation, dissolved air flotation, microbubble generation and electrolytic flotation.

Dispersed air flotation: This technique requires the use of a high speed mechanical agitator for bubble formation and an air injection system as shown in Figure 8 [99]. The gas mixes with the liquid as it is introduced at the top of the vessel and is allowed to pass a disperser that creates bubbles ranging in diameter from 700 to 1500 μm [100]. These bubbles are a magnitude larger (1000 μm in diameter) than those produced using dissolved air flotation technique.

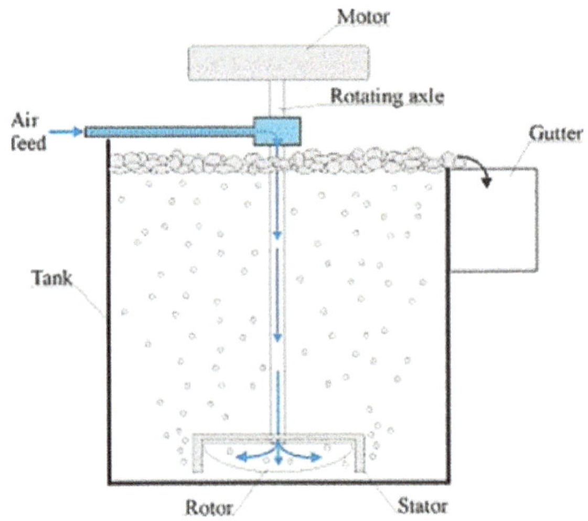

Figure 8: Dispersed air flotation technique [99].

Chen et al. [101] used dispersed air flotation for dewatering of microalgae Scenedesmus quadricauda. They used three varying surfactants in order to remove the microalgal cell: non-ionic X-100, cationic N-Cetyl-N-N-N-trimethulammonium bromide and anionic sodium dodecylsulfate. They also found that surfactants played a role in increasing the integrity of the bubble avoiding rupturing, and noted that this removal method was most successful with the use of cationic N-Cetyl-N-N-N-trimethulammonium bromide surfactant.

Yan and Jameson [102] noted that the dispersed air flotation technology resulted in 98% microalgae removal efficiency. Xu et al. [103] reported a 93.6% recovery efficiency of B. braunii using dispersed air flotation technique. Kurniawati et al. [104] harvested Chlorella vulgaris and Scenedesmus obliquus using dispersed air flotation assisted with chitosan and achieved a recovery efficiency greater than 93%.

Dissolved air flotation: The dissolved air flotation technique is the most widely used flotation technique in the treatment of industrial effluent [98,100]. This method requires a reduction in water pressure that is presaturated with air. The liquid is then injected into the flotation tank at atmospheric pressure. Bubbles are generated from the diffuser nozzles and rise through the liquid carrying microalgae cells in the suspended media to the surface of the tanks as shown in Figure 9 [21]. The cumulated biomass at the surface can be skimmed off and collected. The clarified liquid portion is saturated with air and recycled back into the flotation tank [100]. The supply of air into the system can be controlled by changing the saturator pressure or by changing the ratio that is recycled back into the tank. The size of the bubbles formed can be controlled by the saturator, operated above atmospheric pressure and by the injection flow rate [16,105]. The flow rate must be great enough to prevent backflow, provide a pressure drop and allow for bubble growth on the pipes surface [16]. Small bubbles ranging in size from 10 to 100 μm (with a mean size of 40 μm) are desirable [105].

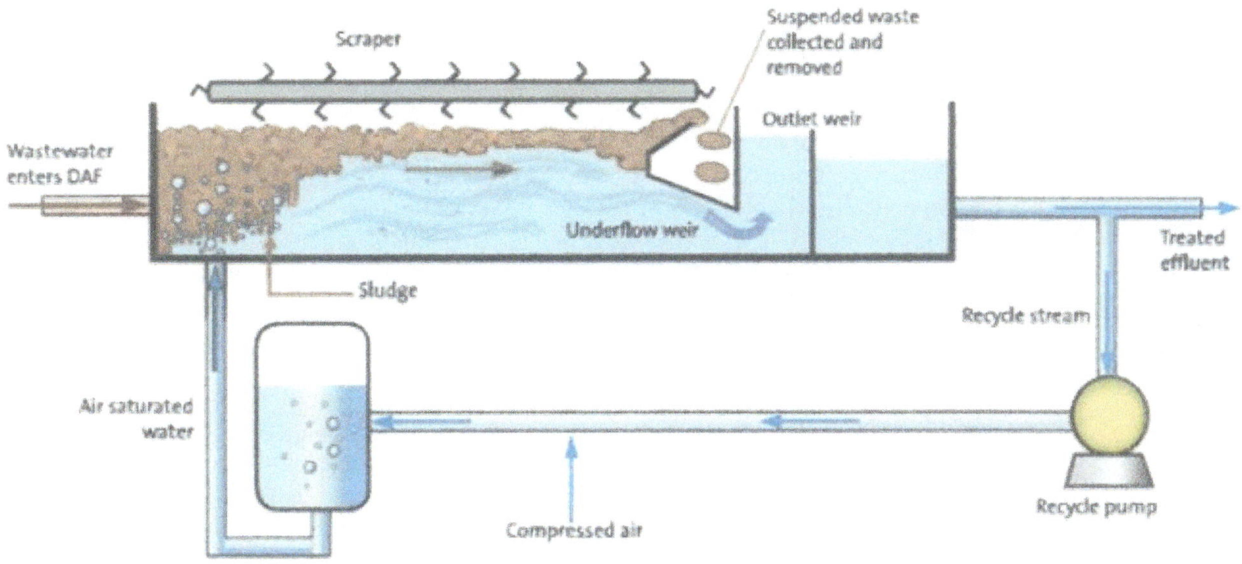

Figure 9: Dissolved air flotation apparatus [21].

The dissolved air floatation microalgae separation is usually coupled with the use of a chemical flocculation process. Edzwald [28] investigated the use of dissolved air flotation for microalgae recovery and noted that this method required pretreatment by flocculation, but was more successful than the settling technique. Wiley et al. [106] used dissolved air flotation for microalgae harvest and noted a suspended solids concentrate of 5% with an energy consumption of 7.6 kWh/m^3. Goh [91] noted that this method was effective in harvesting microalgae from pig slurry when coupled with the alum flocculent with a high dosage of 0.3 g/L.

Fluidic oscillation: The recovery of microalgae can also be achieved by micro-bubble generation through fluidic oscillation as shown in Figure 10 [107]. This method works by converting a continuous air supply into oscillatory flow with a regular frequency, generating bubbles that are the size of the exit pores [108]. The miniature bubbles are formed by fitting a diffuser to the bi-stable valve which ensures that the bubbles formed are approximately 10 times smaller than those originally dispersed in flotation methods [107]. Fine bubbles that are the size of the exit pores are generated by use of a fluidic oscillator [108]. The bubbles formed attach to the hydrophobic cells suspended in solution and carry them to the surface. At the surface the bubble raptures leaving the cells behind [109].

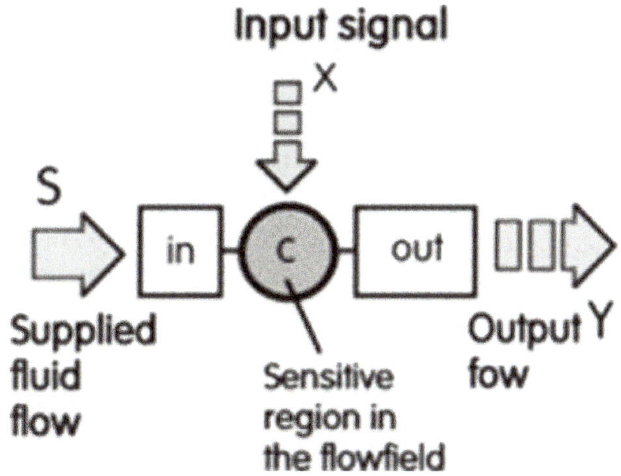

Figure 10: Principle of fluidic oscillation - The continuous flow of fluid (S) is integrated with a weak input signal (X) which causes the change in output flow (Y) [107].

Hanotu et al. [108] reported a microalgae recovery efficiency of 99.2% using microbubble generation at a pH of 5 with the aid of ferric chloride coagulant (150 mg/L). Elder [110) noted a removal efficiency greater than 95% using micro-bubble flotation. Yap et al. [111] noted a removal efficiency greater than 95% for removal of Microcystis and filamentous Cylindrospermospsis using micro-bubble flotation. However, not much research has been performed using this technique for microalgae recovery to deem it economically suitable for large scale recovery of microalgae cells [20].

Factors Affecting Air Flotation

The factors affecting the efficiency of harvesting microalgae using the air flotation technique include: pH, air flow rate, alkalinity, recycle rate, hydraulic loading and time. The velocity of the skimmer determines the concentration of the slurry formed and the height above the surface of the water [16].

pH: One of the most important parameters that affects the flotation processes is the pH. Lin and Liu [112] stated that the pH affects the reaction routes and the interfacial properties. Chen et al. [101] noted that maintaining the pH in the range of 5-8 using anionic sodium dodecylsulfate had little to no effect on the removal efficiency (95%) of microalgae *Scendesmus quadricauda.* They attributed this to the electrostatic interactions between the positively charged microalgae surface and the surfactant that are facilitated by pH values less than 8. However, at pH values above 8 the interaction between the algae surface and surfactant was weak, which did not allow for efficient removal. Kwon et al. [113] reported that flocculent assisted air flotation technique required a pH of 7-8 for organic and 5-6 for inorganic flocculants in order to reach a removal efficiencies greater than 90%. Zhang et al. [114] reported a harvesting efficiency greater than 90% for *Chlorella zofingiensis* using an air flotation technique by adjusting the pH to 6.2. Hanotu et al. [108] reported that acidic conditions were optimal for effective removal of microalgae.

Alkalinity: Besson and Guiraud [115] noted that the microalgae flotation recovery efficiency was improved by using sodium hydroxide at a concentration of 0.0085 mol/L (over the tested range of 0.000 to 0.025 mol/L). Schlesinger et al. [116] noted that calcium hydroxide is most effective in flocculating microalgae that can be recovered using flotation techniques. Thangavel and Sridevi [117] reported that carbonate salts can be used for microalgae flocculation to enhance the recovery efficiency of the microalgae. Chen et al. [101] reported that the change in alkalinity from 0 to 50 mg/L using $NaHCO_3$ did not have any effect on the removal efficiency (95%) of Scenedesmus quadricauda species.

Air flow rate: Reports in the literature show that varying the air flow rate from 68 to 206 ml/min did not have any effect on the removal efficiency (92%) of the *Chlorella sp.* species (Liu et al. [118]. Coward et al. ([119] harvested *Chlorella sp.* using flotation technique at a flow rate of 100 L/h. Dassey and Theegala ([120] reported that an increase in the air flow rate beyond 160 ml/min did not significantly increase the microbubble production and as such no increase in harvesting efficiency was noted. Hanotu et al. [108] noted that the air flow rate of 85 L/min was effective in the separation of microalgae form the media.

Medium: Zhang et al. [121] found that the algal media influences the rate of harvesting efficiency. They noted that the algal cells that were nitrogen deprived had lower concentrations of surface functional groups as they went from the exponential to the stationary and declining growth phases. They noted that these functional groups are made up of proteins and polysaccharides that are important for stabilizing the surface charge which affects the adhesion onto the bubbles in dissolved air flotation technique. They also noted that more Alum coagulant was required to achieve higher harvesting efficiency (90%) in the exponential phase compared to the stationary and declining phases as more alum is required to destabilize the surface of the generated cells in the exponential phase. There was a linear correlation between the concentration of surface functional groups and alum dosage for a given harvesting efficiency. Coward et al. [119] stated th cells at the highest flotation harvesting efficiency was observed during high growth phase as a result of higher zeta potential. Schenk et al. [11] noted that nutrient limitation in the media can improve the harvesting efficiency of microalgae flotation methods.

Recycle rate: The recycle rate reported in the literature varied from 5 to 10%. Edzwald and Wingler [122] obtained a 97-99% removal efficiency of *Chlorella vulgaris* species with a recycle rate of 8%. Vlaski et al. [123] reported a removal efficiency of 94.5% for Microcystics with a recycle rate of 5-10%. Kempeneers et al. [124] achieved a removal efficiency of 80% for *Melosira cyclotella* with a recycle rate of 6%. Teixeira and Rosa [125] reported a 92-98% recovery efficiency of bluegreen microalgae using a recycle rate of 8% and stated that the recycle system was vital for effective particle recovery but recycle rates past 8% illustrated little improvements. They also found that the addition of pressurized recycle system did not improve the recovery rate of the cells. This phenomena is attributed to the lack of particle destabilization since particle destabilization is vital to the effectiveness of dissolved air flotation as opposed to floc size. Gregory and Edwald [126] reported a recovery efficiency of 90-99% using dissolved air flotation with a recycle rate of 10%.

Hydraulic loading: The hydraulic loading rate for industrial air flotation applications ranges from 0.504 m/h to 40 m/h [127, 128]. Edzwald [129] reported that high rate dissolved air flotation techniques can be performed at hydraulic loadings of 20-40 m/h. Haarhoff and Rykaart [130] noted that increasing the hydraulic loading lowered bubbles formation. Dassey and Theegala [120] stated that increased hydraulic loadings decreased the time for air to dissolve which results in poor bubble productions.

Time: The time required for effective air floatation reported in the literature varied from 3 to 30 min. Edzwald and Wingler [122] noted that the time required to remove 96-99% of *Chlorella vulgaris* using air flotation was 5 min. Vlaski et al. [123] used an air flotation technique to concentrate the species *Microcystis aeruginosa* to 94.5% in 8 min. Kempeneers et al. [124] achieved an 80% removal efficiency of *Melosira cyclotella* using air flotation in 3 min. Xu et al. [103] noted a 93.6% recovery efficiency for *B. braunii* using air floatation in 14 min. Coward et al. [119] noted that the air flotation technique effectively harvested microalgae within 30 min.

Energy Consumption

The flotation technique is an energy intensive process as a result of the high pressure required [108]. Algae harvest using dissolved air flotation is an efficient method but has high operational costs that are associated with the use of energy intensive compressor that function at pressures of 390 kPa [131-133]. However, the fluidic oscillator does not require much energy for operation and has been noted to consume 2-3 orders less energy than dissolved and dispersed air flotation methods [107,108].

Advantages and Disadvantages

Compared to sedimentation, flotation method is much faster and more effective for harvesting of microalgae [96]. Mohn [58] reported that dewatering microalgae using flotation methods (7% concentration) is much more rapid and efficient than the use of sedimentation (1.5% concentration). However, the flotation method has only been reported to be effective in microalgae harvesting on a bench scale and is not suitable for recovering microalgae cells on large scale [43]. In addition, this method is an energy intensive [16], has high operational costs and high energy is required for small bubble generation [58]. The cost of flotation has been reported to be as high as or even higher than the centrifugation method.

CHEMICAL HARVESTING METHODS

On the basis of energy requirement, chemical flocculation as a dewatering method seems to be the most promising for large scale utilization [16,134, 135]. These flocculation methods work by concentrating the cells (coagulation) followed by settlement to the bottom of the cultivating apparatus due to the increased density of the concentrate as shown in Figure 11 [19].

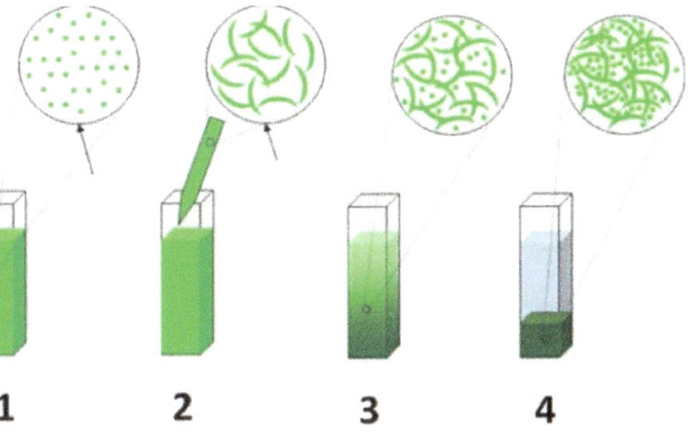

Figure 11: Flocculation harvesting process of microalgae [19].

Type of Chemical Flocculation

There are two types of chemicals (Table 1) that can induce flocculation: inorganic and organic polymers [136]. The effects of these polymers on microalgae are shown in Table 2 [11,17,24,137,139,141]. Typically, cationic, anionic and non-ionic polyelectrolytes are used to flocculate the microalgae cells [16].

Table 1: Chemical flocculent type and the effect on microalgae

Chemical Type	Flocculent	Reference
Inorganic	Ferric sulphate	Kown et al. [113]
	Ferric chloride	Papazi et al. [137]
	Aluminium chloride	Molina-Grima et al. [17]
	Aluminium sulfate	Oh et al. [136]
Organic	Magnafloc LT 25	Knuckey et al. [25]
	Zetag	t'Lam et al. [138]
	Praestol	Pushparaj et al. [139]
	Chitosan	Chen et al. [140]

Table 2: Chemical flocculent type and the effect on microalgae

Flocculent Type	Effect on Microalgae	Reference
Inorganic	Toxic to the cells	Papazi et al. [137]
	Alter the color of the medium	Schenk et al. [11]
	Alter the pH of the media which may make it unsuitable for reuse	Molina Grima et al. [17]
Organic	High cell viability (>75%)	Harith et al. [24]
	No inhibitory effect the cells	Pushparaj et al. [139]
	Non-toxic	Vandamme et al. [141]

Inorganic Compounds: Microalgae flocculation using inorganic compounds works by charge neutralization [142]. The flocculation process using these compounds works in low pH environments in order to form cationic hydrolysis products [143]. Under optimal pH, these flocculants form polyhydroxy complexes. A large number of chemicals (ferric sulphate, ferric chloride, aluminum chloride, aluminum sulfate) have been tested with the microalgae inorganic flocculation process, the most effective of which was aluminum sulfate [136]. In wastewater treatment, multi-valent metal salts such as ferric sulphate, ferric chloride and aluminium chloride have been used to remove algae [16].

Papazi et al. [137] achieved a harvesting efficacy for *Chlorella minutissma* species greater than 85% using ferric salts. Kown et al. [113] reported a flocculation efficiency of 85.6% for Tetraselmis sp. using ferric sulfate (a dose of 0.7 g/L) at a pH of 4-8. Wyatt et al. [144] reported a harvesting efficacy greater than 90% for the *Chlorella zofingiensis* species using a ferric chloride concentration of 40% (w/v) at a pH of 4.0. Xuan [145] achieved a 90% removal efficacy for Nannochloropsis sp. using ferric chloride administered at 0.18 mg/l. Sukenik et al [146] achieved a flocculation efficiency greater than 80% for marine microalgae using ferric chloride. Bintisaarani [147] found the ferric chloride flocculation to be the most effective for harvesting *Nannochloropsis* species and reported a removal efficiency of 89% using a ferric chloride concentration of 0.9 M at a pH of 7.5.

Aluminium (alum) salts have been noted to effectively flocculate the microalgae species *Chlorella* and *Scenedesmus* [17]. Papazi et al. [137] found aluminum salts to be more effective in flocculating *Chlorella* species than ferric salts. Shelef et al. [43] noted that alum was a superior flocculating agent compared to ferric sulfate in terms of pH, amount of flocculent and the quality of the final water slurry. Kown et al. [113] reported a flocculation efficiency of 92.6% for *Tetraselmis sp.* using aluminium sulfate dose of 1.2 g/L at a pH of 5-6. Millamena et al. [148] stated that alum was effective in flocculating *Chaetoceros calcitrans*, *Tetraselmis chui,*

Skeletonema costarum and *Isochrysis goibana*species at a pH of 6.5. Aragon et al. [149] used aluminium sulfate to harvest a culture made up of Scenedesmus acutus (80%) and *Chlorella vulgaris* (20%) using a dosage of 30-50 mg/L at a pH of 6-6.5.

Organic Compounds

Organic polymers (chitosan) or polyelectrolyte (polyelectolyte polyamine) flocculants are known as polymeric flocculants (synthetic and natural) that consist of both ionic and non-ionic species. The use of organic compounds for flocculation works by combining both particle bridging and charge neutralization. The charge density and polymer chain length determines the extent to which each is used. The process begins by the attachment of the polymer onto the microalgal surface through chemical or electrostatic forces.

The polymer is able to attach to the surface of the cells through Coulombic (charge-charge), dipole-dipole, van der Waals or hydrogen interactions as shown in Figure 12 [150-153]. Coulmbic force attraction works by having unlike charges on the surface of the polymer and the microalgae attach to one another, following the notion that like charges repel one another and unlike charges attract one another. Dipole-dipole interactions occur when two polar molecules approach one another and the partially negative portion bonds to the partially positive one. Van der Waals forces are the attraction of intermolecular forces between molecules. Hydrogen bonding is a type of dipole-dipole attraction in which a hydrogen atom is bonded to a highly electronegative atom nearby [154]. In this manner, the polymer attaches to the surface leaving its tail out into the solvent forming loops. The loops and tails of the polymer allow it to attach to other cells to from bridges between them [155].

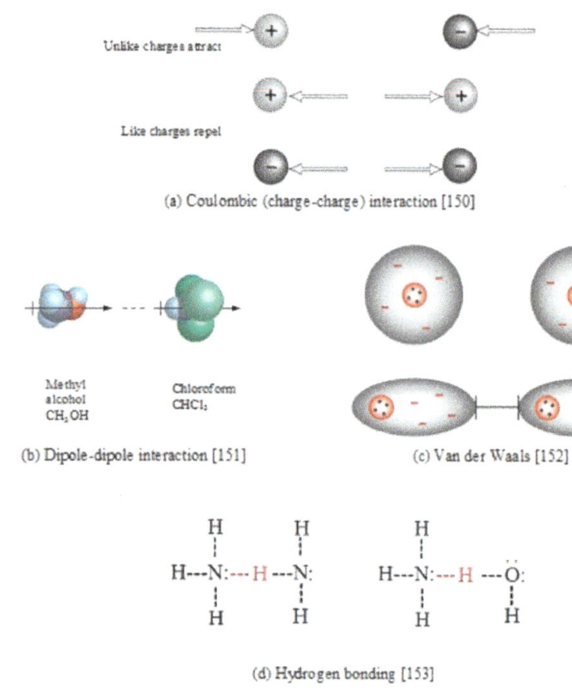

Figure 12: Methods of attachment of polymer to cell surface.

The efficiency of this flocculation process depends on the degree in which the microalgae cells cover the polymer. If the attachment of the polymer to the cell surface is less than the optimum amount, then it may not be able to withstand shear forces as a result of agitation. On the other hand, excess coverage of the polymer onto the cell surface can cause static hindering of the bridging process [156].

Recent studies have revealed that cationic polyelectrolytes flocculent agents are the most effective for microalgae recovery [16,30]. Granadoes et al. [30] noted that inorganic flocculants were less efficient in the flocculation of *Muriellopsis sp.* species than organic agents. Knuckey et al. [25] reported that adding 0.5 mg/L of non-ionic polymer Magnafloc LT25 (anionic polyacrylamide from BASF chemical company) to a medium with a pH adjusted to 10-10.6 effectively concentrated and settled a wide range of microalgae species at rates that are 200-800 times higher than the control. Harith et al. [24] maintained a high microalgae cell viability (75%) for the *Chaetoceros* species using Magnafloc LT25 flocculent at a dosage of 1 mg/L. They stated that increasing the Magnafloc LT 25 and Magnafloc LT 27 dosage did not increase the flocculation efficiency but increased the settling rates.

t'Lam et al. [138] reported a 98% flocculation efficiency for the species *P. tricorntum* using Zetag flocculent at 10 ppm, but only achieved a 52% recovery for the *N. oleoabundans* species. Udom et al. [157] found that using Zetag at a dosage of 34 mg/l flocculated microalgae and resulted in a 98% recovery efficiency. Buelna et al. [158] noted a 95-100% removal efficiency for *Chlorella culture* using 5 mg/L Zetag 63 at a pH of 6-9.

Pushparaj et al. [139] flocculated Teraselmis and Spirulina with a 70% biomass recovery efficiency using praestol (a cationic organic polyacrylamide based flocculent) with no inhibitory effects on recycled and reused media. However, inhabitation of flocculation has been noted for organic cationic polymers in environments with salinity above 5 g/L [17,25]. Sukenik et al. [146] found that the amount of flocculent required to remove 90% of microalgae from liquid suspension increased linearly with increased salinity. Danquah et al. [15] noted that the amount of energy required to achieve 15% (w/v) microalgae concentration using polymer flocculation was 14.81 kWh/m^3.

Chen et al. [140] reported that the general dosage range of chitosan required to effectively flocculate microalgae species was 5-200 mg/L. Xu et al. [159] noted a 99% clarification efficiency for *Chlorella sorokiniana* using chitosan at an optimal dosage of 10 mg/g dried microalgae and pH values below 7. Ahmed et al. [160] achieved a 99% flocculation efficiency in 20 minutes for *Chlorella sp.* with a chitosan dosage of 10 ppm. Chang and Lee [161] reported a flocculation efficiency of 99% for *Chlorella vulgaris* using chitosan at a dosage of 200 mg/L and a pH of 8.7. Sirin et al. [162] reported a flocculation efficiency of 92% in 10 min for the *Phaeodactylum tricornutum* species using chitosan (20 mg/L). Morales et al. [163] noted a 100 % flocculation efficiency for *Chlorella sp.* using chitosan at a concentration of 40 mg/L. Beach et al. [164] compared the chitosan flocculation, centrifugation and filtration methods for microalgae harvesting and noted that chitosan flocculation was the least energy consuming method of the three.

Factors Affecting Chemical Flocculation

Inorganic Flocculation

The factors affecting inorganics flocculation include: concentration of the flocculent, pH and the surface charge of the flocculent.

Flocculent concentration: The concentration at which the flocculent is administered into the system has been noted to affect the efficiency of the microalgae recovery. Rakesh et al. [165] used aluminium sulphate concentrations ranging from 50 to 300 mg/L for the recovery of *Chlorella sp.*, *Chlorococcum sp.* and *Chlorella sorokiniana* and found 50 mg/L to be the most effective dose. Garzon-Sanabria et al. [166] evaluated the recovery efficiency of N. oculata using aluminum chloride at concentrations in the range of 50-100 mg/L and found 50 mg/L to be the most effective dose. Ferriols and Aguilar [167] reported on the use of calcium chloride and sodium hydroxide at concentrations of 100-200 mg/L for the recovery of *Tetraselmis terrahele* and achieved the highest recovery efficiency at 200 mg/L. Wyatt et al. [144] noted that in media with low algae concentrations, the concentration of ferric chloride required to flocculate *Chlorella zofingiensis* increases linearly with cell concentration.

pH: Varying the pH of the medium using inorganic flocculants can promote cell aggregation. Knuckey et al. [25] noted effective flocculation (>80%) of *Chaetoceros calcitrans, Chlorella muelleri, Thalassiosira pseudonana, Attheya septentrionalis, Nitzschia closterium, Skeletonema sp., Tetraselmis suecica* and *Rhodomonas salina* by altering the pH with the addition of sodium hydroxide. Garzon-Sanabria et al. [166] used aluminum chloride (50-100 mg/L) to modify the pH (4-7) and achieved the highest recovery efficiency of N. oculata using a dosage of 50 mg/L (pH =5.3). Lee et al. [168] noted that changing sodium hydroxide concentration affected the flocculation efficiency of *Botryococcus braunii* as a result of pH change in the medium. Sanyano et al. [169] successfully flocculated *Chlorella sp.* using ferric chloride at a pH of 8.1.

Surface charge: Microalgae surface cells are negatively charged, indicating that a positively charged flocculent would be required to bond the cells to one another [144,169]. Algal coagulation is induced by the attraction of the positively charged flocculent onto the negatively charged cell surface and the attachment of another algal cell onto the positively charged flocculent [161]. The efficiency of the flocculation is depended on the amount of flocculent available to bridge the algae to one another [144,170]. Wyatt et al. [144] noted that the positive nature of ferric chloride induced microalgae flocculation with a recovery efficiencies of 90% at a pH above 4.1 and below 8. Knuckey et al. [25] flocculated microalgae with an efficiency of 80% using Fe^{3+} ions. Garzon-Sanabria et al. [166] recovered *Nannocloris oculata* with a 90% efficiency using aluminium chloride to counteract the surface charge of the microalgae cells at a pH of 5.3. Sanyano et al. [169] successfully flocculated*Chlorella sp.* using ferric chloride. Lee et al. [171] achieved 100% flocculation efficiency in *Chlorella sp.* using synthesized cationic aluminum and magnesium organoclays.

Organic Flocculation

The factors affecting organic flocculation include: pH, charge on polymer, dosage and salinity.

pH: Some microalgae species can flocculate together by adjusting the pH [17]. Uduman et al. [16] stated that the pH and the chemical composition of the microalgal medium impact the amount of flocculent required. They noticed less electrostatic repulsion between colloids at low pH levels resulting in increased amounts of bridging since the polymer chains are longer. They also found the dose of polymeric flocculent required to vary with microalgae concentration, because of the charge in surface area of algae. Knuckey et al. [25] reported that the non-ionic polymer Magnafloc LT25 settled a wide range of microalgae species effectively using a rate of 0.5 mg/L in a pH adjusted media to 10- 10.6. Tenney et al. [172] noted the most effective flocculation resulted when using cationic polyelectrolytes at low pH levels. Ras et al. [47] noted that the *Chlorella* species flocculated when the pH was increased to 11-12. Lee et al. [173] stated that extreme pH levels can result in cell death or impairment.

Charge on polymer: The polyelectrolyte charge plays an important role in the flocculation process of microalgae. Anionic polyelectrolytes are not effective flocculent agents on their own due to the negatively charged microalgae cell surface because like charges repel one another and/or the length of the polymer is not sufficient enough to bridge the particles together [142,172]. It is for this reason that cationic polyelectrolytes are found to be much more effective in the flocculation of microalgae. Morrissey et al. [174] noted that the N,Ndimethylaminopropyl acrylamid polymer (positive character) resulted in recovery efficiencies of microalgae greater than 99% at a pH of 7 and increasing the pH to 13 (activating negative functionality) resulted in flocculation efficiencies of less than 12%. Chang et al. [161] noted that the positively charged surface of chitosan resulted in a 99% removal efficiency for *Chlorella vulgaris*.

Flocculent concentration: The amount of cationic flocculent required for effective bridging between the cells depend on the amount of negative charges present in the system, the surface charge density, the cell counts per volume, the total cell surface area and the charge density of the positively charged polyelectrolyte. The negative charge on the surface is induced by the functional groups (carboxyl groups) present on the microalga cells which have been noted to affect the isoelectric point of the cells [142]. Uduman et al. [16] reported that the growth phase and the metabolic conditions of the microalgal cells dictate the concentration and the reactivity of these functional groups. Granados et al. [30] showed that the species *Chlorella* and *Scenedesmus* were effectively flocculated to a concentration of 2 g/L after 15 min using polyelectrolyte dosages of 2-25 mg/g. Tenney et al. [172] stated that the cationic polyelectrolyte polyamine was effective in flocculating the microalgae cells at a dosage of 2.5 mg/l. Sukenik et al. [146] reported that marine microalgae Isochrysis *galbana* and *Chlorella stigmatophora* require 5-10 times more flocculent dosages than those required by freshwater microalgae.

Salinity: The salinity level affects the organic flocculation of microalgae. Bilanovic and Shelef [175] noted that the polyelectrolyte flocculent was inhibited in the marine medium due to its high salinity and observed effective flocculation at salinity levels below 5 g/l. This was attributed to the fact that high ionic strength causes the polymer to shrink in dimension, thus failing to form a bridge to link the microalgal cells. Schlesinger et al. [116] reported that the addition of alkali to *Chlamydomonas* did not result in rapid flocculation in the saline medium.

Advantages and Disadvantages

In comparison with other methods, chemical flocculation is considered to be one of the best methods for cell harvesting because it can handle large amounts of microalgae, it can be used with a wide range of species, it is reliable and it is cost-effective [16,176].

The costs of inorganic flocculants are much less than those of organic ones [138,177,178]. However, the higher amounts required using inorganic flocculants can result in higher costs per unit of microalgae than the more expensive organic flocculants [58]. Sukenik et al. [146] reported that the optimal dosage of inorganic flocculent required to flocculate marine microalgae was 5-10 times higher than that required to flocculate freshwater microalgae. Shammas [177] also noted that the higher cost of organic coagulants can be offset by the low dosages required compared to those of inorganic flocculants.

Microalgae harvesting techniques using chemical flocculation are not environmentally friendly because they introduce chemicals into the system which increase the dissolved solids and change color [179]. Inorganic flocculants can be toxic and can also have negative effects on microalgae by modifying their growth media and changing their color which prevents the reuse and recycling of water [11,17,137]. Hee-Mock et al. [180]) and Vandamme et al. [181] stated that chemical flocculants that are toxic and carcinogenic and are not suitable for harvesting microalgae biomass that is being processed for food supplements and food additives. Therefore, selection of flocculent should be based on cost, toxicity and reusability of the media [17].

Auto-flocculation Harvesting Methods

Some microalgae species can flocculate spontaneously, a process known, in a response to certain environmental stresses. This phenomenon is known as autoflocculation.

Types of Environmental Stress

There are several factors that affect the efficiency of autoflocculation, which include: pH, dissolved oxygen content, nitrogen concentration and the amount of calcium and magnesium ions in solution.

pH

When the pH of the medium is increased the cells come together and settle by gravitational force. The addition of more bases into the medium increased the formation of dense flocs which result in less settling times. However, it is important to note that not all species flocculate with increased pH levels [17,182].

Harith et al. [24] noted that at pH values less than 10 only slight separations between the microalgae and liquid media occurred after 4 h and increasing the pH from 8 to 10 using NaOH and KOH increased the flocculation efficiency from 13 to 82% and from 35 to 78% in 4 h, respectively. Wu et al. [182] noted that a pH of 10.5 resulted in 90% flocculation efficiency for the freshwater species Chlorella vulgaris,*Scenedesmus sp.* and *Chlorococcum sp.* and a pH of 9.0-9.3 resulted in 90% flocculation efficiency for the marine species *Nannochloropsis*sp. and *Phaeodactylum tricornutum.* Horiuchi et al. [183] noted a 96% flocculation efficiency in the marine species *Dunaliella tertiolecta* when the pH was adjusted to 8.6. Millamena et al. [148] also noted effective flocculation of microalgae when the pH was maintained above 10 in salt water.

Dissolved Oxygen

Uduman et al. [16] noted that flocculation in some microalgae species can occur naturally with changes in dissolved oxygen concentration. Schenk et al. [11] reported that dissolved oxygen stress can result in microalgae flocculation. Liao et al. [184] reported that increased dissolved oxygen in solution triggers autoflocculation by increasing the binding sites available on the cell surface. Greater binding sites result in bulk formation of the cells which increases the weight of the flocs and increases the settling rate). They also noted that increased photosynthetic activity by the microorganisms increases the dissolved oxygen content and the formation of dense flocs. Wilen and Balmer [185] noted that large flocs can be generated when the dissolved oxygen concentration is high in the media. Koopman et al. [186] noted that dissolved oxygen concentrations of 14-16 mg/l promoted flocculation in the system.

Nitrogen

Auto-flocculation in microalgae cells may be triggered naturally as a result of environmental stress caused by nitrogen concentration [11.16]. Sukenik and Shelef [187] noted that certain species of microalgae flocculate with one another as a result of nitrogen stress in the media. Becker [188] noted that microalgae cells can aggregate with one another as a result of nitrate assimilation. Assimilation of nitrate nitrogen increases the pH of the medium and promotes cell flocculation [182,189]. Nurdogen and Oswald [190] also noted that nitrate assimilation resulted in auto-flocculation in microalgae species.

Addition of Ca^{2+} and Mg^{2+}

Autoflocculation occurs in the culture media spontaneously as a result of coprecipitation of calcium and magnesium salts present in the media which results in change of the pH of the medium [43]. Smith and Davis [192] evaluated Mg^{2+}, Ca^{2+} and CO_3 2- ions for their effectiveness in flocculating and settling of microalgae cells and found that Mg^{2+} ion with high pH levels resulted in effective flocculation and rapid sedimentation. They achieved settling rates that were 100- fold higher than those achieved with sedimentation. The reason for this phenomenon is that magnesium hydroxide flocs are positively charged while calcium carbonate flocs are negatively charged [193]. Thus, destabilization of the negatively charged microalgae cells is greater using magnesium as opposed to calcium. The optimal pH for autoflocculation is strain specific [187]. Nguyen et al. [191] reported that the species *Chlorella vulgaris* autoflocculated with and efficiency of 90% by addition of $Ca^{2+}$and Mg^{2+} at concentrations of 120 mg/l and 1000 mg/l, respectively. Vandamme et al. [194] noted that addition of Mg^{2+} in *Chlorella vulgaris* culture induced autoflocculation.

Advantages and Disadvantages

The advantages of this harvesting technique are the simplicity and low costs. The process can be reverted by pH adjustment using HCl to decrease the pH back to 7.5-8 [25]. However, autoflocculation does not occur in all species making it an unreliable process [39].

Using pH flocculation is beneficial since pH induced flocculated cells were identical to non-flocculated microalgae cells. This means that this autoflocculation technique has low-shear force on the cells when compared to centrifugation [195]. Knuckey et al. [25] found that the chlorophyll a from T. pseudonana cultures were intact using pH-induced methods, but centrifuged microalgae cells only has slight chlorophyll a peaks. This is necessary when the microalgae biomass is required for use as feed diets or for use of extraction of certain compounds (such as chlorophyll a) from the cell. They also noted that the harvested microalgae species (used for oyster feed) using pH induced flocculation were a better diet choice compared to those harvested by centrifugation and much better than those harvested using ferric chloride flocculation processes. D'Souza et al. [196] also reported that pH induced harvesting of C. muelleri for feed to tiger prawn *P. mondon* were only slightly slower in developmental rate compared to those using fresh *C. muelleri.*

Bio-Flocculation Harvesting Methods

The use of microorganisms for the recovery of microalgae biomass has been investigated (Table 3) [11,19,114,136,173,197,198]. This method works by the addition of microorganisms to the culture which adhere to the microalgae cells causing the weight to increase and resulting in settlement of the cells to the bottom of the vessel. The supernatant containing the culture medium is decanted and washed with water in order to reduce the salinity [145,199].

Table 3: Bio-flocculation of microalgae by use of fungi and bacteria microorganisms

Microorganism	Bio-Flocculated Microalgae	Reference
Bacteria		
Bacillus licheniformis	*Desmodesmus sp.*	Ndikubwimana et al. [197]
P. stutzeri & B.cereus	*Pleurochrysis carterae*	Lee et al. [173]
Paenibacillus sp.	*Chlorella vulgaris*	Oh et al. [136]
Paenibacillus polymixa	*Scenedesmus* sp.	Kim et al. [198]
Bacillus subtilis	*Chlorella vulgaris*	Zheng et al. [114]
Fungi		
Ankistrodesmus falcatus	*Chlorella vulgaris*	Salim et al. [19]
Scenedesmus obliquus	*Chlorella vulgaris*	Salim et al. [19]
Tetraselmis suecica	*Nannochloropsis oleabundans*	Salim et al. [19]
Skeletonema	*Nannochloropsis*	Schenk et al. [11]

Molina Grima et al. [17] noted the effective flocculation of *Chlorella* using bio-flocculent from bacteria species. Oh et al. [136] successfully harvested *Chlorella vulgaris* using the bio-flocculent bacterium Paenibacillus sp. Kim et al. [200] noted effective flocculation of the species Scenedesmus sp. using the bio-flocculent *Paenibacillus polymixa*. Ndikubwimana et al. [197] harvested the microalgae species Desmodesmus sp. using the bacterium *Bacillus licheniformis* with a 98% removal efficiency. Zhang and Hu [201] co-cultured the species *Chlorella vulgaris* with different filamentous fungi and extracted the microbial oil for transesterification into biodiesel.

Factors Affecting Bio-flocculation

The factors affecting bio-flocculation include: concentration of the bio-flocculent, pH and the selectivity of the microorganism.

Bio-flocculent Concentration

The rate at which bio-flocculation is achieved depends on the ratio of the bio-flocculent to the non-flocculating microalgae species. Bioflocculent concentrations that are greater than the concentration of non-flocculating microalgae increase the rate of sedimentation [19]. Lee et al. [173] found that the addition of bacteria to the non-flocculating microalgae culture increased the rate of sedimentation. Salim et al. [19] successfully harvested non flocculating microalgae by addition of bioflocculating species and noted that the addition of bioflocculating microalgae induced the microalgae sedimentation and increased the efficiency of harvesting. Oh et al. [137] reported that the flocculation efficiency of *C. vulgaris* using the bacterium *Paenibasillus sp.* decreased with increasing dilutions of the bacterium. Zheng et al. [114] reported that the flocculation efficiency of *C. vulgaris* using the bio-flocculent B. subtilis increased with increasing concentrations of C. vulgaris biomass. Lee et al. [168] noted that *C. vulgaris* flocculation efficiency increased with increasing bacteria (*Flavobacterium, Terrimonas and Sphingobacterium*) concentration in the culture.

pH

The efficiency of bio-flocculation was noted to be affected by the pH of the medium. The pH alters the surface charge of the molecules in the medium which dictate the degree of attraction/repulsion. Oh et al. [136] reported that the flocculation efficiency of C. vulgaris using the bacterium *Paenibasillus sp.* increased with increasing pH from 5 to 11. Ndikubwimana et al. [197] noted that the flocculation efficiency of*Desmodesmus sp.* increased from 43 to 98% using the bacterium *Bacillus licheniformis* as the pH decreased from 7.2 to 3. Lee et al. [168] noted that pure *C. vulgaris* cultures showed no flocculation as the pH was increased from 3-11, but cultures with bacteria demonstrated increased flocculation efficiencies (from 43 to 94%) with increases in the pH over the range of 3-11. However, Zheng et al. [114] noted that the bio-flocculation efficiency using *B. subtilis* with microalgae species *Chlorella vulgaris* and *Chlorella protothecoides* were noted effected by pH.

Species Selectivity

the bacterium *Paenibasillus* resulted in a flocculating efficiency of 83% and 95% with Chlorella vulgaris and *Scenedesmus sp.*, respectively.

Advantages and Disadvantages

The advantages of using bio-flocculants include their biodegradability, non-toxic nature and the intermediates formed during degradation are not secondary pollutants [203]. Salim et al. [19] noted that a two step harvesting process using naturally flocculating microalgae to induce non-flocculating microalgae followed by centrifugation reduced the energy use from 13.8 MJ/kg (dry weight) to less than 2 MJ/kg (dry weight).

The disadvantages of this technique are that the microorganisms used to flocculate the algae are species-specific, and the recycling and recovery of these organisms can be difficult [17,204]. Oh et al. [136] stated that the bio-flocculants used to dewater the microalgae cells should be tested for acute oral toxicity in order for the retrieved biomass to be used in food additives and feed supplement. Vandamme et al. [181] indicated that the use of fungi or bacteria as flocculating agents results in microbiological contamination of the microalgae biomass, which needs to be assessed before use in feed or food products.

Electrophoresis Harvesting Methods

The electrolytic methods is used to eliminate the use of costly and toxic chemicals since microalgae behave much the same as colloid particles which allows for their separation from a water based medium by electric flied [16].

Types of Electrophoresis

The main electrophoresis methods that can be used for harvesting microalgae are: electrolytic coagulation, electrolytic flocculation and electrolytic flotation.

Electrolytic Coagulation

This type of electrophoresis method requires the use of both physical and chemical stimuli for the effective separation of microalgae biomass. The coagulation process is induced by the generation of current from an iron or aluminum electrode as shown in Figure 13 [205]. The amount of electrical current passing through the water medium dictates the amount of metal ions dissolved into the liquid suspension [206]. The metal ions released into the solution are metal hydroxides that contribute to the destabilization of colloid suspension and coagulate the biomass. Flocculation is achieved by the linking of the positively charged metal to the negatively charged microalgae cell and the movement toward the anode as a result of electrophoretic motion [179,206]. Rapid coagulation results from high current densities but the cost associated with the method is high.

Uduman et al. [207] used an aluminium electrode set at 5 V and achieved electrocoagulation efficiencies of 93.3% and 87.3% in 600 s for the species Tetraselmis sp. and *Chloroccum sp.*, respectively. Azarian et al. [208] recovered 99.5% of total suspended solids in 15 minutes by electrocoagulation using a power source of 550 W with aluminium anode. They also noted that a power supply of 100 W required 30 minutes to achieve similar results. Ghernaout et al. [209] achieved a 99% removal of *Escherichia coli* in 20 min by electrocoagulation using aluminium electrode with a power supply of 12 W.

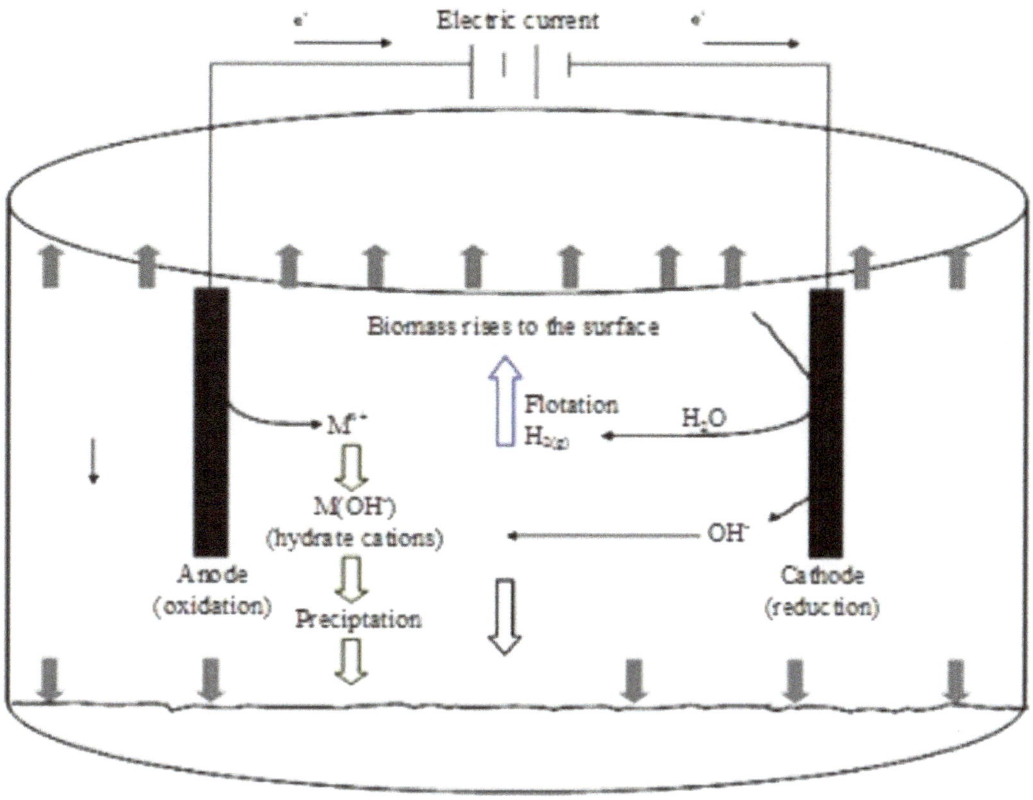

Figure 13: Dewatering microalgae through electrolytic coagulation [205].

Electrolytic Flocculation

Electrolytic flocculation works by movement of negatively charged cells toward the anode as shown in Figure 14 [16]. At the anode, the cells lose their charge forming flocs that can be lifted to the surface by adhering to the bubbles formed by the electrolysis of the water [210].

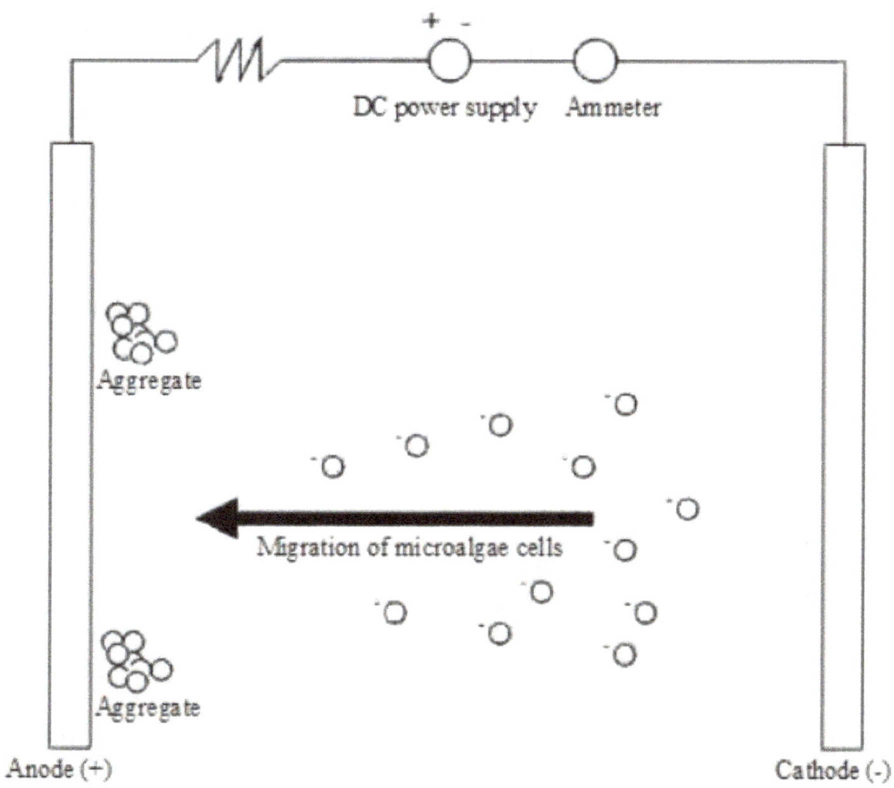

Figure 14: Dewatering microalgae through electrolytic flocculation [16].

Poelman et al. [210] tested the effectiveness of electrolytic flocculation in a 100 L vessel equipped with vertical electrodes and noted a removal efficiency of 80-95% in 35 min. They also noted that the rate of microalgae removal decreased with decreasing voltages and less energy was consumed when the total surface area of the electrodes was decreased and/or the distance between the electrodes was decreased.

Xu et al. [103] achieved a 93.6% recovery efficacy of *Botryococcus branii* after 30 minutes using electrolytic flocculation technique with a power supply of 6 W. Lee et al [168] noted a marine microalgae recoveries of 85% and 95% after 60 minutes using electrolytic flocculation with a power supply of 5 V and 5.2 V, respectively. Zenouzi et al. [211] reported a 97.4% removal efficiency of *Dunaliella salina* after 3 min using electrolytic flocculation.

Electrolytic Flotation

This technique is similar to electrolytic coagulation in that active metal anodes are used to flocculate the microalgae cells. The difference between the two techniques is that the cathode in electrolytic flotation is made from an inactive metal (steel) that is electrochemically nondepositing as shown in Figure 15 [212]. The inactive metal forms hydrogen bubbles from the electrolysis of the water. The particulates in the suspension attach to the gaseous bubbles and are lifted to the surface of the vessel [208,213].

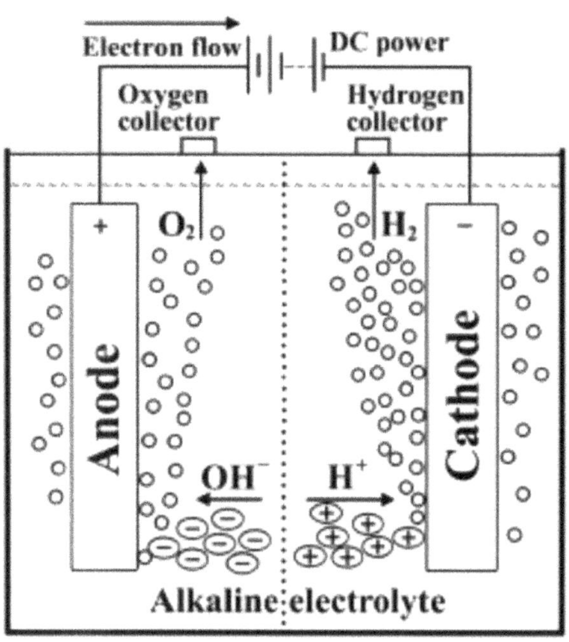

Figure 15: Dewatering microalgae through electrolytic flotation [212].

Alfafara et al. [213] investigated the use of electrolytic floatation of microalgae cells using a polyvalent aluminium anode and titanium alloy cathode and found that increasing the electrical power decreased the electrolysis time and increased the rate of chlorophyll a removal. They also noted that the amount of chlorophyll measured was related to the concentration of microalgae removed by electrolysis. The usage of high electrical power is limited by the increase in heat and the increase in pH.

De Carvalho Neto et al. [214] reported a chlorophyll a removal efficiency of 99% using an electroflotation method running for 140 min with a power supply of 60 W. Ghernaout et al. [215] reported a microalgae removal efficiency of approximately 100% after 140 minutes in Ghrib Dam water using electroflotation method. Shelef et al. [43] noted that electroflotation technique resulted in total suspended solids in the range of 3-5%. Brennan and Owende [48] noted that there is little proof of the economic feasibility of this recovery method.

Energy Consumption

Dewatering microalgae by electrophoresis techniques requires less energy (0.2-2.1 Wh/g) compared to other harvesting methods (0.18-35.62 Wh/g). Vandamme et al. [141] reported that the amount of energy required to flocculate the freshwater microalgae *C. vulgaris* and the marine P. tricornutum species was 2.1 Wh/g and 0.2 Wh/g (dry weight), respectively. Gonzalez-Fernandez and Ballesteros [45] noted that the marine microalgae required less energy for harvesting, because the marine medium allows for a higher conductivity that favors the electrocoagulation process. Kim et al. [200] reported that the electrical energy consumption of polarity exchange using two types of electrodes ranged from 1.19 to 1.23 Wh/g for 99% harvesting recovery of microalgae after 15 min. Bektas et al. [216] noted that dewatering of microalgae using electrocoagulation by 0.8-1.5 Wh/g of microalgae culture whereas cross flow filtration, pressure filters, vacuum filters, centrifugation and flocculation using polymers have consume 3.47, 0.18, 1.19, 1.67 and 35.62 Wh/g, respectively [15,17].

Advantages and Disadvantages

The advantages of the electrophoresis harvesting technique include: versatility, energy efficiency, safety, selectivity, environmental compatibility and cost effectiveness [206]. Minimum energy is consumed when using optimum potential difference (0.331 kWh/m^3) by controlling the electrode surface area and distance between the electrodes. There are no added costs associated with flocculent products [16]. The costs associated with dewatering microalgae via electrolytic methods were significantly less than other harvesting methods such as sedimentation with flocculants, centrifugation and flotation with flocculants [210]. This indicates that although higher electrical energy is consumed using electrolytic methods, the cost to harvest is much lower than other harvesting techniques.

Some of the drawbacks associated with this harvesting technique include: cathode fouling and change in cell composition [210]. The current intensity decreases by 5-10% upon reuse of the cathode due to internal resistance [141] and changes in cell composition can be induced using high current densities [45].

COMBINATION OF METHODS

Cost effective methods for cell harvesting are vital for the economics of biodiesel production. Harvesting processes account for 20-30 % of the total biomass production costs [17]. Selection of harvesting technique is dependent on the size and density of the microalgae cells, conditions of the culture and concentration of biomass and target product value [13].

Schenk et al. [11] reviewed various harvesting methods and noted that the combination of flocculation with sedimentation or flotation with filtration or flotation with centrifugation to be the most economical alternatives. Brennan and Owende [48] and Uduman et al. [16] reported that energy can be conserved and cost can be reduced by harvesting the microalgae in a two-step process using two techniques. Initially, the microalgae are concentrated to 2-7% total suspended solids by the process of flocculation and the cells are further concentrated into a paste (suspended solid concentration of 15-25%) by a secondary harvesting step such as filtration or electrophoresis. Funk et al. [217] integrated dissolved air flotation with chemical flocculation (ferric sulfate) and noted increased recovery efficiency from 88% to 95% for *Chlorella vulgaris.*

Kim et al. [200] reported that a new and innovative technique for improving the economics of the electrophoresis processes is the combination of electrolytic coagulation and electrolytic flotation into continuous electrolytic microalgae. In this method, the current direction (polarity exchanges) is exchanged for the continuous harvest of microalgae and their cultivation. The current direction creates two phases by using a pair of electrodes. The first phase works to destabilize the negatively charged microalgae cells forming flocs. The formation of flocs is mediated by metal ions that are released from the electrode dissolving in solution. In the second phase, the metal ion generation is halted and the bubbles formed from both electrodes lift the flocs to the top of the solution causing them to float.

Xu et al. [131] used electroflocculation integrated with dispersed air flotation and noted a harvesting efficiency of 98.9% in 14 min. They noted that the cell aggregate increased with the integrated system as opposed to those observed with electroflocculation. The use of dispersed air flotation increased the rate of aggregate formation. However, the stress from continued air supplementation into the system disturbed the up-floated flocs into algal aggregates. Thus disturbance was avoided by halting the supplementation of air into the system once the aggregate size reached its peak value.

COMPARATIVE ANALYSIS

Selection of Criteria

Eight criteria (Table 4) were used for the evaluation of microalgae harvesting techniques: (a) dewatering efficiency, (b) cost, (c) toxicity (d) suitability for large scale use, (e) time, (f) species specificity, (g) reusability of media and (h) maintenance. These criteria were selected based on the information reported in the literature about these microalgae harvesting methods. The comparative analysis was performed using these criteria to determine the most efficient, cost effective and environmentally friendly dewatering technique for a wide array of microalgae species that is suitable for large scale application.

Assigning Score to Each Criterion

Each of the selected criteria was assigned a score from 7 to 15 which was determined by the degree of importance of the criterion (Table 4). Higher values were given to the criteria that were deemed most important for development of an efficient and economic large scale dewatering method for microalgae. Lower values were given to criteria that were deemed necessary for determining a suitable method but were considered less important. These values were then used to determine the effectiveness of each harvesting method as shown in Tables 5-20.

Table 4: Criteria used for the comparative analysis of different harvesting techniques

Criteria	Importance	Description
Dewatering Efficiency	15	The system should be able to effectively concentrate and remove high percentage of the cells from their surrounding liquid media
Cost	15	The operational costs of the process should be low in order to reduce the total processing costs associated with microalgae recovery
Toxicity and health and environmental impact	15	The method should be non-toxic so that the retrieved algae biomass maybe processed for a number of value added products including ones for human consumption It should also be environmentally friendly in order to reduce the amount of toxic wastes produced
Suitability for Large Scale Use	15	The method should effective in handling large volumes for industrial production
Time	15	The rate of harvest should be quick to ensure the sustainability purposes
Species Specificity	10	The method should not be species or strain specific
Reusability of Media	8	The media should be recycled for reuse in order to minimize costs
Maintenance	7	Costs for maintaining the method should be low

Table 5: Evaluation of sedimentation

Criteria	Description	Score
Dewatering Efficiency (15)	Settlement is based on density and since microalgae density is similar to that of water media, efficiency is low without the use of flocculants	5
Cost (15)	Minimum energy costs are required for this technique as gravitational forces are cost free	15
Toxicity and health and environmental impact (15)	This method is nontoxic to the cells, since it works by gravitational forces	15
Suitability for Large Scale (15)	Unsuitable for large scale use because of the long periods required for the process and it only works for microalgae cells with higher densities	5
Time (15)	Long periods of time are required to achieve settlement of microalgae cells through gravitational forces	2
Species Specificity (10)	Highly dependent on the type of species used. Species should have a higher density than that of water	4
Reusability of Media (8)	Method does not introduce any chemicals or alter the composition of the species/media	8
Maintenance (7)	No maintenance costs are required	7
Total (100)		**61**

Table 6: Evaluation of vacuum filtration

Criteria	Description	Score
Dewatering Efficiency (15)	Effective recovery of microalgae cells Depends on filter size and the size of microalgae cells	13
Cost (15)	Costs associated with pump and replacement of filters	9
Toxicity and health and environmental impact (15)	Cell composition remains intact and toxic chemicals are not required	15
Suitability for Large Scale (15)	Large pump and large filters are required for large scale	10
Time (15)	Rapid cell recovery	12
Species Specificity (10)	Dependent on the microalgae cell size	6
Reusability of Media (8)	Liquid media can be recycled	8
Maintenance (7)	Frequent filter replacement as a result of clogging	2
Total (100)		**75**

Table 7: Evaluation of pressure filtration

Criteria	Description	Score
Dewatering Efficiency (15)	Effective in dewatering the microalgae. Suspended solids in the filtrate are low	13
Cost (15)	Costs associated with pump to create pressure and with filter replacements	9
Toxicity and health and environmental impact (15)	This method is non-toxic and cell composition is not altered	15
Suitability for Large Scale (15)	Suitable for large volumes but requires large filters and large pump	10
Time (15)	Relatively rapid cell recovery	12
Species Specificity (10)	Dependent on the size of the species	5
Reusability of Media (8)	Filtrate can be recycled and reused again for microalgae growth	8
Maintenance (7)	Costs associated with filter replacement	2
Total (100)		**74**

Table 8: Evaluation of cross flow filtration

Criteria	Description	Score
Dewatering Efficiency (15)	Complete removal of microalgae cells from the media	15
Cost (15)	Costs associated with pump and membrane	12
Toxicity and health and environmental impact (15)	This method is non-toxic	15
Suitability for Large Scale (15)	Suitable for large volumes of microalgae Smaller microalgae result in membrane clogging	10
Time (15)	Rapid cell recovery	12
Species Specificity (10)	Wide range of cell sizes can be used	8
Reusability of Media (8)	Filtrate can be recycled and reused again for microalgae growth	8
Maintenance (7)	Costs associated with filter replacement	4
Total (100)		**84**

Table 9: Evaluation of disc stack centrifuge

Criteria	Description	Score
Dewatering Efficiency (15)	Effective separation of solid particles from liquid suspensions	13
Cost (15)	Large amounts of energy are required for operation	8
Toxicity and health and environmental impact (15)	The use of toxic materials is not required Cell composition is not altered	15
Suitability for Large Scale (15)	Suitable for large volumes of microalgae	12
Time (15)	Rapid	15
Species Specificity (10)	No dependence on the type of species	10
Reusability of Media (8)	Supernatant can be easily recovered and recycled	8
Maintenance (7)	Not much maintenance is required	6
Total (100)		**87**

Table 10: Evaluation of decanter centrifuge

Criteria	Description	Score
Dewatering Efficiency (15)	Effective separation of solid particles from liquid suspensions Solid concentrates are much more dense than those recovered using disc type	15
Cost (15)	More energy is required for operation compared to disc type	6
Toxicity and health and environmental impact (15)	The use of toxic materials is not required Cell composition is not altered	15
Suitability for Large Scale (15)	Suitable for large volumes of microalgae	12
Time (15)	Rapid	15
Species Specificity (10)	Suitable for larger species only	5
Reusability of Media (8)	Supernatant can be easily recovered and recycled	8
Maintenance (7)	Not much maintenance is required	6
Total (100)		**82**

Table 11: Evaluation of dispersed air flotation

Criteria	Description	Score
Dewatering Efficiency (15)	Effectiveness depends on the likelihood of the cells coming into contact with the air bubbles in order to float to the surface Cell may rupture	12
Cost (15)	High costs are required for high speed agitation in order to produce the bubbles. Additional surfactants increase the costs	10
Toxicity and health and environmental impact (15)	The use of toxic materials is not required	12
Suitability for Large Scale (15)	Large volumes of microalgae can be used	13
Time (15)	The time required is dependent on the rate of agitation	10
Species Specificity (10)	Species should have high tolerance to avoid rupturing	5
Reusability of Media (8)	Media may be recycled for further use	8
Maintenance (7)	Not much maintenance is required	7
Total (100)		**77**

Table 12: Evaluation of dissolved air flotation

Criteria	Description	Score
Dewatering Efficiency (15)	Effective with the use of additional flocculants	10
Cost (15)	Operational costs are high, large amounts of energy would be required and the cost of flocculants for effective recovery is also high	9
Toxicity and health and environmental impact (15)	Inorganic flocculants are toxic	8
Suitability for Large Scale (15)	Large volumes of microalgae can be harvested	13
Time (15)	Dependent on the likelihood of the cells interacting with air bubble	10
Species Specificity (10)	Wide range of species can be used but species ability to adhere onto gas bubble is key	8
Reusability of Media (8)	Chemicals are not used and the medium can be recycled after it is saturated with air	5
Maintenance (7)	Not much maintenance is required	7
Total (100)		**70**

Table 13: Evaluation of fluidic oscillation

Criteria	Description	Score
Dewatering Efficiency (15)	Effective with the use of additional flocculants	9
Cost (15)	Large amounts of energy would be required, but operational costs are much lower than those of dispersed and dissolved air techniques	10
Toxicity and health and environmental impact (15)	Coagulants are required for improving recovery effectiveness	10
Suitability for Large Scale (15)	Large volumes of microalgae can be harvested,	13
Time (15)	Dependent on the likelihood of the cells interacting the with air bubble	10
Species Specificity (10)	Can be used on a wide range of species but species ability to adhere onto gas bubble is key	8
Reusability of Media (8)	Recycled after it is saturated with air	6
Maintenance (7)	Not much maintenance is required	7
Total (100)		**73**

Table 14: Evaluation of inorganic flocculation

Criteria	Description	Score
Dewatering Efficiency (15)	Cell concentration in liquid is low and depends on the position of the flocculent on the cell	10
Cost (15)	Large amounts of flocculants are required Does not require high amounts of energy	11
Toxicity and health and environmental impact (15)	Flocculating agents are toxic and not suitable for food additive and pharmaceutical products	5
Suitability for Large Scale (15)	Large volumes of microalgae suspensions can be used	15
Time (15)	Relatively fast	10
Species Specificity (10)	Wide range of species can be used but the process is dependent on the type of species used and how well the flocculent attaches to the cells	5
Reusability of Media (8)	The pH of the media left after harvest of microalgae is low which is not suitable for some microalgae species	2
Maintenance (7)	No maintenance required	7
Total (100)		**65**

Table 15: Evaluation of organic flocculation

Criteria	Description	Score
Dewatering Efficiency (15)	Effectiveness is dependent on the position of the flocculent on the cell and the cell surface charge	11
Cost (15)	Expenses are associated with cost of flocculent Less amounts are required when compared to inorganic agents	11
Toxicity and health and environmental impact (15)	Organic compounds are non-toxic and can be used in the formation of addible and cosmetic by-products	15
Suitability for Large Scale (15)	Large volumes of microalgae suspensions can be used	15
Time (15)	Relatively fast	10
Species Specificity (10)	Process is dependent on the type of species used and how well the flocculent agent attaches to the cells	6
Reusability of Media (8)	Organic agents are non-toxic and the media can be recycled Changes in pH of media occur with flocculent addition and can affect the microalgae species	5
Maintenance (7)	No maintenance required	7
Total (100)		**80**

Table 16: Evaluation of auto-flocculation

Criteria	Description	Score
Dewatering Efficiency (15)	Flocculation is induced by pH change and separation depends on the response of the species to the environment	11
Cost (15)	Cost of chemicals purchased is relatively reasonable	12
Toxicity and health and environmental impact (15)	Chemicals used to induce flocculation can be toxic	8
Suitability for Large Scale (15)	Large volumes of microalgae suspension can be used Prolonged periods of time for sufficient settling is not suitable for large scale use	8
Time (15)	Long periods are required for the settling of the cells	8
Species Specificity (10)	Dependent on the response of the cell to the pH altered environment Is not suitable for all species	4
Reusability of Media (8)	The pH of the media is altered to induce the flocculation making the media unsuitable for reuse	2
Maintenance (7)	No maintenance required	7
Total (100)		**60**

Table 17: Evaluation of bio-flocculation

Criteria	Description	Score
Dewatering Efficiency (15)	Effectiveness is dependent on the linkage of the bio-flocculent to the cells in order to increase their density	11
Cost (15)	Costs are associated with the purchasing the microorganisms and the maintenance of the culture	10
Toxicity and health and environmental impact (15)	Bio-flocculent species are non-toxic	15
Suitability for Large Scale (15)	Effective on large volumes of microalgae species	15
Time (15)	Based on the ability of the cells to link onto the microorganisms in order be more dense and improve settling time	10
Species Specificity (10)	Wide array of species can be used	8
Reusability of Media (8)	Microorganisms can be harvested and the medium can be recycled	3
Maintenance (7)	Maintenance of microorganism culture is required	4
Total (100)		**76**

Table 18: Evaluation of electrolytic coagulation

Criteria	Description	Score
Dewatering Efficiency (15)	Effective concentration of cells in liquid suspension	13
Cost (15)	High electrical power is required	11
Toxicity and health and environmental impact (15)	The addition of toxic chemicals is not required Cell composition can be altered	12
Suitability for Large Scale (15)	Large volumes require large power inputs which deem this method unsuitable for large scale	7
Time (15)	Rapid	15
Species Specificity (10)	Dependent on the charge of the species Conductivity of the water (marine water requires less energy)	3
Reusability of Media(8)	Microalgae can be harvested and the media can be recycled	5
Maintenance (7)	Effectiveness of electrode is reduced with continued use Frequent replacement of electrode maybe necessary	3
Total (100)		**69**

Table 19: Evaluation of electrolytic flocculation

Criteria	Description	Score
Dewatering Efficiency (15)	Effective in cumulating the cells together with one another	13
Cost (15)	Energy is required for rapid accumulation of the cells	11
Toxicity and health and environmental impact (15)	The addition of toxic chemicals is not required Cell composition can be altered	12
Suitability for Large Scale (15)	Unsuitable for large scale production because of energy requirement and the alteration of cell composition	8
Time (15)	Rapid	15
Species Specificity (10)	Dependent on the charge neutralization of the cell Conductivity of the water (marine water requires less energy)	5
Reusability of Media (8)	Media can be reused Cell composition can be altered	4
Maintenance (7)	Effectiveness of electrode is reduced with continued use Frequent replacement of electrode maybe necessary	3
Total (100)		**71**

Table 20: Evaluation of electrolytic flotation

Criteria	Description	Score
Dewatering Efficiency (15)	Effective in accumulating the cells together with one another Effectiveness is based on the likelihood that the cell comes in contact with the bubble in order to float to the surface	13
Cost (15)	High costs are associated with power supple	11
Toxicity and health and environmental impact (15)	The addition of toxic chemicals is not required Cell composition can be altered	12
Suitability for Large Scale (15)	Unsuitable for large scale production because of the energy required and the alteration of cell composition	8
Time (15)	Dependent on the likelihood that the cell adheres to a bubbles in order to float to the surface	10
Species Specificity (10)	Dependent on the charge neutralization of the cell Conductivity of the water (marine water requires less energy)	4
Reusability of Media (8)	Media can be reused	4
Maintenance (7)	Effectiveness of electrode is reduced with continued use Frequent replacement of electrode maybe necessary	3
Total (100)		**65**

Analysis

The sum of the scores obtained for each method are presented in Table 21. The results indicated that of the 16 methods used for microalgae harvesting evaluated in this study, 4 had scores of 80/100 or above and are, therefore, deemed suitable for harvesting a wide array of microalgae species at the industrial scale. These methods are disc stack centrifuge (87/100), cross flow filtration (84/100), decanter centrifugation (82/100) and organic flocculation (80/100).

Table 21: Comparative analysis

Criteria	Physical									Chemical		AF	BF	Electrophoresis		
	S	VF	PF	CFF	DSC	DC	DAF	DVF	FO	IF	OF			EC	EFC	EFT
Dewatering Efficiency (15)	5	13	13	15	13	15	12	10	9	10	11	11	11	13	13	13
Cost (15)	15	9	9	12	8	6	10	9	10	11	11	12	10	11	11	11
Toxicity and health and environmental impact (15)	15	15	15	15	15	15	12	8	10	5	15	8	15	12	12	12
Suitability for large scale (15)	5	10	10	10	12	12	13	13	13	15	15	8	15	7	8	8
Time (15)	2	12	12	12	15	15	10	10	10	10	10	8	10	15	15	10
Species specificity (10)	4	6	5	8	10	5	5	8	8	5	6	4	8	3	5	4
Reusability of Media (8)	8	8	8	8	8	8	8	5	6	2	5	2	3	5	4	4
Maintenance (7)	7	2	2	4	6	6	7	7	7	7	7	7	4	3	3	3
Total (100)	**61**	**75**	**74**	**84**	**87**	**82**	**77**	**70**	**73**	**65**	**80**	**60**	**76**	**69**	**71**	**65**

S: Sedimentation
VF: Vacuum filtration
PF: Pressure filtration
CFF: Cross flow filtration
DSC: Disc stack centrifugation
DC: Decanter centrifugation
DAF: Dispersed air flotation
DVF: Dissolved air flotation
FO: Fluidic oscillation
IF: Inorganic flocculation
OF: Organic flocculation
AF: Auto-flocculation
BF: Bio-flocculation
EC: Electrolytic coagulation
EFC: Electrolytic flocculation
EFT: Electrolytic flotation

Three of the nine physical harvesting methods (disk stack centrifugation, cross flow filtration and decanter centrifugation) were considered suitable for large scale harvesting of microalgae because they are highly effective in removing microalgae biomass from the liquid medium, non-toxic and rapid and the medium can be reused. The organic flocculation is suitable for microalgae harvesting on a large scale because it can be used with a wide range of microalgae, the organic chemicals are not toxic and the medium can be recycled.

The other 12 harvesting methods were deemed unsuitable for harvesting microalgae at the industrial scale because they did not meet the evaluation criteria (suitability for dewatering a wide array of microalgae species, suitability for large volumes, low operation costs and low maintenance). The other 6 physical methods were not as effective in removing the algae biomass, required long time, were not suitable for large scale, required high maintenance and were not effective for a wide array of microalgae. Autoflocculation techniques are unsuitable for large scale use because the chemicals used for pH change are toxic, are species specific, require long time and recycling of the medium requires additional costly treatment. The bio-flocculation technique depends on the desirable end product, since the microorganisms used for flocculation are harvested with the cells, the flocculating cultures must be adjusted for viable growth and the process is costly. Electrophoresis methods were deemed unsuitable for large scale microalgae harvesting because of difficulty in scaling up and the disruption of the cells can affect the quality and yield of the desired end product.

CONCLUSIONS

The major obstacle for using microalgae biomass on an industrialscale for production of value added products is the dewatering step which accounts for 20-30% of the total costs associated with the process. A comparative analyses of 16 harvesting techniques that included 9 physical methods (sedimentation, vacuum filtration, pressure filtration, cross flow filtration, disc stack centrifugation, dispersed air floatation, dissolved air flotation, fluidic oscillation), 2 chemical methods (inorganic flocculation, organic flocculation), autoflocculation method, bio-flocculation method and 3 electrophoresis methods (electrolytic coagulation, electrolytic flocculation, electrolytic floatation) were undertaken. Selection of the most suitable harvesting methods was based on the effectiveness, cost, toxicity, processing time, species specificity, maintenance and suitability for operating on large scale. Each of the selected criteria was assigned a score ranging from 7 to 15 depending on the degree of importance of the criterion.

The results indicated that of the 16 methods evaluated, 4 scored values of 80/100 or above and were deemed suitable for harvesting microalgae on an industrial scale. Three of which were physical techniques (disc stack centrifuge (87/100), cross flow filtration (84/100), decanter centrifugation (82/100)) and the forth was the organic flocculation (80/100) method. These techniques were deemed suitable for large scale use because of their effective dewatering ability, low operational costs, suitability for numerous species, rapidness, require minimal maintenance and being environmentally friendly. The other methods were deemed unsuitable because they are not suitable for dewatering a wide array of microalgae species, not suited for large volumes, costly and require high maintenance.

Although any of these four techniques is deemed suitable for harvesting of microalgae and can be used alone, a combination of methods can also be used to further enhance the recovery efficiency and improve the economics. The use of flocculation as an initial harvesting step to concentrate the algae suspension allows for effective removal of algae biomass from large liquid media. The costs associated with energy intensive centrifugation or filtration techniques can be reduced by using these methods as secondary techniques since less volume of microalgae suspension will be required to undergo the secondary treatment. It is, therefore, recommended that the use of centrifugation or filtration microalgae harvesting techniques be coupled (as secondary techniques) with organic flocculation (as an initial dewatering step) in order to improve the economics of the overall process.

REFERENCES

1. Wahlen BD, Willis RM, Seefeldt LC(2011)Biodiesel production by simultaneous extraction and conversion of total lipids from microalgae, cyanobacteria, and wild mixed-cultures.BioresourcTechnol 102: 2727-2730.
2. Slade R,Bauen A(2013)Microalgae cultivation for biofuels: Cost, energy balance, environmental impacts and future prospects. Biomass Bioenergy 53: 29-38.
3. Spolaore P, Joannis-Cassan C, Duran E, Isambert A(2006) Commercial applications of microalgae.J BiosciBioengg 101: 87-96.
4. Bishop WM,Zubeck HM(2012)Evaluation of microalgae for use as nutraceuticals and nutritional supplements.J Nutrition Food Sci2: 147-183.
5. Luiten EEM, Akkerman I, Koulman A, Kamermans P, Reith H, et al (2003)Realizing the promises of marine biotechnology.BiomolEngg 20: 429-39.
6. Soontornchaiboon W, JooSS, Kim SM(2012) Anti-inflammatory effects of violaxanthin isolated from microalga Chlorella ellipsoidea in RAW 264.7 macrophages.Biological Pharmaceutical Bulletin 35: 1137-1144.
7. Chen P, Min M, Chen Y, Wang L, Li Y, et al(2009) Review of the biological and engineering aspects of algae to fuels approach. IntJ AgriBiolEngg 2: 1-30.
8. Harun R, Singh M, Gareth MF, Michael KD(2010) Bioprocess engineering of microalgae to produce a variety of consumer products. Renew Sustain Energy Rev 14: 1037-1047.
9. Becker W (2004)Microalgae in human and animal nutrition in Handbook Microalgal Culture: Biotechnology and Applied Phycology. (edn A), Blackwell Publishing Ltd, Oxford, UK.
10. Guil-Guerrero JL, Navarro-Juárez R, López-Martínez JC, Campra-Madrid P, Rebolloso-Fuentes MM(2004) Functional properties of the biomass of three microalgal species.J Food Engg 65: 511-517.
11. Schenk P, Thomas-Hall S, Stephens E, Marx U, Mussgnug J, et al(2008)Second generation biofuels: high-efficiency microalgae for biodiesel production. Bioenerg Res1: 20-43.
12. Demirbas MF, BalatM,Balat H(2011)Biowastes-to-biofuels.Energy Convers Manag52: 1815-1828.ISSN: 0196-8904.
13. Demirbas A (2010) Use of algae as biofuel sources. Energy Convers Manag 51: 2738-2749.
14. Pulz O (2001)Photobioreactors: production systems for phototrophic microorganisms. ApplMicrobiolBiotechnol 57: 287-293.
15. Danquah MK, Gladman B, MoheimaniN, Forde GM(2009)Microalgal growth characteristics and subsequent influence on dewatering efficiency. ChemEngg J 51: 73-78.
16. http: //xa.yimg.com/kq/groups/13238265/955710696/

name/RSE012701.pdf

17. Molina Grima E,Belarbi EH,Acien-Fernandez FG, Robles-Medina A,Yusuf C(2003)Recovery of microalgal biomass and metabolites: process options and economics. BiotechnolAdv 20: 491-515.
18. Zitelli GC, Rodolfi L, BiondiN,Tredici MR(2006) Productivity and photosynthetic efficiency of outdoor cultures of Tetraselmissuecica in annular columns. Aquaculture 261: 932-943.
19. Salim S, Bosma R, VermuëMH,Wijffels RH(2011) Harvesting of microalgae by bio-flocculation. JApplPhycol 23: 849-855.
20. Milledge JJ, Heaven S(2013) A review of the harvesting of micro-algae for biofuel production.Rev Environ SciBiotechnol12: 165-178.
21. Trevi (2014)Dissolved air flotation (DAF).Trevinv, 9050 Gentbrugge, Belgium.2014.
22. Generalic E (2014)Sedimentation. Croatian-English Chemistry Dictionary & Glossary.
23. Mohn FH (1980)Experiences and strategies in the recovery of biomass from mass cultures of microalgae. In Algae Biomass. (G. Schelef and C. J. Soeder ed.), Elsevier, Amsterdam: 547-571.
24. Harith ZT,YusoffFM, Mohammed MS, Shariff M, Din M, et al (2009) Effect of different flocculants on the flocculation performance of microalgae, Chaetoceroscalcitrans, cells.African J Biotechnol 8: 5971-5978.
25. Knuckey RM, Brown MR, Robert R, Frampton DMF(2006) Production of microalgal concentrates by flocculation and their assessment as aquaculture feeds. Aquaculture Engg 35: 300-313.
26. Waite AM, Thompson PA, Harrison PJ(1992) Does energy control the sinking rates of marine diatoms. Limnol Oceanography37: 468-477.
27. Van LerlandET,Peperzak L(1984)Separation of marine seston and density determination of marine diatoms by density gradient centrifugation. J Plankton Res6: 29-44.
28. Edzwald JK (1993) Algae, bubbles, coagulants, and dissolved air flotation. Water SciTechnol 27: 67-81.
29. Reynolds SC (1994)The ecology of freshwater phytoplankton. Cambridge University Press, Cambridge, United Kingdom.
30. Granados MR, Acién FG, Gómez C, Fernández-SevillaJM,Grima EM(2012)Evaluation of flocculants for the recovery of freshwater microalgae. BioresourceTechnol118: 102-110.
31. Murphy CF, Allen DT(2011) Energy-water nexus for mass cultivation of algae.Environ SciTechnol45: 5861-5868.
32. Cole TM,Buchak EM(1995)CE-QUAL-W2: A two dimensional, laterally averaged, hydrodynamic and water quality model, version 2.0: user manual. Instruction Report EL-95-1, US Army Engineer Waterways Experiment Station, Vicksburg, Mississippi.
33. Peperzak LF, Koeman CR,GieskesWWC,Joordens JCA(2003) Phytoplankton sinking rates in the Rhine region of freshwater influence.J Plankton Res 25: 365-368.
34. Yang MD, Sykes RM,Merry CJ(2000) Estimation of algal biological parameters using water quality modeling and SPOT satellite data.Ecological Modelling, 125: 1-13.
35. Choi SK, Lee JY, Kwon DY, Cho KJ(2006)Settling characteristics of problem algae in the water treatment process.Water sciTechnol: A jIntAssociation Water Res 53: 113-119.
36. Choi SK, Lee JY, Kwon DY, Cho KJ(2006)Settling characteristics of problem algae in the water treatment process.Water sciTechnol: A jIntAssociation Water Res 53: 113-119.
37. Davis EM, Downs TD, Shi Y,Ajgaonkar AA(1995) Recycle as an alternative to algal TSS and BOD removal from an industrial waste stabilization pond system. Industrial Waste Conference 50: 65-73.
38. Greenwell HC, Laurens LML, Shields RJ, LovittRW, Flynn KJ(2010)Placing microalgae on the biofuels priority list: a review of the technological challenges.J Royal Society Interface7: 703-726.
39. Manheim D, Nelson Y(2013) Settling and bioflocculation of two species of algae used in wastewater treatment and algae biomass production.Environ Prog SustainEnergy 32: 946-954.
40. Schlenk D, Batlet G, King C, Stauber J, Adams M, et al(2007)Effects of light on microalgae concentrations and selenium uptake in bivalves exposed to selenium-amended sediments.Arch EnvironContamination Toxicology 53: 365-370.
41. Park JBK, CraggsRJ,Shilton AN(2011)Recycling algae to improve species control and harvest efficiency from a high rate algal pond. Water Res45: 6637-6649.
42. Griffiths MJ, Van Hille RP, Harrison STL(2012) Lipid productivity, settling potential and fatty acid profile of 11 microalgal species grown under nitrogen replete and limited conditions.J ApplPhycol 24: 989-1001.
43. Wang ZJ, Hou, Bowden D,Belovich JM(2014)Evaluation of an inclined gravity settler for microalgae harvesting.J ChemTechnolBiotechnol89: 714-120.
44. Shelef G, SukenikA, Green M (1984) Microalgae harvesting and processing: a literature review.Solar Energy Res InstGolden, Colorado.
45. Taher S, Al-Zuhair S, Al-Marzouqi, HaikY,Farid MM(2011)A review of enzymatic transesterification of microalgal oil-based biodiesel using supercritical technology.Enzyme Research 468292: 25.doi: 10.4061/2011/468292
46. Gonzalez-Fernandez C, Ballesteros M(2013)Microalgae Autoflocculation: an alternative to high-energy consuming harvesting methods.J Applied Phycology, 25: 991-999.
47. Mata TM, Martins AA, Caetano NS(2010) Microalgae for biodiesel production and other applications: a review. Renew Sustain Energy Rev14: 217-232.
48. Ras M, Lardon L, Bruno S, BernetN,Steyer JP(2011) Experimental study on a coupled process of production and anaerobic digestion of Chlorella vulgaris.

BioresourcTechnol 102: 200-206.

49. Brennan L, Owende P(2010) Biofuels from microalgae a review of technologies for production, processing, and extractions of biofuels and co-products.Renew Sustain Energy Rev14: 557-577.

50. Green FB (2008) Harvesting microalgae: challenges and achievements.Microalgae Biomass Summit, Algal Biomass Organization, Seattle, Washington, USA.

51. http: //www.dicalite-europe.com/filtrat.htm

52. http: //www.chem.arizona.edu/

53. Sparks T (2012) Solid-liquid filtration: understanding filter presses and belt filters.Filtration + Separation, 49: 20-24.

54. http: //www.komline.com/docs/rotary_drum_vacuum_filter.html

55. Stucki S, Vogel F, Ludwig C, HaiducAG,Brandenberger M(2009) Catalytic gasification of algae in supercritical water for biofuel production and carbon capture.Energy Environ Sci 2: 535-541.

56. Ferrentino JM, FaragIH,Jahnke LS(2006) Microalgae oil extraction and in-situ transesterification.AIChE Annual Meeting, Conference Proceedings, San Francisco, California, USA.

57. 57.Umesh BV (1984) Performance of two new filtration devices for harvesting of Spirulina alga. BiotechnolLetters6: 309-312.

58. Gudin C, Chaumont D(1991) Cell fragility, the key problem of microalgae mass production in closed photobioreactors. BioresourcTechnol 38: 141-51.

59. Umesh BV (1984) Performance of two new filtration devices for harvesting of Spirulina alga. BiotechnolLetters6: 309-312.

60. Mohn F (1988) Harvesting of micro-algal biomass. In: Micro-algal biotechnology (LJBorowitzka and MABorowitzkaedn.), Cambridge University Press, Cambridge, UK.ISBN: 0521323495

61. Petrusevski B, van Breemen AN, Alaerts GJ, Bolier G(1995) Tangential flow filtration: a method to concentrate freshwater algae. Water Research 29: 1419-1424.

62. Amaro HM, GuedesAC,Malcata FX(2011) Advances and perspectives in using microalgae to produce biodiesel. Appl Energy 88: 3402-3410.

63. Pittman JK, Dean AP,Osundeko O(2011)The potential of sustainable algal biofuel production using wastewater resources. BioresourcTechnol102: 17-25.

64. Wang B, Li Y, Wu N, Lan CQ(2008)CO2 bio-mitigation using microalgae. ApplMicrobiolBiotechnol79: 707-718.

65. Arar EJ, Collins GB(1997) In-vitro determination of chlorophyll a and pheophytina in marine and freshwater algae by fluorescence.National Exposure Research Laboratory, US Environmental Protection Agency, Cincinnati, Ohio .

66. Rossi N, Jaouen O, LegentilhommeP, Petit I(2004) Harvesting of cyanobacteriumArthrospiraplatensis using organic filtration membranes. Food and Bioproducts Processing 82: 244-250.

67. Velamakanni BV, Lange FF(1991) Effect of interparticle potentials and sedimentation on particle packing density of bimodal particle distributions during pressure filtration.J American ChemSociety 74: 166-172.

68. Rushton A (1996) Solid liquid filtration and separation technology (Rushton A, ASWard, RG Holdrichedn.) Wiley-VCH, Verlag GmbH, Weinheim, Germany: 1-32.

69. Raghvan GSV,Sanga ECM(2003) Encyclopedia of Agricultural, Food, and Biological Engineering.(DR Heldmanedn.) CRC Press, Boca Raton, Florida: 329

70. Show KY, Lee DJ, Chang GS(2013) Algal biomass dehydration.BioresourcTechnol 135: 720-729.

71. Nagle N, Lemke P(1990) Production of methyl ester fuel from microalgae.ApplBiochemBiotechnol 25: 355-361.

72. Spellman FR (2008) Handbook of water and wastewater treatment plant operations.(2ndedn.)CRC Press, Boca Raton, Florida.

73. http: //www.novasep.com/biomolecules/BioIndustries/Bioindustries-technologies.asp

74. Rossignol N, Vandanjon L, JaouenP,Quemeneur F(1999) Membrane technology for the continuous separation microalgae/culture medium: compared performances of cross-flow microfiltration and ultrafiltration. Aquacultural Engineering 20: 191-208.

75. Zhang X, Hu Q, Sommerfeld M, Puruhito E, Chen Y(2010)Harvesting algal biomass for biofuels using ultrafiltration membranes.BioresourceTechnol 101: 5297-5304.

76. Rossi N, Petit I, Jaouen P, LegentilhommeP,Derouiniot M(2005)Harvesting of cyanobacteriumArthrospiraplatensis using inorganic filtration membrane. Separation SciTechnol40: 3033-3050.

77. Rose PD, Maart BA, Phillips TD, Tucker SL, Cowan AK, et al.(1992)Cross-flow ultrafiltration used in algal high rate oxidation pond treatment of saline organic effluents with the recovery of products of value.Water Science Technol25: 319-327.

78. Walsh DT, Petrovits EJ, Kraus RA, Wuetrich LM, WithstandleyCA, et al. (1986)An evaluation of tangential flow filtration for microalgal cell harvesting.National Science Foundation, Washington.

79. Ahmed AL, Yasin NHM, Derek CJC, Lim JK(2012) Cross flow microfiltration of microalgae biomass for biofuel production.Desalination 302: 65-70.

80. Crittenden JC, Trussell RR, Hand DW, Howe KJ(2012) MWH water treatment: principles and design.Hoboken, New Jersey, USA.

81. Borowitzka MA, Moheimani NR(2013) Algae for biofuels and energy.Springer Dordrecht Heidelberg, New York, London.

82. http: //www.esi.info/detail.cfm/Euroby/Disc-stack-centrifuges-with-selfcleaning-bowl/_/R-35093_BN133BX

83. Mannweiler K, Hoare M(1992) The scale-down of an industrial disk stack centrifuge. Bioprocess Engg8: 19-

25.

84. Heasman M, Diemar J, O'Connor W, Sushames T, Foulkes L, et al.(2000)Development of extended shelf-life microalgae concentrate diets harvested by centrifugation for bivalve molluscs—a summary. Aquaculture Res31: 637-659.

85. Sim T, Goh A, Becker E(1988) Comparison of centrifugation, dissolved air flotation and drum filtration techniques for harvesting sewage-grown algae.Biomass 16: 51-62.

86. Vasudevan V, Stratton RW, Pearlson MN, Jersey GR, Beyene AG, et al. (2012) Environmental performance of algal biofuel technology options.Environ SciTechnol 46: 2451-2459.

87. Mackay D (1996)Downstream processing of natural products: A practical handbook (MS Verraledn.) John Wiley and Sons: 11-40.

88. Chojnacka K, ChojnackiA,Górecka H(2005)Biosorption of Cr3+, Cd2+ and Cu2+ ions by blue-green algae Spirulina sp.: kinetics, equilibrium and the mechanism of the process. Chemosphere 59: 75-84.

89. Smith J, Charter E(2009) Functional food product development.Wiley-Blackwell, Etobicoke, Ontario.

90. Rees F, LeenheerJ, Ranville J(1991) Use of a single-bowl continuous-flow centrifuge for dewatering suspended sediments: effect on sediment physical and chemical characteristics. Hydrological Processes 5: 201-214.

91. Dassey AJ,Theegala CS(2013) Harvesting economics and strategies using centrifugation for cost effective separation of microalgae cells for biodiesel applications. BioresourceTechnol128: 241-245.

92. www.alfalaval.com/solution-finder/products/px-series/pages/documentation.aspx

93. Goh A (1984) Production of microalgae using pig waste as a substrate.Algal Biomass Workshop, University of Colorado, Boulder, USA.

94. Sharma KK, Garg S, Li Y, Malekizadeh A, Schenk PM(2013)Critical analysis of current microalgae dewatering techniques.Biofuels, 4: 397-407.

95. Tarleton S,Wakeman R(2006)Solid/liquid separation: equipment selection and process design.Elsevier Science, Sydney, Australia.

96. Metcalf & Eddy (2003) Wastewater Engineering: Treatment and Reuse, (TchobanoglousG, Burton FL,Stensel HD, edn.) McGraw-Hill Education, New York, USA.

97. Packer M (2009)Algal capture of carbon dioxide; biomass generation as a tool for greenhouse gas mitigation with reference to New Zealand energy strategy and policy. Energy Policy 37: 3428-3437.

98. Singh A, Nigam PS, Murphy JD(2011) Mechanism and challenges in commercialisation of algal biofuels. BioresourceTechnol 102: 26-34.

99. Schofield T (2001) Dissolved air flotation in drinking water production. Water SciTechnol43: 9-18.

100. Matis KA, GalliosGP, Kydros KA(1993) Separation of fines by flotation techniques. Separations Technol 3: 76-90.

101. Kaartinen J,Koivo H(2014)Machine vision based measurement and control of zinc flotation circuit. Helsinki University of Technology, Helsiniki, Finland.

102. Rubio J, Souza ML, Smith RW(2002) Overview of flotation as a wastewater treatment techniques.Minerals Engineering, 15: 139-155.

103. Chen YM, Liu JC, Ju YH(1998) Flotation removal of algae from water.Colloids and Surfaces, B: Biointerfaces 12: 49-55.

104. Yan YD, Jameson GJ(2004) Application of the Jameson Cell technology for algae and phosphorus removal from maturation ponds.Int J Mineral Process73: 23-28.

105. Xu L, Wang F, Li HZ, Hu ZM, Guo C, et al.(2010) Development of an efficient electroflocculation technology integrated with dispersed-air flotation for harvesting microalgae.J ChemTechnolBiotechnol 85: 1504-1507.

106. Kurniawati HA, IsmadjiS,Liu JC(2014) Microalgae harvesting using natural saponin and chitosan. BioresourcTechnol 166: 429-434.

107. Edzwald JK (2010) Dissolved air flotation and me.Water Res44: 2077-2106.

108. Wiley PE.,BrennemanKJ, Jacobson AE(2009) Improved algal harvesting using suspended air flotation.Water EnvironRes 81: 702-708.

109. Zimmerman WB, Hewakandamby BN, Tesar V, Bandulasena HCH,Omotowa OA(2009) On the design and simulation of an airlift loop bioreactor with microbubble generation by fluidic oscillation. Food Bioproducts Processing, 87: 215-227.

110. Hanotu J, BandulasenaHCH, Zimmerman WB(2012) Microflotation performance for algal separation. BiotechnolBioengg 109: 1663-1673.

111. Dai Z, FornasieroD, Ralston J(2000)Particle-bubble collision models--a review. Adv Colloid InterfSci 85: 231-256.

112. http: //digitalcommons.usu.edu/cgi/viewcontent.cgi?article=2050&context=etd

113. Yap RKL, Whittaker M, Diao M,Stuetz RM,Jefferson B,et al.(2014)Hydrophobically-associating cationic polymers as micro-bubble surface modifiers in dissolved air flotation for cyanobacteria cell separation.Water Res 61: 253-262.

114. Lin MC, Liu JC(1996) Adsorbing colloid flotation of as (V)—feasibility of utilizing streaming current detector. Separation SciTechnol31: 1329-1641.

115. Kown H, Lu M, Lee EY,Lee J(2014)Harvesting microalgae using flocculation combined with dissolved air flotation.Biotechnol Bioprocess Engg 19: 143-149.

116. Zheng H, Gao Z, Yin J, Tang X, JiX, et al.(2012) Harvesting of microalgae by flocculation with poly (gamma-glutamic acid).BioresourceTechnol 112: 212-220.

117. Besson A,Guiraud P(2013) High-pH-induced flocculation-flotation of the hypersaline microalga Dunaliellasalina.BioresourcTechnol 147: 464-470.

118. Schlesinger A,Eisenstadt D, Bar-Gil A, Carmely H, EinbinderS, et al.(2012) Inexpensive non-toxic flocculation of microalgae contradicts theories; overcoming a major hurdle to bulk algal production. BiotechnolAdv 30: 1023-1030.
119. Thangavel P,Sridevi G(2012) Environmental Sustainability.Springer, New Delhi, India.
120. Liu JC, Chen YM, Ju YH(1999) Separation of algal cells from water by column flotation.Separation SciTechnol 34: 2259-2272.
121. Coward T, Lee JGM, Caldwell GS(2014) Harvesting microalgae by CTAB-aided foam flotation increases lipid recovery and improves fatty acid methyl ester characteristics.Biomass Bioenergy 67: 354-362.
122. Dassey A Theegala C(2012) Optimizing the air dissolution parameters in an unpacked dissolved air flotation system.Water 4: 1-11.
123. Vlaski A, van BreemenAN,Alaerts GJ(1996) Optimisation of coagulation conditions for the removal of cyanobacteria by dissolved air flotation or sedimentation. Aqua, 45: 253-261.
124. Zhang S, Amendola P, Hewson JC, SommerfeldM, Hu Q(2012)Influence of growth phase on harvesting of Chlorella zofingiensis by dissolved air flotation. BioresourcTechnol 116: 447-484.
125. Edzwald JK,Wingler BJ(1990) Chemical and physical aspects of dissolved air-flotation for the removal of algae. Aqua, 39: 24-35.
126. Vlaski A, van BreemenAN,Alaerts GJ(1996) Optimisation of coagulation conditions for the removal of cyanobacteria by dissolved air flotation or sedimentation. Aqua, 45: 253-261.
127. Kempeneers S, MenxelFV,Gille L(2001)A decade of large scale experience in dissolved air flotation.Water SciTechnol43: 27-34.
128. Teixeira MR, Rosa MJ(2006) Comparing dissolved air flotation and conventional sedimentation to remove cyanobacterial cells of Microcystisaeruginosa: Part I: The key operating conditions. Separation PurifTechnol 52: 84-94.
129. Gregory R,Edzwald JK (2010) Water Quality and Treatment (6thedn.) McGraw Hill, New York.
130. Metcalf and Eddy (1991) Wastewater Engineering: Treatment, Disposal, and Reuse.(3rdedn.)Mcgraw-Hill College Division New York.
131. Corbitt RA (1999) Standard handbook of environmental engineering.(2ndedn.)McGraw-Hill, New York, New York.
132. Edzwald JK (2007) Development of high rate dissolved air flotation for drinking water treatment.J Water Supply 56: 399-409.
133. Haarhoff J, Rykaart E(1995) Rational design of packed saturators. Water SciTechnol 31: 179-190.
134. Fe´ris LA, Rubio J(1999) Dissolved air flotation (DAF) performance at low saturation pressures. FiltraSeparat 36: 61-65.
135. Haarhoff, J, Steinbach S(1996) A model for the prediction of the air composition in pressure saturators. Water. Res30: 3074-3082.
136. Green FB, Lundquist TJ, Oswald WJ(1995) Energetics of advanced integrated wastewater pond systems. Water SciTechnol31: 9-20.
137. Araujo GS, Matos LJBL, Fernandes JO, Cartaxo SJM, Goncalves LRB, et al.(2013)Extraction of lipids from microalgae by ultrasound application: prospection of the optimal extraction method.UltrasonicsSonochemistry 20: 95-98.
138. Wijffels RH, Barbosa MJ(2010)Microalgae for the production of bulk chemicals and biofuels.Biofuels BioproductsBiorefining, 4: 287-295.
139. Oh HM, Lee SJ, Park MH, Kim HS, Kim HC, et al.(2001) Harvesting of Chlorella vulgaris using a bio-flucclant form Paenibacillus sp. AM49.BiotechnolLetters 23: 1229-1234.
140. Papazi A, Makridis P,Divanach P(2010) Harvesting Chlorella minutissima using cell coagulants.J ApplPhycol 22: 349-355.
141. t'Lam GP, Vermue MH, Olivieri G, van den Broek LAM, Barbosa MJ, et al. (2014)Cationic polymers for successful flocculation of marine microalgae. BioresourceTechnol169: 804-807.
142. Pushparaj B, Pelosi E, TorzilloG, Materassi R(1993) Microbial biomass recovery using a synthetic cationic polymer. BioresourceTechnol43: 59-62.
143. Chen G, Zhao L, Qi Y, Cui YL(2014) Chitosan and its derivatives applied in harvesting microalgae for biodiesel production: an outlook.J Nanomaterials 2014: 1-9.
144. Vandamme D, Vieira Pontes SC, Goiris K, Foubert I, Jozef L, et al (2011)Evaluation of electro-coagulation-flocculation for harvesting marine and freshwater microalgae. BiotechnolBioengg, 108: 2320-2329.
145. Bernhardt H, Clasen J(1991) Flocculation of micro-organisms. JWater Supply: Res Technol-Aqua 40: 76-87.
146. Briley DS,Knappe DRU(2002) Optimizing ferric sulphate coagulation of algae with streaming current measurements.American Water Works Association 94: 80-90.
147. Wyatt NB,Gloe LM, Brady PV, Hewson JC, Grillet AM, et al.(2012)Critical conditions for ferric chloride-induced flocculation of freshwater algae.BiotechnolBioengg 109: 493-501.
148. www.nus.edu.sg/nurop/2009/FoE/U067436X.PDF
149. Sukenik A, BilanovicD,ShelefG(1988)Flocculation of microalgae in brackish and sea waters. Biomass 15: 187-199.
150. http: //umpir.ump.edu.my/3643/1/CD6374_NUR_ANIS_SAARANI.pdf
151. 151.http: //eng.thesaurus.rusnano.com/wiki/article619
152. 152.http: //www.zmescience.com/science/physics/direct-measurement-of-van-der-waals-force-made-for-the-first-time/
153. Millamena O, AujeroE,Borlongan I(1990) Techniques on algae harvesting and preservation for use in culture as larval food.AquaculturalEngg 9: 295-304.

154. Aragon AB, Prdilla RB,Ros de Ursinos JAF(1992) Experimental study of the recovery of algae cultured in effluents from the anaerobic biological treatment of urban wastewaters. ResourcConservRecycl6: 293-302.

155. http: //www.wikipremed.com/01physicscards.php?card=628

156. http: //eng.thesaurus.rusnano.com/wiki/article619

157. http: //www.zmescience.com/science/physics/direct-measurement-of-van-der-waals-force-made-for-the-first-time/

158. http: //chemwiki.ucdavis.edu/Physical_Chemistry/Intermolecular_Forces/Hydrogen_Bonding

159. Bondi A (1964) Van der Waals volumes and radii.J PhysChem68: 441-451.

160. Hunter RJ (1993) Introduction to Modern Colloidal Science.Oxford University Press, New York.

161. Hogg R (1999)The role of polymer adsorption kinetics in flocculation.Colloids and Surfaces A: Physicochemical and Engineering Aspects 146: 253-263.

162. Udom I, Zaribaf BH, Halfhide T, Gillie B, Dalrymple O, et al.(2013)Harvesting microalgae grown on wastewater. BioresourcTechnol 139: 101-106.

163. Buelna G, Bhattarai KK, DelanoueJ,Taiganides EP(1990) Evaluation of various flocculants for the recovery of algae biomass grown on pig-waste. Biological Wastes 31: 211-222.

164. Xu Y, PurtonS,Baganz F(2013)Chitosan flocculation to aid the harvesting of the microalga Chlorella sorokiniana. BioresourcTechnol129: 296-301.

165. Ahmad AL, Mat Yasin NH, Derek CJC, Lim JK(2011) Optimization of microalgae coagulation process using chitosan.ChemEngg J173: 879-882.

166. Chang, YR, Lee DJ(2012)Coagulation-membrane filtration of Chlorella vulgaris at different growth phases. Drying Technol30: 1317-1322.

167. Sirin S, Trobazo RC, Ibanez,Salvado J(2012) Harvesting the microalgae Phaeodactylumtricornutum with polyaluminum chloride, aluminium sulphate, chitosan and alkalinity-induced flocculation. J ApplPhycol 24: 1067-1080.

168. Morales J, de la Noue J, Picard G(1985)Harvesting marine microalgae species by chitosan flocculation. AquaculturalEngg 4: 257-270.

169. Beach ES, Eckelman MJ, Cui Z, BrentnerL, Zimmerman JB(2012) Preferential technological and life cycle environmental performance of chitosan flocculation for harvesting of the green algae Neochlorisoleoabundans. BioresourcTechnol 121: 445-449.

170. Rakesh S, Saxena S, Dhar DW, PrasannaR,Saxena AK(2014)Comparative evaluation of inorganix and organix amendments for their flocculation efficiency of selected microalgae.JApplPhycol 26: 399-406.

171. Garzon-Sanabria AJ, Davis RT,Nikolov ZL(2012) Harvesting Nannochlorisoculata by inorganic electrolyte flocculation: effect of initial cell density, ionic strength, coagulant dosage, and media pH.BioresoureTechnol 118: 418-424.

172. Ferriols VNE, Aguilar RO(2012) Efficiency of various flocculants in harvesting the green microalgae Tetraselmistetrahele (Chlorodendrophyceae: Chlorodendraceae).Aquaculture, Aquarium, Conservation and Legislation, 5: 265-273.

173. Lee YC., Kim B, Farooq W, Chung J, Han JI, et al. (2013) Harvesting of oleaginous Chlorellasp by organoclays. BioresourcTechnol 132: 440-445.

174. Sanyano N, ChetpattananondhP,Chongkhong S(2013) Coagulation-flocculation of marine Chlorellasp for biodiesel production. BioresourcTechnol 147: 471-476.

175. Ching HW, Tanaka TS,Elimelech M(1994) Dynamics of coagulation of kaolin particles with ferric chloride. Water Res28: 559-569.

176. Lee SJ, Kim SB, Kim JE, Kwon GS, Yoon BD, et al. (1998)Effects of harvesting method and growth stage on the flocculation of the green alga Botrycococusbraunii. Letters in ApplMicrobiol27: 4-18.

177. 177.Shammas NK (2005) Coagulation and flocculation. In: Physicochemical treatment processes, Handbook of environmental engineering. The Humana Press, Totowa, New Jersey, 3: 103-139.

178. Tenney MW, Echelberger WF, SchuesslerRG,Pavoni JL(1969) Algae flocculation with synthetic organic polyelectrolytes. ApplMicrobiol 18: 965-971.

179. Lee A, Lewis D, Ashman P(2009)Microbial flocculation, a potentially low-cost harvesting technique for marine microalgae for the production of biodiesel. J ApplPhycol 21: 559-567.

180. Morrissey KI,He C,Wong MH,Zhao X,Chapman RZ, et al.(2014)Charge-tunable polymer as reversible and recyclable flocculants for the dewatering of microalgae. Biotechnol Bioengineering 112: 74-83.

181. Bilanovic D,Shelef G(1988)Flocculation of microalgae with cationic polymers: effects of medium salinity. Biomass 17: 65-76.

182. Benemann J, Koopman B, Weissman J, Eisenberg D, Goebel R(1980)Development of microalgae harvesting and high-rate pond technologies in California, in Algae Biomass: production and use.Elsevier/North-Holland Biomedical Press, Amsterdam, Netherlands.

183. Shammas NK (2005) Coagulation and flocculation. In: Physicochemical treatment processes, Handbook of environmental engineering. The Humana Press, Totowa, New Jersey, 3: 103-139.

184. Zhou FS, Wang X, Zhou L, Liu Y(2014)Preparation and coagulation behavior of a novel multiple flocculent based on cationic polymer-hydroxy aluminium-clay minerals. Adv Materials SciEngg, Article ID: 581051.

185. BukhariAA(2008)Investigation of the electrocoagulation treatment process for the removal of total suspended solids and turbidity from municipal wastewater. BioresourcTechnol 99: 914-921.

186. Koopman B, Lincoln EP, Nordstedt RA(1982) Anaerobic-photosynthetic reclamation of swine waste. Proceedings-Water Reuse Symposium,Washington, DC, II: 924-934.

187. Hee-Mock O, Lee SJ, Park MH, Kim HS, Kim HC,

et al.(2001)Harvesting of Chlorella vulgaris using a bioflocculant from Paenibacillus sp. AM49.Biotechnol Letters 23: 1229-1234.

188. Vandamme D, FoubertI, Muylaert K(2013) Flocculation as a low-cost method for harvesting microalgae for bulk biomass production.Trades in Biotechnol 31: 233-239.

189. Wu Z, Zhu Y, Huang W, Zhang C, Li T, et al(2012) Evaluation of flocculation induced by pH increase for harvesting microalgae and reuse of flocculated medium. BioresourcTechnol 110: 496-502.

190. Horiuchi J, Ohba I, Tada K, Kobayashi M, KannoT, et al.(2003)Effective cell harvesting of the halotolerant microalga Dunaliellatertiolecta with pH Control. J BiosciBioengg, 95: 412-415.

191. Liao BQ, Lin HJ, Langevin SP, Gao WJ,Leppard GG(2011)Effects of temperature and dissolved oxygen on sludge properties and their role in bioflocculation and settling.Water Res 45: 509-520.

192. Wilén BM, Balmér P(1999)The effect of dissolved oxygen concentration on the structure, size and size distribution of activated sludge flocs. Water Res 33: 391-400.

193. SukenikA, Shelef G(1984) Algal autoflocculation-verificationandproposedmechanism.BiotechnolBioengg 26: 142-147.

194. Becker EW (1994) Microalgae: Biotechnology and Microbiology.Cambridge University Press, Cambridge, United Kingdom.

195. Uusitalo J (1996) Algal carbon uptake and the difference between alkalinity and high pH (“alkalization”), exemplified with a pH drift experiment.Scientia Marina 60: 129-134.

196. Nurdogen Y, Oswald WJ(1995)Enhances nutrient removal in high-rate ponds.Water SciTechnol31: 33-43.

197. Nguyen TDP, Frappart M, Jaouen P, PruvostJ,Bourseau P(2014) Harvesting Chlorella vulgaris by natural increase in pH: effect of medium composition.Environ Technol 35: 1378-1388.

198. Smith BT, Davis RH(2012) Sedimentation of algae flocculated suing naturally-available, magnesium-based flocculants.Algal Res 1: 32-39.

199. Ayoub GM, Lee SL,Koopman B(1986)Seawater induced algal flocculation. Water Res 20: 1265-1271.

200. VandammeD, FoubertI, FraeyeI, MeesschaertB,Muylaert K(2012) Flocculation of Chlorella vulgaris induced by high pH: Role of magnesium and calcium and practical implications. BioresourcTechnol 105: 114-119.

201. Knuckey RM(1998) Australian microalgae and microalgal concentrates for use as aquaculture feeds. Ph.D. Thesis. University of Tasmania, Hobart, Tasmania.

202. D'Souza FML, Knuckey RM, HohmannS, Pendrey RC(2002)Flocculated microalgae concentrates as diets for larvae of the tiger prawn PenaeusmonodonFabricius. Aquaculture Nutrition 8: 113-120.

203. Ndikubwimana T, Zeng, X, Liu Y, Chang JS, Lu Y(2014) Harvesting of microalgae Desmodesmus sp. F51 by bioflocculation with bacterial bioflocculant.Algal Res 6: 186-193.

204. Kim DG,La HJ, Ahn CY, Park YH, Oh HM(2011) Harvest of Scenedesmus sp. with bioflocculant and reuse of culture medium for subsequent high-density cultures. BioresourcTechnol 102: 3163-3168.

205. Hanifa T, Al-Zuhair S, Al-Marzouqi AH, HaikY,Farid MM (2011) A review of enzymatic transesterification of microalgal oil-based biodiesel using supercritical technology.Enzyme Res 468292: 1-25.

206. Kim J, Tyu BC, Kim K, Kim BK, Han JI, et al.(2012) Continuous microalgae recovery using electrolysis: effect of different electrode pairs and timing of polarity exchange.BioresourcTechnol 123: 164-170.

207. Zhang J, Hu B(2012) A novel method to harvest microalgae via co-culture of filamentous fungi to form cell pellets.BioresourcTechnol 114: 529-535.

208. Grossart HP, Kiorboe T, Tang KW, Allgaier M, Yam EM, et al.(2006) Interactions between marine snow and heterotrophic bacteria: aggregate formation and microbial dynamics.Aquatic Microbial Ecology 42: 19-26.

209. SalehizadehH, Shojaosadati S (2001) Extracellular biopolymeric flocculants: recent trends and biotechnological importance.BiotechnolAdv19: 371-385.

210. Shen Y, Yuan W, Pei ZJ, Wu Q, Mao E(2009) Microalgae mass production methods. Transactions of the ASABE, 52: 1275-1287.

211. Cerqueira AA, Marques MRC(2012)Electrolytic treatment of wastewater in the oil industry.New Technologies in the Oil and Gas Industry, ISBN: 978-953-51-0825-2.

212. Mollah MYA, Morkovsky P, Gomes JGA, Kesmez M, PargaJ, Cocke DL(2004) Fundamentals, present and future perspectives of electrocoagulation. J Hazardous Materials 114: 199-210.

213. Uduman N, Lee H, DanquahMK,Hoadley AFA(2011) Electrocoagulation of marine microalgae.Engineering a Better World: Sydney Hilton Hotel, NSW, Australia.

214. Azarian GH, MesdaghiniaAR,Vaezi F(2007) Algae removal by electro-coagulation process, application for treatment of the effluent from an industrial wastewater treatment plant. Iranian J Public Health 36: 57-64.

215. Ghernaout D, Badis A, KellilA, Ghernaout B(2008) Application of electrocoagulation in Escherichia coli culture and two surface waters.Desalination 219: 118-125.

216. Poelman E, de PauwN,Jeurissen B(1997) Potential of electrolytic flocculation for recovery of micro-algae. Research, Conservation and Recycling, 19: 1-10.

217. Zenouzi A, Ghobadian B, Hejzai MA,Rahnemoon P(2013)Harvesting of microalgae Dunaliellasalina using electroflocculation.J AgriSciTechnol 15: 879-888.

CHAPTER 19

MICROALGAE OIL EXTRACTION PRE-TREATMENT METHODS

Mariam Al hattab and Abdel Ghaly

Keywords: Microalgae, Oil extraction, Lipid

INTRODUCTION

One of the most important resources for humans today is energy and its sustainability [1]. Today, petroleum products contributes 80% of the world's energy need as a fuel source for transportation and other energy demanding sectors [2,3]. Millions of years were required for the formation of fossil fuels and the diminishing crude oil reserves, increasing fuel prices and environmental concerns associated with fossil fuels usage make them unsustainable energy sources [4-6]. Renewable biofuels are regarded as an important energy resource in many countries globally, but only account for 10% of the total energy consumption [6].

There are three main generations of liquid biofuels [7]. The first generation of liquid biofuels includes both bioethanol and biodiesel production from food crops (corn, vegetable oils and sugarcane). This first generation of liquid fuels presents a problem with food supply and increased cost of food crops, thus limiting their use as energy sources. The second generation of liquid biofuels uses waste cooking oils, animal fats and non-edible plant seeds [7,8]. This second generation of biofuels solve the problem associated with the first generation of biofuels as these sources are not edible, but maintaining a consistent feedstock is a challenge. This inconsistency with feed stock supply gave rise to a third generation of liquid biofuels from aquatic algae [9].

Biodiesel from algae is a green alternative that reduces CO, CO_2 and hydrocarbon emissions compared to the currently used diesel fuel [10,11]. However, the problem associated with the use of algae for biodiesel production is the high costs associate with the algae production and oil extraction processes. In order for microalgae biodiesel to be economically viable, the processing needs to be made more economic by improving the methods of algae production, harvesting and oil extraction. This can be done by optimizing the algae production and harvesting processes and the oil extraction process for use on large scale. In particular, reducing the costs associated with lipid recovery and maximizing the recovery of the lipids in the biomass is vital for economically viable biodiesel production process, since these lipids are transesterified into biodiesel [12].

The objectives of this study were (a) to review the current methods used for pre-treatment of microalgae for enhancing oil extraction (b) to review the current methods used for oil extraction from microalgae and (c) perform a comparative analyses in order to determine the most economically viable and efficient algae pre-treatment and oil extraction methods from microalgae at a large scale.

Microalgae Lipids

Microalgae biomass is composed of proteins, carbohydrates, lipids and nucleic acid that vary widely in proportion (Table 1) depending on the species cellular response [13-15]. There are numerous applications in which these components can be used such as health food additives for human consumption, animal feed, plant fertilizers and/or biofuels [4,13].

Citation: Hattab MA, Ghaly A (2015) Microalgae Oil Extraction Pre-treatment Methods, doi: 10.4172/20904541.1000172.

Table 1: Microalgae composition [15]

Species of sample	Proteins	Carbohydrates	Lipids	Nucleic acid
Scenedesmus obliquus	50–56	10–17	12–14	3–6
Scenedesmus quadricauda	47	-	1.9	-
Scenedesmus dimorphus	8–18	21–52	16-40	-
Chlamydomonas rheinhardii	48	17	21	-
Chlorella vulgaris	51–58	12–17	14–22	4–5
Chlorella pyrenoidosa	57	26	2	–
Spirogyra sp.	6–20	33–64	11–21	-
Dunaliella bioculata	49	4	8	-
Dunaliella salina	57	32	6	-
Euglena gracilis	39-61	14-18	14-20	-
Prymnesium parvum	28-45	25-33	22-38	1-2
Tetraselmis maculata	52	15	3	-
Porphyridium cruentum	28-39	40-57	9-14	-
Spirulina platensis	46–63	8–14	4–9	2–5
Spirulina maxima	60–71	13–16	6–7	3–4.5
Synechoccus sp.	63	15	11	5
Anabaena cylindrica	43–56	25–30	4–7	-

Lipids are contained as small spherical droplets in the chloroplast and between the thylakoid membranes as shown in Figure 1 [16]. They function in the structural support for the cell, the metabolic organelles in photosynthesis metabolism, the growth process of the cell and in the synthesis of lipoprotein membranes contained in the chloroplast [17]. Biodiesel production is generated through the conversion of lipids into fatty acids methyl ester using transesterification process as shown in Figure 2 [18-20]. Thus, lipids must be extracted from the microalgae biomass in order to avoid product contamination from other cellular components.

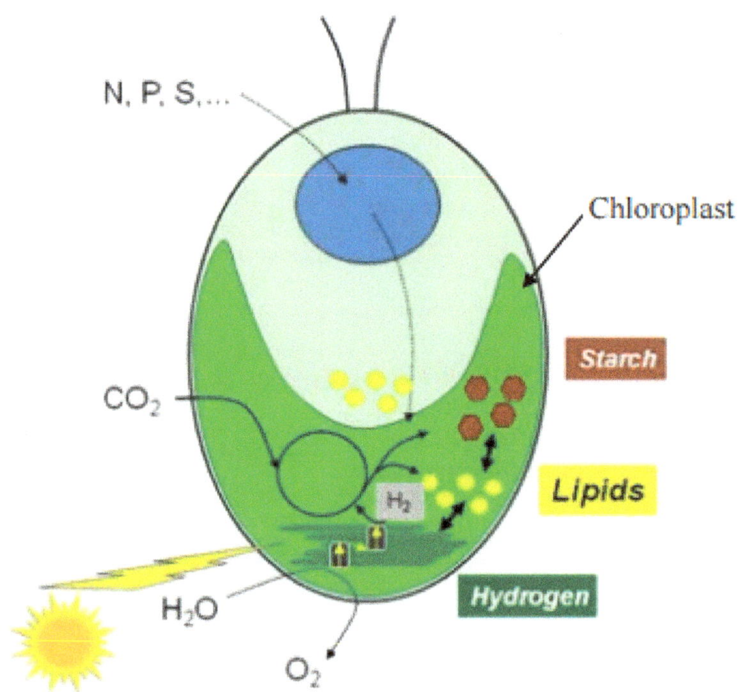

Figure 1: Schematic depicting the location of lipids microalgae cells and the process of formation [16].

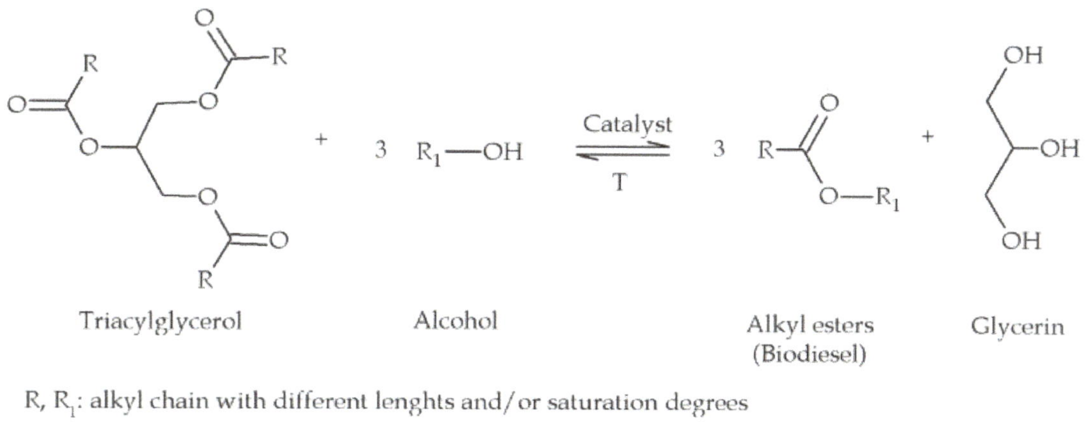

Figure 2: Transesterification of algal lipid into biodiesel [20].

MECHANICAL PRE-TREATMENT OF MICROALGAE

Mechanical pre-treatment of algae to disrupt the cell wall and enhance the efficiency of the lipid extraction process by enhancing the solvent/lipid contact. Disruption of the cellular wall allows for easier recovery of the intracellular lipids resulting in rapid and increased efficiencies in lipid extraction [10,21-23]. Mechanical pre-treatment methods include beadmilling and ultrasound.

Bead Milling

Bead milling is a process that works to disturb the extracellular wall of microalgae by grinding and agitation of the cells on a solid surface of glass beads [24]. Exciting the beads using a bead mill produces a high shear force that can destroy the microalgal cell walls [25]. The optimal diameter size of the beads for effective microalgae cell wall disruption is 0.3-0.5 mm [26,27]. These beads can be made of zirconia-silica, zirconium oxide or titanium carbide [27]. Figure 3 and 4 depicts the difference between a disrupted microalgae cell with one that has not been pre-treated [23,26].

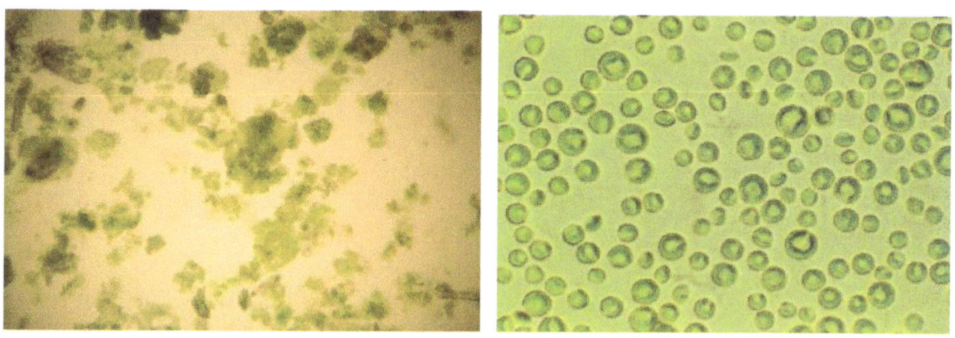

(a) Bead Milling pre-treatment (b) No pre-treatment

Figure 3: *Chlorella* microalgae species before and after treatment [26].

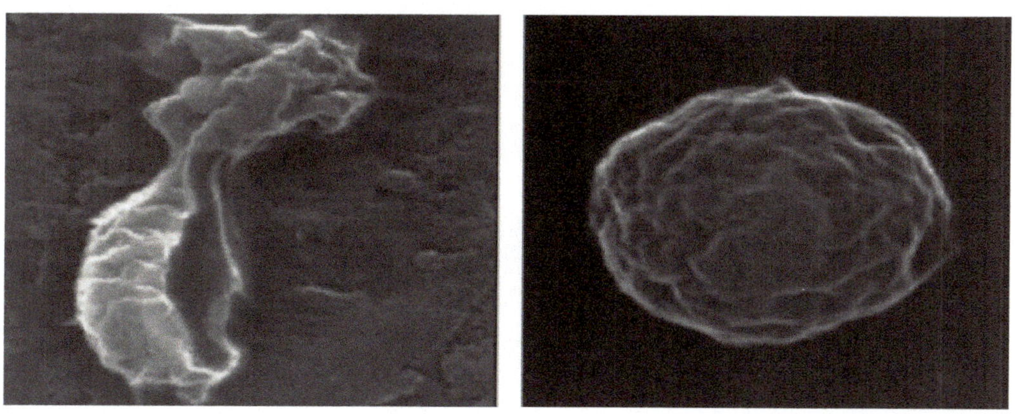

(a) Grinding pre-treatment (b) without pre-treatment

Figure 4: *Nannochlorpisis* sp. cells [23].

Types of bead milling: There are two types of bead milling vessels: shaking vessels and agitated vessels.

Shaking vessels: The shaking vessel works to disrupt the cell walls through the shaking of the entire culture vessel [28]. The vessel or multiple vessels are placed on a platform that vibrates allowing the beads to move and collide with the cells. The use of this type of bead mill is limited to laboratory scale and is not as effective in damaging the cell walls as agitated beads method [28,29]. Figure 5 shows a schematic of shaking vessel [30] Zheng et al. [29] used a bead milling vessel to extract lipids from *Chlorella vulgaris* and noted a recovery of 11%, which was lower than other methods tested. Shen et al. [31] noted that the highest lipid recovery content of 18.8% from *C. protothecoides* was achieved using bead beater shaking vessel. Lee et al. [10] noted that bead beating of *Botryococcus* sp. cells resulted in a lipid extraction of 28%. Prabakaran and Ravindran [32] reported a lipid content recovery of 25-30% from Chlorella sp., *Nostoc* sp. and *Tolypothrix* sp. species using bead beating. Ryckebosch et al. [33] noted a lipid recovery efficiency of 40% from *P. tricronutum* using a shaking beat beater.

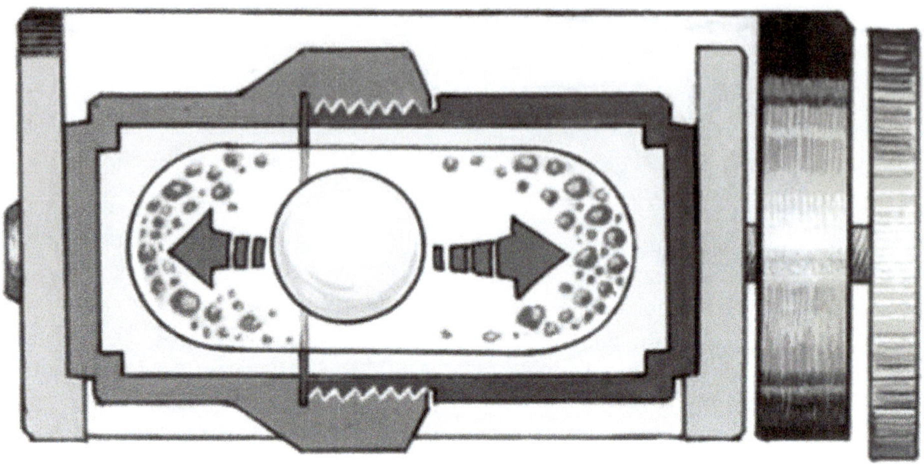

Figure 5: Schematic of bead shaking vessel, jar is held in place inside the vessel and agitated [30].

Agitated beads: In this method, the beads and the culture are both agitated. A rotating agitator inside the vessel supplies the heat. Agitating the beads provides better disruption of the cell walls which increases the extraction efficiency. Figure 6 illustrates a schematic of the bead milling system [34]. For heat-sensitive molecules, the vessel is equipped with cooling jackets. This technique provides agitation, collision and grinding of the biomass which results in effective disruption efficiencies [35].

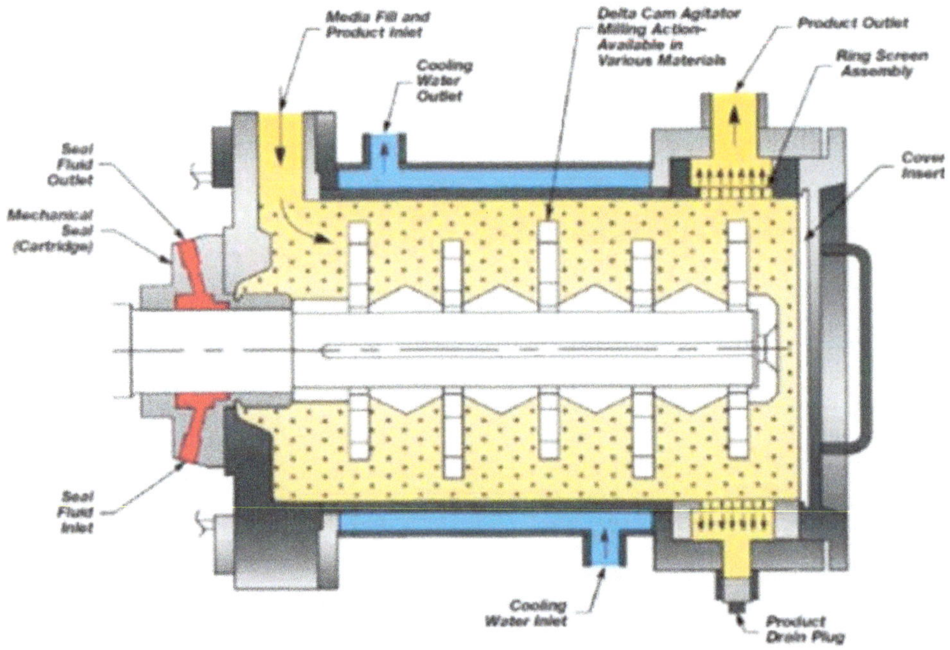

Figure 6: Agitator bead mill unit [34].

Gouveia et al. [36] extracted 33 g of oil from 100 g of *Nannochloropsis* sp. using bead mill assisted techniques. Halim et al. [37] found that agitated bead beating resulted in the disruption of 17.5% of *Chlorococcum* sp. microalga. Baldev et al. [38] reported a 2 fold increase in lipid yield in *Scenedesmus* sp. using agitated bead beater and mechanical grinding of the cells. Lee et al. [10] used bead milling and reported an oil yield range from 7.9-8.1 g/L from *Botryococcus* sp., *Chlorella vulgaris* and *Scenedesmus* sp. Shen et al. [31] noted an oil recovery of 20.5% using bead milling pre-treatment from Chlorella protothecoides. Ceron et al. [39] found that bead milling was the best method for lipid recovery from *Scenedesmus* almeriensis.

Factors affecting bead milling: The degree of disruption of cells is depended on the strength of the microalgae cell walls, contact between the cells and the beads, as well as the shape, size and composition of the beads [26]. However, the biomass concentration, residence time and agitator speed have been identified to be most influential on the cell wall degradation efficiency, processing time and on the energy consumption [22,40].

Biomass concentration: The biomass concentration is one of the primary factors affecting the efficiency of microalgae cell wall disruption [40]. Increasing the biomass concentration in a pre-treatment bead mill leads to a higher fraction of microalgae cell disintegration as shown in Figure 7 [40]. The authors found that increasing the biomass concentration in bead milling of *Chlorella vulgaris* over the range of 25-145 g/L resulted in higher disintegration efficiency.

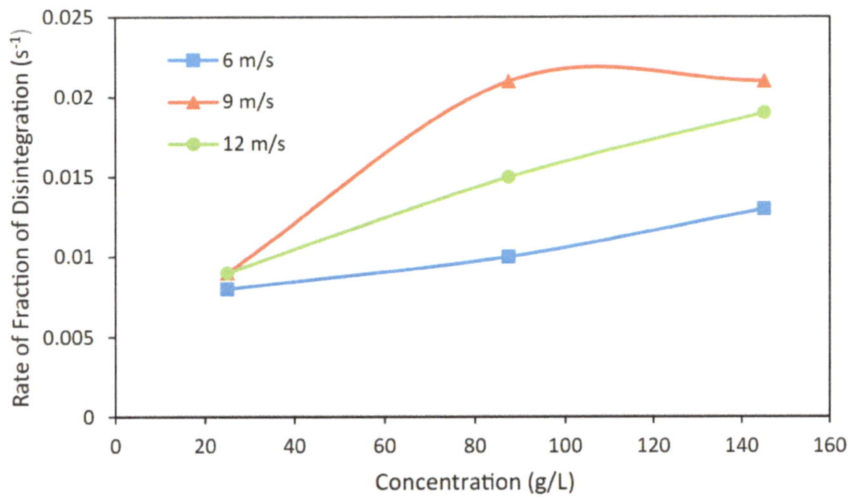

Figure 7: Fraction of cell disintegration as a function of concentration and agitation speed [40].

Doucha and Livansky [26] reported that increasing the Chlorella biomass concentration in bead milling resulted in increased disintegration of microalgae cell wall. Mogren et al. [41] also noted that disintegration rates increased with increasing concentration of yeast biomass in bead milling pre-treatment. Greenwell et al. [22] recommended that the cell concentrations should range between 100 to 200 g/L for high efficiencies and suitable economics, energy-wise.

Agitator speed: The agitator speed is another primary determinant of the efficiency of microalgae cell wall disruption. Postma et al. [40] stated that increasing the speed of agitation increases the impact force and the frequency which results in higher breakdown of cell wall, to a certain degree as shown in Figure 7 [40]. The authors found that increasing the agitator speed of bead milling in *Chlorella vulgaris* culture over the tested range of 9 - 12 m/s resulted in higher disintegration efficiency, but also noted an optimum speed in the range of 9-10 m/s. Doucha and Livansky [26] noted that increasing the agitator speed in bead milling processing of microalgae resulted in increased cell wall destruction. Hedenskog et al. [42] reported cell destruction efficiency in *Scenedesmus* quadricauda biomass of 55% using a speed of 2800 rpm. Lee et al. [10] noted lipid recovery efficiencies of 7.9-8.1 g/L from *Botryococcus* sp., Chorella vulgaris and Scendesmus sp. using a rotational speed of 2800 rpm. Bert et al. [43] noted that increasing the agitator speed from 8 to 14 m/s increased the dispersion efficiency of the agglomerate.

Residence time: The residence time is another primary determinant of the efficiency of microalgae cell wall disruption [22]. The fraction of microalgae cell disintegration as a function of time is depicted in Figure 8 [40]. Increasing the time of bead mill processing leads to higher cell disintegration until maximum disintegration is reached.

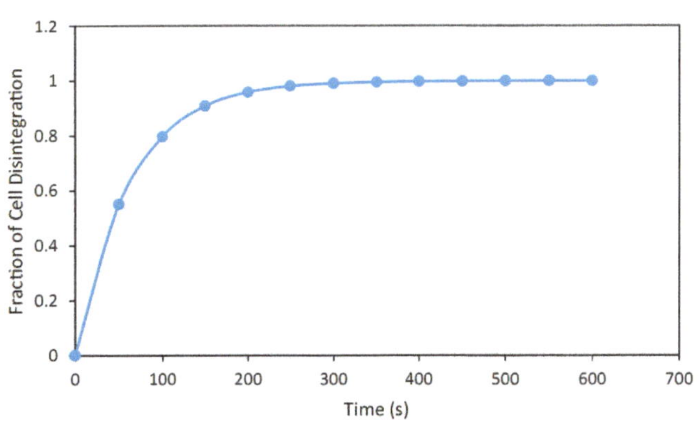

Figure 8: Microalgae fraction of cell disintegration as a function of time [40].

Doucha and Livansky [26] found that the breakdown of microalgae cell wall using bead milling increased from 67% to 95% as the treatment time increased from 30 min to 90 min, respectively. Postma et al. [40] noted that 500s of bead mill pre-treatment of *Chlorella vulgaris* biomass resulted in 99% disintegration, but 90-95% disruption was achieved at 200-250 s. Hedenskog et al. [42] reported cell destruction efficiency in*Scenedesmus* quadricauda biomass of 87% in 5 min. Safi et al. [44] reported that increasing the bead milling time increased the cell wall disintegration which lead to greater protein and pigment recovery efficacies from *Chlorella vulgaris*, up to the maximum reached at 40 min.

Bead density: The bead filling volume plays a role in the effectiveness of cell wall disruption. Increasing the contact between the particles and the cells by increasing the bead density enhances the dispersion upto a limit of 85% [43]. Doucha and Livansky [26] reported that the cell wall destruction values of 65, 83 and 85% were achieved from Chlorella using chamber bead volumes of 60, 75 and 80%, respectively. Hedenskog et al. [42] noted that 33 and 50% bead filling of the vessel resulted in a 55% and 90% cell disintegration of *Scenedesmus* quadricauda biomass, respectively. Liang et al. [45] extracted 38% of lipids from Chlorella vulgaris using 71% bead filling volume. Schwenzfeier et al. [46] used a bead filling volume of 65% for the algae Tetraselmis sp. biomass for effective cell disruption for protein recovery. Bert et al. [43] noted that the dispersion efficiency was improved with increasing bead filling volumes over the tested range of 70-85%, but bead fill volumes greater than 85% significantly increased the wear of the mill.

Bead type and size: Varying beads and bead sizes have different impact on the dispersion efficiency [43]. Hedenskog et al. [42] noted a cell disintegration of 55% using ballotini beads (0.35-0.5 mm) in *Scenedesmus* quadricauda biomass. Doucha and Livansky [26] found that zirconium dioxide beads (0.5 mm) and glass beads (0.42-0.58 mm) resulted in Chlorella biomass cell disintegration efficiencies of 98.5 and 99.9%, respectively. Lee et al. [10] noted lipid recovery efficiencies of 7.9-8.1 g/L from *Botryococcus* sp., *Chlorella vulgaris* and Scendesmus sp. using glass bead (0.1 mm diameter).

Zheng et al. [29] used glass beads (0.4-0.6 mm diatmeter) to recover 10% of lipids in *Chlorella vulgaris* biomass. Lee et al. [47] reported a lipid recovery efficiency of 28.6% from *Botryococcus* braunii using 1 mm glass beads. Bert et al. [43] tested the effect of varying ceramic bead sizes (0.6-1.6 mm) on the agglomerate size and found that 0.8 mm beads were the most effective in achieving the lowest agglomerate size of 41 μm and ceramic beads significantly improved the dispersion efficiency compared to glass ones.

Energy consumption: Effective disruption of microalgae cell wall using bead mill pre-treatment has been noted to consume 35-810 Wh/kg of power [22,26,40]. The large variation in power consumption is a result of varying microalgae species and operation parameters.

Lee et al. [10] disrupted the microalgae *Botryococcus*, Chlorella and *Scenedesmus* using agitated bead mill with an energy consumption of 140 Wh/kg. Doucha and Livansky [26] noted that the power consumption for rupturing Chlorella using an agitated bead mill was in the range of 35-250 Wh/kg. Greenwell et al. [22] reported that the energy consumption for disruption of cell wall using a bead beater ranged from 300-400 Wh/kg. Postma et al. [40] noted an energy consumption of 810 Wh/kg for bead milling of Chlorella sp. biomass.

Advantages and disadvantages: The advantages of using bead beating are the simplicity, rapidness of the method, reproducibility of results and low labor intensity requirement [48,49]. However, pre-treatment of microalgae cells using bead milling can be difficult to scale up and requires the use of a cooling jacket in order to prevent the degradation of the desired product [10,25,48,50]. Additionally, bead mill pretreatment is not a selective product recovery technique, which requires further processing to remove the undesired compounds [49]. The biomass undergoing bead disruption techniques must be dry and concentrated in order to achieve high disruption efficiency [49,51].

Ultrasonication

Ultrasonication is another mechanical method that can be used for pre-treatment of microalgae prior to lipid extraction. In this method, algae are exposed to high intensity ultrasonic waves, creating tiny cavitation bubbles around the cells. The bubbles collapse and emit shockwaves that shatter the cell walls causing the intracellular lipids to enter the bulk of the solution as shown in Figures 9 and 10 [24,52-54]. Ultrasonic assisted microalgae lipid extraction has been noted to significantly increase the yields and reduce the extraction time [10,24,55]. Pernet and Tremblay [56] concluded that the ultrasonic method for oil extraction from Chaetoceros gracilis increased the extraction rate which affects the recovery of lipid extracts. There are some contradictions in the literature regarding scale up. Halim et al. [37] noted that this technique is moderately suitable for scale up whereas Mercer and Armenta [24] stated that ultrasound maybe difficult for upscale.

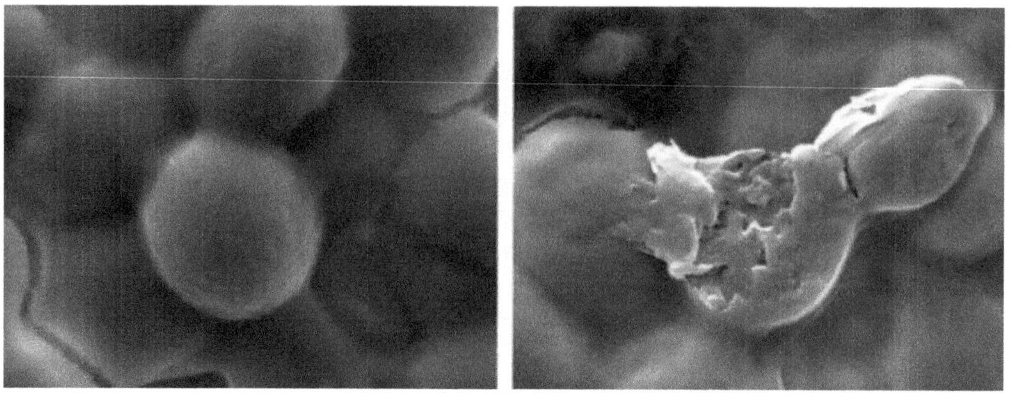

Figure 9: Micrographs of microalgae before and after ultrasound treatment [53].

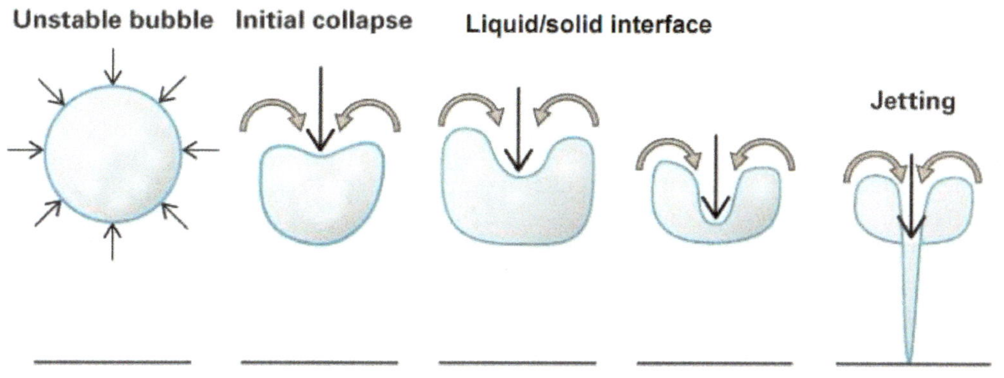

Figure 10: Schematic of cell disruption using ultrasound energy [54].

Wiltshire et al. [57] reported a 90% extraction efficiency of fatty acids and pigments from the species *Scenedesmus* obliquus using ultrasound extraction. Ranjan et al. [58] reported that ultrasound assisted microalgae lipid extraction demonstrated more distorted clusters of biomass on micrographs, in comparison to cells with solvent penetration. Cravotto et al. [59] noted that ultrasound assisted lipid extraction from Crypthecodinium cohnii resulted in an increase in lipid yield of 21.1% as opposed hexane solvent extraction.

Lee et al. [10] tested various cell disruption techniques (autoclave, bead milling, microwave and sonication) on lipid extraction from*Botryococcus* sp. and found that sonication was the least efficient in lipid extraction, but the yields were higher than that with sole solvent extraction. Shen et al. [31] reported that the lipid content for Chlorella protothecoides was least using sonication treatment compared to methods of bead beating and press. Zheng et al. [29] tested treatment of *Chlorella vulgaris* biomass using microwave, ultrasound, enzyme lysis and bead beating, which resulted in lipid yields of 18, 15, 22 and 10%, respectively. Prabakaran and Ravindran [32] found that the lipid recovery from Chlorella sp., *Nostoc* sp. and *Tolypothrix* sp. was highest using sonication pre-treatment compared with auto-calving, bead beating and microwave. De Souza Silva [60] tested the pre-treatment of microalgae culture using microwave, autoclaving and ultrasonication technology for lipid extraction and found that ultrasound resulted in the lowest yields. Koberg et al. [61] reported a lipid yield of 18.9% and 32.8% in Nannochlorpsis species using microwave and ultrasonication pre-treatment, respectively.

Ultrasound types: There are two types of ultrasound units: horns and baths [62]. Both of these types are used for batch operations, but the addition of flow cells can alter them into continuous operational modes [63,64].

Horn: Horns type ultrasonic technology use a piezoelectric generator that is composed of lead zirconate titanate crystals that vibrate with amplitude ranging from 10 μm to 15 μm (Figure 11) [65]. The formed vibrations travel down the titanium metal horn or probe, increasing in amplitude ranging from 100 μm to 150 μm at the tip. The power at the tip should be of high intensity in order to create cavitation with sufficient disruptive force since the energy dissipates rapidly with distance. The use of horn type sonicators is limited to laboratory scale handling 10- 100 mL volumes. Setting the horn to vibrate laterally would increase the area of contact for larger cavitation, but larger area would decrease the intensity of the cavitation.

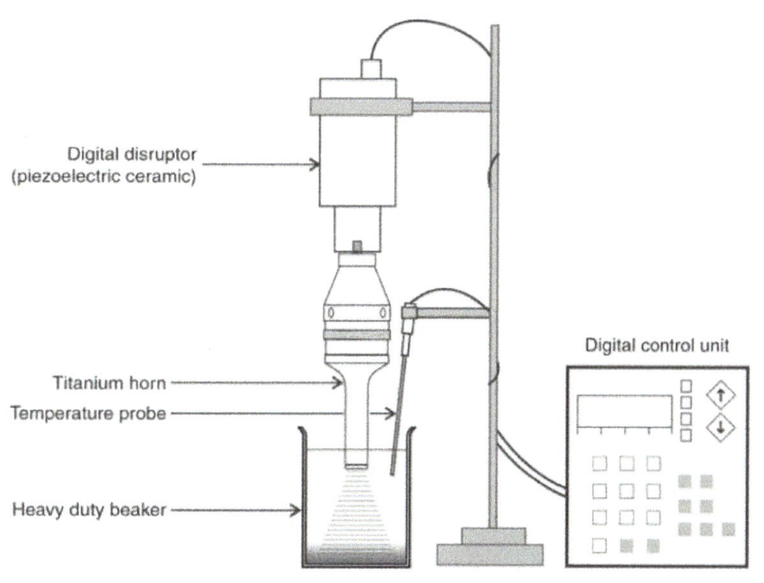

Figure 11: Horn sonication unit [65].

The scale-up of horns sonication would require the use of multiple systems and the use of continuous flow of the cells [35]. Jeon et al. [66] stated that disruption of microalgae biomass using horn sonicators is not suitable because the cavitation is localized. Wang et al. [67] noted that the relative lipid increase rate for *S. dimorphus* (30 ml) using horn sonication was lower than that using bath sonication at operational times of 1 and 5 min. Cravotto et al. [59] noted that horn sonication of the marine microalgae Crypthecodinium cohnii (50 ml) was effective in increasing the lipid recovery. Menendez et al. [55] reported that horn ultrasonic treatment did not increase the lipid yields in *Nannochloropsis* gaditana microalgae, compared to conventional extraction techniques.

Bath: Ultrasonic baths use transducers, placed as the bottom of the reactor, to generate the ultrasonic waves (Figure 12). The number and the arrangement of the transducers vary with the capacity of the reactor and the shape. Sonicator baths have a larger capacity (up to 3 L), but the reactor size is limited by the rapid rate of sonic energy dissipation with distance. Baths can be modified in flow cells for continuous industrial operation and higher efficiencies may be achieved using multiple transducers operating with 2 or 3 varying frequencies in the cell. Wiyarno et al. [69] noted a lipid recovery of 26% from Nannochlorpsis using ultrasonic bath treatment. McMillan et al. [70] investigated the disruption efficiency of *Nannochloropsis* oculata biomass using microwave, mechanical force and ultrasonic bath treatments and found that ultrasound was the least effective. Neto et al. [71] treated *Chlorella minutissima*, *Thalassiosira fluviatillis* and *Thalassiosira pseudonana* with ultrasonic bath waves and noted that ultrasonic was a necessary step for cell disruption in order to increase the lipid recoveries. Piasecka et al. [72] noted that sonication bath treatment of *Chlorella protothecoides* biomass resulted in an increase in lipid yields of 37%.

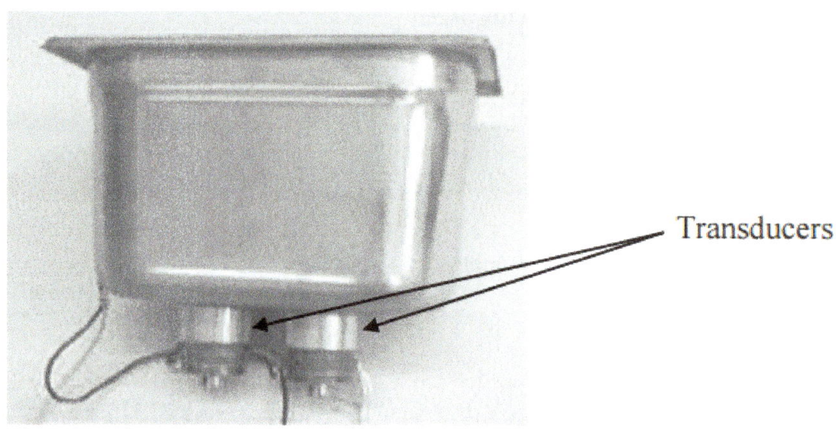

Figure 12: Ultrasonic bath transducers [68].

Factors affecting ultrasonic assisted extraction: The pretreatment time, temperature of the reaction, cell concentration and microalgae type have been reported to influence the lipid yield in microalgae lipid extraction assisted with ultrasound technology.

Extraction time: Increasing the time in which the cells are exposed to ultrasound can affect the recovery rate. Longer treatment time allows for increased cell disruption as a result of additional energy input [35]. Menendez et al. [55] reported that an increase in extraction time from 5 to 20 min increased the lipid yield from 31 to 36%, respectively. Adam et al. [73] noted that increasing the treatment time resulted in higher lipid recovery efficiencies. Tang et al. [74] found that increasing the ultrasonic treatment time over the range of 15 to 90 min has no significant effect on the microalgae lipid recovery. Wiyarno et al. [69] reported that an increase in sonication time results in higher lipid yields. McMillan et al. [70] found that increasing the sonication time resulted in greater cell disruption efficiencies. However, Prommuak et al. [75] found that increasing the reaction time from 15 to 30 min resulted in decreased lipid yields from *Chlorella vulgaris* and *Haematococcus pluvials* species as a result of lipid oxidation with prolonged treatments.

Reaction temperature: Altering the reaction temperature during ultrasonic treatment has been noted to influence the lipid recoveries from microalgae biomass. Prommuak et al. [75] found that an increase in temperature from 30 to 40°C recovered slightly higher lipids after 5-10 min. Adam et al. [73] reported that increasing the temperature from 1 to 35°C in ultrasound assisted oil extraction from *Nannochloropsis*oculata resulted in increased oil yields by a factor of 1.5. Wiyanrno et al. [69] noted that increasing the reaction temperature from 23 to 60°C resulted in increases in lipid yields.

Cell concentration: The effectiveness of microalgae cell disruption using sonication technique has been noted to change with cell concentrations. Lee et al. [35] stated that increased cell fragments build up during treatment reduces the efficiency of the process and concluded that sonication of concentrated biomass is less efficient in dilute suspensions. Nowotarski et al. [76] reported that sonication of the microalgae Dunnaliella salina was most effective at low densities. Adam et al. [73] found that the optimal concentration of microalgae cells for lipid extraction using sonication was 5% over the tested range of 5-30%. However, Gerde et al. [77] found that varying the microalgae concentration over the tested range of 1.5-14.1 g cells/L did not require higher sonication energy for complete disruption. Natarajan et al. [78] reported that the cell disruption efficiency increased with an increase in cell concentration up to 6.84 g/L over the tested range of 0.07 g/L to 12.22 g/L.

Microalgae type: The components of the microalgae cell wall coating are species specific and play an important role in the disruption efficiency using sonication techniques. Diatoms have siliceous cell wall coating called frustules and the degree of silicification of the frustules varies with cells of the same species as a result of nutrient availability [79]. Neto et al. [71] reported that for lipid extraction from diatom microalgae, sonication greatly impacted the cell disruption and improved the oil recovery. Joyce et al. [80] reported that suspensions of*Nannochloropsis* oculata was unaffected by the ultrasound treatment as a result of varying microalgae cell wall thickness but Dunaliella salina and Chlorella concordia resulted in complete cell disruption with 4 min and 16 min sonication treatments, respectively. Nowotarski et al. [76] reported that sonication of Dunnaliella salina was effectively disrupted at low concentrations while *Nannochloropsis* oculata was much more resistant to sonication even at long sonication exposures. Natarajan et al. [78] found that the Chlorella sp. cells released the lipids into the liquid suspension after sonication treatment but Tetraselmis suecica and *Nannochloropsis* sp. retained the lipids in the membrane after undergoing ultrasonication treatment.Takeda [81] noted that lipid extraction from Chlorella minutissima did not improve with sonication treatment as a result of cell fragility compared to diatoms. Sostaric et al. [82] reported an increased in lipid yield of 6.02% in *Chlorella vulgaris*biomass using ultrasonic bath treatment because the Chlorella species possess a thin cell wall made up of sugar [83]. Kaiwan-arporn et al. [84] compared lipid yields from Synechocystis aquatilis using grinding and sonication techniques and found that the yields were 21.3% and 10.2%, respectively. Shen et al. [31] noted that the lipid yield using sonication is affected by cell size, shape and structure and reported lipid yields of 10.7 and 21.2% for *Chlorella protothecoides* and *Scenedesmus* dimorphus, respectively.

Ultrasonic power: The ultrasonic power has been noted to have negligible effects on the oil yield of varying microalgae species. Natarajan et al. [78] tested ultrasonic power of 500, 750 and 1000 W on Tetraselmic suecica and Chlorella sp. oil yields and found that over the tested range no variations in yields were noted. Tang et al. [74] tested the effects of varying ultrasonic power on the lipid yield of microalgae over the range of 80-200 W and found no significant difference in the oil yields with varying power. Cravotto et al. [59], Singh and Gu [52] and Adam et al. [73] have also reported that the power of sonication has little effect on the oil yields of microalgae species.

Energy consumption: Menendez et al. [55] reported that the energy consumption using ultrasound assisted lipid extraction from*Nannochloropsis* gaditana increased from 1.7 to 6.7 Wh/g dried biomass when the reaction time was increased from 5 to 20 min which corresponded to increases in the lipid yields of 31 and 36%. Guldhe et al. [85] reported that the energy consumption using ultrasound assisted lipid extraction from *Scenedesmus* sp. was 119 Wh/g. Adam et al. [73] reported an energy utilization of 16.66 Wh/g for lipid extraction from the microalga *Nannochloropsis* oculata. Halim et al. [37] noted that the sonication energy consumption for *Chlorococcum* sp. was 36.66 Wh/g of dried biomass, but the lipid yields were only slightly higher than that from untreated biomass. Bigelow et al. [86] found that sonication of Chlamydomonas reinhardtii microalgae biomass consumed 36 Wh/g. Wang et al. [67] noted that sonication of *Scenedesmus* dimorphus and*Nannochloropsis* oculata consumed energy in the range of 17.3-86.8 Wh/g.

Advantages and disadvantages: Pre-treating microalgae biomass using ultrasound has several advantages which include: reduced extraction time, less solvent requirement, higher yields as a result of easier cell penetration, the biomass does need to be dry and easier release of intracellular components to the bulk of the solvent [87-90]. Chemat et al. [89] and Wang and Weller [90] stated that ultrasound assisted extraction can be operated at low temperatures (less thermal denaturation of biomolecules) and is much more economical compared to other conventional extraction methods. However, ultrasound assisted techniques consume large amounts of power and pose scale up difficulties [87,88]. The use of high-intensity sonication can result in pressurize/heat that is damaging to the cells or tissue [67]. This technique can be species specific since it has been noted to be ineffective in the destruction of diatom species with thick cellular coating [76,78,80].

Thermal Pre-Treatment of Microalgae

Thermal methods can be used to enhance the lipid recovery from microalgae by disrupting the cell wall which allows for the release of the intercellular lipids. Thermal pre-treatment methods of microalgae include steam explosion, autoclaving and freeze drying.

Steam Explosion

Steam explosion may also be used to disturb the microalgae cell wall so that the intracellular components maybe easily recovered. The treatment works by exposing the biomass to high temperatures that vary from 160°C to 260°C [91-93] and vapor pressure in the range of 1.03 and 3.45 MPa, followed by the return to ambient conditions through depressurization. The drop in pressure to ambient conditions results in cell wall rupture of the biomass [94-98]. Pre-treating microalgae biomass with steam technology should be performed at lower temperatures in order to prevent lipid degradation [95].

A batch unit setup for steam pre-treatment of microalgae is illustrated in Figure 13 [94]. The unit consists of a 4 L steam generator, 2 L reactor and a collection vessel. The sudden depressurization takes place in the collection tank which is equipped with a flash valve to allow the return to atmospheric conditions. This form of cell wall disruption is efficient and economically viable method [95].

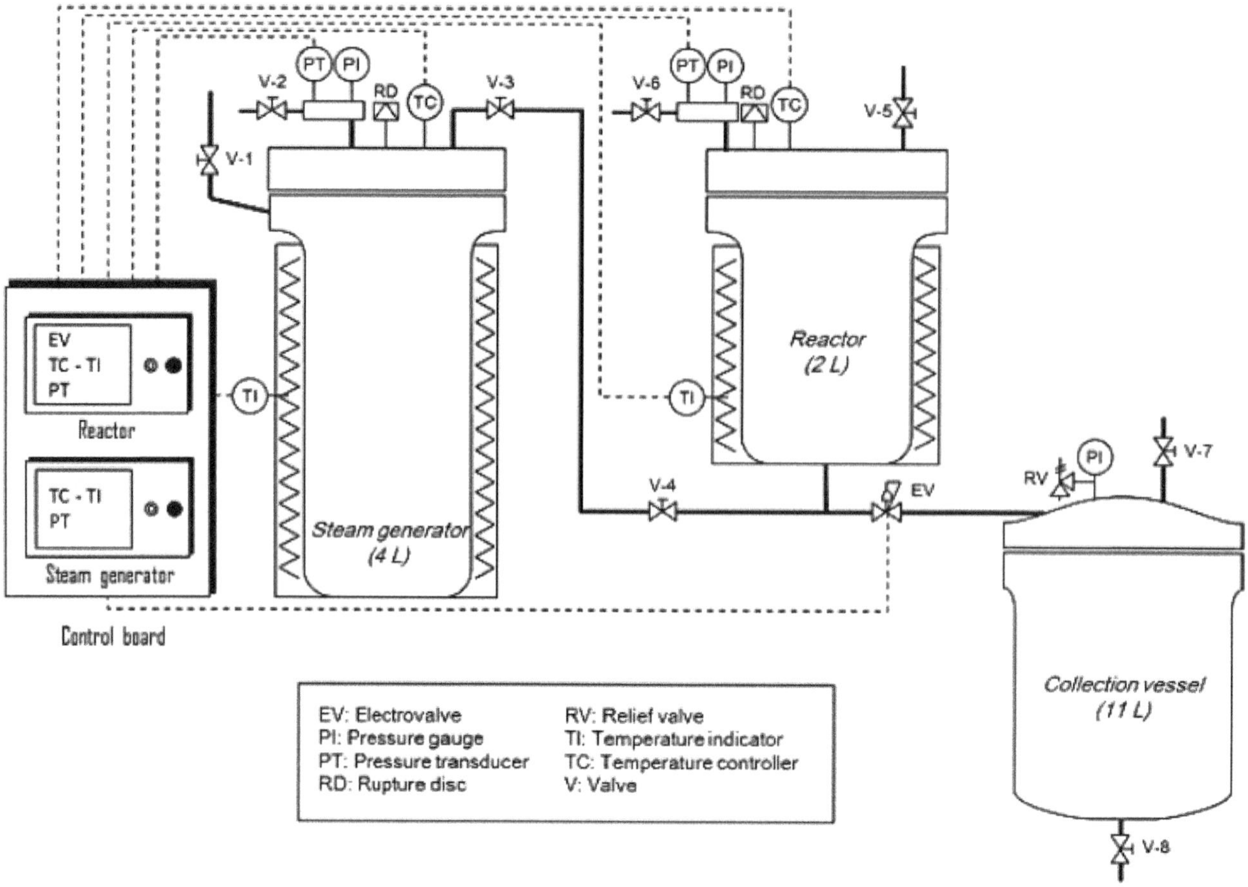

Figure 13: Schematic of steam explosion apparatus [94].

Lorente et al. [94] reported that steam explosion pre-treatment of three types of microalgae biomass resulted in the higher lipid extraction efficiencies compared to autoclaving, ultrasound and microwave techniques. Nurra et al. [95] noted that the steam explosion pretreatment of*Nannochloropsis* gaditana biomass resulted in an increased lipid recovery from 0.3% to 3.6%. Mosier et al. [99] and Montane et al. [100] reported that steam explosion is an economically efficient pretreatment technique for fractionating and modifying lignocellulosic materials which improves the feedstock quality for downstream processing.

Factor affecting steam explosion: The factors which play an important role are steam explosion cell wall destruction of microalgae is temperature, pressure and microalgae species.

Temperature: The temperature plays an important role in the effectiveness of microalgae cell wall rupture [95]. The optimum temperature for hydrolysis of microalgae biomass depends on the microalgae species [101].

Lorente et al. [94] found that steam explosion of *Nannochloropsis* gaditana at 120°C and 150°C resulted in an increase in lipid extraction yields of 8.1 and 8.4%, respectively. Nurra et al. [95] noted that increasing the temperature in steam explosion pre-treatment of *Nannochloropsis*gaditana biomass from 120 to 180°C increased the lipid yields from 3.6 to 8.8%. Du et al. [102] noted that pre-treatment of microalgae biomass using steam explosion at a temperature of 150°C for 50 min was not sufficient in destruction of the cell wall. Alzate et al. [101] reported that increasing the temperature from 110°C to 170°C resulted in increased microalgae biomass biodegradability (10% to 27%). Mendez et al. [103] reported an increase in protein solubilisation in *Chlorella vulgaris* biomass of 16% at temperature of 140°C and an increase in temperature in the range of 160 to 180°C resulted in solubilisation of 45%. Aguirre and Bassi [104] noted that an increase in temperature from 104°C to 210°C in *Chlorella vulgaris* increased the extraction efficiencies from 24 to 77%.

Pressure: Altering the temperature in the steam explosion treatment vessel, changes the pressure of the system which effects the solubilisation of the microalgae biomass. Aguirre and Bassi [104] stated that temperature and pressure are correlated in steam treatment. Lorente et al. [105] noted a slight increase in microalgae lipid recovery from 17.9 to 18.2% using steam explosion with increases in pressure from 2 to 4.7 bars which was achieved by increasing the temperature from 120 to 150°C. Nuraa et al. [95] reported an increase in lipid recovery from 3.6 to 8.8% using steam explosion pre-treatment of microalgae biomass as the pressure was increased from 2 to 10 bar by increasing the temperature from 120°C to 180°C. Mendez et al. [103] reported that with increasing pressure from 3 to 10 bars in steam explosion of *Chlorella vulgaris* biomass the carbohydrate solubilisation increased. Robles Medina et al. [105] noted that carotenoid extraction from Haematococcus biomass increased by 15% when treated with high pressure compared to no pretreatment techniques.

Microalgae species: The lipid recovery from microalgae species using steam explosion varies with the type of species as a result of different cellular composition. Lorente et al. [94] found that the lipid recovery using steam explosion resulted in lipid recovery increases of 8.7%, 9.5% and 2.1% compared to Bligh and Dyer extraction from *Nannochloropsis* gaditana, Chlorella sorkiniana and Phaeodactylum tricornutum, respectively. Alzate et al. [101] noted that steam explosion treatment for the production of biogas using Chlamydomonas, *Scenedesmus* and Nannocloropsis resulted in a biodegradability of 70% while Acutodesmus obliquus, Oocystis, Phormidium and Nitzschia sp. species resulted in a biodegradability of 30% using steam explosion treatment. Pandey et al. [106] stated that the effectiveness of steam pre-treatment of microalgae is dependent on the cell wall composition which varies among microalgae species.

Cell concentration: The biomass concentration has been noted to influence the effectiveness of the steam explosion technique. Aguirre and Bassi [104] found that the highest lipid yields from *Chlorella vulgaris* treated with steam, were achieved at concentrations lower than 5 g/L. Kita et al. [107] found that *Botryococcus* braunii treated with steam resulted in hydrocarbon recoveries of 90% or greater when operating at thermal temperatures of above 85°C with a biomass concentration of 1.5 g/L, while Frenz et al. [108] noted a recovery of less than 1% using cell concentrations of 1 g/L of B. braunii, thus concentrations above 1 g/L were recommended. Menedez et al. [55] reported that high pressure steam treatment at temperatures above 160°C of *Chlorella vulgaris* at concentrations of 16 g/L only solubilized 45% of the proteins, thus lower concentrations are required to achieve a higher degree of solubilisation.

Energy consumption: Very little work has been done on the investigation of steam explosion as a pre-treatment technique on microalgae biomass. Keymar et al. [109] noted that the energy consumption for using steam explosion pre-treatment on microalgae biomass consumed 2.92 Wh/g of volatile solid. Ko et al. [110] reported an energy consumption of 1.11 Wh/g for steam explosion pre-treatment of Ma bamboo for the production of 1 L of ethanol. Zhu and Pan [111] found that the steam pre-treatment of wood biomass consumed 0.98 Wh/g.

Advantages and disadvantages: Microalgae steam explosion pretreatment is advantageous as it disrupts that cellular wall as a result of sudden pressure release, making the lipids accessible for rapid recovery [95] without the release of hazardous wastes [91,112]. Other advantages to steam explosion pre-treatment are its relatively low energy consumption [111,112], low maintenance costs and its low corrosion potential [112]. However, the effectiveness of this technique is species specific [106]. To date steam explosion pre-treatment of microalgae biomass has only been recently studied and is under investigation at laboratory scale and has been primarily used for the production of biogas product [106].

Autoclaving

Autoclaving (Figure 14) microalgae biomass is a form of thermal treatment operating at a temperature of 121°C and pressure of 15 lbs [113,114]. High thermal stress causes the cell walls to rapture forcing the release of the intracellular lipids [32,37].

Figure 14: Autoclave [115].

Surendhiran and Vijay [113] noted that autoclave pre-treatment of *Nannochloropsis* oculata biomass resulted in higher lipid yields than those without initial treatment. Lee et al. [10] also found that maximum oil recoveries from *Chlorella vulgaris* were achieved with autoclave pre-treatment. Prabakaran and Ravindran [32] reported that the lipid yield recovered from *Chlorella vulgaris* was higher with autoclave pretreatment than that without pre-treatment, but the lipids were not higher than microwave treated biomass. De Souza Silva et al. [60] noted that autoclave and microwave pre-treatment of microalgae biomass resulted in lipid yields of 15.4 and 33.7%, respectively.

Factors affecting autoclave assisted extraction: The efficiency of autoclave technology is dependent on time elapsed of the treatment as well as the type of microalgae.

Time: Altering the autoclave treatment duration has been reported to effect the effectiveness of microalgae lipid recoveries. Surendhiran and Vijay [113] pretreated *Nannochloropsis* oculata biomass using autoclave operating at 121°C and a pressure of 15 lbs at 10, 20 and 30 min and achieved the highest lipid recovery of 29.34% at 30 min treatment. Prabakaran and Ravindran [32] noted an increase in lipid content of 22% in*Nannochloropsis* oculata with autoclave pre-treatment at 121°C for 5 min, but higher yields were achieved using microwave pre-treatment. Lee et al. [10] reported an increase in lipid recovery of 4% in *Botryococcus* when autoclaved for 5 min. Rakesh et al. [115] found that the autoclave pre-treatment of *Botryococcus* sp. for 15 min was effective in increasing the lipid recovery from the cells.

Microalgae type: The species type has been noted to effect the autoclaves ability to effectively disrupt the cell wall. Rakesh et al. [115] reported that autoclaving microalgae biomass for 15 min effectively increased the lipid recoveries in *Botryococcus* sp., but resulted in a 50% reduction in total lipid recovered from Chlorella sorokinana species. Prabakaran and Ravindran [32] reported higher lipid recoveries in Chlorella sp., *Nostoc* sp. and *Tolypothrix* when autoclaved, however higher recoveries were achieved using microwave, sonication and bead beating since they are more abrasive destruction techniques. Lee et al. [10] found that autoclave pre-treatment was effective in increasing the lipid recovery in *Scenedesmus* sp., but was not effective on *Botryococcus* sp. and *Chlorella vulgaris* compared to other pre-treatment methods investigated. Yu et al. [116] reported that autoclaving Chlorella sorokiniana species resulted in the lowest lipid yields compared to other techniques such as bead beating, microalgae and sonication. Miranda et al. [117] noted that autoclave treatment of *Scenedesmus* biomass was the most efficient technique for sugar recovery. The effectiveness of autoclaving treatment on different microalgae spices varies as a result of different cell wall structures that can be tough and unaffected by autoclave disruption techniques [117,118].

Energy consumption: The energy consumption for autoclave pretreatment of microalgae biomass is high as a result of high operational temperatures and pressures [119]. This technique has been shown to consume higher amounts of energy than those of bead beating techniques [10]. Lopata et al. [120] reported that the costs of autoclave machinery range from high to very high. Panasonic [121] reported that their laboratory size (50 L) autoclave (Model No. MLS-3751L-PA) has a power consumption of 1900 watts. SciCan [122] noted a power consumption of 1700 watts for their Bravo laboratory scale (17 L) autoclave. Market Forge STM-E autoclave, higher capacity of 85 L has a power consumption of 9000 watts [123].

Advantages and disadvantages: The use of autoclave pre-treatment of microalgae biomass prior to lipid extraction is advantages because it disturbs the extracellular cell membrane allowing for easier recovery of lipids due to increased penetration [119]. Autoclaving cells at high temperatures over a short duration can reduce the degradation of the desired product [124]. However, this technique is difficult to upscale, long duration of time required, and large scale use would require high energy consumptions [119].

Freeze drying

Freeze drying (Figure 15) microalgae biomass is one of the most preferred drying techniques as a result of its mild operating conditions and the ease of lipid extraction after pre-treatment [50,125]. Figure 16 illustrates the cell structure of microalgae species that have been freeze dried prior to lipid extraction [126]. The solvent extraction of lipids from wet biomass can be difficult, thus prior freeze drying of the biomass will overcome this difficulty [50]. However, microalgae lipids are susceptible to degradation under thermal drying techniques and can result in evaporative loss of lipids [127]. Further milling of the freeze dried microalgae would enhance the efficiency of lipid extraction as a result of increased surface area for biomass-solvent contact and diminishes the diffusion pathway [37].

Figure 15: Industrial freeze dryer [126].

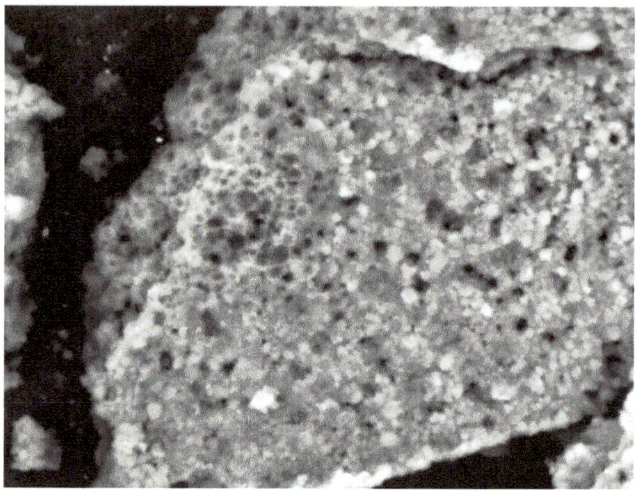

Figure 16: Scanning electron microscope micrograph depiction of freeze-dried microalgae biomass [127].

Guldhe et al. [85] stated that water needs to be removed from microalgae biomass in order to increase the lipid extraction efficiency. Pasquet et al. [126] found that freeze drying of the microalga species Dunaliella tertiolecta preserved the integrity of the cell. Arora [128] noted that freeze dried microalgae biomass is more effective in the recovery of microalgae lipids as opposed to the solvent recovery using wet biomass. Belarbi et al. [129] reported that solvent extraction of oil from wet biomass can be difficult, but prior freeze drying of the biomass can extract lipids readily. However, Guldhe et al. [85] noted that solvent lipid extraction from *Scenedesmus* sp. dried using freeze-dryer; oven and sun drying techniques did not significantly differ from one another. Balasubramanian et al. [130] also noted that the drying type (freezedrying, oven and sun drying) of *Nannochloropsis* sp. biomass has no significant influence on the lipid recovery using solvent extraction. Fajardo et al. [131] used freeze dried biomass in the solvent recovery of oil from P. tricornutum microalga species.

Energy consumption: The energy consumption reported in the literature for freeze-drying of microalgae biomass has only been investigated on a small laboratory scale. Guldhe et al. [85] noted that the energy consumption for freeze-drying *Scenedesmus* sp. biomass was 17.7 kWh for every gram lipid recovered. Green and Perry [132] reported that the energy consumptions for drying microalgae using a freeze dryer of 5.56 Wh/g. Bennion [133] found that the energy consumption for freeze drying microalgae biomass was 5.27 Wh/g. Ratti [134] calculated that the industrial freeze drying of biomass would require 38.88 Wh/g of biomass. Variation in the energy usage is a result of varying operational parameters and operational duration.

Advantages and disadvantage: Freeze drying of microalgae biomass breaks up the biomass cells and turns them into fine powder, eliminating the need for homogenization [135]. Freeze drying is one of the most commonly used technique for the production of high value products because it is gentle and does not have adverse effects on the cellular components [33]. Recovery of lipids from freeze-dried microalgae biomass does not require the use of a prior pre-treatment technique [129]. However, the process can be time consuming and the costs can be high [35,136]. Freeze drying has been deemed unsuitable for large scale operation as a result of high costs associated with scale up [50,136].

ELECTROMAGNETIC RADIATION

Lee et al. [10] tested bead milling and microwave cell disruption

Microwave

Microwave pretratment processes enhance the kinetics of the lipid extraction process by disruption of the cellular wall [90,103]. This technique provides large amounts of thermal energy from electromagnetic radiation with a certain frequency to the cells [137,138]. A schematic of the microwave processing system is illustrated in **Figure 17** [137]. The process consists of a feeding tank, microwave unit, water bath and a peristaltic pump that generates feed circulation through the system. Tubes with a diameter of 0.953 cm were used to carry the microalgae biomass from the feed-tank to the microwave unit. In the microwave chamber the mixture is heated using microwave radiation within seconds. Varying the radiation time alters the cell wall disruption efficiency. After treatment, the processed fluid is released into a 50 mL beaker that is held in a constant temperature water bath [137].

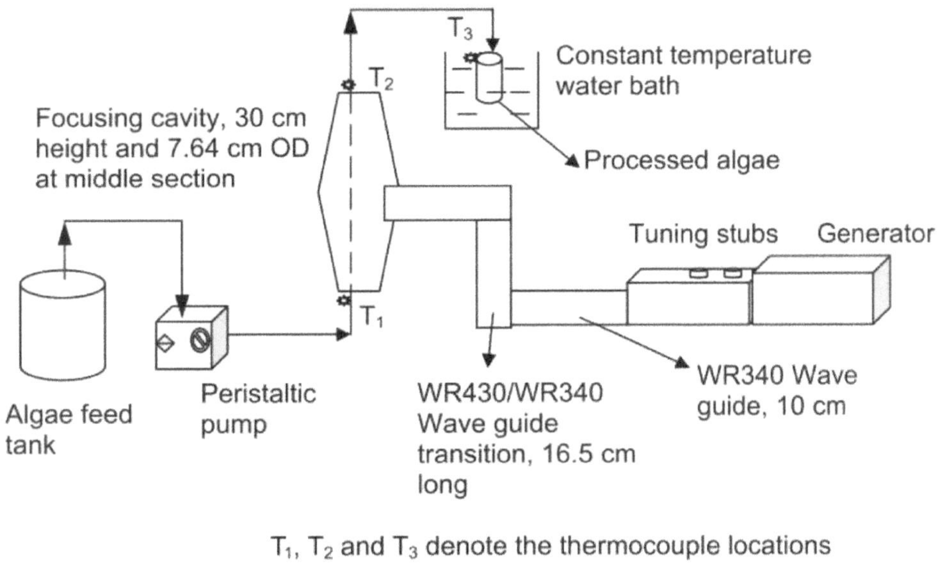

Figure 17: Microwave processing system [139].

The thermal energy is a result of frictional forces that are caused by inter and intra-molecular movements [139]. This energy causes temperature and pressure effects on the cell wall that result in cell wall rapture [37]. Disruption of the cell walls allows for the cell components to be released which increases the efficiency of the extraction process by overcoming the concentration gradient associated with solvent extraction [10,140].

Balasubramanian et al. [137] investigated the use of microwave assisted solvent extraction for lipids from S. obliquus and noted that higher oil yields were achieved with solvent system hexane compared to the hexane extraction without microwave radiation at all temperatures and reaction times investigated.

Techniques on lipid extraction from *Botryococcus* sp., Chlorella vulgaris and *Scenedesmus* sp. with chloroform/ methanol (1/1 v/v) solvent system and found that microwave assisted method resulted in the highest the lipid yield for all three species. Sostaric et al. [82] reported that pre-treatment of microalgae biomass using microwave irradiation results in higher oil yields. Cheng et al. [140] noted that the use of microwave irradiation resulted in a 31% increase in oil yields. Sostaric et al. [82] postulated that pre-treatment of microalgae biomass resulted in higher bio-oil yields due to micro-cracks present in the cell wall.

Factors affecting microwave assisted extraction: Temperature, reaction time and energy input can impact the efficiency of microalgae lipid extraction using microwave pre-treatment technology.

Temperature: High amounts of thermal energy in the form of heat disrupt the cellular wall of microalgae organisms and allows for easier recovery of intracellular lipids [90,103]. Figure 18 depicts the effects of microwave pre-treatment on microalgae cells [141]. Increases in temperature result in rapid destruction of the cellular wall which results in increased lipid recoveries [75,103]. Figure 19 depicts the effect of microwave pre-treatment temperature on the lipid recoveries of varies microalgae species [51].

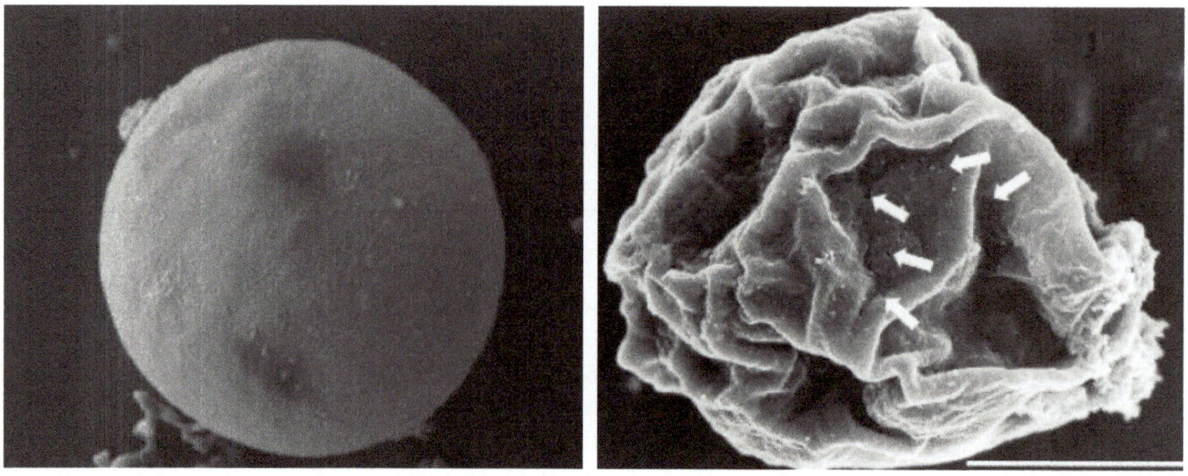

(a) Without pre-treatment (b) Microwave pre-treatment

Figure 18: Microwave treatment of microalgae biomass [143].

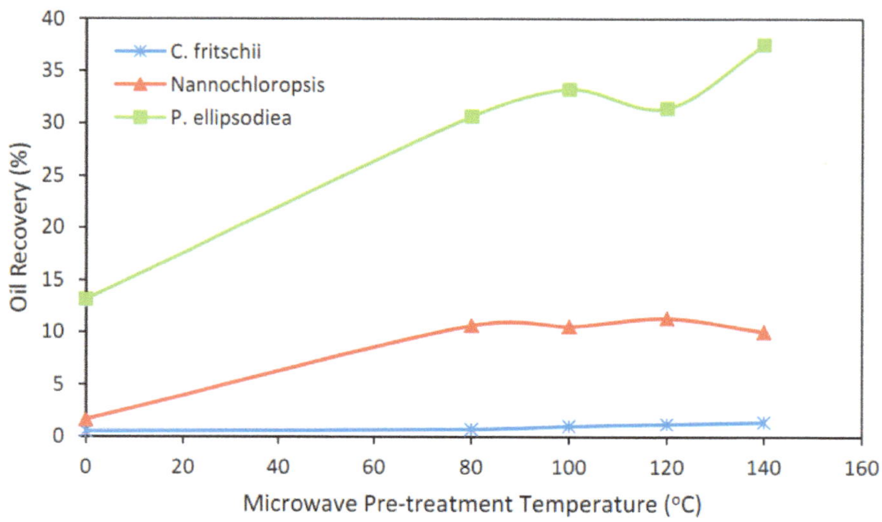

Figure 19: Effect of increasing microwave treatment temperature on the lipid extraction of microalgae [51].

Menendez et al. [103] noted that increasing the reaction temperature from 60 to 90°C in microwave assisted lipid extraction from*Nannochloropsis* gaditana resulted in increased extraction efficiency from 29 to 40%, respectively. Prommuak et al. [75] found that the lipid extracts from *Chlorella vulgaris* and Haematococcus pluvialis using microwave assisted techniques at high temperatures resulted in higher lipids. Balasubramanian et al. [137] reported that increasing the temperature from 80 to 95°C in microwave assisted lipid extraction resulted in higher lipid yields (23-30% dry weight) from Scendesmus obliquus. Passos et al. [142] noted increased solubilisation of microalga biomass with increasing temperatures over the tested range of 50 to 98°C. Cheng et al. [140] reported an increase in microalgae cell wall destruction with microwave pre-treatment as the temperatures increased from 80-120°C.

Time: The length of time in which cells are exposed to microwave radiation has significant influence on the effectiveness in the destruction of the cell wall, which determines the recovery efficiency of the lipids present in microalgae biomass [143]. Figure 20 depicts the relationship between varying the length of microwave treatment with the oil recovery [137].

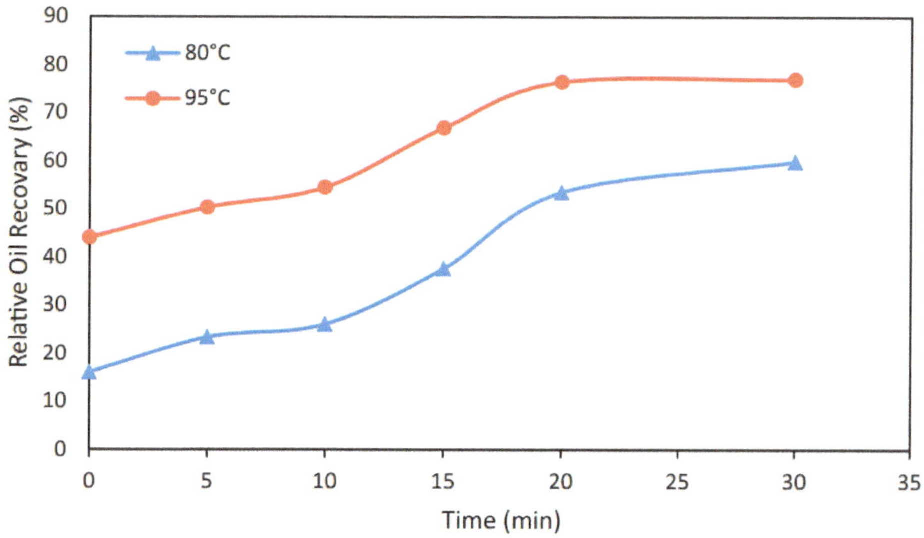

Figure 20: Varying microwave treatment time on *Scenedesmus obliquus* biomass with oil recovery efficiencies [139].

Patil et al. [143] noted that the reaction time has a significant effect on the effectiveness of the microwave assisted pre-treatment and that increased duration of microwave exposure increased the oil yields. Menendez et al. [103] found that increasing the reaction time from 10 min to 20 min in microwave assisted lipid extraction from *Nannochloropsis* gaditana at a temperature of 60°C resulted in increased extraction efficiency from 29 to 40%. Balasubramanian et al. [137] reported that increasing the extraction time from 10 to 20 min in microwave assisted lipid extraction resulted in an increased lipid yield from 10% to 22% for Scendesmus obliquus. Dai et al. [144] noted that increased microwave extraction time from 10 to 40 min resulted in increased microalgae lipid recovery 14 to 18%.

Applied energy: The amount of microwave energy applied has been noted to effect the microalgae solubilisation. Dai et al. [144] noted an increase in microalgae lipid recovery with increasing microwave treatment power from 400 to 1000 W. Qv et al. [145] reported an increase in microalgae extraction efficiency with increases in microwave power from 140 to 560 W. However, the authors noted that further increase to 700 W decreased the extraction recovery. Biller et al. [51] found that increasing the microwave power from 25 to 61 Wh/g resulted in increased lipid yield from *Nannochloropsis* biomass from 1.6 to 10%. Passos et al. [142] noted that increasing the microwave energy from 300 to 900 W resulted in increased microalgae biomass solubilisation.

Energy consumption: The energy consumption of microwave pretreatment depends on the temperature and duration of the treatment. Menendez et al. [103] reported that the energy consumption using microwave assisted lipid extraction from *Nannochloropsis* gaditana increased from 0.9 Wh/g to 1.6 Wh/g of dried biomass, as the reaction temperature increased from 60°C to 90°C which corresponded to lipid yields 29% and 40%. Balasubramanian et al. [137] extracted 76-77% of the oil from *Scenedesmus* obliquus using microwave radiation with an energy consumption of 60 Wh/g of dried biomass. Guldhe et al. [85] found that the energy consumption for microalgae lipid extraction from*Scenedesmus* sp. was 98 Wh/g of dried biomass at a temperature of 100°C. Bermudez et al. [146] noted that the energy consumption for microwave assisted lipid extraction from microalgae over 60 min period increased from 441 Wh/g fatty acid to 588 Wh/g fatty acid, as the temperature increased from 60°C to 120°C; additionally, they found that reducing the treatment time to 20 min at 90°C consumed 204 Wh/g fatty acid and did not vary significantly from a higher treatment time. Passos et al. [142] noted an increase in microwave power from 300 W to 900 W resulted in increased temperatures from 50°C to 98°C and that increasing the time from 3 to 9 min at 300 W resulted in temperature increases from 50 to 95°C. Biller et al. [51] found that increasing temperature from 80°C to 140°C resulted in increased microwave power consumption from 25 Wh/g to 61 Wh/g. Azcan and Yilmaz [147] also noted that an increase in microwave temperature radiation consumed higher amounts of energy.

Advantages and disadvantages: Microwave assisted lipid extraction from microalgae is one of the simplest methods and most effective amongst other extraction methods [10]. This technique is moderately suitable for scale up [37]. Microwave microalgae assisted lipid extractions have been noted to be the most applicable method for large scale use due to its simplicity and effectiveness [10]. The rapid extraction time, high heating rates, low operating costs, environmentally friendly nature, lesser solvent requirements, high product purity and high efficiency make it an attractive method for microalgae lipid recovery [28,82,137,138,148]. However, the disadvantages include the maintenance costs associated with large scale applications, an additional step required after pre-treatment for lipid recovery, high temperatures can result in lipid degradation and lengthy cooling times are required to avoid lipid loss [28,149].

BIOLOGICAL PRE-TREATMENT OF MICROALGAE

Enzyme hydrolysis

Enzymes are used to hydrolyze the microalgae cell walls, releasing the intercellular components (such as lipids) into the bulk of the medium, making their recovery much more rapid and effective [24]. Numerous types of enzymes have been reported to be effective in the cell wall degradation of microalgae. These include sanilase, cellulase, neutral protease, alkaline protease, papain and lysozyme (Table 2). Addition of these enzymes to the biomass will work to degrade the polymers on the cell surface which will allow easier recovery of the lipids from the biomass [78,157]. A schematic depicting the microalgae cell wall hydrolysis process is illustrated by Figure 21 [159].

Table 2: Enzymes-assisted hydrolysis of microalgae for lipid recovery

Enzyme	Microalgae	Lipid Yield (%)	Reference
Snailase	*Chlorella vulgaris* *Scenedesmus dimorphus* *Nannochloropsis* sp.	34	Liang et al.[150]
Trypsin		34	
Cellulase		17	
Neutral protease		11	
Alkaline protease		8	
Neutral protease	*Neochloris oleoabundans*	24.3	Wang et al. [151]
Cellulase Neutral protease	*Chlorella vulgaris*	66	Liu et al. [152]
Neutral protease	*Chlorella vulgaris*	85	Ying et al. [153]
Alkaline protease		80	
Cellulase Novozyme 188	*Chlorella vulgaris*	73	Cho et al. [154]
Papain	*Spirogyra* sp.	18	Reddy and Majumder [155]
Cellulase	*Chlorella* sp.	63	Fu et al. [156]
Cellulase Papain	*Neochloris oleoabundans*	10	Wang et al. [151]
Lysozyme	*Scenedesmus* sp.	16.6	Taher et al. [157]
Cellulase		15.4	
Lysozyme Cellulase	*Scenedesmus* sp.	12	Taher et al. [157]
Cellulase	*Nannochloropsis oculata*	32.7%	Surendhiran & Vijay [113]
Cellulase	*Chlorella vulgaris*	65	Liu et al. [152]
Cellulase Alkali protease	*Chlorella vulgaris*	66	Liu et al. [152]
Cellulase	*Chlorella vulgaris*	35	Zheng et al. [158]
Sanilase	*Chlorella vulgaris*	7	Zheng et al. [29]
Cellulase		24	
Lysozyme		22	

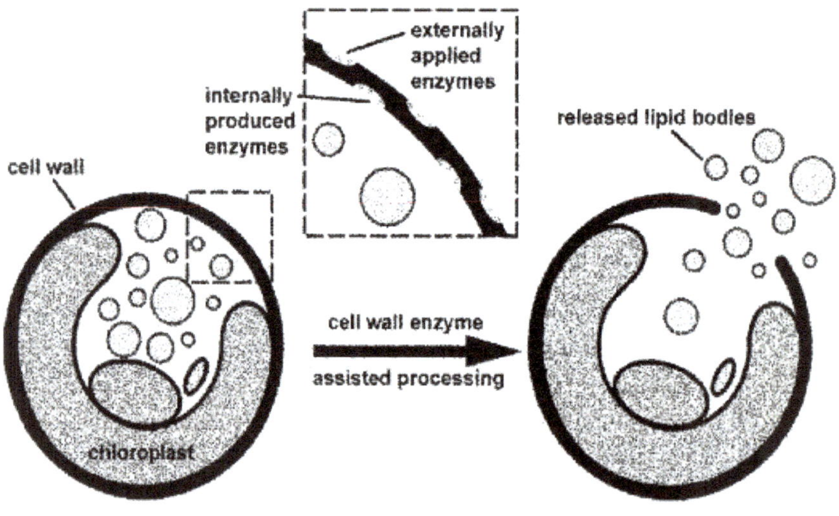

Figure 21: Schematic depicting enzyme assisted cell wall hydrolysis [161].

Freshwater microalgae possess highly resistant, non-hydrolyzable aliphatic biopolymers that are made up of even carbon numbered longchain (30 to 34 carbon atoms) ω^9-unsaturated ω-hydroxy fatty acid monomers [160]. Intermolecular ester links of the monomers form linear chains that act as the starting position of ether cross-linking. These algaenans are highly resistant against degradation as a result of their polyether nature [160-162]. Blokker et al. [160] predicted the structure of these algaenans consisting of linear polyester chains that are cross-linked by ether-bonds as that depicted in Figure 22 [160].

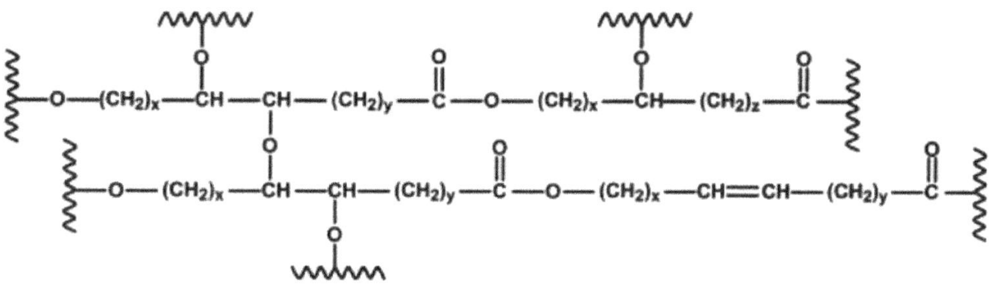

Figure 22: Predicted algaenans structure [162].

Zheng et al. [29] tested the effectiveness of sanilase assisted cell hydrolysis on *Chlorella vulgaris* biomass and noted a lipid recovery of 7% which was much lower than that achieved using lysozyme and cellulase. Liang et al. [150] reported a lipid recovery of 34% using the enzyme snailase on microalgae biomass consisting of Chlorella vulgaris, *Scenedesmus* dimorphus and *Nannochloropsis* sp. species. Lu et al. [163] reported that the snailase enzyme was more effective in disrupting *Chlorella protothecoides* cell wall than cellulase as a result of higher protoplast concentrations achieved. Cheng et al. [8] also noted effective recovery of Schizochytrium protoplasts using the combination of snailase and cellulase enzymes for cell wall hydrolysis

Liang et al. [150] noted that lipid recoveries greater than 30% were achieved from *Chlorella vulgaris* using trypsin and snailase enzymes while lower yields resulted from neutral protease, alkaline protease and cellulase. Sander and Murthy [164] noted that effective cell wall degradation in microalgae cells can be achieved using enzymatic cellulase. Cho et al. [154] noted that the presence of cellulose enzymes increased the cell hydrolysis which resulted in higher solvent lipid recoveries. Fu et al. [156] reported that cellulose enzyme hydrolyses of Chlorella sp. increased the lipid recovery to 56% compared to 32% prior to treatment. Liu et al. [152] reported a lipid recovery of 65% from *Chlorella vulgaris*using cellulase assisted enzyme hydrolysis. Zheng et al. [29] reported a lipid recovery of 24% from *Chlorella vulgaris* using cellulase cell wall hydrolysis.

Liang et al. [150] noted that a lipid recovery of 11% using neutral protease microalgae hydrolysis of *Chlorella vulgaris*, *Scenedesmus*dimorphus and *Nannochloropsis* sp. Liu et al. [152] reported a lipid recovery of 66% from *Chlorella vulgaris* using neutral protease enzyme hydrolysis. Wang et al. [151] noted a cell disintegration that ranged from 10.9% to 24.3% in Neochloris oleoabundans using neural protease corresponding to enzyme concentrations of 1% to 6%. Ying et al. [153] tested neutral protease enzyme extraction of lipids from Chlorella vulgaris biomass and noted a recovery efficiency of 85%.

Khomova et al. [165] noted a complete recovery of lipids (37%) from Monochrysis biomass using alkaline protease enzyme treatment. Liu et al. [152] obtained a lipid recovery of 66% from *Chlorella vulgaris* biomass using a mixture of alkali protease and cellulase. Liang et al. [150] noted a lipid recovery of 8% when a mixture of microalgae biomass (Chlorella vulgaris, *Scenedesmus* dimorphus and *Nannochloropsis* sp.) was treated with alkaline protease. Ying et al. [153] reported an extraction efficiency of 80% using enzymatic alkaline protease in *Chlorella vulgaris* biomass.

Wang et al. [151] reported disintegration efficiency in Neochloris oleoabindans of up to 45.2% using papain enzyme for cell wall hydrolysis. Horst et al. [166] noted that the papain enzyme assisted cell hydrolysis of Phaeodactylum tricornutum biomass was effective in lipid extraction. Morris et al. [167] tested used the enzyme papain to hydrolyze *Chlorella vulgaris* biomass and found that it resulted in a 16% disintegration efficiency. Reddy and Majumder [155] used papain to hydrolyze Spirogyra sp. biomass and noted a lipid recovery of 18%.

Gerken et al. [168] noted that the enzyme lysozyme was effective in the digestion of *Chlorella vulgaris* cell wall. Taher et al. [157] reported a lipid recovery from Scenedsmus sp. of 16.6% using the enzyme lysozyme for cell wall hydrolysis. Zheng et al. [29] recovered a lipid concentration of 22% from *Chlorella vulgaris* biomass using lysozyme enzyme hydrolysis. Natarajan et al. [78] reported that microalgae cell wall can be effectively degraded using lysozyme enzyme hydrolysis. Cuellar-Bermudez et al. [169] reported that lysozyme was effective in disrupting microalgae cell wall.

Factors affecting enzyme assisted extraction: A number of factors have been noted to influence the efficiency of enzymatic pre-treatment for microalgae lipid extraction. These factors include: treatment time, enzyme concentration, cell concentration, pH and temperature.

Time: Prolonged exposure of enzyme to microalgae biomass allows for increased cell wall degradation, Figure 23 [150]. The rupture of microalgae cell wall allows for easier recovery of the intracellular lipids [113,156].

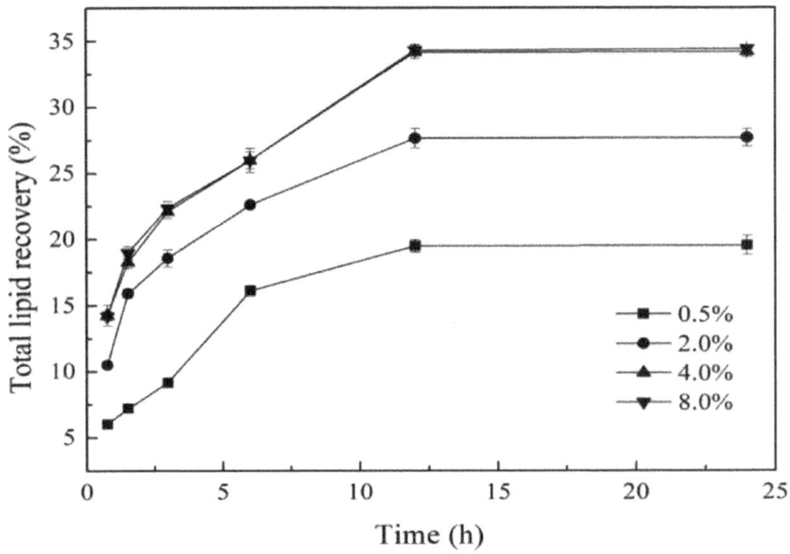

Figure 23: Effect of total lipid recovery as a function of reaction time and enzyme concentration in Chlorella vulgaris biomass [152].

Liang et al. [150] found that lipid yields increased with increasing reaction time from 2 to 12 h for *Chlorella vulgaris*, but reached a plateau at 12 h. Surendhiran and Vijay [113] noted that increasing the reaction time from 8 to 12 h resulted in increased lipid yield from N. oculata species. Reddy and Majumder [155] found that increasing the reaction time of enzyme hydrolysis from 1 to 4 h resulted in increased oil yield from 12 to 18%. However further increase in incubation time to 8 h did not show any increases in oil yield and 10 h incubation time resulted in reduction of yield to 17%. Cho et al. [154] noted that increasing the enzymatic reaction time from 1 to 72 h resulted in increased *Chlorella vulgaris* oil yield from 18 to 85%. Fu et al. [156] found that the microalgae lipid content increased with increasing reaction time from 1 to 70 h.

Enzyme concentration: The enzyme concentration affects the efficiency of microalgae cell wall degradation. Greater enzyme concentration allows for rapid cell wall degradation by increasing the enzyme to biomass content, Figure 19 [150].

Liang et al. [150] noted that the lipid recoveries from Chlorella vulgaris using Trypsin and snailase enzymes at concentrations of 4 to 8% did not vary from one another, but increasing the enzyme concentrations from 0.5 to 2% significantly increased oil yields. Wang et al. [151] noted that increasing the enzyme concentration from 1 to 6% in Neochloris oleoabundans resulted in increased cell disintegration from 48.2% to 64.4%, 38.1% to 45.2% and 10.9% to 24.3% using cellulose, papain and neutral protease, respectively. Morris et al. [170] noted that increasing the enzyme concentration from 0.55 to 4.5% resulted in increased cell hydrolysis from 10 to 25%. Kose and Oncel [173] found that the highest protein recovery, as a result of effective cell wall hydrolysis, was achieved at an enzyme to biomass ratio of 8% over the rested range of 1-20%.

Cell concentration: The microalgae biomass concentration also plays a role in the efficiency of microalgae cell wall degradation. The greater the cell concentration the lower the contact of enzyme to biomass [154].

Cho et al. [154] found that increasing the microalgae cell concentration while holding the concentration of enzyme, resulted in decreased hydrolysis yields in *Chlorella vulgaris* as a result of increased mass transfer with higher microalgae concentrations. Fu et al. [156] achieved the highest cell wall hydrolysis in Chlorella species at the lowest cell concentration of 20 g/l (20 to 40 g/l). Morris et al. [170] noted that the highest protein yield, as a result of cell wall hydrolysis, resulted at a biomass concentration of 10 or 15% over the tested range of 5-20%. Kose and Oncel [171] found that the increasing the biomass concentration (lowering the enzyme to biomass ratio) resulted in lower protein yields which is attributed to lower cell wall hydrolysis. Ho et al. [172] noted that increasing the microalgae biomass concentration (10 - 40 g/L) while holding the enzyme concentration consistent, resulted in decreased glucose production yields.

pH: The pH of the reaction medium plays an important role in the effectiveness of the enzyme (Table 3) as the function of the enzyme is strongly influenced by the pH, Figure 24 [150].

Table 3: Optimal pH for enzyme assisted microalgae cell wall hydrolysis

Enzyme	Microalgae	Tested pH range	Optimal pH	Reference
Cellulase Novozyme 188	*Chlorella vulgaris*	3.8-5.8	4.8	Cho et al. [154]
Snailase Trypsin	*Chlorella vulgaris* *Scenedesmus dimorphus* *Nannochloropsis* sp.	4-9	4	Liang et al. [150]
Cellulase	*Chlorella* sp.	3.6-7.6	4.6	Fu et al. [156]
Cellulase	*Chloroccum* sp.	2.5-7.5	4-6	Harun and Danquah [173]
Papain	*Spirogyra* sp.	6	-	Reddy and Majumder [155]
Cellulase Papain	*Neochloris oleoabindans*	6.5	-	Wang et al. [151]
Lysozyme Cellulase	*Scenedesmus* sp.	7.48	-	Taher et al. [157]

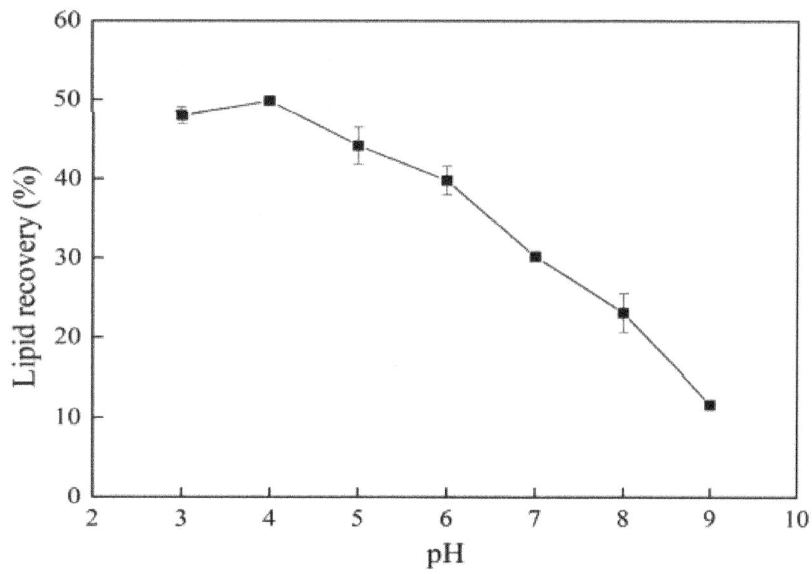

Figure 24: Effect pH on the lipid recovery from Chlorella vulgaris biomass [152].

Cho et al. [154] noted that an optimal pH exists for maximum enzymatic (Cellulase and Novozyme 188) hydrolysis of Chlorella vulgaris of 4.8 over the tested range of 3.8-5.8. Liang et al. [150] found that snailase and trypsin enzymatic lipid extraction from microalgae was pH dependent and that the increases in pH above 4 (4-9) resulted in lower lipid yields. Fu et al. [156] achieved the highest cell wall hydrolysis in Chlorella species at a pH of 4.6 using cellulase over the tested range of 3.6 to 7.6. Harun and Danquah [173] investigated the hydrolysis of Chloroccum sp. using cellulase over the pH range of 2.5 to 7.5 and found an optimal pH between 4 to 6, as a result of enzyme impairment under acidic and alkaline conditions. Reddy and Majumder [155] used papain assisted enzyme cell wall hydrolyses on Spirogyra sp. and found it effective at a pH of 6.

Temperature: The temperature of the reaction is crucial for efficient microalgae cell wall degradation. Low temperatures can slow down the reaction rate and high temperatures can result in lipid degradation. Thus the temperature of the reaction must be optimized for the enzymatic activity in order to achieve high lipid recoveries.

Cho et al. [154] noted that a temperature of 50°C resulted in the highest hydrolysis efficiency in *Chlorella vulgaris* over the tested range of 40-60°C. Reddy and Majumder [155] found that increasing the reaction temperature in enzyme hydrolysis from 30 to 60°C resulted in increased oil yields from 14 to 27%, but further increase to 70°C resulted in a decreased oil yield (25%). Fu et al. [156] found that the cell wall hydrolysis rate was highest at a temperature of 50°C over the tested range of 40-60°C in Chlorella species. Harun and Danquah [173] found that an optimal temperature of 40°C existed for cellulase activity during over the tested range from 28 to 60°C. Saha and Cotta [174] and Mtui et al. [175] also reported an optimum temperature range of 30-45°C for cellulase cell wall hydrolysis.

Advantages and disadvantages: Microalgae lipid extraction using enzymes is a highly specific and rapid which make it desirable for specific biproducts [150,176]. This method requires low operational temperatures and has high specificity/selectivity of the lipid class and no corrosion issues associated, making it a desirable technique for microalgae cell wall hydrolysis compared to chemical and physical methods [157,177]. However, efficiency is affected by the lipid composition and the type of microalgae [150] and enzymes can be high in cost [28] which limits the use of enzyme hydrolysis on industrial scale [156,178].

COMPARATIVE ANALYSIS

Selection of Criteria

Eight criteria (Table 4) were used for the evaluation of microalgae oil extraction pre-treatment techniques: (a) cell wall disruption efficiency, (b) cost, (c) time, (d) suitability for large scale use, (e) toxicity and health, (f) environmental impact, (g) reusability and (h) maintenance. These criteria were selected based on the information reported in the literature regarding the technique. Comparative analysis was performed using these criteria in order to determine the most efficient, cost effective and environmentally friendly microalgae oil extraction pre-treatment technique that is suitable for large scale application.

Table 4: Criteria used for the comparative analysis of different oil extraction pre-treatment techniques

Criteria	Importance	Description
Cell well disruption efficiency	15	The system should be able to effectively disrupt the cell wall of microalgae in order to increase the oil extraction efficiency
Cost	15	The operational costs of the pre-treatment process should be low, so that the additional inquired pre-treatment costs can be justified
Time	15	The rate of microalgae cell wall degradation should be quick to ensure the sustainability purposes
Suitability for Large Scale Use	15	The method should effective in handling large volumes for industrial production
Toxicity and Health	10	The method should be non-toxic so that the retrieved algae biomass maybe processed for a number of value added products for animal and/or human consumption
Environmental Impact	10	Method should be environmentally friendly with to no toxic wastes produced
Reusability	10	The pre-treatment method should be resalable in order to reduce costs associated with equipment
Maintenance	10	Costs for maintaining the method should be low

Assigning a Score to Each Criterion

Each of the selected criterions was assigned a score from 10 to 15 which were determined by the degree of importance of the criterion (Table 4). Higher values were given to the criteria that were deemed most important for development of an efficient and economic large scale pre-treatment technique for microalgae biomass. Lower values were given to criteria that were deemed necessary for determining a suitable method but were considered less important. These values were then used to determine the effectiveness of each of the investigated pretreatment method on microalgae biomass as shown in Tables 5-13.

Table 5: Evaluation of shaking vessel bead mill microalgae pre-treatment

Criteria	Description	Score
Cell wall disruption efficiency (15)	Effective cell wall disruption of microalgae biomass, but dependent on the likelihood of the contact between cell and bead.	10
Cost (15)	Costs associated with drying of biomass before use and the high energy required for shaking/agitation. Requires cooling jacket in order to avoid degradation of desired product	6
Time (15)	Rapid	15
Suitability for Large Scale Use (15)	Difficulty in scale up	3
Toxicity and Health (10)	No hazardous substances are used	10
Environmental Impact (10)	No release of harmful compounds or disruption to the environment	10
Reusability (10)	Method can be reused	10
Maintenance (10)	Maintenance costs associated with bead replacement	6
Total (100)		**70**

Table 6: Evaluation of agitated bead mill microalgae pre-treatment

Criteria	Description	Score
Cell wall disruption efficiency (15)	Effective cell wall disruption of microalgae biomass, but dependent on the likelihood of the contact between cell and bead	12
Cost (15)	Costs associated with drying of biomass before use and the high energy required for shaking/agitation Requires cooling jacket in order to avoid degradation of desired product	6
Time (15)	Rapid	15
Suitability for Large Scale Use (15)	Difficulty in scale up	3
Toxicity and Health (10)	No hazardous substances are used	10
Environmental Impact (10)	No release of harmful compounds or disruption to the environment	10
Reusability (10)	Method can be reused	10
Maintenance (10)	Maintenance costs associated with bead replacement	6
Total (100)		**72**

Table 7: Evaluation of horn sonication microalgae pre-treatment

Criteria	Description	Score
Cell wall disruption efficiency (15)	Effective cell wall destruction method	15
Cost (15)	Operational costs are associated with high amounts of power Multiple units required for large scale use	4
Time (15)	Relatively rapid	12
Suitability for Large Scale Use (15)	Upscale difficulty Cavitation area is limited Multiple systems required for scale up	4
Toxicity and Health (10)	No hazardous substances are used	10
Environmental Impact (10)	No release of harmful compounds or disruption to the environment	10
Reusability (10)	Method can be easily reused	10
Maintenance (10)	Minimal costs associated with maintenance	8
Total (100)		**73**

Table 8: Evaluation of bath sonication microalgae pre-treatment

Criteria	Description	Score
Cell wall disruption efficiency (15)	Effective cell wall destruction method	15
Cost (15)	Operational costs are associated with high amounts of power	8
Time (15)	Relatively rapid	12
Suitability for Large Scale Use (15)	Upscale difficulty	8
Toxicity and Health (10)	No hazardous substances are used	10
Environmental Impact (10)	No release of harmful compounds or disruption to the environment	10
Reusability (10)	Method can be easily reused	10
Maintenance (10)	Minimal costs associated with maintenance	8
Total (100)		**81**

Table 9: Evaluation of steam explosion microalgae pre-treatment

Criteria	Description	Score
Cell wall disruption efficiency (15)	Effective cell wall disruption, resulting in increased lipid recovery Effectiveness is dependent on the species	12
Cost (15)	Relatively low energy requirement Low maintenance costs are associated	13
Time (15)	Rapid treatment	14
Suitability for Large Scale Use (15)	Method is suitable for use on industrial scale	15
Toxicity and Health (10)	No hazardous substances are used	10
Environmental Impact (10)	No release of harmful compounds or disruption to the environment	10
Reusability (10)	Method can be easily reused	10
Maintenance (10)	Little maintenance is required	9
Total (100)		**93**

Table 10: Evaluation of freeze drying microalgae pre-treatment

Criteria	Description	Score
Cell wall disruption efficiency (15)	Integrity of the cell wall is weakened	12
Cost (15)	High energy consumption	6
Time (15)	Prolonged periods of time required	3
Suitability for Large Scale Use (15)	Upscale difficulty	8
Toxicity and Health (10)	No hazardous substances are used	10
Environmental Impact (10)	No release of harmful compounds or disruption to the environment	10
Reusability (10)	Method can be easily reused	10
Maintenance (10)	Costs associated with pump maintenance	5
Total (100)		**64**

Table 11: Evaluation of autoclave microalgae pre-treatment

Criteria	Description	Score
Cell wall disruption efficiency (15)	Integrity of the cell wall is weakened, however not as effective as other cell distribution methods	10
Cost (15)	High costs associated with high energy consumption required for high heat and pressure	6
Time (15)	Prolonged periods of time required for effective disruption	3
Suitability for Large Scale Use (15)	Upscale difficulty	8
Toxicity and Health (10)	No hazardous substances are used	10
Environmental Impact (10)	No release of harmful compounds or disruption to the environment	10
Reusability (10)	Method can be easily reused	10
Maintenance (10)	Costs associated with maintenance can be restively low	8
Total (100)		**65**

Table 12: Evaluation of microwave microalgae pre-treatment

Criteria	Description	Score
Cell wall disruption efficiency (15)	Effective cell wall disruption, resulting in increased lipid recovery	13
Cost (15)	Costs associated with thermal energy are relatively low	11
Time (15)	Rapid treatment	15
Suitability for Large Scale Use (15)	Suitable for industrial scale	15
Toxicity and Health (10)	No hazardous substances are used	10
Environmental Impact (10)	No release of harmful compounds or disruption to the environment	10
Reusability (10)	Treatment process can be easily reused	10
Maintenance (10)	High maintenance costs	3
Total (100)		**87**

Table 13: Evaluation of enzyme microalgae pre-treatment

Criteria	Description	Score
Cell wall disruption efficiency (15)	Effective cell wall hydrolysis	15
Cost (15)	High costs associated with enzyme purchase	7
Time (15)	Prolonged time durations	5
Suitability for Large Scale Use (15)	Suitable for large scale use	15
Toxicity and Health (10)	No hazardous substances are used	15
Environmental Impact (10)	Enzymes must be disposed of after use	6
Reusability (10)	Enzymes can be reused, however efficiency may decrease	5
Maintenance (10)	Maintenance is required for preservation	6
Total (100)		**74**

Analysis of Microalgae Pre-Treatment Techniques

The sum of the scores obtained for the different microalgae pretreatment techniques are presented in Table 14. The results indicate that of the 9 microalgae pre-treatment methods investigated, 1 mechanical, 1 thermal and 1 electromagnetic technique resulted in scores of 80/100 or greater deeming them efficient and economically suitable for industrial scale. These methods were bath sonication (81), steam explosion (93) and microwave radiation (87). Bath sonication method is effective in cell wall degradation, nontoxic, rapid technique that requires minimal maintenance, but it does however suffer from high operational costs and difficulty in scale up for industrial use. Microwave assisted technology is rapid, effective in cell wall disruption, non-toxic, can be used for large volumes of biomass and the medium maybe reused, but it suffers from high costs associated with maintenance. Steam explosion microalgae pre-treatment is an effective technique for microalgae cell wall rupture allowing the releasing intracellular components, rapid, reusable, relatively low in costs, environmentally friendly and can be used on an industrial scale. However, this method is species specific making it effective for certain microalgae species. Overall, the negative aspects of these three techniques are outweighed by their effectiveness, rapidness and relatively low costs when compared to other pre-treatment techniques. Other mechanical extraction methods suffered from high operational costs, lengthy treatment times, high maintenance costs and the scale up difficulty. Freeze drying and autoclave techniques were deemed unsuitable microalgae pre-treatment techniques because of the high costs, scale up difficulty and long processing times associated. Biological pre-treatment technique are deemed unsuitable as a result of high costs associated with purchasing of enzymes and recovery/separation difficulty after undergoing treatment, long treatment times and high maintenance of enzymes for high efficiency.

Table 14: Comparative analysis of microalgae pre-treatment methods

Criteria	Mechanical				Thermal			Electromagnetic	Biological
	SBM	ABM	HS	BS	SE	FD	AC	M	EZ
Cell well disruption efficiency (15)	10	12	15	15	12	12	10	13	15
Cost (15)	6	6	4	8	13	6	6	11	7
Time (15)	15	15	12	12	14	3	3	15	5
Suitability for Large Scale Use (15)	3	3	4	8	15	8	8	15	15
Toxicity and Health (10)	10	10	10	10	10	10	10	10	15
Environmental Impact (10)	10	10	10	10	10	10	10	10	6
Reusability (10)	10	10	10	10	10	10	10	10	5
Maintenance (10)	6	6	8	8	9	5	8	3	6
Total (100)	**70**	**72**	**73**	**81**	**93**	**64**	**65**	**87**	**74**

SBM= Shaking vessel bead mill SE= Steam explosion ABM= Agitated bead mill FD= Freeze-Drying HS= Horn sonication AC= Autoclave BS= Bath sonication EZ = Enzyme M= Microwave

CONCLUSIONS

Microalgae biomass can be used to produce numerous value added products such as biodiesel, bioethanol, biogas, biohydrogen, fish feed, animal feed, human food supplements and skin care products. Production of value added products from microalgae biomass requires the growth and recovery of the algae biomass, extraction and downstream processing of the desired product. One of the major obstacles for using microalgae biomass on an industrial-scale, for the production of biodiesel, is the high processing costs. Increasing the lipid recovery efficiency from the microalgae biomass would result in greater product yields (biodiesel). Thus, the aim of this study was to review the current methods used for microalgae pre-treatment and perform a comparative analysis in order to determine the most economically efficient method for large scale use. The effectiveness of the pre-treatment methods investigated was evaluated based on (a) cell wall disruption efficiency, (b) cost, (c) toxicity (d) suitability for large scale use, (e) time, (f) reusability and (g) maintenance. The results indicated that of the 9 microalgae methods investigated 1 mechanical, 1 thermal and 1 electromagnetic radiation technique were suitable. These methods were bath sonication (81), steam explosion (93) and microwave radiation (87). Microwave assisted microalgae pre-treatment technique is rapid, effective in cell wall disruption, non-toxic, can be used for large volumes and the medium maybe reused, but it does however suffer from high maintenance costs. Bath sonication technique is effective in the degradation of cell wall, nontoxic, rapid technique with minimal maintenance required, but suffers from high costs and difficulty in scale up for industrial use. Steam explosion pre-treatment is effective in degrading microalgae cell wall, releasing intracellular components, rapid, reusable, relatively low in costs, environmentally friends and reusable, but is species specific. Overall, the negative aspects of these three techniques are outweighed by their effectiveness, rapidness and relatively low costs when compared to other pre-treatment techniques. Other mechanical extraction methods suffer from high operational costs, lengthy treatment times, high maintenance costs and the scale up difficulty. Freeze drying and autoclave techniques were deemed unsuitable microalgae pre-treatment techniques because of the high costs, scale up difficulty and long processing times associated. Biological pre-treatment technique were deemed unsuitable as a result of high costs associated with purchasing of enzymes, difficulty in recovery/ separation after treatment, long treatment time, and high maintenance required for high efficiency.

REFERENCES

1. Huang Z, OngSL, Ng HY(2011) Submerged anaerobic membrane bioreactor for low-strength wastewater treatment: effect of HRT and SRT on treatment performance and membrane fouling. Water Res45: 705-713.
2. Sathish A, Smith BR, Sims RC(2013) Effect of moisture on in situ transesterification of microalgae for biodiesel production. JChemTechnolBiotechnol 89:137-142.
3. Demirbas A (2009) Production of biodiesel from algae oils. Energy Source Part A 31:163-168.
4. Chisti Y (2007) Biodiesel from microalgae. BiotechnolAdv25: 294-306.
5. Lardon L, H´elias A, Sialve B, SteyerJP, Bernard O(2009) Lifecycle assessment of biodiesel production from microalgae. Environ SciTechnol 43: 6475–6481
6. Demirbas A, Demirbas MF(2011) Importance of algae oil as a source of biodiesel. Energy Convers Manag 52: 163-170.
7. Nagarajan S, Chou SK, Cao SY, Wu C, Zhou Z(2012) An updated comprehensive techno-economic analysis of algae biodiesel. BioresourceTechnol 145: 150-156.
8. Cheng R, Ma R, Li K, Rong H, Lin X, et al. (2012) Agrobacterium tumefaciens mediated transformation of marine microalgae Schizochytrium. Microbiol Res 167:179-186.
9. Jones CS,Mayfieldt SP(2012) Algae Biofuels: Versatility for the Future of Bioenergy. Current Opinion in Biotechnol23: 346-351.
10. Lee JY, Yoo C, Jun SY, AhnCY, OhHM(2010) Comparison of several methods for effective lipid extraction from microalgae. BioresourcTechnol101:S75–S77.
11. Valente OS, Silva MJD, Pasa VMD, BelchiorCRP,Sodr´e JR(2010) Fuel consumption and emissions from a diesel power generator fuelled with castor oil and soybean biodiesel. Fuel 89: 3637–3642.
12. Halim R, Gladman B, DanquahMK, Webley PA(2011) Oil extraction from microalgae for biodiesel production. BioresourceTechnol102:178–185.
13. Becker EW (2007) Micro-algae as a source of protein. BiotechnolAdv25: 207–210.
14. Pimentel D, Marklein A, Toth MA, Karpoff M,Paul GS, et al. (2008) Biofuel impacts on world food supply: Use of fossil fuel, land and water resources. Energies, 1:41-78.
15. Demirbas A (2011) Biodiesel from oilgae, biofixation of carbon dioxide by microalgae: A solution to pollution problems.Appl Energy 88:3541-3547.
16. Li-Beisson Y (2015)Triacylglyserol biosynthesis in eukaryotic microalgae. The AOCS Lipid Library, Plant Lipid Biochemistry, Urbana, Illinois, USA.
17. Hu Q, Sommerfeld M, Harvis E, Ghirardi M, Posewitz M, et al.(2008) Microalgaltriacylglycerols as feedstocks for biofuel production: perspectives and advances. The Plant J54: 621-639.
18. Spolaore P, Joannis-Cassan C, Duran E, Isambert A (2006) Commercial applications of microalgae. J Bioscience Bioengg 101: 87-96.
19. Gavrilescu M, Chisti Y(2005) Biotechnology—a sustainable alternative for chemical industry. BiotechnolAdv23: 471–499.
20. Feltes MMC, de Oliveira D, NinowJL, de Oliveira JV(2011)An overview of enzyme-catalyzed reactions and alternative feedstock for biodiesel production. Alternative Fuel 21-46.
21. Gouveia L, Nobre BP, Marcelo FM, Mrejen S, Cardoso MT, et al.(2007) Functional food oil colored by pigments extracted from microalgae with supercritical CO2. Food Chem 101:717–723.
22. Greenwell HC, Laurens LML, Shields JR, LovittRW, Flynn KJ(2010) Placing microalgae on the biofuels priority list: A review of the technological challenges. J Royal SocietyInterf 7: 703-726.
23. Cooney M, Young G, Nagle N(2009) Extraction of bio-oils from microalgae. Sep Purification Rev 38: 291–325.
24. Mercer P,Armenta RE(2011) Developments in oil extraction from microalgae. European J Lipid SciTechnol113: 539-547.
25. Munir N, Sharif N, Naz S, SaleemF,Manzoor F(2013) Harvesting and processing of microalgae biomass fractions for biodiesel production (a review). J SciTechnol Develop 23: 235-243.
26. Doucha J, L ´vansky´ K(2008) Influence of processing parameters on disintegration of Chlorella cells in various types of homogenizers. ApplMicrobiolBiotechnol81: 431-440.
27. Hopkins TR (1991) Physical and chemical cell disruption for the recovery of intracellular protein.(Seetharam R, Sharma S Edn) Purification and analysis of recombinant proteins. Marcel Dekker, New York, USA: 57-84.
28. Kumar RR, RaoPH,Arumugam M(2015) Lipid extraction methods from microalgae: a comprehensive review. Frontiers Energy Res2:1-9.
29. Zheng H1, Yin J, Gao Z, Huang H, Ji X, et al.(2011) Disruption of Chlorella vulgaris cells for the release of biodiesel-producing lipids: a comparison of grinding, ultrasonicaiton, bead milling, enzymatic lysis, and microwaves. ApplBiochemBiotechnol 164: 1215-1224.

30. Goldberg S (2008) Mechanical/physical methods of cell disruption and tissue homogenization. Methods MolBiol424: 3-22.

31. Shen Y, Pei Z, Yuan W, Mao E(2009) Effect of nitrogen and extraction method on algae lipid yield. IntJ Agri Biological Engg2:51-57.

32. Prabakaran P, Ravindran RD(2011) A comparative study on effective cell distruption methods for lipid extraction from microalgae. LettApplMicrobiol 53: 150-154.

33. Ryckebosch E, MyuylaertK,Foubert I(2011) Optimisation of an analytical procedure for extraction of lipids from microalgae. J AmrOil Chemists' Society 89:189–198.

34. Schilling R (2008) Media milling advances. Union Process Inc., Akron, Ohio.

35. Lee AK, Lewis DM, Ashman PJ (2012) Disruption of microalgal cells for the extraction of lipids for biofuels: Processes and specific energy requirements. Biomass Bioenergy 46: 89-101.

36. Gouveia L, Batista PA, Nobre PB, Marques AP, Moura P, et al. (2012) Exploring Scenedesmusobliquus and Nannochloropsis sp. potential as a sustainable raw material for biofuels and high added value compounds. 1-CIAB 1stCongresoIberoamericanoSobreBiorefinerias: 707-707.

37. Halim R, Harun R, DanquahMK, Webley PA(2012) Microalgal cell disruption for biofuel development. Appl Energy, 91(1): 116-121.

38. Baldev E, MubarakAli D, Dhivya M, Kanimozhi M, Shakena-Fatima M, et al.(2014) Facile and novel strategy for methods of extraction of biofuel grade lipids from microalgae an experimental report. Int J Biotechnol Wellness Ind3: 121-127.

39. Ceron MC, Campos I, Sanchez JF, Acien FG, Molina E, et al.(2008) Recovery of lutein from microalgae biomass: development of a process for Scenedesmusalmeriensis biomass. J Agri Food Chem56:11761-11766.

40. Postma PR, Miron TL, Olivieri G, Barbosa MJ, WijffelsRJ, et al.(2015) Mild disintegration of the green microalgae Chlorella vulgaris using bead milling. BioresourcTechnol 184: 297-304.

41. Mogren H, LindblomM, Hedenskog G(1974) Mechanical disintegration of microorganisms in an industrial homogenizer. BiotechnolBioengg, 16: 261-274.

42. Hedenskog G, Enebo L, VendlováJ,Prokes B(1969) Investigation of some methods for increasing the digestibility in vitro of microalgae. BiotechnolBioengg 11: 37–51.

43. Bert C, Ropers B,Sertain E(2011) Dispersion in a pressurized mill with electrofused ceramic beads. SEPR Research Center. Mountainside, New Jersey.

44. Safi C, Frances C, Ursu AV, Laroche C, Pouzet C, et al.(2015) Understanding the effect of cell disruption methods on the diffusion of Chlorella vulgaris proteins and pigments in the aqueous phase. Algal Res8: 61-68.

45. Liang Y, SarkanyN, Cui Y(2009) Biomass and lipid productivities of Chlorella vulgaris under autotrophic, heterotrophic and mixotrophic growth conditions. BiotechnolLett31: 1043-1049.

46. Schwenzfeier A, WierengaPA,Gruppen H(2011) Isolation and characterization of soluble Protein from the green microalgae Tetraselmis sp. BioresourcTechnol 102:9121–9127.

47. Lee SJ, Yoon BD, Oh HM(1998) Rapid method for the determination of lipid from the green algae Botryococcusbraunii. BiotechnolTechniques 7:553–556.

48. Kim J, Yoo G, Lee H, Lim J, Kim K, et al.(2013) Methods of downstream processing for the production of biodiesel from microalgae. BiotechnolAdv 31: 862-876.

49. Gunerken E, D'Hondt E, Eppink MHM, Garcia-Gonzalez L, ElstK, et al. (2015) Cell disruption for microalgae biorefineries, BiotechnolAdv33: 243-260.

50. Grima ME, Belarbi EH, Fernandez FGA, Medina AR,Chisti Y(2003) Recovery of microalgal biomass and metabolites: process options and economics. BiotechnolAdv20:491–515.

51. Biller P, Ross AB(2012) Hydrothermal processing of algal biomass for the production of biofuels and chemicals. Biofuels 3:603–623.

52. Singh J, Gu S(2010) Commercialization potential of microalgae for biofuels production. Renew Sustain Energy Rev 14: 2596-2610.

53. Keris-Sen UD, Sen U, Soydemir G,Gurol MD(2014) An investigation of ultrasound effect on microalgal cell integrity and lipid extraction efficiency. BioresourcTechnol152: 407-413.

54. Acoustical Society of America (2013) Ultrasound 'making waves' for enhancing biofuel production. Biomass Magazine, BBI International, Grand Forks, North Dakota.

55. Menendez JMB, Arenillas A, Diaz JAM, Boffa L, Mantegna S, et al.(2013) Optimization of microalgae oil extraction under ultrasound and microwave irradiation. J ChemTechnolBiotechnol89: 1779-1784.

56. Pernet F, Tremblay R(2003) Effect of ultrasonication and grinding on the determination of lipid class content of microalgae harvested on filters. Lipids 38: 1191–1195.

57. Wiltshire KH, Boersma M, Moller A, Buhtz H(2000) Extraction of pigments and fatty acids from the green alga Scenedesmusobliquus (Chlorophyceae). Aquatic Ecology 34: 119–126.

58. Ranjan A, Patil C, Moholkar SV(2010) Mechanistic assessment of microalgal lipid extraction. Industrial EnggChem Res 49: 2979–2985.

59. Cravotto G, Boffa L, Mantegna S, Perego P(2008) Improved extraction of vegetable oils under high-intensity ultrasound and/or microwaves. UltrasonicsSonochem15: 898–902.

60. De Souza Silva APF, Costa MC, Lopes AC, Neto EFA, Leitao RC, et al.(2014) Comparison of pretreatment methods for total lipids extraction from mixed microalgae. Renew Energy, 63: 762-766.

61. Koberg M, Cohen M, Ben-AmotzA,Gedanken A(2011) Bio-diesel production directly from the microalgae biomass of Nannochloropsis by microwave and

ultrasound radiation. BioresourcTechnol 102:265-4269.

62. Hosikian A, Lim S, Halim R,Danquah MK(2010) Chlorophyll extraction from microalgae: A review on the process engineering aspects. IntJChemEngg391632: 1-11.

63. Borthwick KAJ, Coakley WT, McDonnell MB, Nowotny H, Benes E, et al. (2005) Development of a novel compact sonicator for cell disruption. J Microbiol Methods 60: 207e16.

64. Gogate PR (2007) Application of cavitational reactors for water disinfection: current status and path forward. J Environ Manag85:801-8015.

65. Minoletti F, HermosoM, Gressier V(2008) Seperation of sedimentary micron-sized particles for palaeoceanography and calcareous nannoplankton biogeochemistry. Nature Protocols 4: 14-24.

66. Jeon BH, Choi JA, Kim HC, Hwang JH, Abou-Shanab EA, et al.(2013) Ultrasonic disintegration of microalgal biomass and consequent improvement of bioaccessibility/ bioavailability in microbial fermentation.Biotechnol Biofuels 6:37-48.

67. Wang M, Yuan W, Jiang X, Jing Y, Wang Z(2014) Disruption of microalgal cells using high-frequency focused ultrasound. BioresourcTechnol153:315-321.

68. Tovatech(2015)How ultrasonic cleaning works. Tovatech Inc., South Organge, New Jersey, USA.

69. Wiyarno B, YunusRM, Mel M(2011) Extraction of algae oil from nannochloropsis sp.: A study of soxhlet and ultrasonic-assisted extractions. J ApplSci 11: 3607-2011.

70. McMillan JR, Watson IA, Ali M, Jaafar W(2013) Evaluation and comparison of algal cell disruption methods: microwave, waterbath, blender, ultrasonic and laser treatment. Appl Energy 103: 128-134.

71. Neto AMP, de Souza RAS, Leon-Nino RD, da Costa JDA, Tiburcio RS, et al.(2013) Improvement in microalgae lipid extraction using a sonication-assisted method. Renew Energy 55: 525-531.

72. Piasecka A, KrzeminskaI, Tys J(2014) Physical methods of microalgal biomass pre-treatment. IntAgrophys 28: 341-348.

73. Adam F, Abert-Vian M, Peltier,Chemat F(2012) "Solvent-free" ultrasound-assisted extraction of lipids from fresh microalgae cells: A green, clean and scalable process. BioresourcTechnol114: 457-465.

74. Tang S, Qin C, Wang H, Li S,Tian S(2011) Study on supercritical extraction of lipids and enrichment of DHA from oil-rich microalgae. The J Supercritical Fluids 57: 44-49.

75. Prommuak C, Pavasant P, Quitain AT, Goto M,Shotipruk A(2012)Microalgal lipid extraction and evaluation of single-step biodiesel production. Engg J 16: 157-166.

76. Nowotarski K, King PM, Joyce EM, Mason TJ(2012) Ultrasonic distruption of algae cells. AIP Congress Proceedings 1433:237-240.

77. Gerde JA, Montalbo-Lomboy M, Yao L, GrewellD, Wang T(2012) Evaluation of microalgae cell disruption by ultrasonic treatment. BioresourcTechnol 125:175-181.

78. Natarajan R, Chen X, Lau R(2014) Ultrasound applications in lipid extraction from microalgae. Springer, 4: 117-139.

79. Tomas CR (1997) Identifying marine Phytoplankon. Academic Press, Inc., San Diego, California, USA: 5-385.

80. Joyce EM, King PM, Mason TJ(2013)The effect of ultrasound on the growth and viability of microalgae cells. J ApplPhycol26: 1741-1748.

81. Takeda H (1991) Sugar composition of the cell wall and the taxonomy of Chlorella (Chlorophyceae). J Phycol27:224-232.

82. Sostaric M, Klinar D, Bricelj M, Golob J, Berovic M, et al.(2012) Growth lipid extraction and thermal degradation of the microalga Chlorella vulgaris. New Biotechnol 29: 325–331.

83. Bicudo CEM, Menezes M(2006)Gêneros de algas de águascontinentaisdoBrasil. (2ndedn.) São Carlos, SP: Rima, Brazil.

84. Kaiwan-arporn P, Hai PD, Thu NT,Annachhatre AP(2012) Cultivation of cyanobacteria for extraction of lipids. Biomass Bioenergy 44:142-149.

85. Guldhe A, Singh B, Rawat I, RamluckanK, Bux F(2014) Efficacy of drying and cell disruption techniques on lipid recovery from microalgae for biodiesel production. Fuel 128: 46-52.

86. Bigelow TA, Xu J, Stessman DJ, Yao L, Spalding MH (2014)Lysis of Chlamydomonasreinhardtii by high-intensity focused ultrasound as a function of exposure time. UltrasonicsSonochem 21:258-1264.

87. Luque-Garcı´a JL, de Castro MD(2003) Ultrasound: A powerful tool for leaching. TrAC Trends in Analytical Chem 22: 41-47.

88. Lee S, Shah YT(2012) Biofuels and bioenergy: processes and technologies. CRC Press, London, England.

89. Chemat F, ZillEH, Khan MK(2011) Applications of ultrasound in food technology: processing, preservation and extraction. UltrasonicsSonochemistry 18: 813-835.

90. Wang L, Weller CL(2006) Recent advances in extraction of nutraceuticals from plants. Trends Food SciTechnol 17:300–312.

91. Cheng JJ, Timilsina GR(2011) Status and barriers of biofuel technologies: A review. Renew Energy 36:3541-3549

92. Salvado J, Velasquez JA, Ferrando F(2003)Binderless fiberboard from steam exploded MiscanthusSinensis: optimization of pressing and pretreatment conditions. Wood SciTechnol 37:279–286.

93. Velasquez JA, Ferrando F, FarriolX,Salvado J(2003)Binderless fiberboard from steam exploded Miscanthussinensis. Wood SciTechnol37: 269–278.

94. Lorente E, FarriolX, Salvado J(2015) Steam explosion as a fractionation step in biofuel production from microalgae. Fuel ProcTechnol131: 93-98

95. Nurra C, Torras C, Clavero E, Rios S, Rey M, et al.(2014)Biorefinery concept in a microalgae pilot plant. Culturing, dynamic filtration and steam explosion

fractionation. BioresourcTechnol 163: 136-142.

96. Morgan-Sagastume F, Pratt S, Karlsson A, Crine D, LantP, et al.(2011) Production of volatile fatty acids by fermentation of waste activated sludge pre-treated in full-scale thermal hydrolysis plants. BioresourcTechnol 102:3089-3097.

97. Pérez-Elvira SI, Nieto P,Fdz-Polanco F(2006) Sludge minimization technologies. Rev EnvironSciBiotechnol 5:375–398.

98. Kepp U, Machenbach I, WeiszN, Solheim OE(2000) Enhanced stabilisation of sewage sludge through thermal hydrolysis- three years of experience with full-scale plant. Water SciTechnol 42: 89-96.

99. Mosier N, Wyman C, Dale B, Elander R, Lee YY, et al.(2005) Features of promising technologies for pretreatment of lignocellulosic biomass. BioresourcTechnol 96: 673–686.

100. Montane D, Farriol X, Salvado J, JollezP, Chornet E(1998) Application of steam explosion to the fractionation and rapid vapor-phase alkaline pulping of wheat straw. Biomass Bioenergy 14: 261–276.

101. Alzate ME, Munoz R, Rogalla F, Fdz-Polanco F, Perez-Elvira SI(2012) Biochemical methane potential of microalgae: influence of substrate to incoculum ratio, biomass concentration and pre-treatment. BioresourcTechnol123:488-494.

102. DuZ, MohrM, MaX, ChengY, LinX(2012) Hydrothermal pre-treatment of microalgae for production of pyrolytic bio-oil with a low nitrogen content. BioresourcTechnol 120: 13-8.

103. Mendez L, Mahdy A, Demuez M, Ballesteros M, Gonzalez-Fernandez C(2014) Effect of high pressure thermal pretreatment on Chlorella vulgaris biomass: organic matter solubilisation and biochemical methane potential. Fuel 117: 674-679.

104. Aguirre AM,Bassi A(2014) Investigation of high pressure steaming (HPS) as a thermal treatment for lipid extraction from Chlorella vulgaris. BioresourcTechnol 164: 136-142

105. MedinaAR, GrimaEM, GimenezAG, GonzalezMJ(1998) Downstream processing of algal polyunsaturated fatty acids. BiotechnolAdv 16:517-580.

106. Pandey A, Negi S, BinodP, Larroche C(2014) Pretreatment of biomass: processes and technologies. Academic Press, San Diego, California, USA.

107. Kita K, Okada S, Imou K, Yokoyama S, Amano T(2010) Thermal pre-treatment of wet microalgae harvest for efficient hydrocarbon recovery. Appl Energy 87: 2420-2423.

108. Frenz J, Largeau C, Casadevall E, KollerupF, Daugulis AJ(1989) Hydrocarbon recovery and biocompatibility of solvents for extraction from cultures of Botryococcusbraunii. BiotechnolBioengg, 34: 755–762.

109. Keymar P, RuffellI, Pratt S,Lant P(2013) High pressure thermal hydrolysis as pre-treatment to increase the methane yield during anaerobic digestion of microalgae. BioresourcTechnol 131:128-133.

110. Ko CH, Wang YN, Chang FC, Lee CY, Chen WH, et al.(2013) Energy analysis of bioethanols produced from Dendrocalamuslatiflorus. ChemEngg Transactions 24: 109-114.

111. Zhu JY, Pan XJ(2010) Woody biomass pretreatment for cellulosic ethanol production: technology and energy consumption evaluation. BioresourcTechnol 101: 2992-5002.

112. Silva ASA, Teixeira RSS, de Oliveira Moutta R, Ferreira-Leitao VS, Olivieri de Barros RR, et al.(2013)Surgarcane and woody biomass pretreatments for ethanol production, sustainable degradation of lignocellulosic biomass - techniques, applications and commercialization. Intech, USA.

113. Surendhiran D, Vijay M(2014) Effect of various pretreatment for extraction intracellular lipid from Nannochloropsisoculata under nitrogen replete and depleted conditions. ISRN Chemical Engg 2014 536310: 1-9.

114. USGS (2015)Quality assurance/ quality control manual: Ohio water microbiology laboratory. U.S. Department of the Interior, Reston, Virginia, USA.

115. Rakesh S, Dhar DW, Prasanna R, Saxena AK, Saha S, et al.(2015) Cell disruption methods for improving lipid extraction efficiency in unicellular microalgae. EnggLife Sci 15: 443-447.

116. Yu X, Dong T, Zheng Y, Miao C, Chen S(2015) Investigations on cell distruption of oleaginous microorganisms: Hydrochloric acid digestion is an effective method for lipid extraction. Europ J Lipid SciTechnol117: 730-737.

117. Miranda JR, PassarinhoPC, Gouveia L(2012) Pre-treatment optimization of Scenedesmusobliquus microalga for bioethanol production. BioresourcTechnol 104:342–348.

118. Russell BL (1995) Determination of factors limiting enzymatic hydrolysis of the "Chlorella sorokiniana" cell wall. Unpublished PhD Thesis., University of Florida, Gainesville, Florida, USA.

119. King PM (2014) The use of ultrasound on the extraction of microalgal lipids. Unpublished PhD Thesis. Coventry University, Coventry, United Kingdom.

120. Lopata VJ, Saunders CB, Singh A, Janke CJ, WrennGE, et al.(1999) Electron-beam-curable epoxy resins for the manufacture of high-performance composites. RadPhyChem 56: 405-415.

121. Panasonic (2014) Sterilization and Drying. Panasonic Healthcare Company of North America, Wood Dale, Illinois.

122. SciCan(2015) Bravo autoclave from SciCan. SciCan, Canonsburg, Pennsylvania, USA.

123. BME (2015) Market Forge STM-E sterilmatic autoclave-New Chamber, 95-2678. Booth Medical Equipment, Alexander, Arkansas.

124. Luo J, Xiao X,Luo SL(2010)Biosorption of cadmium (II) from aqueous solu-tions by industrial fungus Rhizopuscohnii. Transactions of Nonferrous Metals Society of China 20:1104-1111.

125. MacLab (2015) Freeze drying. Maclab, Victoria, Australia.

126. Pasquet V, Cherouvrier JR, Farhat F, Thiery V, Piot JM, et al. (2011) Study on the microalgal pigments extraction process: Performance of microwave assisted extraction. Process Biochemistry 46: 59-67.

127. Pourmortazavi SM, Hajimirsadeghi SS (2007) Supercritical fluid extraction in plant essential and volatile oil analysis — review. J Chromatogr A 1163: 2-24.

128. Arora R (2012) Microbial biotechnology: Energy and environment. CABI, Oxfordshire, United Kingdom.

129. Belarbi EH, Molina E, Chisti Y (2000) A process for high yield and scaleable recovery of high purity eicosapentaenoic acid esters from microalgae and fish oil. Enzyme Microb Tech 26: 516-529.

130. Balasubramanian RK, Yen Doan TT, Obbard JP (2013) Factors affecting cellular lipid extraction from marine microalgae. ChemEng J, 216: 929–936.

131. Fajardo AR, Cerdan LE, Medina AR, Fernandez FGA, Moreno PAG, et al. (2007) Lipid extraction from the microalga Phaedactylumtricornutum. Eur J Lipid SciTechnol, 109: 120–126.

132. Green DW Perry HR (2008) Perry›s Chemical Engineers› Handbook. McGraw-Hill, New York, USA.

133. Bennion EP (2014) Life cycle assessment of microalgae to biofuel: thermochemical processing through hydrothermal liquefaction or pyrolysis. All Graduate Theses and Dissertations: 2341.

134. Ratti C (2001) Hot air and freeze-drying of high-value foods: a review. J Food Eng 49: 311–319.

135. Ahlgren G, Merino L (1991) Lipid analysis of freshwater microalgae: a method study. ArchivFürHydrobiol 121: 295–306.

136. Taher H, Al-Zuhair S, Al-Marzouqi AH, Haik Y Farid MM (2011) A review of enzymatic transesterification of microalgal oil-based biodiesel using supercritical technology. Enzyme Res 1-25.

137. Balasubramanian S, Allen JD, Kanitkar A Boldor D (2011) Oil extraction from Scenedesmusobliquus using a continuous microwave system—design, optimization, and quality characterization. BioresourcTechnol 102: 3396-3403.

138. Hemwimon S, Pavasant P, Shotipruk A (2007) Microwave-assisted extraction of antioxidativeanthraquinones from roots of Morindacitrifolia. Sep PurifTechnol 54: 44–50.

139. Amarni F, Kadi H (2010) Kinetics study of microwave-assisted solvent extraction of oil from olive cake using hexane: comparison with the conventional extraction. IFSET 11: 322-327.

140. Cheng J, Yu T, Li T, Zhou J, Cen K (2013) Using wet microalgae for direct biodiesel production via microwave irradiation. Bioresourtechnol 131: 531-535.

141. Ma YA, Ming Y, Cheng, Huang JW, Jen JF, et al. (2014) Effects of ultrasonic and microwave pretreatments on lipid extraction of microalgae. Bioprocess BiosystemsEngg, 37: 1543-1549.

142. Passos F, Sole M, Garcia J, Ferrerl(2013) Biogas production from microalgae grown in wastewater: Effect of microwave pretreatment. Appl Energy, 108: 168-175.

143. Patil PD, Gude VG, Mannarswamy A, Cooke P, McGee SM, et-al.(2011) Optimization of microwave-assisted transesterification of dry algal biomass using response surface methodology. BioresourcTechnol 102: 1399-1405.

144. Dai YM, Chen KT, Chen CC (2014) Study of the microwave lipid extraction from microalgae for biodiesel production. Chemical Engineering Journal, 250: 267-273.

145. Qv XY, Zhou QF, Jiang JG (2014) Ultrasound-enhanced and microwave-assisted extraction of lipid from Dunaliellatertiolecta and fatty acid profile analysis. J Sep Sci 37: 2991-2999.

146. Bermudez JM, Boffa L, Mantegna S, Arenillas A, Menendez JA (2012) Energetic evaluation of microalgae extraction under ultrasound and microwave irradiation. Proceedings of International Congress of Chemical Engineering, ICCE 2012, Sevilla, Spain.

147. Azcan N, YilmazO(2012) Microwave irradiation application in biodiesel production from promising biodiesel feedstock: microalgae (Chlorella protothecoides). Proceedings of the World Congress on Engineering and Computer Science, San Francisco, USA

148. Refaat AA, Sheltawy EL, Sadek KU (2008) Optimum reaction time, performance and exhaust emissions of biodiesel produced by microwave irradiation. Int J Environ Sci 5:315–322.

149. Wang LJ (2010) Advances in extraction of plant products in nutraceutical processing. CRC Press, Taylor & Francis, Boca Raton, Florida, USA 2:15–52.

150. Liang K, Zhang Q, Cong W (2012) Enzyme-assisted aqueous extraction of lipid from microalgae. J Agri Food Chem 60:11771-11776.

151. Wang D, Li Y, Hu X, Weimin Su, Zhong M (2015) Combined enzymatic and mechanical cell disruption and lipid extraction of green alga Neochlorisoleoabundans. Int J MolSci 16: 7707-7722.

152. Liu Y, Zhang X, Tan T (2013) Enzyme-aided extraction of lipid from wet microalgae Chlorella vulgaris. J Bioprocess EnggdBiorefiner 2: 193-197.

153. Ying L (2012) The research on extraction microalgae lipid by enzymatic hydrolyzation. Beigin University of Chemical Technology, Masters Thesis.

154. Cho, H-S., Y-K. Oh, S-Ch. P. J-W. Lee and J-Y. Park, 2013. Effects of enzymatic hydrolysis on lipid extraction from Chlorella vulgaris. Renewable Energy, 54: 156-160.

155. Reddy A, Majumder AB (2014) Use of a combined technology of ultrasonication, three-phase partitioning, and aqueous enzymatic oil extraction for the extraction of oil from Spirogyra sp. J Engg, 740631: 1-6.

156. Fu CC, Hung TC, Chen JY, Su CH, Wu WT (2010) Hydrolysis of microalgae cell walls for production of reducing sugar and lipid extraction. BioresourcTechnol 101: 8750-8754.

157. Taher H, Sulaiman AZ, Marzouqi AH, Haik Y, Farid MM (2014) Effective extraction of microalgae lipids from wet biomass for biodiesel production. Biomass Bioenergy 66: 159–167.

158. Zheng G, Li C, Guo L, Ruo W, Wang S (2012) Purificaiton of extracted fatty acids from the microalgae Spirulina. J Amer Oil Chem Society 89: 561-566.

159. Knoshaug EP, Donohoe BS, Gerken H, Laurans L, Wychen SR (2013) Disruption of cell walls for enhanced lipid recovery. Google Patents: US20130171721 A1.

160. Blokker P, Schouten S, Van den Ende H, De Leeuw JW, Hatcher PG, et al. (1998) Chemical structure of algaenans from the fresh water algae Tetraedron minimum, Scenedesmuscommunis and Pediastrumboryanum. Organic Geochem 29: 1453–1468.

161. Burczyk J, Śmietana B, Termińska-Pabis K, Zych M, Kowalowski P (1999) Comparison of nitrogen content amino acid composition and glucosamine content of cell walls of various chlorococcalean algae. Phytochem 51: 491–497.

162. Allard B, Templier J (2000) Comparison of neutral lipid profile of various trilaminar outer cell wall (TLS)-containing microalgae with emphasis on algaenan occurrence. Phytochem 54: 369–380.

163. Lu Y, Kong R, Hu L (2011) Preparation of protoplasts from Chlorella protothecoides. World J MicrobiolBiotechnol 28: 1827-1830.

164. Sander K, Murthy GS (2009) Enzymatic degradation of microalgal cell wall. ASABE Paper 1035636, St. Joseph, Michigan.

165. Khomova TV, Gusakova SD, Mezhlum'yan LG, Glushenkova AI (1993) Lipids of microalgae. I. lipids and a protein hydrolysate from Monochrysislutheri. Chem Natural Comp 24: 507-511.

166. Horst I, Parker BM, Dennis JS, Howe CJ, Scott SA (2012) Treatment of Phaeodactylumtricornutum cells with papain facilitates lipid extraction. J Biotechnol 162: 40-49.

167. Morris HJ, Carrillo OV, Almarales A, Bermudez RC, Alonso ME, et al. (2009) Protein hydrolysates from the alga Chlorella vulgaris 87/1 with potentialities in immunonutrition. BiotechologiaAplicada 26: 162-165.

168. Gerken HG, Donohoe B, Knoshaug EP (2013) Enzymatic cell wall degradation of Chlorella vulgaris and other microalgae for biofuels production. Planta 237: 239-253.

169. Cuellar-BermudezSP, Aguilar-Hernandez I, Cardenas-Chavez DL, Ornelas-Soto N, Romero-Ogawa MA, et al. (2014) Extraction and purification of high-value metabolites from microalgae: essential lipids, astaxanthin and phycobiliproteins. Microbial Biotechnol 8: 109-209.

170. Morris HJ, Almarales A, Carrillo O, Bermudez RC(2008) Utilisation of Chlorella vulgaris cell biomass for the production of enzymatic protein hydrolysates. BioresourcTechnol 99:7723-7729.

171. Kose A,Oncel SS(2015) Properties of microalgal enzymatic protein hydrolysates: Biochemical composition, protein distribution and FTIR characteristics. Biotechnol Reports, 6: 137-143.

172. HoSH, Huang SW, Chen CY, HasunumaT, Kondo A(2013) Bioethanol production using carbohydrate-rich microalgae biomass as feedstock. BioresourceTechnol 135: 191-198.

173. HarunR,Danquah MK(2011) Influence of acid pre-treatment on microalgal biomass for bioethanol production. Process Biochem46: 304-309.

174. Saha BC, Cotta MA(2007) Enzymatic hydrolysis and fermentation of lime pretreated wheat straw to ethanol. J ChemTechnolBiotechnol 82: 913–919.

175. Mtui GYS (2009) Recent advances in pretreatment of lignocellulosic wastes and production of value added products. African J Biotechnol 8: 1398–1415.

176. Bhat MK (2000)Cellulases and related enzymes in biotechnology. BiotechnolAdv 18:355–383.

177. Balat M, BalatH, Oz C(2008) Progress in bioethanol processing. Prog Energy Combustion Sci 34: 551–573.

178. Dincer A, Telefoncu A(2007) Improving the stability of cellulase by immobilization on modified polyvinyl alcohol coated chitosan beads. J MolCatalysis B-Enzymatic 45:10–14.

CHAPTER 20

CHALLENGES AND OPPORTUNITIES FOR HYDROGEN PRODUCTION FROM MICROALGAE

Melanie Oey, Anne Linda Sawyer, Ian Lawrence Ross and Ben Hankamer,

Keywords: Algae, fuel, hydrogen, renewable energy, solar, water

INTRODUCTION

The global economy is valued at ~US$100 Tn pa (CIA, 2014) and is powered by the $6 Tn pa energy sector (Dittrick and Izundu, 2008), with ~80% provided in the form of fuel and ~20% as electricity (IEA, 2009). While renewable electricity generation systems are being deployed rapidly, renewable fuel technologies are ~10–20 years behind on the development curve. This is due to the challenges of reducing fuel production costs towards ~$100 per barrel of oil equivalent to compete with fossil fuels, and the need to improve 'Energy Return on Energy Invested' (ERoEI – a measure of process efficiency) and reduce greenhouse gas (GHG) emissions.

Additionally, by 2050 the human population is forecast to increase from 7.3 billion to over 9 billion people (USCB, 2013), and together with continued global economic growth, this is projected to require 50% more fuel (IEA, 2010), 70% more food (FAO, 2009), 50% more fresh water (OECD, 2014) and cuts in CO_2emissions of 80% (IPCC, 2007; Stocker, 2013) to maintain political, social, fuel and climate security. It is therefore critical to fast-track the development of 'commercially-ready' CO_2 neutral fuel systems that do not compete with food and water needs. The importance of this challenge is highlighted by the common vision of the Group of 7 (G7) nations to achieve GHG reductions of 70% by 2050 based on 2010 figures (G7, 2015).

Citation: Oey, M., Sawyer, A.L., Ross, I.L. and Hankamer, B. (2015) Challenges and opportunities for hydrogen production from microalgae. Plant Biotechnol. J., doi: 10.1111/pbi.12516.

SOLAR FUEL OPTIONS

The production of clean fuels at a globally significant level requires a renewable energy source sufficiently large to drive this process. Solar energy is by far the largest energy source available to us, with 5500 ZJ (1 ZJ = 10^{15} J) (Smil, 2008) reaching the Earth›s atmosphere every year. Of this, 1300 ZJ is photosynthetically active radiation (PAR) available at the Earth›s surface (Oey*et al.*, 2013). This annual irradiance level (1300 ZJ/year) is ~2300× the total global energy demand (0.56 ZJ/year) (BP, 2011) and dwarfs the combined energy from all reported oil, coal, gas and uranium reserves (Stephens *et al.*, 2010b). Photosynthetic organisms have over 3 billion years, evolved intricate photosynthetic systems capable of tapping into this abundant solar resource for the production of food, fuel and atmospheric oxygen (O_2) and not only support life on Earth, but provide a basis for next-generation solar fuel production.

First and second generation crop-based biofuels include sugar and corn ethanol, as well as lignocellulosic ethanol from woody biomass and agricultural residues. However, these technologies require arable land and fresh water, and so ultimately result in '*food vs fuel*' or '*forest vs fuel*' competition (e.g. palm oil). To achieve a transition from '*food/forest vs fuel*' scenarios to a '*food and fuel*' future, next-generation biofuel systems will need to expand photosynthetic capacity into urban areas, onto nonarable land (~25% of the Earth's surface), or into the oceans to utilize alternative water sources and conserve arable land (~3% of the Earth's surface) (Stephens *et al.*, 2012). Microalgae and cyanobacteria systems can be located on nonarable land (and potentially in urban areas and oceans), cultivated at least in part in saline and/or waste water, facilitating increased nutrient and water recycling, and achieve higher yields than crop plants due to the ability to optimize light distribution, CO_2 supply and production conditions and thus offer advantages over crop-based biofuel feed stocks (Pulz and Gross, 2004).

It can be calculated that at a 2% solar to biomass conversion efficiency, which is already achievable at demonstration scale for biomass production in microalgal systems, ~2.15% of the Earth›s surface would be required to supply the current global energy demand in the form of biomass at average illumination levels (0.56 ZJ/year/1300 ZJ/year ×100/2 = 2.15%). Subsequent production of fuels (e.g. oil, bio-diesel, ethanol, methane and H_2) reduces this efficiency, but this is offset, at least in part, by the fact that natural photosynthesis can attain theoretical efficiencies of ~6% (Melis and Happe, 2001; Zhu *et al.*, 2008, 2010). Engineered microalgae and artificial systems, particularly those capable of H_2production, can theoretically achieve higher efficiencies, but require further development (Blankenship *et al.*, 2011; Kruse *et al.*, 2005b; Melis, 2009). This article reviews advances in the development of microalgae-based solar-driven H_2 production processes from water, with a particular focus on the challenges that must be overcome to deliver them and potential solutions that they offer.

Microalgal H_2 Production

The increasing requirement for carbon emission reduction makes H_2 more attractive as a fuel as its combustion yields only H_2O. Furthermore, H2 has a 2.7–3.09 higher energy density (~120 (LHV) – 140 (HHV) kJ/g) than other hydrocarbon fuels (Gupta et al., 2013). Conventional, fossil fuel-based H_2production methods (e.g. steam reformation of natural gas (50%), industrial oil and naphtha reforming (30%), coal gasification (18%) and fossil fuel driven water electrolysis (3.9%) (Gupta *et al.*, 2013; Kalamaras and Efstathiou, 2013)) are costly, energy intensive and emit high levels of CO_2. In comparison, renewable bio-H_2 production systems (such as microalgae and cyanobacteria systems) have the potential to be carbon negative and less energy intensive, as they operate at ambient temperature and pressure (Karthic and Joseph, 2012).

Hydrogen production using oxygen evolving photosynthesis (oxygenic photosynthesis) is the most advanced biological H_2 production approach and the focus of this article. It uses solar energy to split water into protons (H^+), electrons (e^-) and O_2 (Figure 1a) and recombines the derived H^+ and e^- by either hydrogenase (Figure 1b) or nitrogenase enzymes to produce H_2. The utilization of H^+ and e^- derived from water photolysis for H_2 production has only been reported in green algae and cyanobacteria, and provides the basis for the development of bio-inspired artificial photosynthetic systems (Blankenship*et al.*, 2011). The water photolysis process can be divided into indirect and direct pathways.

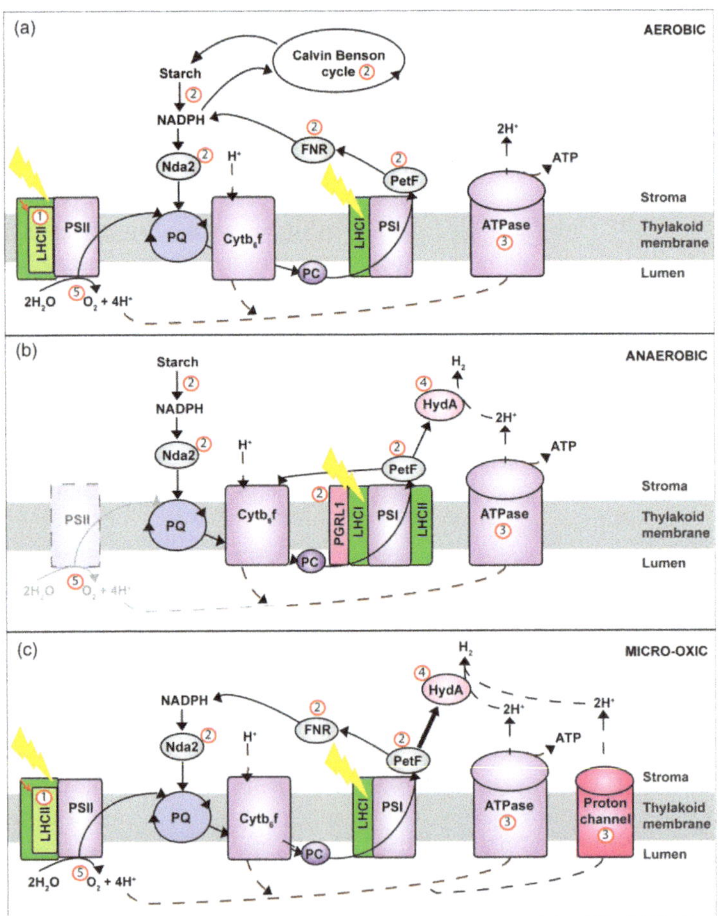

Figure 1: Overview of the photosynthetic microalgal H2 production from water. Aerobic (a) and anaerobic (b) stages of the two phase H2 production and Micro-oxic continuous H2 production (c). Proton (H^+) flow is marked with dashed lines, electron flow with continuous lines. Potential engineering targets for improved H_2-production are marked in red (labels #1–5) and described in detail in section '*Biological challenges and bottlenecks for microalgal H2-production*'. *LHCII*: light-harvesting antenna proteins of photosystem II; *PSII*: photosystem II reaction

centre; *PQ*: plastoquinone pool; *Cytb6f*: cytochrome b_6f complex; *PC*: plastocyanin; *PGRL1*: proton gradient regulation like1 protein;*LHCI*: light-harvesting antenna proteins of photosystem I; *HydA*: hydrogenase A; *PETF:* ferredoxin; *FNR*: *ferredoxin* (flavodoxin)-NADP(H) reductase; *NDH*: NADPH-dehydrogenase.

Indirect Water Photolysis H_2 Production Process

The indirect process can take place in both microalgae and cyanobacteria (Kruse *et al.*, 2005a; Mathews and Wang, 2009; Melis and Happe, 2001; Melis*et al.*, 2000; Rathore and Singh, 2013). Solar energy is first converted into chemical energy in the form of carbohydrates, which are then used as substrates for H_2 production. Cyanobacteria, depending on the species, utilize both nitrogenases and hydrogenases for H_2 production (Dutta *et al.*, 2005), whereas microalgae rely solely on hydrogenases (Oncel *et al.*, 2015). Nitrogenases have the advantage that they act unidirectionally, whereas hydrogenases are bidirectional (Melnicki *et al.*, 2012). However, the high O_2sensitivity of both enzymes requires the separation of H_2 evolution and CO_2fixation. In cyanobacteria, the reactions are either temporally separated during light and dark periods, or physically separated, with H_2 production occurring in heterocysts and CO_2 fixation in the vegetative cells (Dutta *et al.*, 2005). The low ratio of heterocysts to vegetative cells (about 1:10) limits H_2 production levels (Bandyopadhyay *et al.*, 2010). High H_2 production rates have been reported using cyanobacteria as catalysts (i.e. through the inhibition of cell replication) (Bandyopadhyay *et al.*, 2010; Melnicki *et al.*, 2012). However, the energy requirement for the nitrogenase reaction (4 ATP per H_2) reduces H_2 production and conversion efficiencies to less than half of the efficiencies expected from hydrogenase-based indirect photolysis (LaurentPilon and Halil, 2014). In algal systems, the most common, but not ideal practice, is the temporal separation of the O_2 and H_2 production phase (aerobic and anaerobic phase, Figure 1) to circumvent oxygenic inactivation of the hydrogenase (further described below) (Kruse *et al.*, 2005a; Melis and Happe, 2001; Melis *et al.*, 2000).

Direct Water Photolysis H_2 Production Process

Direct photolysis has only been reported in microalgae, and involves the funnelling of e^- derived from the light-driven water splitting reaction of photosystem II (Figure 1c) directly to a H_2-producing hydrogenase (Melis and Happe, 2001; Melis *et al.*, 2000). While direct solar-driven H_2 production from water offers the highest theoretical photon conversion efficiency, achieving this is technically challenging due to the high O_2 sensitivity of the hydrogenase. It requires customized cell lines and production systems which allow the simultaneous production of H_2 and O_2 as this co-production is tightly down-regulated in native strains (Melis *et al.*, 2000).

To date, the highest bio-H_2 production efficiencies have been reported for microalgae (Scoma *et al.*, 2012; Volgusheva *et al.*, 2013). In microalgae, the first step of photosynthesis is the capture of solar energy by the light-harvesting complex (LHC) proteins (Figure 1a). The LHC proteins are classified as LHCI or LHCII types depending on their predominant interaction with either photosystem I (PSI) or II (PSII), respectively. The LHC proteins belong to a large gene family, which in the green alga, *Chlamydomonas reinhardtii,* consists of over 20 members (Dittami *et al.*, 2010). The excitation energy transferred to PSII by LHCII (Figure 1a) drives the photosynthetic water splitting reaction which converts H_2O into H^+, e^- and O_2 (Eq. (1)).

The photons captured by LHCI and LHCII excite e^- derived from H_2O, and drive their transfer along the photosynthetic electron transport chain (e^- flow is indicated by solid black lines in Figure 1a) from the PSII-LHCII supercomplex via plastoquinone (*PQ*), cytochrome b_6f (*Cyt*), the PSI-LHCI supercomplex and ferredoxin (*Fd*) (via the ferredoxin-NADP$^+$ oxido-reductase; FNR) to finally reduce NADP$^+$ to NADPH. NADPH, together with ATP, is used in the Calvin–Benson cycle and subsequent biochemical pathways to fix (i.e. reduce) CO_2 to the sugars, starch, oils and other bio-molecules which collectively form biomass.

Simultaneously, H^+ are released into the thylakoid lumen by PSII and the PQ/PQH_2 cycle (H^+ flow is indicated by black dashed lines in Figure 1). These H^+generate an electrochemical H^+ gradient across the thylakoid membrane, which drives ATP production via the flow of H^+ back to the stroma through the ATP synthase.

Under normal aerobic conditions (Figure 1a), photosynthesis produces the carbohydrates that fuel mitochondrial respiration and cell growth. However, under anaerobic light conditions (Figure 1b), mitochondrial oxidative phosphorylation is largely inhibited by the lack of oxygen, leading to chloroplast over-reduction and a slowing of the electron transport chain, which can ultimately result in photo-damage and reduced ATP generation. Under these conditions, the H^+ and e^- extracted from water by the remaining PSII or starch can be fed to the hydrogenase (HYDA) via the electron transport chain, which recombines the H^+ and e^- to produce H_2 (Eq. (2)).

Overall therefore, Eqs (1) and (2) describe solar-driven H_2 production from H_2O. The hydrogenase essentially acts as a H^+/e^- release valve by recombining H^+(from the medium) and e^- (from reduced ferredoxin) to produce H_2 gas that is excreted from the cell.

$$2H_2O \rightarrow 4H^+ + 4e^- + O_2 \tag{1}$$

and

$$4H^+ + 4e^- \rightarrow 2H_2 \tag{2}$$

The high H_2 production efficiency of microalgae is partly due to the high efficiency of the algal [FeFe]-hydrogenase, which is 100-fold higher than that of other hydrogenases (turnover rate of up to 10^4 H_2 molecules/s) (Lubitz *et al.*, 2014; Volgusheva *et al.*, 2013, 2015). The overall process is also attractive for sustainable applications as it occurs at ambient temperature and pressure. The volatility of H_2 has the advantage that it prevents feedback inhibition and its separation into the gas phase assists harvesting, although there are technical challenges in terms of H_2 containment and its separation from contaminating oxygen.

Chlamydomonas reinhardtii is one of the best studied microalgae species with respect to H_2 production processes, availability of genetic tools and structural biology knowledge of the photosynthetic machinery (Nield *et al.*, 2004; Posten and Walter, 2012; Tokutsu *et al.*, 2012). It is not only able to produce H_2 via oxygenic photolysis (direct and indirect), but also via dark fermentation (see below), in a process similar to that of other micro-organisms (Eroglu and Melis, 2011; Stripp and Happe, 2009; Tolleter *et al.*, 2011; Volgusheva *et al.*, 2013; Wecker and Ghirardi, 2014). Of these pathways, light-dependent H_2 production is the most direct photon-to-fuel conversion process. However, the natural hydrogenase is only expressed and active under micro-oxic or anaerobic conditions (Ghirardi *et al.*, 1997) and is rapidly inactivated by the gradual accumulation of photosynthetically produced O_2. While this provides the basis for a ~95% pure H_2 stream (Kruse *et al.*, 2005a) and potentially protects against the formation of explosive O_2/H_2 mixtures, under natural conditions it only allows for transient H_2 production.

Currently, most algal systems are based on a simple process of temporal separation to circumvent the oxygenic inactivation of the hydrogenase (Meliset *al.*, 2000). This involves first growing algae aerobically for biomass accumulation, ideally by CO_2 fixation, during which oxygen is released and the reducing power stored, typically as carbohydrates (Figure 1a). This step is followed by sulphur (S) starvation. Among other changes, S deprivation reduces the rate of repair of a PSII reaction centre protein (D1), which contains the S-containing amino acid methionine. Photo-damage then reduces PSII to ~ 25% of the original level (Volgusheva *et al.*, 2013). This drops the O_2 production below the rate of mitochondrial respiration and achieves the anaerobic/micro-oxic conditions which induce the expression of an active hydrogenase (Figure 1b). While anaerobiosis initially blocks the e^- transport from the remaining PSII in the phase preceding H_2 production, the block is subsequently released and direct e^- transport to the hydrogenase occurs, with ~80% of the H_2 being derived directly from PSII activity (Volgusheva *et al.*, 2013). Under these conditions, the reactions described in Eqs (1) and (2) are effectively coupled, resulting in the overall process described in Eq. (3). However, they also limit the overall efficiency of the process.

To overcome the limitations of S deprivation, other approaches aim to develop O_2-tolerant hydrogenases or balance O_2 production with metabolic O_2utilization (Oey *et al.*, 2013). A further approach is the use of S microdosing (Figure 1c) (Kosourov *et al.*, 2005) to allow PSII repair, the production of H^+ and e^- from water and to balance O_2 evolution with O_2 utilization through mitochondrial respiration (Kruse *et al.*, 2005a). This micro-oxic approach has the advantage of being more efficient and can provide a basis for a continuous H_2 production process. Another, recently published strategy is magnesium deprivation, which results in a decreased photosynthetic activity but increased respiration and starch accumulation allowing a prolonged H_2 production period (>7 days). Under these conditions, functional PSII, which appears to be crucial for H_2 production, was only reduced by 20%, implying a higher e^- availability to the hydrogenase (Volgusheva *et al.*, 2015). Furthermore, the co-cultivation of algae and bacteria has been presented as a new approach to achieve algal anaerobiosis via increased bacterial respiration (Lakatos *et al.*, 2014). It is also of note that two algae strains are reported to be able to produce low levels of H_2 under aerobic conditions (Hwang *et al.*, 2014).

$$2H_2O \rightarrow 2H_{2(gas)} + O_{2(\text{extracted by respiration})} \text{(Typically } H_2 \text{ gas is} \sim 95\% \text{ pure(Kruse et al., 2005a))} \quad (3)$$

BIOLOGICAL CHALLENGES AND BOTTLE-NECKS FOR MICROALGAL H_2 PRODUCTION

In addition to *Chlamydomonas*, several other algal species, including species of*Chlorella* (Rashid *et al.*, 2011), *Scenedesmus* (Schulz *et al.*, 1998) and*Tetraselmis* (D'Adamo *et al.*, 2014), have also been reported to produce H_2 but at lower levels. The *C. reinhardtii* system has the advantage that all three genomes (nuclear, chloroplast and mitochondrial) have been sequenced and transformed, and much more detailed information is available on the organization of photosynthetic complexes during H_2 production (Hemschemeier and Happe, 2011; Tolleter *et al.*, 2011). Transformation techniques are well established and include particle bombardment (Debuchy*et al.*, 1989; Kindle *et al.*, 1989; Mayfield and Kindle, 1990), glass bead or silicon-carbide whisker methods (Dunahay, 1993; Kindle, 1990), electroporation (Shimogawara *et al.*, 1998) and *Agrobacterium tumefaciens*-mediated gene transfer (Kumar *et al.*, 2004). *Chlamydomonas* therefore remains the best characterized model organism for microalgal H_2 production, associated genetic engineering and the identification of molecular bottlenecks.

Optimizing Light Capture Efficiency

The LHC antennae systems have a dual role: to capture photons and dissipate excess light energy to provide photoprotection (Niyogi, 1999; Pascal *et al.*, 2005; Takahashi *et al.*, 2006). Biomass production efficiency, at least in the laboratory, can be improved through LHC antenna reduction (Figure 1 – label #1), as this enhances light distribution through the bioreactors, enables the use of increased operational cell concentrations and can yield improved overall photosynthetic efficiencies of these systems (Beckmann *et al.*, 2009; Meliset *al.*, 1999; Mussgnug *et al.*, 2007; Oey *et al.*, 2013; Polle *et al.*, 2003). Furthermore, *C. reinhardtii* antenna mutants have also been reported to exhibit an earlier onset of H_2 production under S deprivation conditions (Oey *et al.*, 2013), which is likely due to three factors: (i) improved light distribution leading to higher photon conversion efficiencies of the overall culture, (ii) the ability to lower the dissolved O_2 concentration through the use of higher cell densities, which balance O_2 production with metabolic load, and (iii) altered photo-inhibition and stabilization of PSII required for subsequent H_2 production (Volgusheva *et al.*, 2013). Collectively these properties resulted in cultures exhibiting higher H_2 production rates under the conditions tested (Oey *et al.*, 2013).

The precise engineering of the LHC antenna requires a detailed understanding of the structural complexity and dynamic response of algae to light in larger scale production systems (de Mooij *et al.*, 2015). To date, the engineering of the nuclear encoded antenna genes has utilized chemical or random insertion mutagenesis (Polle *et al.*, 2002, 2003), the manipulation of antenna regulation proteins (e.g. NAB1 (Beckmann *et al.*, 2009)) or *RNAi* knock-down approaches (Mussgnug *et al.*, 2007; Oey *et al.*, 2013).

The least specific method of engineering antenna mutants involves the introduction of foreign DNA into the nuclear genome via *random insertion*(Zhang *et al.*, 2014). As clear phenotypes are needed for the rapid screening of mutants, the expected light green phenotypes of putative antenna mutants seems an attractive selection criteria. However, while this approach identifies mutants impaired in antenna or chlorophyll synthesis, it does not select for specific multiple mutations likely required for optimal H_2 production.

A more targeted approach to engineer antenna cell lines is via the indirect route of manipulating antenna regulation proteins. An example of this is the over-expression of the translational repressor NAB1, which results in the specific down-regulation of specific LHC proteins (Beckmann *et al.*, 2009). This approach relies on cellular regulatory mechanisms and is therefore limited in its target scope. Another approach is *RNAi* mediated knock-down (De Riso *et al.*, 2009; Moellering and Benning, 2010; Molnar *et al.*, 2007, 2009; Rohr *et al.*, 2004; Schmollinger *et al.*, 2010; Sun *et al.*, 2008; Zhao *et al.*, 2007, 2009) of specific target genes such as *LHC* genes, or genes involved in chlorophyll biosynthesis. A challenge for precise antenna engineering is that the coding regions of the *LHC* genes are highly homologous (Natali and Croce, 2015), which complicates the specific down-regulation of target *LHC* genes. Additionally, the need to maintain the *RNAi* expressing mutants, typically through ongoing selective pressure, makes this approach less than ideal for industrial scale up. *RNAi* is therefore best suited for proof-of-principle studies to identify target genes.

The most elegant technologies facilitate precise and permanent genome editing enabling the fine tuning of antenna genes and adjustments to the pigment content, expanding the available solar spectrum (Blankenship and Chen, 2013; Chen and Blankenship, 2011). Several genome editing systems have been developed in recent years, including zinc finger nucleases (ZFN) (Sizova *et al.*, 2013), transcription activator-like effector nucleases (TALENs) (Gao *et al.*, 2014) and the CRISPR/Cas systems (Cho *et al.*, 2013; Mali *et al.*, 2013), all of which rely on nuclease-induced DNA strand breaks and endogenous cell repair mechanisms to obtain mutants with specifically edited genomes. This allows processes such as photosynthesis to be fine-tuned for biotechnological applications. Another attraction of these techniques is the potential to develop mutants which are nongenetically modified organisms due to the transient introduction of the required nucleases. This would be a significant advance for large-scale applications such as biofuel production, particularly given current legislative limitations on the use of GMO strains.

While ZFN (Sizova *et al.*, 2013) and TALENs (Gao *et al.*, 2014) have been used for genome editing in *Chlamydomonas,* they still involve labour-intensive cloning steps. In contrast, the CRISPR/Cas system (Cho *et al.*, 2013; Mali *et al.*, 2013) promises to be a simpler technique only requiring the expression of an RNA guided nuclease, Cas9, to introduce strand breaks. The template for the guide RNA is fairly short (~23 bp), assisting with the targeted engineering of specific antenna genes given their high level of homology. Although successful CRISPR/Cas usage in *Chlamydomonas* is yet to be reported, it is anticipated that this technique will be possible in *Chlamydomonas* in the near future (Jiang*et al.*, 2014), having already been successfully demonstrated in mammalian cells (Cong *et al.*, 2013; Mali *et al.*, 2013; Park *et al.*, 2014), plants (Fauser*et al.*, 2014; Feng *et al.*, 2013; Xie and Yang, 2013) and zebra fish (Hwang*et al.*, 2013). A patent application for its usage in the algal genus*Nannochloropsis* is also encouraging (Patent US 20140220638 A1).

Electron Supply to the Hydrogenase and Availability of Reduced PETF

Limited electron flow to the hydrogenase is another potential bottleneck for sustainable H_2 production (Hallenbeck and Benemann, 2002) (Figure 1 – label #2). This can occur due to the limited availability of reduced ferredoxin (PETF) as a result of other competing pathways (Winkler *et al.*, 2011) (e.g. ferredoxin-$NADP^+$ reductase (FNR), sulphite reductase, nitrate reductase, glutamate synthase and fatty acid desaturases (reviewed in Hemschemeier and Happe, 2011)). The hydrogenase (HYDA) can accept e^- via a direct (PSII-dependent; 2 photons per electron to HYDA) or indirect (PSII-independent from starch; 3 photons per electron to HYDA) route (Chochois *et al.*, 2009; Fouchard *et al.*, 2005). To improve electron flow, PETF, FNR and the hydrogenase have all been engineered (Long *et al.*, 2009; Lubner *et al.*, 2011; Rumpel *et al.*, 2014; Sun*et al.*, 2013; Wittenberg *et al.*, 2013; Yacoby *et al.*, 2011), with a particular focus on the improvement of the affinity of the hydrogenase to PETF, the reduction of the affinity of PETF for FNR and the fusion of PETF and PSI with the hydrogenase. However, the identified candidates have so far only been tested*in vitro* and are yet to be assessed for their performance *in vivo*. Photosynthetic, H_2 producing algal cells supply an excellent scaffold to carry out such proof of principle studies. While most engineering efforts have focused on the chloroplast-localized nuclear encoded genes, recent efforts have been made towards the *in situ* overexpression of the hydrogenase in the plastid to uncouple it from its native control system (Reifschneider-Wegner *et al.*, 2014). Engineering has also focused on indirect targets, including electron competitors such as RuBisCo (Pinto *et al.*, 2013), cyclic electron flow (Kruse *et al.*, 2005a; Tolleter *et al.*, 2011), starch degradation (Chochois *et al.*, 2009) and respiration (Ruehle *et al.*, 2008), with mutants in all of these pathways reportedly yielding increased H_2 production levels. Finally, additional components have been added to the electron transfer pathway, including a plastid-expressed NAD(P)H dehydrogenase (Baltz *et al.*, 2014), a hexose transporter (Doebbe *et al.*, 2007) and exogenous hydrogenases (Chien *et al.*, 2012).

Alteration of the Thylakoid Proton Gradient

Proton supply to the hydrogenase is another potential bottleneck (Figure 1 – label #3). The transport of e^- from water to PETF via the photosynthetic electron transport chain is involved in establishing a proton gradient across the thylakoid membrane which drives ATP synthesis. While ATP production is essential during CO_2 fixation, the ATP requirement drops during H_2 production (Das *et al.*, 2014) and electron transport is reduced at the point of *Cytb6f*(Antal *et al.*, 2009; Burgess *et al.*, 2011), leading to an impaired dissipation of the proton gradient and therefore decreased proton availability for the hydrogenase and reduced H_2 production. One strategy to improve H_2production is to artificially dissipate the proton gradient to increase H_2production transiently in the presence of the chemical uncoupler carbonyl cyanide *m*-chlorophenyl hydrazine (CCCP), which causes an efflux of H^+ from the thylakoid lumen into the stroma (Kruse *et al.*, 2005a; Lee, 2013; Lee and Greenbaum, 2003). This suggests that the integration of a proton channel into the thylakoid membrane could more permanently restore proton and electron flow to the hydrogenase. Such a proton channel, however, would need to be inducibly expressed, as the addition of the uncoupler prior to anaerobiosis was found to abolish hydrogenase activity, suggesting that the proton gradient is important for initial hydrogenase expression (Lee, 2013) and aerobic growth. A similar strategy involves the development of a leaky ATPase to increase proton flow and reduce ATP production (Das *et al.*, 2014; Robertson *et al.*, 1990). Reduced ATP production caused by the introduction of a proton channel or mutated ATPase may additionally reduce reactions competing for reducing equivalents and therefore increase electron supply to the hydrogenase (Kumar and Das, 2013).

Oxygen Sensitivity of the Hydrogenase

Sustained H_2 production under standard growth conditions remains a major challenge, as O_2 sensitivity of the hydrogenase is a multifaceted problem. This is because O_2 oxygen inhibits not just hydrogenase enzyme function, but also transcription and protein maturation (Cohen *et al.*, 2005). Interestingly, two algae strains were recently reported to be able to express the hydrogenase in the presence of more than 21% O_2 and to produce low levels of H_2 at 15% atmospheric O_2 (Hwang *et al.*, 2014). Several genetic engineering approaches have also been tested to reduce the O_2 sensitivity of the hydrogenase (Figure 1– label #4), including random mutagenesis (Flynn *et al.*, 2002) and targeted mutagenesis of the catalytic site to restrict O_2 access (Stiebritz and Reiher, 2012). While engineering approaches have been successful for a bacterial [NiFe] hydrogenase (Dementin *et al.*, 2009), this has so far not been the case with the microalgal [FeFe] hydrogenase. Despite this, the identification of hydrogenases with reduced O_2 sensitivities will increase the scope for genetic engineering. A hydrogenase with a reduced O_2 sensitivity would open up a direct path to H_2production from oxygenic water splitting, but would also lead to the co-production of O_2, which then requires subsequent gas separation. However, this approach would overcome the need for a two phase aerobic/anaerobic process, potentially enabling a continuous H_2 production process and eliminating the need for ATP and NADPH, which are required to produce starch- and oil-based feed stocks for alternative fuels. Achieving this in thermophiles would be expected to further improve reaction kinetics.

Oxygen Evolution and Removal

A different approach used to alter O_2 sensitivity and extend H_2 production is to control the O_2 concentration of the culture. Increasing the cell density and respiratory load can result in an earlier onset of anaerobiosis and H_2production (Oey *et al.*, 2013; Schönfeld *et al.*, 2004). Furthermore, the down-regulation of PSII (Surzycki *et al.*, 2007) and engineering of O_2 evolution activity (Makarova *et al.*, 2007; Scoma *et al.*, 2012; Torzillo *et al.*, 2009) have been exploited (Figure 1 – label #5). Leghemoglobins, which are able to sequester O_2, have also been expressed in *C. reinhardtii* (Wu *et al.*, 2010, 2011), and in a different approach, a sulphate permease mutant was developed, which allowed greater control of S deprivation (Chen *et al.*, 2005).

H_2-Production Improvement Strategies - Supportive Targets

Despite the obvious benefit of manipulating targets directly involved in H_2production, engineering more indirect targets to simplify cultivation, analysis and H_2 production is also important. The ability to easily report lumen pH (Benčina, 2013; Demaurex, 2002), hydrogenase expression and key metabolic pathways such as sulphur, starch and lipid metabolism, combined with external control of gene expression levels and physiological mechanisms via the supply of stimuli such as hormones or light, would enable the rapid dissection of the changes in cell state occurring during H_2 production. New genome editing tools (Cong *et al.*, 2013; Fauser *et al.*, 2014; Feng *et al.*, 2013; Hwang *et al.*, 2013; Mali *et al.*, 2013; Park *et al.*, 2014; Xie and Yang, 2013) will provide scope for sophisticated gene regulation and diagnosis similar to those used in mammalian systems including inducible promoters and self-altering genes (Cre-lox (floxed) systems) (Hajdukiewicz *et al.*, 2001; Oey *et al.*, 2009), highly specific reporter systems and further tools for improving research and industrial use.

High-Throughput Screening Methods

The development of new H_2 screening methods can also accelerate the engineering of enhanced H_2-producing algal strains. To date, H_2 mutants have been detected through a chemo-chromic sensing assay, which utilizes a palladium/tungsten oxide film (Seibert *et al.*, 1998) or in screens specific to the created mutants (Kruse *et al.*, 2005a; Ruehle *et al.*, 2008; Stapleton and Swartz, 2010; Tolleter *et al.*, 2011). A recent H_2-detection screen developed (Wecker and Ghirardi, 2014), which uses *Rhodobacter capsulatus* expressing an emerald green fluorescent protein (emGFP) driven by the *hupR* H_2-sensing promoter as a reporter for H_2 production, offers a simple, inexpensive and semi-quantitative screening method. The development of a chlorophyll fluorescence assay (Godaux *et al.*, 2013) and an inorganic hydrogenase active site mimic will also allow for quicker screening of hydrogenase mutants (Berggren *et al.*, 2013; Esselborn *et al.*, 2013).

MICROALGAE: A BLUEPRINT AND EVALUATION PLATFORM FOR BIO-INSPIRED HYDROGEN PRODUCTION

Natural microalgae-based light-driven H_2 production processes serve as a blueprint and testing platform for the development of bio-inspired artificial solar H_2 production systems. For example, at the level of component development, the atomic structure of the light-harvesting antenna systems provides valuable insights into the design of engineered and artificial light capturing pigments and systems (Qin *et al.*, 2015). Another example is the design of new water splitting catalysts based on the manganese cluster of PSII (Brimblecombe *et al.*, 2010; Tsui *et al.*, 2013; Zhang *et al.*, 2015). The arrangement of the redox-active components of the photosystems has also been studied extensively, and other research has led to the development of self-assembling synthetic scaffolds able to optimize the coordination and spacing of the components of artificial photosynthetic systems (Farid *et al.*, 2013; Mancini *et al.*, 2013; Weingarten *et al.*, 2014). In this context, it is notable that the hydrogenase of *C. reinhardtii* can be prepared in large quantities as an apo-protein and be readily reconstituted with its iron sulphur clusters to form an active hydrogenase (Esselborn *et al.*, 2013). Through the use of such components, it is therefore theoretically possible to build light-driven redox-active systems which split water and use the H^+ and e^- for H_2production. However, the economic viability and ERoEI of both algae and bio-inspired artificial systems will have to be determined later in the development cycle.

CHALLENGES OF SCALE UP

Coordinated Biological Engineering

The selection and engineering of high-efficiency microalgae cell lines is central to this process. Lessons learnt from the optimization of microalgal H_2production suggest that the highest efficiency strains will require not single mutations, but significant re-engineering of the H_2 production process. For example, down-regulating the light-harvesting proteins has benefits as light green cell lines enable higher cell densities to be used, resulting in more rapid O_2 scavenging, the early onset of H_2 production and an overall higher H_2 yield (Oey *et al.*, 2013). The yield of H_2 can also be increased by locking the photosynthetic electron transport chain into the State 2 configuration and by reducing cyclic electron transport, which competes with the hydrogenase for e^-(Kruse *et al.*, 2005a; Tolleter *et al.*, 2011). Additionally, lessening the competition between the hydrogenase and FNR for electrons for example, by re-engineering the electron donors of the hydrogenase could further improve H_2 production (Rumpel *et al.*, 2014). The ultimate goal would be the combination of all of these targeted mutations into a single production strain for continuous H_2 production. It would essentially allow the microalgae cells to act as catalysts under micro-oxic conditions where O_2 production and consumption are in balance (Figure 1c), and the water splitting reaction is active at the same time as the hydrogenase. This would enable a 2 photon vs 3 photon electron transfer steps from water to H_2, which represents a 33% improvement in process efficiency.

Next-generation photobioreactor development

The scale up of microalgae systems from the laboratory to commercial facilities theoretically offers the potential to couple H_2 fuel production to CO_2sequestration (e.g. the residual biomass to biochar) or other co-products. To establish the viability of a given process, rigorous techno-economic (TEA) and life cycle analyses (LCA), as well as pilot and demonstration scale trials, are required to confirm positive economic returns and ERoEI, as well as reduced GHG emissions. Achieving these targets will not only involve the engineering of next-generation cell lines as described above, but the design of photobioreactors able to deliver optimized production conditions for the chosen production strains and specific geographic locations (Franz *et al.*, 2012; Slegers, 2014; Slegers *et al.*, 2011).

Other Artificial and Naturally Derived Renewable H_2-Production Strategies

To replace fossil fuel-based H_2 production processes, several renewable H_2production strategies are being explored. In addition to biological systems such as algae, these include biologically based or biologically inspired systems as well as physical systems such as PV, wind and wave power. The ability to use electrolysis at peak electricity production times to generate H_2 for energy storage is already being employed by current European demonstration projects such as HyUnder, INGRID, MYRTE and GRHYD (AREVA, 2013; D›Errico and Screnci, 2012; Darras *et al.*, 2012; ENGIE; HyUnder, 2012; INGRID; Poggi *et al.*,2014). Coupling such energy capture and electrolysis processes is more environmentally friendly than using fossil fuels, and could technically provide workable solutions which can reportedly achieve photon-to-electron conversion efficiencies of 20% (Bonke *et al.*, 2015). Nevertheless, challenges to the scale up of these technologies remain. These include the following: (i) driving down the capital costs of production, (ii) eliminating rare earth metals (Kapdan and Kargi, 2006), (iii) increasing the catalyst longevity and (iv) improving the ERoEI towards a value of 10:1 required to support a civilized society (Hall *et al.*, 2009).

Microalgae systems currently only have photon conversion efficiencies of 3% (Scoma *et al.*, 2012; Volgusheva *et al.*, 2013) and a theoretical maximum of ~12–14% (Melis, 2009; calculated based on the Z-scheme of electron transport and mean energy of PAR), which is lower than that of artificial systems (Bonke*et al.*, 2015). However, they are self-replicating and repairing, have water splitting complexes and hydrogenases based on manganese, calcium and iron (not rare Earth metals) and have been proven to be able to support H_2production over long periods (Laurinavichene *et al.*, 2006). Furthermore, H_2-producing microalgae can be used in co-production strategies which theoretically have the potential to improve economic performance (Stephens*et al.*, 2010a). More detailed TEA and LCA are required to establish the pros and cons of each technology and which system is more competitive economically in terms of ERoEI and GHG emissions.

Biological H_2 production can, apart from the above discussed oxygenic photosynthesis, further be achieved via microbial electrolysis cells, dark fermentation and photo-fermentation (nonoxygenic photosynthesis) (Chandrasekhar *et al.*, 2015).

Microbial Electrolysis Cells

Microbial Electrolysis Cells are designed to produce H_2 from electrogenic micro-organisms which consume and release e^-, H^+ and CO_2. An externally supplied voltage allows the formation of H_2 gas from the released H^+. As the H_2 production process itself does not occur inside the cell, microbial electrolysis cells are strictly speaking not bio-H_2production platforms. Additionally, due to the requirement of the external supply of an electric current and carbon-based feed stocks, the economic viability of scaling up this H_2 production strategy must be carefully considered (Heidrich *et al.*, 2014).

Dark Fermentation

Involves the production of H_2 using facultative or obligate anaerobic bacteria which also consume organic material. However, this is a chemical energy conversion process rather than a renewable energy process. Furthermore, while the process occurs at a higher rate compared to photo-fermentation or photolysis, it has a low H_2 yield: substrate use ratio, mainly because of the formation of various volatile by-products including ethanol, acetic acid, propionic acid and butyric acid (Jones *et al.*, 2015), from which H_2 must be purified (Chandrasekhar *et al.*, 2015). Production factors such as tight control of the pH, carbon/nitrogen ratio of the substrate and high temperature requirements add additional challenges and operating costs (Jones *et al.*, 2015; Wang *et al.*, 2012).

Anoxygenic Photo-Fermentation

It is performed by purple nonsulphur photosynthetic bacteria under anaerobic and anoxic conditions. These organisms use solar energy to convert organic compounds such as organic acids, sugars and glycerol into H_2 and CO_2. While O_2 sensitivity of the key enzyme, nitrogenase, is not of concern as these bacteria conduct nonoxygenic photosynthesis, the conversion of organic substrates to H_2 appears unlikely to be economical or scalable and the abstraction of these substrates (e.g. sugars) from the food chain is undesirable. Furthermore, H_2 production occurs in the absence of NH_4 due to the high sensitivity of the nitrogenase to nitrogen in the medium. The main challenges of this process are bacterial contamination, substrate type and carbon/nitrogen ratio, synthesis of competing by-products such as polyhydroxybutyrate, sensitivity to temperature and light fluctuations, the low catalytic activity of the nitrogenase, expression suppression by NH_4 (and thus required dilution of waste as a substrate) and lower photochemical efficacy, which collectively lead to low H_2 production levels (Chandrasekhar *et al.*, **2015**; Hay *et al.*, 2013).

The bio-H_2 production routes described above have the advantage of being able to utilize carbon sources which would otherwise be lost, for example from waste water, while assisting in waste water management. However, these processes can still be regarded as chemical energy conversion technologies and are not strictly speaking net solar fuel technologies in which the energy from light is captured and stored in the form of fuel.

Microalgae H_2 Systems within a Global Context

It is also important to understand the global context into which this technology fits, as hydrogen sits at the nexus of several global challenges. In 2010, the G20 nations accounted for 83% of global energy demand and similar levels of CO_2emissions (BP, 2013). The highest *national* CO_2 emissions levels were reported in China (8 205.9 Mt CO_2), the US (5 074.1 Mt CO_2) and India (1 954.0 Mt CO_2) (IEA, 2014). *Per capita*, CO_2 emissions of each of these G20 nations (2013) range from the highest in Saudi Arabia (~17.2 tCO_2/year/person) to the lowest in the high 'national emission country' India (1.8 tCO_2/ year/person) (GCP, 2013; PRB, 2013). Not surprisingly, this has resulted in a debate between high (e.g. the US, Saudi Arabia and Australia) and low (such as China and India) *per capita*CO_2 emitters, all of whom are seeking to limit their emission liabilities. Hydrogen as stored chemical energy offers the potential to provide the required *per capita* energy demand without the CO_2 emissions associated with carbon-based fuels. It also offers significant advantages in terms of the development of global distributed systems and networks, as it can be produced using a broad range of technologies making it an important energy currency. However, significant challenges remain. Some of the greatest include market penetration given the dominance of fossil fuels, the high cost of replacing existing carbon-based fuel infrastructure and the technical difficulty of capturing, storing, purifying and compressing H_2 (Bimbo *et al.*, 2013; Hruzewicz-Kołodziejczyk *et al.*, 2012).

In terms of the photosynthetic H_2 production component of this infrastructure, considerable lessons can be learnt from demonstration scale systems for nearer market volatile fuels, for example ethanol, which are reportedly close to commercial production using cyanobacteria for less than $2 per gallon (Perkins,2014). Insights can be gained in terms of capital costs and their potential reduction, strategies to optimize light delivery, as well as methods to capture volatile products as part of a strategy to minimize feedback inhibition.

In terms of H_2 storage and distribution systems, there is as yet no global infrastructure enabling the use of H_2 as a 'drop-in' fuel. Nevertheless, H_2 is widely used in industry and this provides a solid knowledge base to build on (Ramachandran and Menon, 1998). Additionally, during the early stages of deployment, storage and distribution are not critical for the productive use of H_2, as H_2 can be used to reduce CO_2 to yield a range of CO_2 neutral carbon-based drop-in fuels (Centi and Perathoner, 2009; Wang *et al.*, 2011) compatible with existing storage and distribution systems. Notably, forecasts indicate that by 2025, H_2 may contribute 8–10% to the energy market (Gupta *et al.*, 2013), which provides a development driver for H_2 production, storage, distribution and conversion technologies, and these are actively being developed and deployed internationally (see below). Policies such as tax incentives are also gradually being established (FCHEA, 2014). Consequently, an international movement is beginning to connect solar H_2-producing technologies with advancements in storage, distribution and fuel cell technologies.

The Patent Landscape

Analysis of the global H_2 patent landscape provides valuable insights for a developing microalgal H_2 biotechnology industry, in terms of both challenges and opportunities. National expenditure levels on H_2 technologies by the G20 nations are shown in Figure 2a, with Japan, the US and the EU supporting the largest number of developments. Current patent submissions in the G20 countries in the areas of H_2 energy (Figure 2b) and H_2 production (Figure 2c), as well as renewable H_2 production (Figure 1c, absolute patent numbers given above bar), H_2 storage (Figure 2d), H_2 fuel cells (Figure 2e) and H_2-driven transport (Figure 2f), highlight the advances in the delivery of these components of a future H_2 economy.

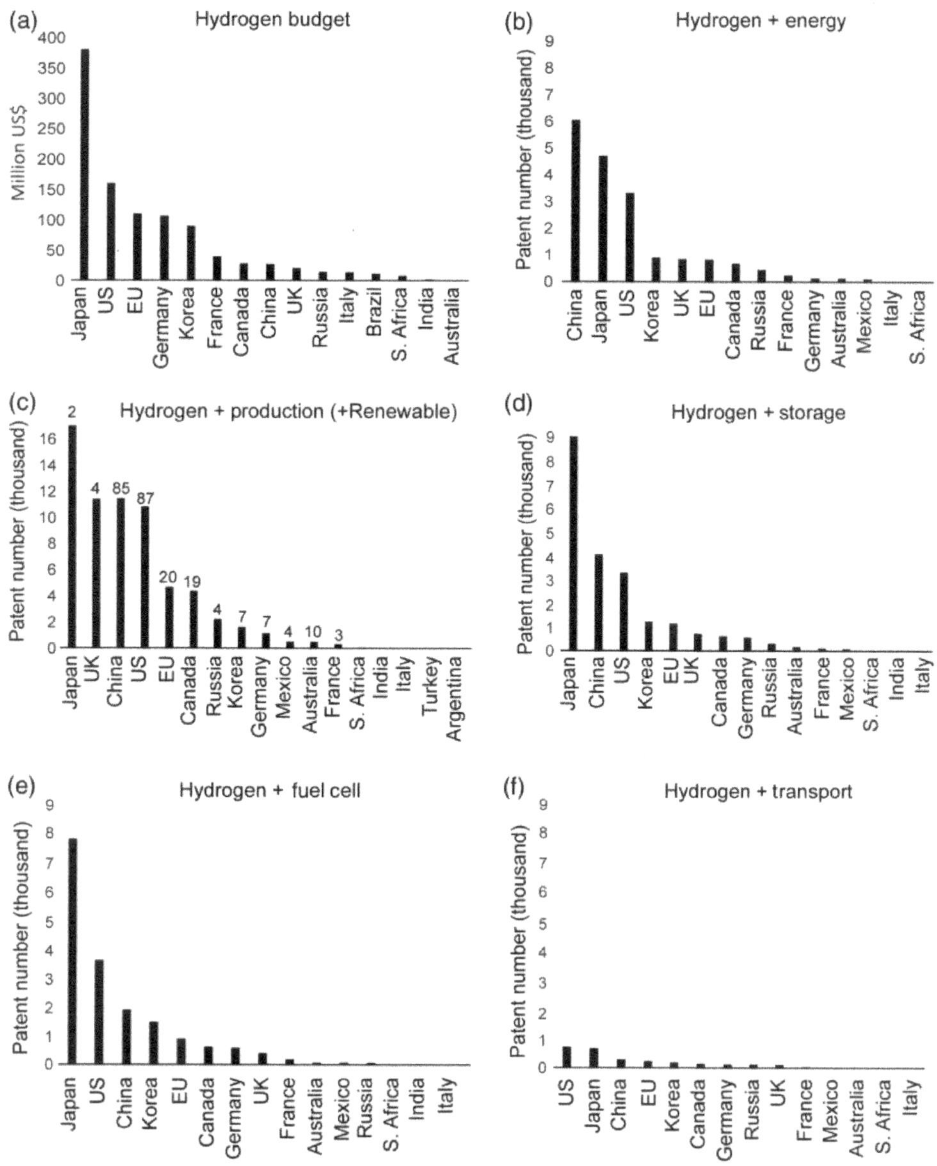

Figure 2: Hydrogen budget and potential market in G20 countries. (a): Allocated budget per country for the development of hydrogen technology. (b–f) Patents filed in the respective countries. Numbers derived from the European Espacenet world wide database keyword search using: Hydrogen + Energy, Hydrogen + Production and Hydrogen + Production + Renewable (renewables are shown as absolute patent numbers above bar), Hydrogen + Storage, Hydrogen + Fuel cell, Hydrogen + Transport.

The majority of patents related to H_2 as a renewable energy source were filed in the US and China (Figure 2c; 87 and 85 patents, respectively), countries which are among the highest *national* CO_2 emitters. Japan reported the highest H_2fuel cell patent submissions (Figure 2e), consistent with its stated aim of transitioning to a H_2 fuel future (GoJ, 2014). These international patenting patterns may be an indicator of early technology adopter hotspots and highlight the potential of developing renewable H_2 production platforms to supply emerging global H_2 markets and distribution networks.

CONCLUSIONS

Solar-driven microalgal H_2 production is both a promising and challenging biotechnology, and will likely play an important role in the global drive to reduce GHG emissions. Advances in molecular engineering, TEA/LCA and pilot scale test facilities and predictive design are providing critical tools for the development of the solar fuel systems required to tap into the huge solar resource available to us and which could be used to drive a sustainable energy economy. Using these tools to develop 'deployment ready' microalgae or artificial solar fuel systems in the next 10 years is of critical importance for climate stabilization, but this can only be achieved within a framework of political leadership which will enable up to 70% emission reductions within the next 10–20 year time window.

ACKNOWLEDGEMENTS

We gratefully acknowledge the funding of the Australian Research Council via grants DP110101699 and DP150100740 and Go8 Australia via grant *'Protein engineering to increase light-to-hydrogen production in algae'*. We have no known conflict of interest to disclose in regard to the presented work.

REFERENCES

1. Antal, T.K., Volgusheva, A.A., Kukarskih, G.P., Krendeleva, T.E. and Rubin, A.B. (2009) Relationships between H2 photoproduction and different electron

transport pathways in sulfur-deprived Chlamydomonas reinhardtii. Int. J. Hydrogen Energy, 34, 9087–9094.

2. AREVA (2013) Renewable energies: AREVA inaugurates an energy storage platform in Corsica. http://www.areva.com/EN/news-9153/renewable-energies-areva-inaugurates-an-energy-storage-platform-in-corsica.html (accessed 29 October 2015).

3. Baltz, A., Kieu-Van, D., Beyly, A., Auroy, P., Richaud, P., Cournac, L. and Peltier, G. (2014) Plastidial expression of type II NAD(P)H dehydrogenase increases the reducing state of plastoquinones and hydrogen photoproduction rate by the indirect pathway in Chlamydomonas reinhardtii. Plant Physiol. 165, 1344–1352.

4. Bandyopadhyay, A., Stöckel, J., Min, H., Sherman, L.A. and Pakrasi, H.B. (2010) High rates of photobiological H2 production by a cyanobacterium under aerobic conditions. Nat. Commun. 1, 139.

5. Beckmann, J., Lehr, F., Finazzi, G., Hankamer, B., Posten, C., Wobbe, L. and Kruse, O. (2009) Improvement of light to biomass conversion by de-regulation of light-harvesting protein translation in Chlamydomonas reinhardtii. J. Biotechnol. 142, 70–77.

6. Benčina, M. (2013) Illumination of the spatial order of intracellular pH by genetically encoded pH-sensitive sensors. Sensors (Basel, Switzerland), 13, 16736–16758.

7. Berggren, G., Adamska, A., Lambertz, C., Simmons, T.R., Esselborn, J., Atta, M., Gambarelli, S. et al. (2013) Biomimetic assembly and activation of [FeFe]-hydrogenases. Nature, 499, 66–69.

8. Bimbo, N., Ting, V.P., Sharpe, J.E. and Mays, T.J. (2013) Analysis of optimal conditions for orptive hydrogen storage in microporous solids. Colloids Surf. A 437, 113–119.

9. Blankenship, R.E. and Chen, M. (2013) Spectral expansion and antenna reduction can enhance photosynthesis for energy production. Curr. Opin. Chem. Biol. 17, 457–461.

10. Blankenship, R.E., Tiede, D.M., Barber, J., Brudvig, G.W., Fleming, G., Ghirardi, M., Gunner, M.R. et al. (2011) Comparing photosynthetic and photovoltaic efficiencies and recognizing the potential for improvement. Science, 332, 805–809.

11. Bonke, S.A., Wiechen, M., MacFarlane, D.R. and Spiccia, L. (2015) Renewable fuels from concentrated solar power: towards practical artificial photosynthesis. Energ Environ. Sci. 8, 2791–2796.

12. BP (2011) Statistical Review of World Energy 2011. http://www.bp.com/sectionbodycopy.do?categoryId=7500&contentId=7068481 (accessed 29 October 2015).

13. BP (2013) Statistical Review of World Energy2013. http://www.bp.com/content/dam/bp/pdf/statistical-review/statistical_review_of_world_energy_2013.pdf (accessed 29 October 2015).

14. Brimblecombe, R., Koo, A., Dismukes, G.C., Swiegers, G.F. and Spiccia, L. (2010) Solar driven water oxidation by a bioinspired manganese molecular catalyst. J. Am. Chem. Soc. 132, 2892–2894.

15. Burgess, S.J., Tamburic, B., Zemichael, F., Hellgardt, K. and Nixon, P.J. (2011) Solar-driven hydrogen production in green algae. In Advances in Applied Microbiology, Vol. 75 (Laskin, A.I., Sariaslani, S. and Gadd, G.M., eds), pp. 71–110. Academic Press: Waltham, USA.

16. Centi, G. and Perathoner, S. (2009) Opportunities and prospects in the chemical recycling of carbon dioxide to fuels. Catal. Today, 148, 191–205.

17. Chandrasekhar, K., Lee, Y.-J. and Lee, D.-W. (2015) Biohydrogen production: strategies to improve process efficiency through microbial routes. Int. J. Mol. Sci. 16, 8266.

18. Chen, M. and Blankenship, R.E. (2011) Expanding the solar spectrum used by photosynthesis. Trends Plant Sci. 16, 427–431.

19. Chen, H.C., Newton, A.J. and Melis, A. (2005) Role of SulP, a nuclear-encoded chloroplast sulfate permease, in sulfate transport and H2 evolution in Chlamydomonas reinhardtii. Photosynth. Res. 84, 289–296.

20. Chien, L.-F., Kuo, T.-T., Liu, B.-H., Lin, H.-D., Feng, T.-Y. and Huang, C.-C. (2012) Solar-to-bioH2 production enhanced by homologous overexpression of hydrogenase in green alga Chlorella sp. DT. Int. J. Hydrogen Energy, 37, 17738–17748.

21. Cho, S.W., Kim, S., Kim, J.M. and Kim, J.-S. (2013) Targeted genome engineering in human cells with the Cas9 RNA-guided endonuclease. Nat. Biotech. 31, 230–232.

22. Chochois, V., Dauvillee, D., Beyly, A., Tolleter, D., Cuine, S., Timpano, H., Ball, S. et al. (2009) Hydrogen production in Chlamydomonas: photosystem II-dependent and -independent pathways differ in their requirement for starch metabolism. Plant Physiol. 151, 631–640.

23. CIA Central Intelligence Agency. (2014) CIA World Factbook 2015. New York, NY: Skyhorse Publishing.

24. Cohen, I., Knopf, J.A., Irihimovitch, V. and Shapira, M. (2005) A proposed mechanism for the inhibitory effects of oxidative stress on rubisco assembly and its subunit expression. Plant Physiol. 137, 738–746.

25. Cong, L., Ran, F.A., Cox, D., Lin, S., Barretto, R., Habib, N., Hsu, P.D. et al. (2013) Multiplex genome engineering using CRISPR/Cas systems. Science, 339, 819–823.

26. D'Adamo, S., Jinkerson, R.E., Boyd, E.S., Brown, S.L., Baxter, B.K., Peters, J.W. and Posewitz, M.C. (2014) Evolutionary and biotechnological implications of robust hydrogenase activity in halophilic strains of Tetraselmis. PLoS ONE, 9, e85812.

27. Darras, C., Muselli, M., Poggi, P., Voyant, C., Hoguet, J.C. and Montignac, F. (2012) PV output power fluctuations smoothing: The MYRTE platform experience. Int. J. Hydrogen Energy, 37, 14015–14025.

28. Das, D., Khanna, N. and Dasgupta, C.N. (2014) Biohydrogen Production: Fundamentals and Technology Advances. Boca Raton, USA: CRC Press.

29. De Riso, V., Raniello, R., Maumus, F., Rogato, A., Bowler, C. and Falciatore, A. (2009) Gene silencing in the marine diatom Phaeodactylum tricornutum. Nucleic Acids Res. 37, e96.

30. Debuchy, R., Purton, S. and Rochaix, J.D. (1989) The argininosuccinate lyase gene of Chlamydomonas reinhardtii: an important tool for nuclear transformation and for correlating the genetic and molecular maps of the ARG7 locus. EMBO J. 8, 2803–2809.

31. Demaurex, N. (2002) pH homeostasis of cellular organelles. Physiology, 17, 1–5.

32. Dementin, S., Leroux, F., Cournac, L., de lacey, A.L., Volbeda, A., Leger, C., Burlat, B. et al. (2009) Introduction of methionines in the gas channel makes NiFe hydrogenase aero-tolerant. J. Am. Chem. Soc. 131, 10156–10164.

33. D'Errico, F. and Screnci, A. (2012) High-capacity hydrogen-based green-energy storage solutions for the grid balancing. In Magnesium Technology 2012 (Mathaudhu, S. N., Sillekens, W. H., Neelameggham, N. R. and Hort, N.eds), pp. 483–488. Hoboken: John Wiley & Sons, Inc.

34. OECD (2014) The Organisation for Economic Co-operation and Development OECD Environmental Outlook to 2050. http://www.oecd.org/env/indicators-modelling-outlooks/oecdenvironmentaloutlookto2050theconsequencesofinaction.htm (accessed 08.12.2015).

35. Dittami, S.M., Michel, G., Collen, J., Boyen, C. and Tonon, T. (2010) Chlorophyll-binding proteins revisited - a multigenic family of light-harvesting and stress proteins from a brown algal perspective. BMC Evol. Biol. 10, 365.

36. Dittrick, P. and Izundu, U. (2008) OTC speakers highlight offshore industry's future. Oil Gas J. 106, 20.

37. Doebbe, A., Rupprecht, J., Beckmann, J., Mussgnug, J.H., Hallmann, A., Hankamer, B. and Kruse, O. (2007) Functional integration of the HUP1 hexose symporter gene into the genome of C. reinhardtii: impacts on biological H2 production. J. Biotechnol. 131, 27–33.

38. Dunahay, T.G. (1993) Transformation of Chlamydomonas reinhardtii with silicon carbide whiskers. Biotechniques, 15, 452–455, 457-458, 460.

39. Dutta, D., De, D., Chaudhuri, S. and Bhattacharya, S.K. (2005) Hydrogen production by cyanobacteria. Microb. Cell Fact. 4, 36–36.

40. ENGIE The GRHYD demonstration project. http://www.gdfsuez.com/en/innovation-energy-transition/smart-energy-management/power-to-gas/the-grhyd-demonstration-project/ (accessed 29 October 2015).

41. Eroglu, E. and Melis, A. (2011) Photobiological hydrogen production: recent advances and state of the art. Bioresour. Technol. 102, 8403–8413.

42. Esselborn, J., Lambertz, C., Adamska-Venkatesh, A., Simmons, T., Berggren, G., Noth, J., Siebel, J. et al. (2013) Spontaneous activation of [FeFe]-hydrogenases by an inorganic [2Fe] active site mimic. Nat. Chem. Biol. 9, 607–609.

43. FAO Food and Agriculture Organization of the United Nations (2009) How to Feed the World 2050. http://www.fao.org/fileadmin/templates/wsfs/docs/expert_paper/How_to_Feed_the_World_in_2050.pdf (accessed 08.12.2015).

44. Farid, T.A., Kodali, G., Solomon, L.A., Lichtenstein, B.R., Sheehan, M.M., Fry, B.A., Bialas, C. et al. (2013) Elementary tetrahelical protein design for diverse oxidoreductase functions. Nat. Chem. Biol. 9, 826–833.

45. Fauser, F., Schiml, S. and Puchta, H. (2014) Both CRISPR/Cas-based nucleases and nickases can be used efficiently for genome engineering in Arabidopsis thaliana. Plant J. 79, 348–359.

46. FCHEA Fuel Cell and Hydrogen Energy Association (2014) International Developments. http://ftp.fchea.org/index.php?id=25 (accessed 29 October 2015).

47. Feng, Z., Zhang, B., Ding, W., Liu, X., Yang, D.-L., Wei, P., Cao, F. et al. (2013) Efficient genome editing in plants using a CRISPR/Cas system. Cell Res. 23, 1229–1232.

48. Flynn, T., Ghirardi, M.L. and Seibert, M. (2002) Accumulation of O2-tolerant phenotypes in H2-producing strains of Chlamydomonas reinhardtii by sequential applications of chemical mutagenesis and selection. Int. J. Hydrogen Energy, 27, 1421–1430.

49. Fouchard, S., Hemschemeier, A., Caruana, A., Pruvost, K., Legrand, J., Happe, T., Peltier, G. et al. (2005) Autotrophic and mixotrophic hydrogen photoproduction in sulfur-deprived Chlamydomonas cells. Appl. Environ. Microbiol. 71, 6199–6205.

50. Franz, A., Lehr, F., Posten, C. and Schaub, G. (2012) Modeling microalgae cultivation productivities in different geographic locations - estimation method for idealized photobioreactors. Biotechnol. J. 7, 546–557.

51. G7 Group of 7 (2015) Leaders' Declaration G7 Summit - Think Ahead. Act Together.

52. Gao, H., Wright, D.A., Li, T., Wang, Y., Horken, K., Weeks, D.P. and Yang, B. (2014) TALE activation of endogenous genes in Chlamydomonas reinhardtii. Algal Res. 1, 52–60.

53. GCP Global Carbon Project (2013) Global Carbon Atlas. http://www.globalcarbonatlas.org/?q=en/emissions (accessed 29 October 2015).

54. Ghirardi, M.L., Togasaki, R.K. and Seibert, M. (1997) Oxygen sensitivity of algal H2-production. Appl. Biochem. Biotechnol. 63–5, 141–151.

55. Godaux, D., Emoncls-Alt, B., Berne, N., Ghysels, B., Alric, J., Remacle, C. and Cardol, P. (2013) A novel screening method for hydrogenase-deficient mutants in Chlamydomonas reinhardtii based on in vivo chlorophyll fluorescence and photosystem II quantum yield. Int. J. Hydrogen Energy 38, 1826–1836.

56. GoJ Government of Japan (2014) The 4th Strategic Energy Plan of Japan - Provisional Translation.

57. Gupta, S.K., Kumari, S., Reddy, K. and Bux, F. (2013) Trends in biohydrogen production: major challenges and state-of-the-art developments. Environ. Technol. 34, 1653–1670.

58. Hajdukiewicz, P.T.J., Gilbertson, L. and Staub, J.M. (2001) Multiple pathways for Cre/lox-mediated recombination in plastids. Plant J. 27, 161–170.

59. Hall, C.A.S., Balogh, S. and Murphy, D.J.R. (2009) What is the minimum EROI that a sustainable society must have? Energies, 2, 25–47.

60. Hallenbeck, P.C. and Benemann, J.R. (2002) Biological hydrogen production; fundamentals and limiting processes. Int. J. Hydrogen Energy 27, 1185–1193.

61. Hay, J.X.W., Wu, T.Y., Juan, J.C. and Md. Jahim, J. (2013) Biohydrogen production through photo fermentation or dark fermentation using waste as a substrate: Overview, economics, and future prospects of hydrogen usage. Biofuels Bioprod. Bior 7, 334–352.

62. Heidrich, E.S., Edwards, S.R., Dolfing, J., Cotterill, S.E. and Curtis, T.P. (2014) Performance of a pilot scale microbial electrolysis cell fed on domestic wastewater at ambient temperatures for a 12 month period. Bioresour. Technol. 173, 87–95.

63. Hemschemeier, A. and Happe, T. (2011) Alternative photosynthetic electron transport pathways during anaerobiosis in the green alga Chlamydomonas reinhardtii. Biochim. Biophys. Acta, 1807, 919–926.

64. Hruzewicz-Kołodziejczyk, A., Ting, V.P., Bimbo, N. and Mays, T.J. (2012) Improving comparability of hydrogen storage capacities of nanoporous materials. Int. J. Hydrogen Energy, 37, 2728–2736.

65. Hwang, W.Y., Fu, Y., Reyon, D., Maeder, M.L., Tsai, S.Q., Sander, J.D., Peterson, R.T. et al. (2013) Efficient genome editing in zebrafish using a CRISPR-Cas system. Nat. Biotech. 31, 227–229.

66. Hwang, J.-H., Kim, H.-C., Choi, J.-A., Abou-Shanab, R.A.I., Dempsey, B.A., Regan, J.M., Kim, J.R. et al. (2014) Photoautotrophic hydrogen production by eukaryotic microalgae under aerobic conditions. Nat. Commun. 5, 3234.

67. HyUnder (2012) HyUnder Project. http://www.hyunder.eu/ (accessed 29 October 2015).

68. IEA International Energy Agency (2009) World Energy Outlook 2009. http://www.iea.org/weo/2009.asp (accessed 29 Otober 2015).

69. IEA International Energy Agency (2010) World Energy Outlook 2010. http://www.worldenergyoutlook.org/publications/weo-2010/ (accessed 29 October 2015).

70. IEA International Energy Agency (2014) World Energy Outlook 2014. http://www.worldenergyoutlook.org/weo2014/ (accessed 08.12.2015)

71. INGRID INGRID Project. http://www.ingridproject.eu/ (accessed 29 October 2015).

72. Jiang, W., Brueggeman, A.J., Horken, K.M., Plucinak, T.M. and Weeks, D.P. (2014) Successful transient expression of Cas9 and single guide RNA genes in Chlamydomonas reinhardtii. Eukaryot. Cell, 13, 1465–1469.

73. Jones, R.J., Massanet-Nicolau, J., Guwy, A., Premier, G.C., Dinsdale, R.M. and Reilly, M. (2015) Removal and recovery of inhibitory volatile fatty acids from mixed acid fermentations by conventional electrodialysis. Bioresour. Technol. 189, 279–284.

74. Kalamaras, C.M. and Efstathiou, A.M. (2013) Hydrogen production technologies: Current state and future developments. Conf. Pap. Energy, 2013, 9.

75. Kapdan, I.K. and Kargi, F. (2006) Bio-hydrogen production from waste materials. Enzyme Microb. Technol. 38, 569–582.

76. Karthic, P. and Joseph, S. (2012) Comparison and limitations of biohydrogen production Processes. Res J. Biotechnol. 7, 59–71.

77. Kindle, K.L. (1990) High-frequency nuclear transformation of Chlamydomonas reinhardtii. Proc. Natl Acad. Sci. USA, 87, 1228–1232.

78. Kindle, K.L., Schnell, R.A., Fernandez, E. and Lefebvre, P.A. (1989) Stable nuclear transformation of Chlamydomonas using the Chlamydomonas gene for nitrate reductase. J. Cell Biol. 109, 2589–2601.

79. Kosourov, S., Makarova, V., Fedorov, A.S., Tsygankov, A., Seibert, M. and Ghirardi, M.L. (2005) The effect of sulfur re-addition on H2 photoproduction by sulfur-deprived green algae. Photosynth. Res. 85, 295–305.

80. Kruse, O., Rupprecht, J., Bader, K.P., Thomas-Hall, S., Schenk, P.M., Finazzi, G. and Hankamer, B. (2005a) Improved photobiological H2 production in engineered green algal cells. J. Biol. Chem. 280, 34170–34177.

81. Kruse, O., Rupprecht, J., Mussgnug, J.H., Dismukes, G.C. and Hankamer, B. (2005b) Photosynthesis: a blueprint for solar energy capture and biohydrogen production technologies. Photochem. Photobiol. Sci. 4, 957–970.

82. Kumar, K. and Das, D. (2013) CO2 sequestration and hydrogen production using cyanobacteria and green algae. In Natural and Artificial Photosynthesis (Reza, R., ed.), pp. 173–215. Hoboken: John Wiley & Sons Inc.

83. Kumar, S.V., Misquitta, R.W., Reddy, V.S., Rao, B.J. and Rajam, M.V. (2004) Genetic transformation of the green alga - Chlamydomonas reinhardtii by Agrobacterium tumefaciens. Plant Sci. 166, 731–738.

84. Lakatos, G., Deak, Z., Vass, I., Retfalvi, T., Rozgonyi, S., Rakhely, G., Ordog, V. et al. (2014) Bacterial symbionts enhance photo-fermentative hydrogen evolution of Chlamydomonas algae. Green Chem. 16, 4716–4727.

85. Laurinavichene, T.V., Fedorov, A.S., Ghirardi, M.L., Seibert, M. and Tsygankov, A.A. (2006) Demonstration of sustained hydrogen photoproduction by immobilized, sulfur-deprived Chlamydomonas reinhardtii cells. Int. J. Hydrogen Energy, 31, 659–667.

86. Lee, J. (2013) Designer transgenic algae for photobiological production of hydrogen from water. In Advanced Biofuels and Bioproducts (Lee, J.W., ed.), pp. 371–404. New York: Springer US.

87. Lee, J.W. and Greenbaum, E. (2003) A new oxygen sensitivity and its potential application in photosynthetic H2 production. Appl. Biochem. Biotechnol. 105, 303–313.

88. Long, H., King, P.W., Ghirardi, M.L. and Kim, K. (2009) Hydrogenase/Ferredoxin charge-transfer complexes: Effect of hydrogenase mutations on the complex association. J. Phys. Chem. A 113, 4060–4067.

89. Lubitz, W., Ogata, H., Ruediger, O. and Reijerse, E. (2014) Hydrogenases. Chem. Rev. 114, 4081–4148.

90. Lubner, C.E., Applegate, A.M., Knörzer, P., Ganago, A., Bryant, D.A., Happe, T. and Golbeck, J.H. (2011) Solar hydrogen-producing bionanodevice outperforms

natural photosynthesis. Proc. Natl Acad. Sci. USA, 108, 20988–20991.

91. Makarova, V.V., Kosourov, S., Krendeleva, T.E., Semin, B.K., Kukarskikh, G.P., Rubin, A.B., Sayre, R.T. et al. (2007) Photoproduction of hydrogen by sulfur-deprived C. reinhardtii mutants with impaired photosystem II photochemical activity. Photosynth. Res. 94, 79–89.

92. Mali, P., Yang, L., Esvelt, K.M., Aach, J., Guell, M., DiCarlo, J.E., Norville, J.E. et al. (2013) RNA-guided human genome engineering via Cas9. Science, 339, 823–826.

93. Mancini, J.A., Kodali, G., Solomon, L.A., Roach, N., Anderson, J.L.R., Esipova, T.V., Vinogradov, S.A. et al. (2013) Light harvesting and light activatable protein maquettes designed from scratch. Biophys. J. 104, 531a–531a.

94. Mathews, J. and Wang, G. (2009) Metabolic pathway engineering for enhanced biohydrogen production. Int. J. Hydrogen Energy 34, 7404–7416.

95. Mayfield, S.P. and Kindle, K.L. (1990) Stable nuclear transformation of Chlamydomonas reinhardtii by using a C. reinhardtii gene as the selectable marker. Proc. Natl Acad. Sci. USA, 87, 2087–2091.

96. Melis, A. (2009) Solar energy conversion efficiencies in photosynthesis: minimizing the chlorophyll antennae to maximize efficiency. Plant Sci. 177, 272–280.

97. Melis, A. and Happe, T. (2001) Hydrogen production. Green algae as a source of energy. Plant Physiol. 127, 740–748.

98. Melis, A., Neidhardt, J. and Benemann, J. (1999) Dunaliella salina (Chlorophyta) with small chlorophyll antenna sizes exhibit higher photosynthetic productivities and photon use efficiencies than normally pigmented cells. J. Appl. Phycol. 10, 515–525.

99. Melis, A., Zhang, L.P., Forestier, M., Ghirardi, M.L. and Seibert, M. (2000) Sustained photobiological hydrogen gas production upon reversible inactivation of oxygen evolution in the green alga Chlamydomonas reinhardtii. Plant Physiol. 122, 127–135.

100. Melnicki, M.R., Pinchuk, G.E., Hill, E.A., Kucek, L.A., Fredrickson, J.K., Konopka, A. and Beliaev, A.S. (2012) Sustained H2 production driven by photosynthetic water splitting in a unicellular cyanobacterium. mBio, 3, e00197–00112.

101. Moellering, E.R. and Benning, C. (2010) RNA interference silencing of a major lipid droplet protein affects lipid droplet size in Chlamydomonas reinhardtii. Eukaryot. Cell, 9, 97–106.

102. Molnar, A., Schwach, F., Studholme, D.J., Thuenemann, E.C. and Baulcombe, D.C. (2007) miRNAs control gene expression in the single-cell alga Chlamydomonas reinhardtii. Nature, 447, 1126–U1115.

103. Molnar, A., Bassett, A., Thuenemann, E., Schwach, F., Karkare, S., Ossowski, S., Weigel, D. et al. (2009) Highly specific gene silencing by artificial microRNAs in the unicellular alga Chlamydomonas reinhardtii. Plant J. 58, 165–174.

104. de Mooij, T., Janssen, M., Cerezo-Chinarro, O., Mussgnug, J., Kruse, O., Ballottari, M., Bassi, R. et al. (2015) Antenna size reduction as a strategy to increase biomass productivity: a great potential not yet realized. J. Appl. Phycol. 27, 1063–1077.

105. Mussgnug, J.H., Thomas-Hall, S.R., Rupprecht, J., Foo, A., Klassen, V., McDowall, A., Schenk, P.M. et al. (2007) Engineering photosynthetic light capture: Impacts on improved solar energy to biomass conversion. Plant Biotechnol. J. 5, 802–814.

106. Natali, A. and Croce, R. (2015) Characterization of the major light-harvesting complexes (LHCBM) of the green alga Chlamydomonas reinhardtii. PLoS ONE, 10, e0119211.

107. Nield, J., Redding, K. and Hippler, M. (2004) Remodeling of light-harvesting protein complexes in Chlamydomonas in response to environmental changes. Eukaryot. Cell 3, 1370–1380.

108. Niyogi, K.K. (1999) Photoprotection revisited: Genetic and molecular approaches. Annu Rev. Plant Phys. 50, 333–359.

109. Oey, M., Lohse, M., Scharff, L.B., Kreikemeyer, B. and Bock, R. (2009) Plastid production of protein antibiotics against pneumonia via a new strategy for high-level expression of antimicrobial proteins. Proc. Natl Acad. Sci. USA, 106, 6579–6584.

110. Oey, M., Ross, I.L., Stephens, E., Steinbeck, J., Wolf, J., Radzun, K.A., Kügler, J. et al. (2013) RNAi knock-down of LHCBM1, 2 and 3 increases photosynthetic H2 production efficiency of the green alga Chlamydomonas reinhardtii. PLoS ONE, 8, e61375.

111. Oncel, S.S., Kose, A. and Faraloni, C. (2015) Chapter 25 - Genetic optimization of microalgae for biohydrogen production. In Handbook of Marine Microalgae (Kim, S.-K., ed.), pp. 383–404. Boston: Academic Press.

112. Park, A., Won, S.T., Pentecost, M., Bartkowski, W. and Lee, B. (2014) CRISPR/Cas9 allows efficient and complete knock-in of a destabilization domain-tagged essential protein in a human cell line, allowing rapid knock-down of protein function. PLoS ONE, 9, e95101.

113. Pascal, A.A., Liu, Z., Broess, K., van Oort, B., van Amerongen, H., Wang, C., Horton, P. et al. (2005) Molecular basis of photoprotection and control of photosynthetic light-harvesting. Nature, 436, 134–137.

114. Perkins, J. (2014) Going commercial. BioFuels J. 12, 60–61.

115. Pilon, L. and Halil, B.l. (2014) Photobiological hydrogen production. In Handbook of Hydrogen Energy (Sherif, S.A., Yogi Goswami, D., Stefanakos, E.K. (Lee), Steinfeld, A. eds), pp. 369–418. Boca Raton: CRC Press.

116. Pinto, T.S., Malcata, F.X., Arrabaça, J.D., Silva, J.M., Spreitzer, R.J. and Esquível, M.G. (2013) Rubisco mutants of Chlamydomonas reinhardtii enhance photosynthetic hydrogen production. Appl. Microbiol. Biotechnol. 97, 5635–5643.

117. Poggi, P., Darras, C., Muselli, M. and Pigelet, G. (2014) The PV-hydrogen MYRTE platform - PV output power fluctuations smoothing. Energy Procedia, 57, 607–616.

118. Polle, J.E.W., Kanakagiri, S., Jin, E., Masuda, T. and Melis, A. (2002) Truncated chlorophyll antenna size of the photosystems - a practical method to improve microalgal productivity and hydrogen production in mass culture. Int. J. Hydrogen Energy, 27, 1257–1264.
119. Polle, J.E.W., Kanakagiri, S.D. and Melis, A. (2003) Tla1, a DNA insertional transformant of the green alga Chlamydomonas reinhardtii with a truncated light-harvesting chlorophyll antenna size. Planta, 217, 49–59.
120. Posten, C. and Walter, C. (2012) Microalgal Biotechnology: Integration and Economy. Walter de Gruyter: Berlin.
121. PRB Population Reference Bureau (2013) 2013 World Population Data Sheet. http://www.prb.org/pdf13/2013-population-data-sheet_eng.pdf (accessed 29 October 2015).
122. Pulz, O. and Gross, W. (2004) Valuable products from biotechnology of microalgae. Appl. Microbiol. Biotechnol. 65, 635–648.
123. Qin, X., Suga, M., Kuang, T. and Shen, J.-R. (2015) Structural basis for energy transfer pathways in the plant PSI-LHCI supercomplex. Science, 348, 989–995.
124. Ramachandran, R. and Menon, R.K. (1998) An overview of industrial uses of hydrogen. Int. J. Hydrogen Energy 23, 593–598.
125. Rashid, N., Lee, K. and Mahmood, Q. (2011) Bio-hydrogen production by Chlorella vulgaris under diverse photoperiods. Bioresour. Technol. 102, 2101–2104.
126. Rathore, D. and Singh, A. (2013) Biohydrogen production from microalgae. In Biofuel Technologies - Recent Developments (Gupta, V.K. and Tuohy, M.G., eds), pp. 317–333. Berlin: Springer, Berlin Heidelberg.
127. Reifschneider-Wegner, K., Kanygin, A. and Redding, K.E. (2014) Expression of the [FeFe] hydrogenase in the chloroplast of Chlamydomonas reinhardtii. Int. J. Hydrogen Energy, 39, 3657–3665.
128. Robertson, D., Boynton, J.E. and Gillham, N.W. (1990) Cotranscription of the wild-type chloroplast atpE gene encoding the CF1/CF0 epsilon subunit with the 3′ half of the rps7 gene in Chlamydomonas reinhardtii and characterization of frameshift mutations in atpE. Mol. Gen. Genet. 221, 155–163.
129. Rohr, J., Sarkar, N., Balenger, S., Jeong, B.-R. and Cerutti, H. (2004) Tandem inverted repeat system for selection of effective transgenic RNAi strains in Chlamydomonas. Plant J. 40, 611–621.
130. Ruehle, T., Hemschemeier, A., Melis, A. and Happe, T. (2008) A novel screening protocol for the isolation of hydrogen producing Chlamydomonas reinhardtii strains. BMC Plant Biol. 8, 107.
131. Rumpel, S., Siebel, J.F., Fares, C., Duan, J., Reijerse, E., Happe, T., Lubitz, W. et al. (2014) Enhancing hydrogen production of microalgae by redirecting electrons from photosystem I to hydrogenase. Energ Environ. Sci. 7, 3296–3301.
132. Schmollinger, S., Strenkert, D. and Schroda, M. (2010) An inducible artificial microRNA system for Chlamydomonas reinhardtii confirms a key role for heat shock factor 1 in regulating thermotolerance. Curr. Genet. 56, 383–389.
133. Schönfeld, C., Wobbe, L., Borgstädt, R., Kienast, A., Nixon, P.J. and Kruse, O. (2004) The nucleus-encoded protein MOC1 is essential for mitochondrial light acclimation in Chlamydomonas reinhardtii. J. Biol. Chem. 279, 50366–50374.
134. Schulz, R., Schnackenberg, J., Stangier, K., Wünschiers, R., Zinn, T. and Senger, H. (1998) Light-dependent hydrogen production of the green alga Scenedesmus obliquus. In BioHydrogen (Zaborsky, O., Benemann, J., Matsunaga, T., Miyake, J. and San Pietro, A., eds), pp. 243–251. New York: Springer US.
135. Scoma, A., Krawietz, D., Faraloni, C., Giannelli, L., Happe, T. and Torzillo, G. (2012) Sustained H2 production in a Chlamydomonas reinhardtii D1 protein mutant. J. Biotechnol. 157, 613–619.
136. Seibert, M., Flynn, T., Benson, D., Tracy, E. and Ghirardi, M. (1998) Development of selection and screening procedures for rapid identification of H2-producing algal mutants with increased O2 tolerance. In Biohydrogen (Zaborsky, O.R., ed.), pp. 227–234. New York: Springer US.
137. Shimogawara, K., Fujiwara, S., Grossman, A. and Usuda, H. (1998) High-efficiency transformation of Chlamydomonas reinhardtii by electroporation. Genetics, 148, 1821–1828.
138. Sizova, I., Greiner, A., Awasthi, M., Kateriya, S. and Hegemann, P. (2013) Nuclear gene targeting in Chlamydomonas using engineered zinc-finger nucleases. Plant J. 73, 873–882.
139. Slegers, P.M. (2014) Scenario Studies for Algae Production. Wageningen: Wageningen University.
140. Slegers, P.M., Wijffels, R.H., van Straten, G. and van Boxtel, A.J.B. (2011) Design scenarios for flat panel photobioreactors. Appl. Energy, 88, 3342–3353.
141. Smil, V. (2008) Energy in Nature and Society: General Energetics of Complex Systems, pp. 480. Cambridge, MA: MIT Press.
142. Stapleton, J.A. and Swartz, J.R. (2010) Development of an in vitro compartmentalization screen for high-throughput directed evolution of FeFe hydrogenases. PLoS ONE, 5, e15275.
143. Stephens, E., Ross, I.L., King, Z., Mussgnug, J.H., Kruse, O., Posten, C., Borowitzka, M.A. et al. (2010a) An economic and technical evaluation of microalgal biofuels. Nat. Biotech. 28, 126–128.
144. Stephens, E., Ross, I.L., Mussgnug, J.H., Wagner, L.D., Borowitzka, M.A., Posten, C., Kruse, O. et al. (2010b) Future prospects of microalgal biofuel production systems. Trends Plant Sci. 15, 554–564.
145. Stephens, E., Wagner, L., Ross, I.L. and Hankamer, B. (2012) Microalgal production systems – Global impact of industry scale up. In Microalgal Biotechnology: Integration and Economy (Posten, C. and Walter, C., eds), pp. 267–301. Berlin/Boston: De Gruyter.
146. Stiebritz, M.T. and Reiher, M. (2012) Hydrogenases and oxygen. Chem. Sci. 3, 1739–1751.

147. Stocker, T.F. (2013) The closing door of climate targets. Science, 339, 280–282.
148. Stripp, S.T. and Happe, T. (2009) How algae produce hydrogen - news from the photosynthetic hydrogenase. Dalton Trans. 45, 9960–9969.
149. Sun, G., Zhang, X., Sui, Z. and Mao, Y. (2008) Inhibition of pds gene expression via the RNA interference approach in Dunaliella salina (Chlorophyta). Mar. Biotechnol. 10, 219–226.
150. Sun, Y., Chen, M., Yang, H., Zhang, J., Kuang, T. and Huang, F. (2013) Enhanced H-2 photoproduction by down-regulation of ferredoxin-NADP(+) reductase (FNR) in the green alga Chlamydomonas reinhardtii. Int. J. Hydrogen Energy, 38, 16029–16037.
151. Surzycki, R., Cournac, L., Peltiert, G. and Rochaix, J.-D. (2007) Potential for hydrogen production with inducible chloroplast gene expression in Chlamydomonas. Proc. Natl Acad. Sci. USA, 104, 17548–17553.
152. Takahashi, H., Iwai, M., Takahashi, Y. and Minagawa, J. (2006) Identification of the mobile light-harvesting complex II polypeptides for state transitions in Chlamydomonas reinhardtii. Proc. Natl Acad. Sci. USA, 103, 477–482.
153. The Intergovernmental Panel on Climate Change (IPCC) (2007) IPCC Fourth Assessment Report (AR4). In. Climate Change 2007: Synthesis Report Contribution of Working Groups I, II and III to the Fourth Assessment Report of the Intergovernmental Panel on Climate Change (Core Writing Team, Pachauri, R.K. and Reisinger, A. eds.), pp. 104. IPCC, Geneva, Switzerland.
154. Tokutsu, R., Kato, N., Bui, K.H., Ishikawa, T. and Minagawa, J. (2012) Revisiting the supramolecular organization of photosystem II in Chlamydomonas reinhardtii. J. Biol. Chem. 287, 31574–31581.
155. Tolleter, D., Ghysels, B., Alric, J., Petroutsos, D., Tolstygina, I., Krawietz, D., Happe, T. et al. (2011) Control of hydrogen photoproduction by the proton gradient generated by cyclic electron flow in Chlamydomonas reinhardtii. Plant Cell, 23, 2619–2630.
156. Torzillo, G., Scoma, A., Faraloni, C., Ena, A. and Johanningmeier, U. (2009) Increased hydrogen photoproduction by means of a sulfur-deprived Chlamydomonas reinhardtii D1 protein mutant. Int. J. Hydrogen Energy, 34, 4529–4536.
157. Tsui, E.Y., Kanady, J.S. and Agapie, T. (2013) Synthetic cluster models of biological and heterogeneous manganese catalysts for O2 evolution. Inorg. Chem. 52, 13833–13848.
158. USCB United States Census Bureau (2013) Total midyear population for the world: 1950-2050. http://www.census.gov/population/international/data/idb/worldpoptotal.php (accessed 29 October 2015).
159. Volgusheva, A., Styring, S. and Mamedov, F. (2013) Increased photosystem II stability promotes H2 production in sulfur-deprived Chlamydomonas reinhardtii. Proc. Natl Acad. Sci. USA, 110, 7223–7228.
160. Volgusheva, A., Kukarskikh, G., Krendeleva, T., Rubin, A. and Mamedov, F. (2015) Hydrogen photoproduction in green algae Chlamydomonas reinhardtii under magnesium deprivation. RSC Adv. 5, 5633–5637.
161. Wang, W., Wang, S., Ma, X. and Gong, J. (2011) Recent advances in catalytic hydrogenation of carbon dioxide. Chem. Soc. Rev. 40, 3703–3727.
162. Wang, A.-J., Cao, G.-L. and Liu, W.-Z. (2012) Biohydrogen production from anaerobic fermentation. In Biotechnology in China III: Biofuels and Bioenergy (Bai, F.-W., Liu, C.-G., Huang, H. and Tsao, G.T., eds), pp. 143–163. Berlin Heidelberg: Springer.
163. Wecker, M.S.A. and Ghirardi, M.L. (2014) High-throughput biosensor discriminates between different algal H2-photoproducing strains. Biotechnol. Bioeng. 111, 1332–1340.
164. Weingarten, A.S., Kazantsev, R.V., Palmer, L.C., McClendon, M., Koltonow, A.R., SamuelAmanda, P.S., Kiebala, D.J. et al. (2014) Self-assembling hydrogel scaffolds for photocatalytic hydrogen production. Nat. Chem. 6, 964–970.
165. Winkler, M., Kawelke, S. and Happe, T. (2011) Light driven hydrogen production in protein based semi-artificial systems. Bioresour. Technol. 102, 8493–8500.
166. Wittenberg, G., Sheffler, W., Darchi, D., Baker, D. and Noy, D. (2013) Accelerated electron transport from photosystem I to redox partners by covalently linked ferredoxin. Phys. Chem. Chem. Phys. 15, 19608–19614.
167. Wu, S., Huang, R., Xu, L., Yan, G. and Wang, Q. (2010) Improved hydrogen production with expression of hemH and lba genes in chloroplast of Chlamydomonas reinhardtii. J. Biotechnol. 146, 120–125.
168. Wu, S., Xu, L., Huang, R. and Wang, Q. (2011) Improved biohydrogen production with an expression of codon-optimized hemH and lba genes in the chloroplast of Chlamydomonas reinhardtii. Bioresour. Technol. 102, 2610–2616.
169. Xie, K. and Yang, Y. (2013) RNA-guided genome editing in plants using a CRISPR–Cas system. Mol. Plant, 6, 1975–1983.
170. Yacoby, I., Pochekailov, S., Toporik, H., Ghirardi, M.L., King, P.W. and Zhang, S. (2011) Photosynthetic electron partitioning between FeFe -hydrogenase and ferredoxin:NADP(+)-oxidoreductase (FNR) enzymes in vitro. Proc. Natl Acad. Sci. USA, 108, 9396–9401.
171. Zhang, R., Patena, W., Armbruster, U., Gang, S.S., Blum, S.R. and Jonikas, M.C. (2014) High-throughput genotyping of green algal mutants reveals random distribution of mutagenic insertion sites and endonucleolytic cleavage of transforming DNA. Plant Cell, 26, 1398–1409.
172. Zhang, C., Chen, C., Dong, H., Shen, J.-R., Dau, H. and Zhao, J. (2015) A synthetic Mn4Ca-cluster mimicking the oxygen-evolving center of photosynthesis. Science, 348, 690–693.
173. Zhao, T., Li, G., Mi, S., Li, S., Hannon, G.J., Wang, X.-J. and Qi, Y. (2007) A complex system of small RNAs in the unicellular green alga Chlamydomonas reinhardtii. Gene Dev. 21, 1190–1203.

174. Zhao, T., Wang, W., Bai, X. and Qi, Y. (2009) Gene silencing by artificial microRNAs in Chlamydomonas. Plant J. 58, 157–164.

175. Zhu, X.-G., Long, S.P. and Ort, D.R. (2008) What is the maximum efficiency with which photosynthesis can convert solar energy into biomass? Curr. Opin. Biotechnol. 19, 153–159.

176. Zhu, X.G., Long, S.P. and Ort, D.R. (2010) Improving photosynthetic efficiency for greater yield. Annu. Rev. Plant Biol. 61, 235–261.

INDEX